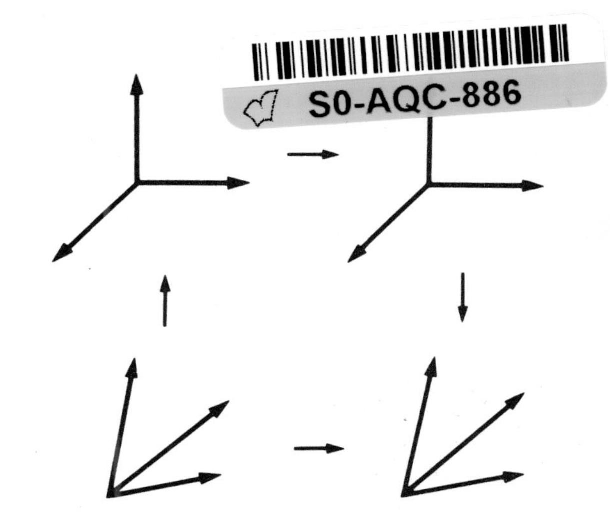

Elementary Linear Algebra with Applications: Concepts and Computations

Marc W. Konvisser

Wayne State University

Address orders and editorial correspondence to:
Ardsley House, Publishers, Inc.
320 Central Park West
New York, NY 10025

ISBN: 0-912675-15-2

Printed in the United States of America

10 9 8 7 6 5 4 3 2

Preface

This book is intended for a first course in linear algebra. It has been extensively class tested in preliminary edition at Wayne State University. Most students will find a background in calculus helpful in the successful use of this text, although it is not essential.

The book treats linear algebra from a geometrically motivated point of view, as recommended by the Committee on the Undergraduate Program in Mathematics. The text not only stresses the computational aspects of the subject, but also gives an equally important role to the abstract ideas of linear algebra. Because many students have difficulty with the more abstract concepts, the need for each new concept is demonstrated before it is introduced.

The book is organized around a core of sections that begin by introducing linear functions and linear transformations in the first lesson. This theme is extended and developed throughout the book, culminating in a discussion of eigenvectors, eigenvalues, and the Spectral Theorem. In addition, applications sections are located throughout the text to emphasize the key concepts as well as to illustrate a variety of uses for those concepts.

The arrangement of topics allows for flexibility. In cases where the students are fairly well prepared, it is advisable to treat the first two chapters rather quickly so that more time can be devoted to the more demanding material in the remaining chapters. Once the desired core sections have been covered, the remaining time can be used to discuss applications and theoretical sections (denoted by *). For those instructors wishing to treat matrix multiplication earlier than Chapter 7, the end of Chapter 3 provides a natural lead-in to that material, as indicated below.

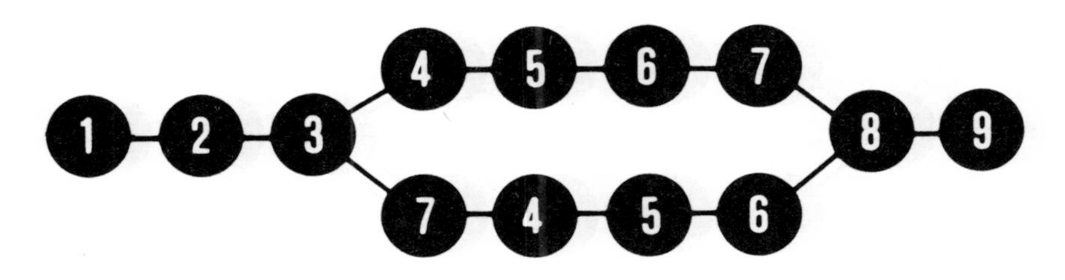

Approximately 1200 exercises and 200 examples are included in the text. Throughout the text, the symbol ★ is also used to denote the more challenging exercises. Many of the exercises at the ends of sections were generously furnished by Professors John Breckenridge, Daniel Frohardt, Togo Nishiura, Claude Schochet, Harold T. Slaby, and Stephen Williams of Wayne State University. I am grateful for their assistance and valuable comments, as well as those of my other colleagues and my students. Special thanks are due to Madelyn Leopold, who first encouraged me to write this book, and to the reviewers: Catherine C. Aust, Clayton Junior College; Ronald D. Baker, University of Delaware; Michael E. Detlefson, Slippery Rock State College; David E. Kullman, Miami University at Oxford, Ohio; Robert M. McConnell, University of Tennessee at Knoxville; David Presser, DePaul University of Illinois; B. David Saunders, Rensselaer Polytechnic University; Ray F. Spring, Ohio University; Harlan R. Stevens, Pennsylvania State University; Joseph Ullman, University of Michigan; Gary L. Walls, University of Southern Mississippi; Cary Webb, Chicago State University; and Edward N. Wilson, Washington University of Missouri. Their comments greatly improved this book.

Letter to Student

This is a letter of introduction — not only to the subject of linear algebra, but also to me, the author. Linear algebra is the study of the relationships between vectors (you can think of them as directed line segments) and a class of functions called linear transformations which map vectors to other vectors. This subject can be applied to solve a huge number of different problems. Some of these applications are discussed in the book. The topics range from how to make up secret codes, to how to win a computer game, to analyzing statistical data. I hope you will find time to read some of these applications sections even if they are not covered in your classroom work.

I decided to write this text because I felt that there was no linear algebra book available which was both at the right level of sophistication and which would provide the student with insight into the logical development of the subject. This book is an outgrowth of my training as a mathematician and my experiences as a teacher. As I was writing, the book seemed to develop a character of its own. I found that I was forced to explain every facet of the subject (this was the mathematician in me), but I was also forced to make each explanation clear so that students who were new at studying mathematical concepts could understand these new ideas (this was the teacher in me).

As you study the material in this book, you may be tempted to ignore the theory and learn only facts. I do not recommend this approach. Theory provides the framework of ideas on which to hang specific facts. If you have a clear understanding of the theory, what is behind the ideas you are taught, you will be able to use and apply these ideas. This will not only make it easier for you to learn, but also help you to assimilate the material more quickly.

While writing this book, I wrestled with a number of mathematical problems trying to understand and explain many rather subtle ideas. To me this was like many of my other mathematical research projects — difficult at the time, but a great source of satisfaction after completion. I hope that you too will share some of this excitement as you learn from my book.

Marc Konvisser

To the first student who read this book, my wife ZIEVA. . .

Who encouraged me when it looked as if this book would never be published.

Who took time from her already too busy life to read and reread the manuscript.

Whose understandings and misunderstandings taught me what was good and what was bad in my writing.

Contents

Basic Concepts

The purpose of this chapter is to introduce some of the basic definitions and techniques of linear algebra. It provides concrete examples of topics to be discussed more abstractly later in the text.

1.1 Functions

The concept of function is of fundamental importance in all branches of mathematics. Linear algebra is the study of a special class of functions called **linear transformations**. Since almost every elementary mathematics text discusses functions, we assume that the student already has some understanding of them. Therefore, the main purpose of this section is to clarify certain properties of functions that are often difficult to understand. We begin by defining a function.

DEFINITION 1.1

A **function** is a correspondence between the elements of one set called the **domain** and another set called the **range** (also called **codomain**). To each element of the domain the function assigns some element of the range (but not the other way around). See Figure 1.1. Not all such correspondences are functions—only those satisfying both of the following conditions:

1. For each element of the domain there is an element of the range corresponding to it.

2. Each element of the domain corresponds to only one element of the range.

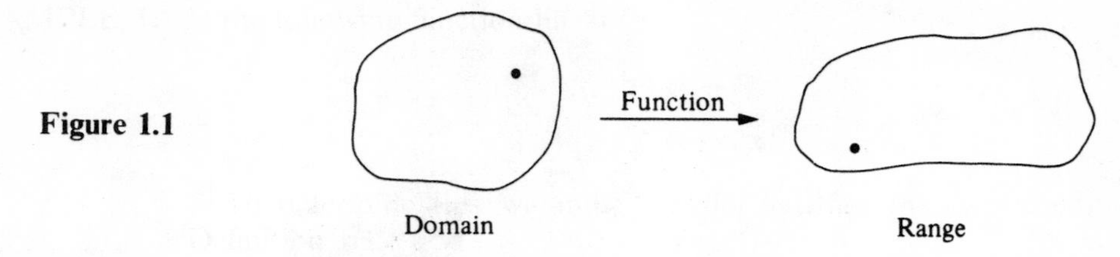

Figure 1.1

EXAMPLE 1 A correspondence often used to identify people in the State of Michigan assigns to each person the number on his or her driver's license. Is this correspondence a function?

The correspondence between the set of all residents of Michigan and the set of integers is not a function because it does not satisfy Condition 1—some residents of Michigan do not have driver's licenses. (This fact causes people who do not have licenses a great deal of difficulty when cashing checks.) ■

Another type of correspondence that is not a function is discussed in the following example.

EXAMPLE 2 The correspondence assigning to each name a person with that name is not a function because it does not satisfy Condition 2. Two people can have the same name. (There are 35 John Smiths in the Detroit telephone directory.) ■

These two examples should help you see how the concept of function can be useful.

To describe a particular function, it is necessary to indicate the domain and range of the function, and which element of the range is assigned to each element of the domain. To specify functions, we give them names. Most functions are called f. A mathematical shorthand is usually used to describe functions. The pattern for this notation is

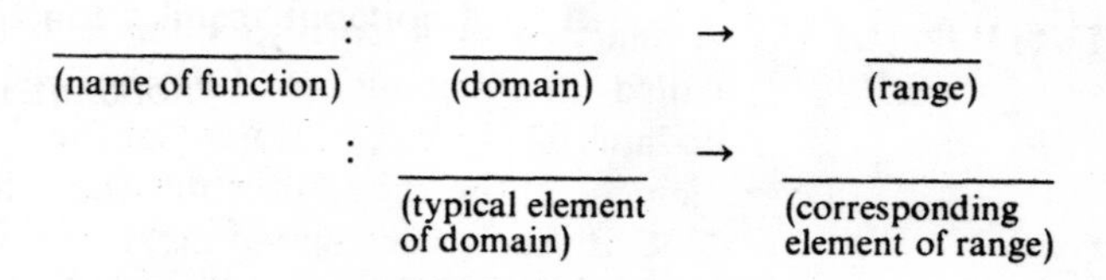

EXAMPLE 3 Describe the domain, range, and action of the function

$$f\colon \text{Integers} \to \text{Integers}$$
$$x \to x^2$$

Both the domain and the range of f are the integers. The action of f is to map each integer onto its square. ■

NOTATION

We indicate the value of a function f at a particular element of its domain by

$$f(\underset{\text{(element of domain)}}{\underline{\qquad\qquad}}) = \underset{\text{(element of range)}}{\underline{\qquad\qquad}}$$

For the function f of Example 3 we can write

$$f(x) = x^2, \quad f(3) = 9, \quad f(y) = y^2, \quad f(x + y) = (x + y)^2$$

Example 3 illustrates two important facts about functions. First, although 2 is an element of the range of f, there is no element x (integer) of the domain of f such that $f(x) = 2$ ($\sqrt{2}$ is not an integer). Thus, although every element of a domain of a function is mapped (or assigned) to some element of a range, not all elements of the range need to be the image of some point of the domain. The second important fact illustrated by Example 3 is that two distinct points of the domain can be mapped onto the same element of the range. For example, $f(2) = 4$ and $f(-2) = 4$.

Now we give two examples of correspondences that are not functions.

EXAMPLE 4 Is the correspondence

$$f\colon \text{Integers} \to \text{Real numbers}$$
$$x \to \sqrt{x}$$

a function?

The correspondence is not a function because it does not satisfy Condition 1 of Definition 1.1. For example, there is no real number corresponding to the integer -1 ($\sqrt{-1}$ is not a real number). ■

EXAMPLE 5 Is the correspondence

$$f\colon \text{Positive real numbers} \to \text{Real numbers}$$
$$x \to \text{A number } y \text{ such that } y^2 = x$$

a function?

The correspondence is not a function because it does not satisfy Condition 2 of Definition 1.1. For example, $f(1)$ can be either $+1$ or -1. ■

Determining the equality of two functions may seem like a trivial problem, but it can be difficult.

DEFINITION 1.2

Let f and g be two functions. Then f and g are **equal** if and only if both

> **1.** The domain of f and the domain of g are the same, and the range of f and the range of g are the same.
>
> **2.** For each element x of the domain of f (or g), $f(x) = g(x)$.

EXAMPLE 6 Are these functions equal?

$$f: \{0, 2\} \to \text{Integers} \qquad g: \{0, 2\} \to \text{Integers}$$

$$x \to 2x \qquad x \to x^2$$

Yes, because they have the same domain and range, and

$$f(0) = 2 \cdot 0 = 0 \quad \text{and} \quad g(0) = 0^2 = 0$$

$$f(2) = 2 \cdot 2 = 4 \quad \text{and} \quad g(2) = 2^2 = 4$$

These functions satisfy both conditions of Definition 1.2 and so they are equal. ■

The purpose of Example 6 is to show that even though two functions may look different, they can be equal. *Two functions are the same if they act the same.*

Exercise Set 1.1

In Exercises 1–5 determine whether the correspondence listed is or is not a function. Explain your answers.

1. f: Integers → Integers

$$x \to \frac{x}{2}$$

2. f: Positive real numbers → Real numbers

x → The number y such that $y^4 = x$

3. $f: \{1, 2, 3, 4\} \to \{1, 2, 3, 4\}$

$$x \to x^2$$

4. f: Integers → Integers

$$x \to -x$$

5. f: Nonzero integers → Real numbers

$$x \to \frac{1}{x}$$

6. Determine whether the functions f and g are equal where

$$f: \{-1, 1\} \to \text{Integers} \qquad g: \{-1, 1\} \to \text{Integers}$$

$$x \to x \qquad x \to \frac{1}{x}$$

7. Determine whether the functions f and g are equal where

$$f: \to \{-1, 1\}\ \text{Integers} \qquad g: \to \{-1, 1\}\ \text{Integers}$$

$$x \to \frac{1}{2}\left(x + \frac{1}{x}\right) \qquad x \to \frac{x}{|x|}$$

8. Consider the following four functions:

f_1: Integers → Integers

$$x \to (2x)^2$$

f_2: Integers → Integers

$$x \to 2x^2$$

f_3: Real numbers → Real numbers

$$x \to (x + 4)^2$$

f_4: Real numbers → Real numbers
$x \rightarrow x^2 + 4^2$

Determine which functions satisfy

a. $f(3) = 18$ b. $f(1) = 2$ c. $f(0) = 0$

d. $f(x^2) = x^4 + 8x^2 + 16$

e. $f(x + y) = 2x^2 + 4xy + 2y^2$

9. Consider the functions

f_1: Real numbers → Real numbers
$x \rightarrow 2x$

f_2: Real numbers → Real numbers
$x \rightarrow 2x + 1$

f_3: Real numbers → Real numbers
$x \rightarrow x^2$

Determine which functions satisfy the following conditions.

a. $f(3x) = 3f(x)$ b. $f(-x) = -f(x)$

c. $f(x + 4) = f(x) + f(4)$

d. $f(x + y) = f(x) + f(y)$

10. Are these functions equal?

$f: \{0, 1, 2, 3, 4, 5\} \rightarrow$ Real numbers
$x \rightarrow (x + 2)^2$

$g: \{0, 1, 2, 3, 4, 5\} \rightarrow$ Real numbers
$x \rightarrow (x^4 + 8x^3 + 24x^2 + 32x + 16)^{1/2}$

11. Consider the following correspondences. Determine which are functions and which are not.

a. f: Set of all books → Integers
Book → Number of pages in that book

b. Integers → Set of all books
n → A book having n pages

c. Set of all cars → {red, blue, yellow, white}
Car → Color of that car

d. Set of all trees → Integers
Tree → The number of leaves on that tree (at a specified time)

1.2 Linear Functions

We begin the study of linear algebra by first defining linear function, the basic concept of linear algebra. Linear functions play important roles in almost every branch of mathematics, and their study is important to many other disciplines. Furthermore, the theory of linear functions serves as a model for investigating many areas of mathematics.

DEFINITION 1.3

A function f that has the real numbers as its domain and range is called a **linear function** if the following two conditions hold:

1. $f(x + y) = f(x) + f(y)$ for all elements x, y in the domain of f.

2. $f(cx) = cf(x)$ for every number c and every element x in the domain of f.

We begin by looking at several examples of functions to determine which are linear. To make things a little easier, we will use the symbol $\mathbb{R}$ to represent the set of real numbers.

EXAMPLE 1 Is the following function linear?

$$f: \mathbb{R} \to \mathbb{R}$$
$$x \to 2x$$

To determine this we must see if f satisfies the two conditions of Definition 1.3.

1. $f(x + y) = 2(x + y) = 2x + 2y$

$f(x) + f(y) = 2x + 2y$ Condition 1 holds

2. $f(cx) = 2(cx) = c(2x)$

$cf(x) = c(2x)$ Condition 2 holds

Thus f is a linear function. ■

This example is a special case of the more general result that every function $f: \mathbb{R} \to \mathbb{R}$ where $f(x) = ax$ for some real number a is a linear function (see Exercise 4 in this section).

EXAMPLE 2 Is the following function linear?

$$f: \mathbb{R} \to \mathbb{R}$$
$$x \to 2x + 1$$

Again, we must see if f satisfies the two conditions of Definition 1.3.

1. $f(x + y) = 2(x + y) + 1 = 2x + 2y + 1$

$f(x) + f(y) = (2x + 1) + (2y + 1) = 2x + 2y + 2$

So $f(x + y) \neq f(x) + f(y)$ and thus Condition 1 doesn't hold

Therefore f is not a linear function. (*Note:* Even though the graph of f is a line, f is not a linear function.) ■

EXAMPLE 3 Is f a linear function?

$$f: \mathbb{R} \to \mathbb{R}$$
$$x \to x^2$$

We check the conditions of Definition 1.3.

1. $f(x + y) = (x + y)^2 = x^2 + y^2 + 2xy$

$f(x) + f(y) = x^2 + y^2$

So $f(x + y) \neq f(x) + f(y)$ and thus Condition 1 doesn't hold

Therefore f is not a linear function. ■

Let us suppose that f is a linear function, $f: \mathbb{R} \to \mathbb{R}$, and we know that $f(1) = 3$. Can we find $f(2)$?

From Condition 2 of Definition 1.3, we see that $f(cx) = cf(x)$. So

$$f(2) = f(2 \cdot 1) = 2f(1) = 2 \cdot 3 = 6$$

Similarly, $\quad f(-4) = f(-4 \cdot 1) = -4f(1) = -4 \cdot 3 = -12$

Using this procedure with the variable x instead of a specific number yields

$$f(x) = f(x \cdot 1) = xf(1) = x \cdot 3 = 3x$$

From this we learn that there is one and only one linear function $f: \mathbb{R} \to \mathbb{R}$ such that $f(1) = 3$, and this function is $f(x) = 3x$. We will use these ideas again.

EXAMPLE 4 Suppose $f: \mathbb{R} \to \mathbb{R}$ is a linear function and $f(1) = \sqrt{2}$. Is there any such linear function? Is there only one such linear function? What is this function? To answer these questions we proceed as before.

Although it is not essential, we begin by finding $f(2)$ to get an idea of what f is like. Using Condition 2 of Definition 1.3, we have

$$f(2) = f(2 \cdot 1) = 2f(1) = 2\sqrt{2}$$

To find the general formula for $f(x)$ we again use Condition 2.

$$f(x) = f(x \cdot 1) = xf(1) = x\sqrt{2}$$

Thus we can conclude that there is one and only one linear function $f: \mathbb{R} \to \mathbb{R}$ such that $f(1) = \sqrt{2}$. This function is $f(x) = \sqrt{2}\,x$. ■

It is now clear that no matter what $f(1)$ is, the procedure will still work. Let us suppose that $f: \mathbb{R} \to \mathbb{R}$ is linear and $f(1) = a$, for some real number a. Using condition 2 of Definition 1.3, we have

$$f(x) = f(x \cdot 1) = xf(1) = xa$$

By using the fact that every function $f: \mathbb{R} \to \mathbb{R}$ with $f(x) = ax$ is linear, we get the result that there is one and only one linear function $f: \mathbb{R} \to \mathbb{R}$ with $f(1) = a$. This function is $f(x) = ax$.

EXAMPLE 5 Suppose $f: \mathbb{R} \to \mathbb{R}$ is a linear function and $f(3) = 12$. Can we find $f(x)$ for every real number x?

It will suffice to find $f(1)$, since, by the result above, $f(x) = f(1)x$. By Condition 2 of Definition 1.3

$$f(3) = f(3 \cdot 1) = 3f(1)$$

So $\quad 12 = f(3) = 3f(1)$

Solving for $f(1)$, $\quad f(1) = f(3)/3 = 12/3 = 4$

Hence $\quad f(x) = f(1)x = 4x$ ■

This procedure works in general. Suppose $f: \mathbb{R} \to \mathbb{R}$ is a linear function and b is a nonzero real number. If $f(b) = c$, find a general formula for $f(x)$. As in Example 5, we begin by finding $f(1)$

$$c = f(b) = f(b \cdot 1) = bf(1)$$

$$f(1) = c/b \qquad \text{(We can divide by } b \text{ since } b \neq 0.)$$

Since $f(x) = f(1)x$, we have

$$f(x) = (c/b)x$$

Thus we have proved the following theorem.

THEOREM 1.1 A linear function $f: \mathbb{R} \to \mathbb{R}$ is completely determined by its value at any nonzero point. Specifically, if $b \neq 0$ and $f(b) = c$, then $f(x) = (c/b)x$.

It should be noted that all linear functions from $\mathbb{R}$ to $\mathbb{R}$ are of the form $f(x) = (c/b)x$; thus $f(0) = 0$, and the graph of the equation $y = f(x)$ is a straight line through the origin.

Exercise Set 1.2

1. Using Definition 1.3, determine which of the following functions are linear.

a. $f: \mathbb{R} \to \mathbb{R}$, $x \to x + 4$ **b.** $f: \mathbb{R} \to \mathbb{R}$, $x \to \sqrt{x}$

c. $f: \mathbb{R} \to \mathbb{R}$, $x \to \dfrac{x}{2}$ **d.** $f: \mathbb{R} \to \mathbb{R}$, $x \to \dfrac{1}{x}$

2. Find all linear functions $f: \mathbb{R} \to \mathbb{R}$ with the property that:

a. $f(1) = 5$ **b.** $f(1/2) = 3$ **c.** $f(2) = 5$
d. $f(0) = 0$ **e.** $f(0) = 1$

3. Graph the functions $y = f(x)$ for the functions f you found in parts a, b, and c of Exercise 2.

4. Show that if a is a real number and

$$f: \mathbb{R} \to \mathbb{R}$$
$$x \to ax$$

then f is a linear function.

5. Let $f, g: \mathbb{R} \to \mathbb{R}$ be linear functions. Show that

a. $(f + g)(x) = f(x) + g(x)$ is a linear function.
b. $f(g(x))$ is a linear function.

6. Find two linear functions $f, g: \mathbb{R} \to \mathbb{R}$ such that $f(x)g(x)$ is not a linear function.

7. Let $f: \mathbb{R} \to \mathbb{R}$ be a linear function such that $f(1) = 0$. Show that $f(x) = 0$ for every real number x.

8. Show that if $f, g\colon \mathbb{R} \to \mathbb{R}$ are linear functions, then $f = g$ if and only if $f(b) = g(b)$ for some number $b \neq 0$.

9. Suppose $f, g\colon \mathbb{R} \to \mathbb{R}$ are linear functions.

a. Show that the function $h(x) = af(x) + bf(y)$ is linear for any real numbers a and b.

b. Show that $f(g(x)) = g(f(x))$.

c. Find nonlinear functions h_1 and h_2 such that $h_1(h_2(x)) \neq h_2(h_1(x))$.

1.3 Linear Functions of Two Variables

In Section 1.2 we analyzed linear functions of one variable. In this section we will consider functions with domain $\mathbb{R}^2$, the set of all ordered pairs of real numbers (x, y), and range $\mathbb{R}$. An example of such a function of two variables follows.

EXAMPLE 1 Find $f(0, 0)$, $f(1, 0)$, and $f(1, 1)$ for the following function.

$$f\colon \mathbb{R}^2 \to \mathbb{R}$$

$$(x, y) \to 3x - 2y$$

We have $f(0, 0) = 0$, $f(1, 0) = 3$, $f(1, 1) = 1$. ■

Now we need to see what it means for a function of two variables to be linear. Definition 1.3 of a linear function of one variable stated f is linear provided that:

1. $f(x + y) = f(x) + f(y)$; or f of the sum of two numbers equals the sum of f of one of the numbers plus f of the other number.

2. $f(cx) = cf(x)$; or f of a multiple of a number equals that multiple of f of that number.

To extend this definition to functions whose domain is the set of all ordered pairs of real numbers instead of just the real numbers, we must define

1. The sum of two ordered pairs of real numbers.

2. The product of a real number and an ordered pair of real numbers.

DEFINITION 1.4

The **sum of two ordered pairs** of real numbers is given by

$$(x_1, y_1) + (x_2, y_2) = (x_1 + x_2, y_1 + y_2)$$

where x_1, y_1, x_2, y_2 are real numbers.

DEFINITION 1.5

The **product of a real number and an ordered pair** of real numbers is given by

$$c(x, y) = (cx, cy)$$

where c, x, and y are real numbers

These definitions have a nice geometrical interpretation. Instead of just looking at a point (x, y), we associate the directed line segment from the origin to that point with the point (see Figure 1.2). These line segments are given a special name and play a central role in the study of linear algebra.

DEFINITION 1.6

With each point (x, y) in $\mathbb{R}^2$, we associate the directed line segment from $(0, 0)$ to (x, y). This line segment is called the **vector** $\begin{bmatrix} x \\ y \end{bmatrix}$.

When we discuss vectors we will often refer to them by a single letter (not their endpoints). To distinguish vectors from real numbers, we set the letter name of the vector in boldface type. For example,

$$\mathbf{v} = \begin{bmatrix} x \\ y \end{bmatrix}.$$

The correspondence between points and vectors (each point is the tip of some vector and each vector represents the point at its tip) can be used to restate Definitions 1.4 and 1.5 in terms of vectors.

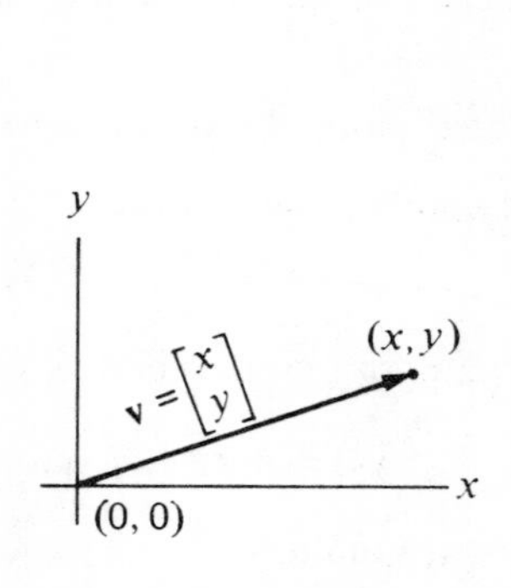

Figure 1.2

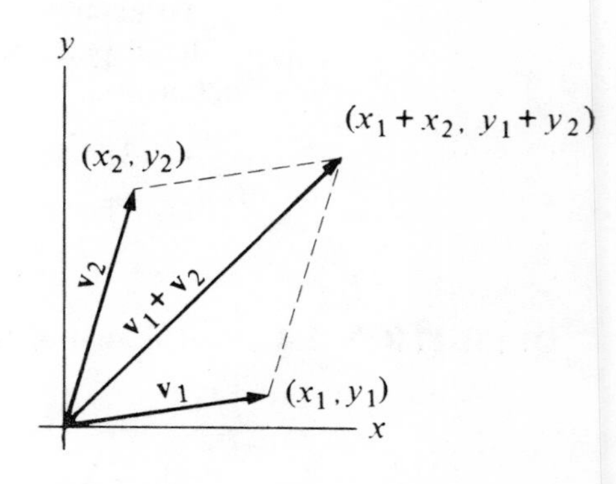

Figure 1.3

DEFINITION 1.7

The **sum of the vectors** $\mathbf{v}_1 = \begin{bmatrix} x_1 \\ y_1 \end{bmatrix}$ and $\mathbf{v}_2 = \begin{bmatrix} x_2 \\ y_2 \end{bmatrix}$ is

$$\mathbf{v}_1 + \mathbf{v}_2 = \begin{bmatrix} x_1 \\ y_1 \end{bmatrix} + \begin{bmatrix} x_2 \\ y_2 \end{bmatrix} = \begin{bmatrix} x_1 + x_2 \\ y_1 + y_2 \end{bmatrix}$$

Figure 1.3 shows the sum of two vectors graphically. The vector $\mathbf{v}_1 + \mathbf{v}_2$ is the diagonal of the parallelogram having sides $\mathbf{v}_1$ and $\mathbf{v}_2$. For this reason it is convenient to think of the sum of two vectors as being obtained by "ungluing" one of the vectors from the origin and attaching its end to the tip of the other vector (without turning it or changing its length). The tip of this "detached vector" is at the tip of the vector that is the sum of the two original vectors. This is just filling in one of the missing sides of the parallelogram (Figure 1.4).

Figure 1.4

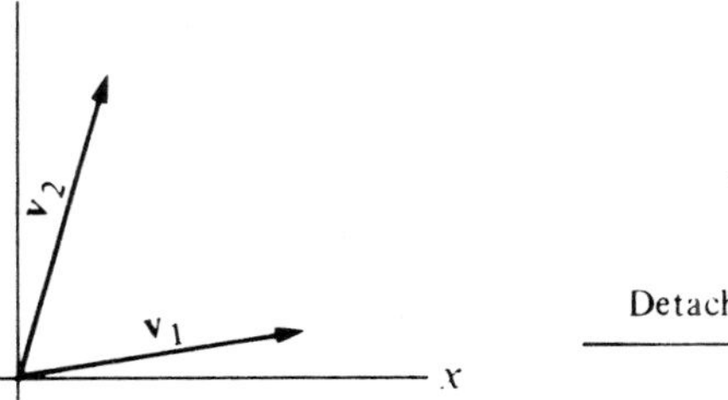

Detach →

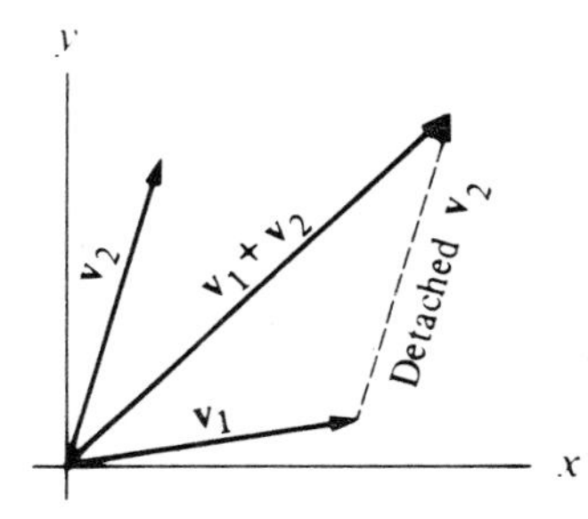

EXAMPLE 2 Find the sum of the vectors $\begin{bmatrix} 1 \\ 2 \end{bmatrix}$ and $\begin{bmatrix} -3 \\ 1 \end{bmatrix}$.

$$\begin{bmatrix} 1 \\ 2 \end{bmatrix} + \begin{bmatrix} -3 \\ 1 \end{bmatrix} = \begin{bmatrix} -2 \\ 3 \end{bmatrix}.$$

The solution is shown graphically in Figure 1.5.

Figure 1.5

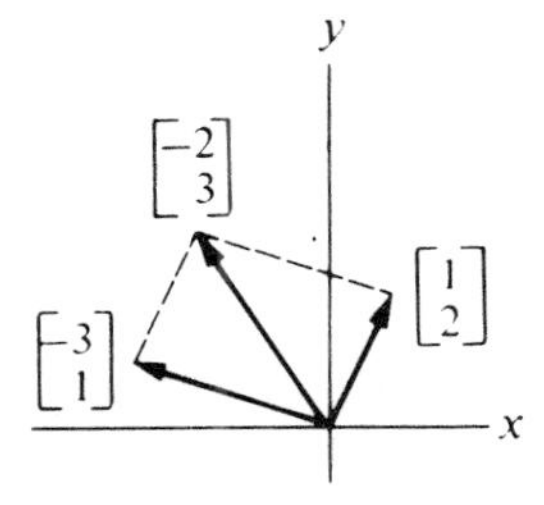

■

DEFINITION 1.8 The **product of a real number** c **and a vector** $\mathbf{v} = \begin{bmatrix} x \\ y \end{bmatrix}$ is

$$c\mathbf{v} = c\begin{bmatrix} x \\ y \end{bmatrix} = \begin{bmatrix} cx \\ cy \end{bmatrix}$$

The geometrical interpretation of multiplication of a vector by a real number c is shown in Figure 1.6.

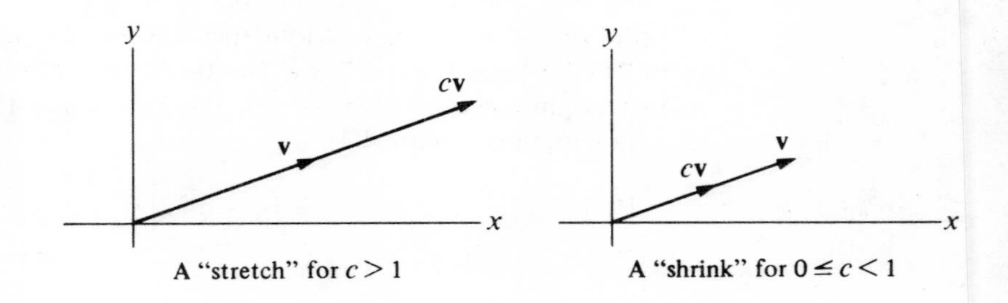

A "stretch" for $c > 1$

A "shrink" for $0 \leq c < 1$

Figure 1.6

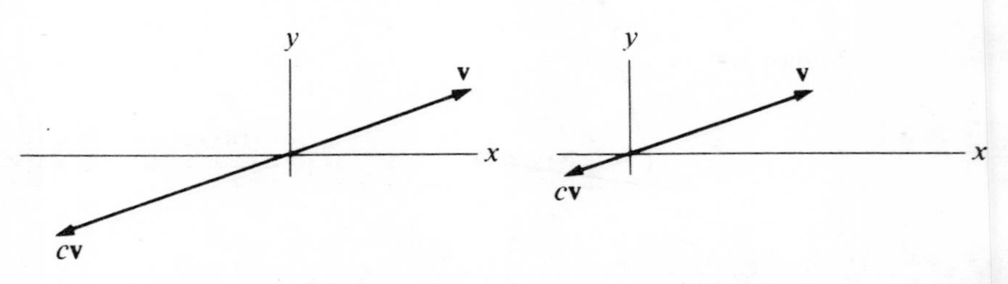

A "reverse and stretch" for $c < -1$

A "reverse and shrink" for $-1 < c < 0$

We can multiply vectors by real numbers, but vectors themselves are not real numbers. To emphasize this difference, real numbers are given a special name—they are called **scalars**. The operation of multiplying a vector by a real number is called **scalar multiplication**.

EXAMPLE 3 The product of a scalar 2 and a vector $\begin{bmatrix} 1 \\ -2 \end{bmatrix}$ is

$$2\begin{bmatrix} 1 \\ -2 \end{bmatrix} = \begin{bmatrix} 2 \\ -4 \end{bmatrix} \quad \blacksquare$$

Now we can define the term linear function for functions with domain $\mathbb{R}^2$ and range $\mathbb{R}$.

DEFINITION 1.9 A function $f\colon \mathbb{R}^2 \to \mathbb{R}$ is **linear** if it satisfies the following two conditions:

1. $f(\mathbf{v} + \mathbf{w}) = f(\mathbf{v}) + f(\mathbf{w})$ for all vectors $\mathbf{v}$ and $\mathbf{w}$
2. $f(c\mathbf{v}) = cf(\mathbf{v})$ for all vectors $\mathbf{v}$ and all scalars c

EXAMPLE 4 Consider the function

$$f\colon \mathbb{R}^2 \to \mathbb{R}$$
$$\begin{bmatrix} x \\ y \end{bmatrix} \to 2x - y$$

To determine if f is a linear function, we must check to see if f satisfies both conditions of Definition 1.9. To check the first condition, let

$$\mathbf{v} = \begin{bmatrix} x_1 \\ y_1 \end{bmatrix} \quad \text{and} \quad \mathbf{w} = \begin{bmatrix} x_2 \\ y_2 \end{bmatrix}$$

be arbitrary vectors. We must see if $f(\mathbf{v} + \mathbf{w}) = f(\mathbf{v}) + f(\mathbf{w})$.

$$f(\mathbf{v} + \mathbf{w}) = f\left(\begin{bmatrix} x_1 \\ y_1 \end{bmatrix} + \begin{bmatrix} x_2 \\ y_2 \end{bmatrix}\right) = f\left(\begin{bmatrix} x_1 + x_2 \\ y_1 + y_2 \end{bmatrix}\right) = 2(x_1 + x_2) - (y_1 + y_2)$$

and

$$f(\mathbf{v}) + f(\mathbf{w}) = f\left(\begin{bmatrix} x_1 \\ y_1 \end{bmatrix}\right) + f\left(\begin{bmatrix} x_2 \\ y_2 \end{bmatrix}\right) = (2x_1 - y_1) + (2x_2 - y_2)$$

Since $2(x_1 + x_2) - (y_1 + y_2) = (2x_1 - y_1) + (2x_2 - y_2)$, f satisfies Condition 1.

To check Condition 2 of Definition 1.9, let c be a scalar and $\mathbf{v} = \begin{bmatrix} x \\ y \end{bmatrix}$. We must see if $f(c\mathbf{v}) = cf(\mathbf{v})$.

$$f(c\mathbf{v}) = f\left(c\begin{bmatrix} x \\ y \end{bmatrix}\right) = f\left(\begin{bmatrix} cx \\ cy \end{bmatrix}\right) = 2(cx) - cy$$

and

$$cf(\mathbf{v}) = c(2x - y) = 2(cx) - cy$$

Thus f also satisfies Condition 2 and is a linear function. ■

EXAMPLE 5 Is the following function linear? $f\colon \mathbb{R}^2 \to \mathbb{R}$

$$\begin{bmatrix} x \\ y \end{bmatrix} \to xy$$

f is not linear, in fact, it satisfies neither condition of Definition 1.9.

Checking Condition 2, with an arbitrary vector $\mathbf{v} = \begin{bmatrix} x \\ y \end{bmatrix}$, and an arbitrary scalar c, yields

$$f(c\mathbf{v}) = f\left(c\begin{bmatrix} x \\ y \end{bmatrix}\right) = f\left(\begin{bmatrix} cx \\ cy \end{bmatrix}\right) = (cx)(cy) = c^2xy$$

but
$$cf(\mathbf{v}) = cf\left(\begin{bmatrix} x \\ y \end{bmatrix}\right) = c(xy)$$

So $f(c\mathbf{v}) \neq cf(\mathbf{v})$. Condition 2 doesn't hold and f is not linear. ■

Exercise Set 1.3

In Exercises 1–3 graph the vectors on the same set of coordinate axes.

1. $\begin{bmatrix} 1 \\ 2 \end{bmatrix}$, $3\begin{bmatrix} 1 \\ 2 \end{bmatrix}$, $-1\begin{bmatrix} 1 \\ 2 \end{bmatrix}$, $1/2\begin{bmatrix} 1 \\ 2 \end{bmatrix}$, $-5/2\begin{bmatrix} 1 \\ 2 \end{bmatrix}$

2. $\begin{bmatrix} 3 \\ 1 \end{bmatrix}$, $\begin{bmatrix} -2 \\ 5 \end{bmatrix}$, $\begin{bmatrix} 3 \\ 1 \end{bmatrix} + \begin{bmatrix} -2 \\ 5 \end{bmatrix}$, $\begin{bmatrix} 3 \\ 1 \end{bmatrix} - \begin{bmatrix} -2 \\ 5 \end{bmatrix}$

3. $\begin{bmatrix} -1 \\ 2 \end{bmatrix}$, $\begin{bmatrix} 2 \\ -1 \end{bmatrix}$, $\begin{bmatrix} 1 \\ 2 \end{bmatrix}$, $\begin{bmatrix} 2 \\ 1 \end{bmatrix}$

4. Solve each of the equations and graph the vectors involved.

a. $\begin{bmatrix} 4 \\ 1 \end{bmatrix} + \mathbf{x} = \begin{bmatrix} 1 \\ 3 \end{bmatrix}$

b. $\begin{bmatrix} 1 \\ 4 \end{bmatrix} + c\begin{bmatrix} 2 \\ 3 \end{bmatrix} = \begin{bmatrix} -3 \\ -2 \end{bmatrix}$

c. $a\begin{bmatrix} 1 \\ 2 \end{bmatrix} + b\begin{bmatrix} -3 \\ 1 \end{bmatrix} = \begin{bmatrix} 1 \\ 0 \end{bmatrix}$

d. $a\begin{bmatrix} 1 \\ 2 \end{bmatrix} + b\begin{bmatrix} -3 \\ 1 \end{bmatrix} = \begin{bmatrix} 0 \\ 1 \end{bmatrix}$

5. Find a missing vertex for the parallelograms having vertices

a. (0, 0), (4, 1), (1, 3)

b. (1, 1), (5, 1), (2, 3)

6. Determine which functions are linear.

a. $f: \mathbb{R}^2 \to \mathbb{R}$

$\begin{bmatrix} x \\ y \end{bmatrix} \to 3x + 5y$

b. $f: \mathbb{R}^2 \to \mathbb{R}$

$\begin{bmatrix} x \\ y \end{bmatrix} \to x^2y$

c. $f: \mathbb{R}^2 \to \mathbb{R}$

$\begin{bmatrix} x \\ y \end{bmatrix} \to x$

d. $f: \mathbb{R}^2 \to \mathbb{R}$

$\begin{bmatrix} x \\ y \end{bmatrix} \to \sqrt{x^2 + y^2}$

$\left(\sqrt{x^2 + y^2} = \text{the length of } \begin{bmatrix} x \\ y \end{bmatrix}\right)$

e. $f: \mathbb{R}^2 \to \mathbb{R}$

$\begin{bmatrix} x \\ y \end{bmatrix} \to ax + by$ for some arbitrary but fixed constants a and b.

7. Graph the following vectors:

a. $\begin{bmatrix} 1 \\ 2 \end{bmatrix}$, $\begin{bmatrix} 1 \\ 2 \end{bmatrix} + \begin{bmatrix} 1 \\ 1 \end{bmatrix}$, $\begin{bmatrix} 1 \\ 2 \end{bmatrix} + 2\begin{bmatrix} 1 \\ 1 \end{bmatrix}$,

$\begin{bmatrix} 1 \\ 2 \end{bmatrix} + 3\begin{bmatrix} 1 \\ 1 \end{bmatrix}$, $\begin{bmatrix} 1 \\ 2 \end{bmatrix} - 1\begin{bmatrix} 1 \\ 1 \end{bmatrix}$, $\begin{bmatrix} 1 \\ 2 \end{bmatrix} - 2\begin{bmatrix} 1 \\ 1 \end{bmatrix}$

b. Connect the tips of the vectors you graphed in part a. What type of curve does this appear to be?

c. What type of curve is traced by the tips of all vectors of the form

$\mathbf{v} = \begin{bmatrix} 1 \\ 2 \end{bmatrix} + t\begin{bmatrix} 1 \\ 1 \end{bmatrix}$ where t is a real number?

8. Graph the set of all points that correspond to solutions of the equation $f\begin{bmatrix} x \\ y \end{bmatrix} = 0$ if

$$f: \mathbb{R}^2 \to \mathbb{R}$$
$$\begin{bmatrix} x \\ y \end{bmatrix} \to 2x - y$$

9. Let $f, g: \mathbb{R}^2 \to \mathbb{R}$ be linear functions. Show that $f + g$ is a linear function.

10. Find two linear functions $f, g: \mathbb{R}^2 \to \mathbb{R}$ such that $f\left(\begin{bmatrix} x \\ y \end{bmatrix}\right)g\left(\begin{bmatrix} x \\ y \end{bmatrix}\right)$ is not linear.

11. Let $f: \mathbb{R}^2 \to \mathbb{R}$ and $g: \mathbb{R} \to \mathbb{R}$ be linear functions.

a. Show that $g(f(\mathbf{v}))$ is a linear function from $\mathbb{R}^2$ to $\mathbb{R}$.

b. Why can't we define $f(g(x))$?

1.4 Classifying Linear Functions of Two Variables

When we investigated real-valued linear functions of one variable, we saw that the value of the function at any nonzero point completely determined the function. The same thing does not hold for functions of two variables. In the following example we give two different linear functions that have the same value at a point other than zero.

EXAMPLE 1

$$f: \mathbb{R}^2 \to \mathbb{R} \qquad \text{and} \qquad g: \mathbb{R}^2 \to \mathbb{R}$$
$$\begin{bmatrix} x \\ y \end{bmatrix} \to x + y \qquad\qquad \begin{bmatrix} x \\ y \end{bmatrix} \to x$$

Both f and g are linear functions and

$$f\left(\begin{bmatrix} 1 \\ 0 \end{bmatrix}\right) = g\left(\begin{bmatrix} 1 \\ 0 \end{bmatrix}\right) = 1$$

However, $f \neq g$ since

$$f\left(\begin{bmatrix} 0 \\ 1 \end{bmatrix}\right) = 1 \quad \text{and} \quad g\left(\begin{bmatrix} 0 \\ 1 \end{bmatrix}\right) = 0 \qquad \blacksquare$$

As we saw in Example 1, knowing the value of a linear function of two variables at one nonzero point does not uniquely determine the function. However, we can get results for linear functions of two variables that are analogous to the results for functions of one variable. We use the fact that if f is linear

$$f\left(\begin{bmatrix} x \\ y \end{bmatrix}\right) = f\left(x\begin{bmatrix} 1 \\ 0 \end{bmatrix} + y\begin{bmatrix} 0 \\ 1 \end{bmatrix}\right) \qquad \text{since } \begin{bmatrix} x \\ y \end{bmatrix} = x\begin{bmatrix} 1 \\ 0 \end{bmatrix} + y\begin{bmatrix} 0 \\ 1 \end{bmatrix}$$

$$= f\left(x\begin{bmatrix} 1 \\ 0 \end{bmatrix}\right) + f\left(y\begin{bmatrix} 0 \\ 1 \end{bmatrix}\right) \qquad \text{Condition 1 of Definition 1.9}$$

$$= xf\left(\begin{bmatrix} 1 \\ 0 \end{bmatrix}\right) + yf\left(\begin{bmatrix} 0 \\ 1 \end{bmatrix}\right) \qquad \text{Condition 2 of Definition 1.9}$$

The following example shows how to use these ideas.

EXAMPLE 2 Suppose $f: \mathbb{R}^2 \to \mathbb{R}$ is a linear function and

$$f\left(\begin{bmatrix}1\\0\end{bmatrix}\right) = 1 \quad \text{and} \quad f\left(\begin{bmatrix}0\\1\end{bmatrix}\right) = -2$$

Find a general expression for f, and use this expression to find $f\left(\begin{bmatrix}3\\4\end{bmatrix}\right)$.

We find

$$\begin{aligned} f\left(\begin{bmatrix}x\\y\end{bmatrix}\right) &= xf\left(\begin{bmatrix}1\\0\end{bmatrix}\right) + yf\left(\begin{bmatrix}0\\1\end{bmatrix}\right) \\ &= x(1) + y(-2) \end{aligned}$$

Using this expression, we have

$$f\left(\begin{bmatrix}3\\4\end{bmatrix}\right) = 3(1) + 4(-2) = -5 \quad \blacksquare$$

We have discovered the general form for all linear functions $f: \mathbb{R}^2 \to \mathbb{R}$.

THEOREM 1.2 If $f: \mathbb{R}^2 \to \mathbb{R}$ is a linear function, then

$$f\left(\begin{bmatrix}x\\y\end{bmatrix}\right) = ax + by$$

where $a = f\left(\begin{bmatrix}1\\0\end{bmatrix}\right)$ and $b = f\left(\begin{bmatrix}0\\1\end{bmatrix}\right)$

This gives a partial answer to our original question, but we still have to find out if knowing the value of f at *any two* nonzero points completely determines f.

EXAMPLE 3 Suppose $f: \mathbb{R}^2 \to \mathbb{R}$ is a linear function with

$$f\left(\begin{bmatrix}1\\-2\end{bmatrix}\right) = 2 \quad \text{and} \quad f\left(\begin{bmatrix}-2\\4\end{bmatrix}\right) = -4$$

Is there only one linear function with these properties?

No. Both

$$f\left(\begin{bmatrix}x\\y\end{bmatrix}\right) = -y \quad \text{and} \quad g\left(\begin{bmatrix}x\\y\end{bmatrix}\right) = 4x + y$$

satisfy these conditions. $\blacksquare$

Comparing Example 3 and Theorem 1.2, we see that if $f: \mathbb{R}^2 \to \mathbb{R}$ is a linear function, then knowing the value of f at the two points $\begin{bmatrix}1\\0\end{bmatrix}$ and $\begin{bmatrix}0\\1\end{bmatrix}$ completely determines the function f, but knowing the value of a linear function at the two points $\begin{bmatrix}1\\-2\end{bmatrix}$ and $\begin{bmatrix}-2\\4\end{bmatrix}$ is not enough to uniquely determine the function. We must now determine which pairs of points can be used to determine linear functions.

EXAMPLE 4 Is there a linear function $f: \mathbb{R}^2 \to \mathbb{R}$ such that

$$f\left(\begin{bmatrix}1\\-2\end{bmatrix}\right) = 2 \qquad \text{and} \qquad f\left(\begin{bmatrix}2\\4\end{bmatrix}\right) = 8$$

If so, is this function unique?

We analyze this problem much like the problem for functions of one variable (Example 5 of Section 1.2). From Theorem 1.2 it is clear that if we can find $f\left(\begin{bmatrix}1\\0\end{bmatrix}\right)$ and $f\left(\begin{bmatrix}0\\1\end{bmatrix}\right)$, we can solve the problem. It is not too difficult to see that

$$\begin{bmatrix}1\\0\end{bmatrix} = (1/2)\begin{bmatrix}1\\-2\end{bmatrix} + (1/4)\begin{bmatrix}2\\4\end{bmatrix}$$

and

$$\begin{bmatrix}0\\1\end{bmatrix} = (-1/4)\begin{bmatrix}1\\-2\end{bmatrix} + (1/8)\begin{bmatrix}2\\4\end{bmatrix}$$

Using the properties of linear functions, we get

$$\begin{aligned}
f\left(\begin{bmatrix}1\\0\end{bmatrix}\right) &= f\left(1/2\begin{bmatrix}1\\-2\end{bmatrix} + 1/4\begin{bmatrix}2\\4\end{bmatrix}\right)\\
&= f\left(1/2\begin{bmatrix}1\\-2\end{bmatrix}\right) + f\left(1/4\begin{bmatrix}2\\4\end{bmatrix}\right) && \text{Condition 1 of Definition 1.9}\\
&= (1/2)f\left(\begin{bmatrix}1\\-2\end{bmatrix}\right) + (1/4)f\left(\begin{bmatrix}2\\4\end{bmatrix}\right) && \text{Condition 2 of Definition 1.9}\\
&= (1/2)(2) + (1/4)(8)\\
&= 3
\end{aligned}$$

and

$$\begin{aligned}
f\left(\begin{bmatrix}0\\1\end{bmatrix}\right) &= f\left(-1/4\begin{bmatrix}1\\-2\end{bmatrix} + 1/8\begin{bmatrix}2\\4\end{bmatrix}\right)\\
&= f\left(-1/4\begin{bmatrix}1\\-2\end{bmatrix}\right) + f\left(1/8\begin{bmatrix}2\\4\end{bmatrix}\right) && \text{Condition 1 of Definition 1.9}
\end{aligned}$$

$$= (-1/4)f\left(\begin{bmatrix}1\\-2\end{bmatrix}\right) + (1/8)f\left(\begin{bmatrix}2\\4\end{bmatrix}\right) \quad \text{Condition 2 of Definition 1.9}$$

$$= (-1/4)\cdot(2) + (1/8)\cdot(8)$$

$$= 1/2$$

Since $$f\left(\begin{bmatrix}x\\y\end{bmatrix}\right) = xf\left(\begin{bmatrix}1\\0\end{bmatrix}\right) + yf\left(\begin{bmatrix}0\\1\end{bmatrix}\right)$$

we have $$f\left(\begin{bmatrix}x\\y\end{bmatrix}\right) = 3x + (1/2)y$$

This is the only linear function that satisfies the given conditions. ■

Note that we could find this function because we could express both $\begin{bmatrix}1\\0\end{bmatrix}$ and $\begin{bmatrix}0\\1\end{bmatrix}$ as sums of scalar multiples of the given vectors $\begin{bmatrix}1\\-2\end{bmatrix}$ and $\begin{bmatrix}2\\4\end{bmatrix}$. In Example 3 the reason we could not find a unique function f satisfying the given conditions was that we could not express $\begin{bmatrix}1\\0\end{bmatrix}$ $\left(\text{or } \begin{bmatrix}0\\1\end{bmatrix}\right)$ as a sum of scalar multiples of the vectors $\begin{bmatrix}1\\-2\end{bmatrix}$ and $\begin{bmatrix}-2\\4\end{bmatrix}$. We will discuss this topic in more detail in Section 1.5.

DEFINITION 1.10

Let $\mathbf{v}_1, \mathbf{v}_2, \ldots, \mathbf{v}_k$ be vectors. If $\mathbf{v}$ is a sum of scalar multiples of $\mathbf{v}_1, \mathbf{v}_2, \ldots, \mathbf{v}_k$, that is,

$$\mathbf{v} = c_1\mathbf{v}_1 + c_2\mathbf{v}_2 + \cdots + c_k\mathbf{v}_k$$

where $c_1, c_2, \ldots, c_k$ are scalars, then $\mathbf{v}$ is said to be a **linear combination** of the vectors $\mathbf{v}_1, \mathbf{v}_2, \ldots, \mathbf{v}_k$. (*Note:* Some or all of the c_i's may be zero.)

Proceeding as in Example 4, we get the following theorem.

THEOREM 1.3

Let $\mathbf{v}$ and $\mathbf{w}$ be vectors in $\mathbb{R}^2$ with the property that both $\begin{bmatrix}1\\0\end{bmatrix}$ and $\begin{bmatrix}0\\1\end{bmatrix}$ can be expressed as linear combinations of $\mathbf{v}$ and $\mathbf{w}$. Then, for any real numbers s and t, there is one and only one linear function f such that $f\colon \mathbb{R}^2 \to \mathbb{R}$ and $f(\mathbf{v}) = s$ and $f(\mathbf{w}) = t$.

EXAMPLE 5 Find the linear function $f\colon \mathbb{R}^2 \to \mathbb{R}$ such that

$$f\left(\begin{bmatrix}1\\2\end{bmatrix}\right) = 4 \quad \text{and} \quad f\left(\begin{bmatrix}-1\\-3\end{bmatrix}\right) = -5$$

Use the facts that

$$\begin{bmatrix}1\\0\end{bmatrix} = 3\begin{bmatrix}1\\2\end{bmatrix} + 2\begin{bmatrix}-1\\-3\end{bmatrix} \quad \text{and} \quad \begin{bmatrix}0\\1\end{bmatrix} = -1\begin{bmatrix}1\\2\end{bmatrix} + -1\begin{bmatrix}-1\\-3\end{bmatrix}$$

Since f is linear,

$$\begin{aligned} f\left(\begin{bmatrix}1\\0\end{bmatrix}\right) &= f\left(3\begin{bmatrix}1\\2\end{bmatrix} + 2\begin{bmatrix}-1\\-3\end{bmatrix}\right) \\ &= 3f\left(\begin{bmatrix}1\\2\end{bmatrix}\right) + 2f\left(\begin{bmatrix}-1\\-3\end{bmatrix}\right) \\ &= 3(4) + 2(-5) \\ &= 2 \end{aligned}$$

Similarly,

$$\begin{aligned} f\left(\begin{bmatrix}0\\1\end{bmatrix}\right) &= f\left(-1\begin{bmatrix}1\\2\end{bmatrix} + -1\begin{bmatrix}-1\\-3\end{bmatrix}\right) \\ &= -1f\left(\begin{bmatrix}1\\2\end{bmatrix}\right) + -1f\left(\begin{bmatrix}-1\\-3\end{bmatrix}\right) \\ &= -1(4) + -1(-5) \\ &= 1 \end{aligned}$$

Using Theorem 1.2, we have

$$\begin{aligned} f\left(\begin{bmatrix}x\\y\end{bmatrix}\right) &= xf\left(\begin{bmatrix}1\\0\end{bmatrix}\right) + yf\left(\begin{bmatrix}0\\1\end{bmatrix}\right) \\ &= 2x + 1y \quad ■ \end{aligned}$$

Exercise Set 1.4

In Exercises 1 to 6 find a linear function $f\colon \mathbb{R}^2 \to \mathbb{R}$ satisfying the conditions listed.

1. $f\left(\begin{bmatrix}1\\0\end{bmatrix}\right) = 5$ and $f\left(\begin{bmatrix}0\\1\end{bmatrix}\right) = -2$

2. $f\left(\begin{bmatrix}1\\0\end{bmatrix}\right) = 17$ and $f\left(\begin{bmatrix}0\\1\end{bmatrix}\right) = 0$

3. $f\left(\begin{bmatrix}1\\2\end{bmatrix}\right) = -7$ and $f\left(\begin{bmatrix}-1\\-3\end{bmatrix}\right) = 4$

4. $f\left(\begin{bmatrix}1\\2\end{bmatrix}\right) = 6$ and $f\left(\begin{bmatrix}-1\\-3\end{bmatrix}\right) = -3$

5. $f\left(\begin{bmatrix}1\\0\end{bmatrix}\right) = 5$ and $f\left(\begin{bmatrix}1\\1\end{bmatrix}\right) = 3$

6. $f\left(\begin{bmatrix}1\\1\end{bmatrix}\right) = 4$ and $f\left(\begin{bmatrix}2\\3\end{bmatrix}\right) = 5$

7. Find two linear functions $f, g: \mathbb{R}^2 \to \mathbb{R}$ such that

$$f\left(\begin{bmatrix}2\\3\end{bmatrix}\right) = g\left(\begin{bmatrix}2\\3\end{bmatrix}\right) = 6$$

and
$$f\left(\begin{bmatrix}2/3\\1\end{bmatrix}\right) = g\left(\begin{bmatrix}2/3\\1\end{bmatrix}\right) = 2$$

8. Express the vectors $\begin{bmatrix}1\\0\end{bmatrix}$ and $\begin{bmatrix}0\\1\end{bmatrix}$ as linear combinations of the two vectors in each of the following sets.

a. $\begin{bmatrix}1\\2\end{bmatrix}$ and $\begin{bmatrix}-3\\4\end{bmatrix}$

b. $\begin{bmatrix}5\\-3\end{bmatrix}$ and $\begin{bmatrix}2\\-3\end{bmatrix}$

c. $\begin{bmatrix}1\\5\end{bmatrix}$ and $\begin{bmatrix}5\\1\end{bmatrix}$

9. Describe the set of all linear combinations of the vectors $\begin{bmatrix}1\\2\end{bmatrix}$ and $\begin{bmatrix}-3\\-6\end{bmatrix}$.

10. Explain why the vector $\begin{bmatrix}1\\0\end{bmatrix}$ cannot be expressed as a linear combination of the vectors $\begin{bmatrix}1\\2\end{bmatrix}$ and $\begin{bmatrix}-3\\-6\end{bmatrix}$. (*Hint:* graph the vectors.)

1.5 Bases for $\mathbb{R}^2$

In the previous section we saw that a linear function on $\mathbb{R}^2$ can be completely determined by its value at two vectors. However, not any pair of vectors in $\mathbb{R}^2$ will do. The condition that such a pair of vectors must satisfy is that both $\begin{bmatrix}1\\0\end{bmatrix}$ and $\begin{bmatrix}0\\1\end{bmatrix}$ can be expressed as linear combinations of these vectors. You should note that this property is a purely vector property and has nothing to do with the particular function in question. Any pair of vectors that satisfies this condition is called a **basis** for $\mathbb{R}^2$. (A more general definition of basis will be given in Section 4.3).

What does it mean for a vector $\mathbf{v}$ to be a linear combination of two vectors $\mathbf{w}_1$ and $\mathbf{w}_2$? This means that there are scalars a and b so that $\mathbf{v} = a\mathbf{w}_1 + b\mathbf{w}_2$. Geometrically, this means that $\mathbf{v}$ is the diagonal of the parallelogram having sides $a\mathbf{w}_1$ and $b\mathbf{w}_2$ (Figure 1.7). We construct this parallelogram beginning with the vectors $\mathbf{v}$, $\mathbf{w}_1$, and $\mathbf{w}_2$. Note that two sides of the parallelogram contain the origin. These are segments of the lines containing the vectors $\mathbf{w}_1$ and $\mathbf{w}_2$. The other two sides lie on the two lines parallel to $\mathbf{w}_1$ and $\mathbf{w}_2$ and contain the endpoint of the vector $\mathbf{v}$. The parallelogram is formed when these four lines are drawn—the vertices of the parallelogram are the points of intersection of these lines (Figure 1.8).

We can conclude that $\mathbf{v}$ is a linear combination of $\mathbf{w}_1$ and $\mathbf{w}_2$ if and only if $\mathbf{v}$ is the diagonal of a parallelogram that has the origin as one vertex and two of its sides parallel to $\mathbf{w}_1$ and the other two sides parallel to $\mathbf{w}_2$.

For a fixed pair of vectors $\mathbf{w}_1$ and $\mathbf{w}_2$, we wish to determine which vectors $\mathbf{v}$ can be expressed as linear combinations of $\mathbf{w}_1$ and $\mathbf{w}_2$.

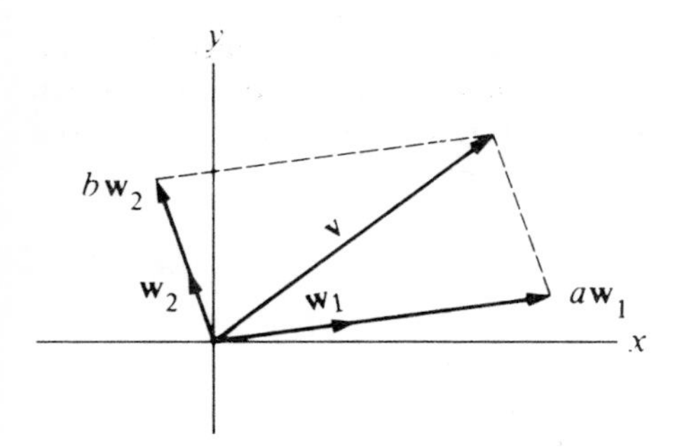

Figure 1.7

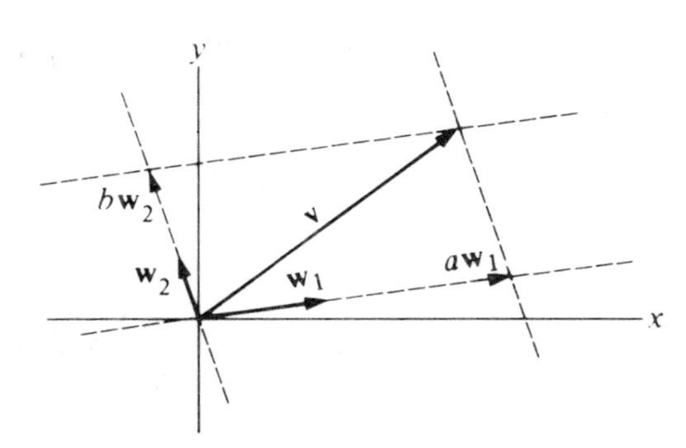

Figure 1.8

If $\mathbf{w}_1$ and $\mathbf{w}_2$ are not parallel (do not lie on the same line through the origin), then for any vector $\mathbf{v}$, we can construct a parallelogram with the origin as one vertex, two sides parallel to $\mathbf{w}_1$, the other two sides parallel to $\mathbf{w}_2$, and $\mathbf{v}$ as the diagonal. In particular both $\begin{bmatrix} 1 \\ 0 \end{bmatrix}$ and $\begin{bmatrix} 0 \\ 1 \end{bmatrix}$ can be written as linear combinations of $\mathbf{w}_1$ and $\mathbf{w}_2$ and form a basis for $\mathbb{R}^2$.

On the other hand, if $\mathbf{w}_1$ and $\mathbf{w}_2$ are parallel (lie on the same line through the origin), then only vectors parallel to $\mathbf{w}_1$ (and $\mathbf{w}_2$) can be expressed as linear combinations of $\mathbf{w}_1$ and $\mathbf{w}_2$ (Figure 1.9). These vectors do not form a basis of $\mathbb{R}^2$. Therefore, we have proved the following theorem.

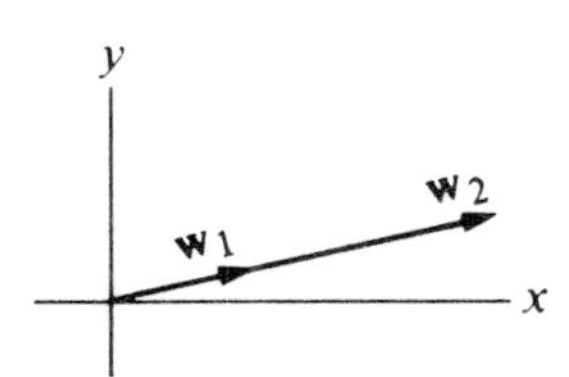

Figure 1.9 $\mathbf{w}_1$ **and** $\mathbf{w}_2$ **are parallel.**

THEOREM 1.4

Let $\mathbf{w}_1$ and $\mathbf{w}_2$ be vectors in $\mathbb{R}^2$. Then $\mathbf{w}_1$ and $\mathbf{w}_2$ form a basis for $\mathbb{R}^2$ if and only if $\mathbf{w}_1$ is not parallel to $\mathbf{w}_2$. (*Note:* The zero vector $\begin{bmatrix} 0 \\ 0 \end{bmatrix}$ is parallel to every vector; this means that neither $\mathbf{w}_1$ nor $\mathbf{w}_2$ is the zero vector.)

Although the geometric interpretation of basis gives a very nice theorem, it is not a practical method for finding which linear combination of a given set of basis vectors equals a particular vector. It is much better to use algebraic methods to solve this kind of problem.

EXAMPLE 1 Express $\begin{bmatrix}1\\0\end{bmatrix}$ as a linear combination of the vectors

$$\mathbf{w}_1 = \begin{bmatrix}2\\1\end{bmatrix} \quad \text{and} \quad \mathbf{w}_2 = \begin{bmatrix}-2\\3\end{bmatrix}$$

Since $\mathbf{w}_1$ and $\mathbf{w}_2$ are not parallel, we know (from Theorem 1.4) that it is possible to write $\begin{bmatrix}1\\0\end{bmatrix}$ as a linear combination of $\mathbf{w}_1$ and $\mathbf{w}_2$. To do this we must find numbers x and y such that

$$\begin{bmatrix}1\\0\end{bmatrix} = x\begin{bmatrix}2\\1\end{bmatrix} + y\begin{bmatrix}-2\\3\end{bmatrix}$$

Hence

$$\begin{bmatrix}1\\0\end{bmatrix} = \begin{bmatrix}2x\\1x\end{bmatrix} + \begin{bmatrix}-2y\\3y\end{bmatrix} = \begin{bmatrix}2x-2y\\x+3y\end{bmatrix}$$

This reduces to solving the system of linear equations

$$1 = 2x - 2y \quad \text{and} \quad 0 = x + 3y$$

which can be solved in several ways. The simplest is to solve the second equation for x

$$0 = x + 3y$$
$$x = -3y$$

and substitute this into the first equation to get

$$1 = 2(-3y) - 2y = -8y$$

Thus $y = -1/8$. Since $x = -3y$, $x = 3/8$. Therefore,

$$\begin{bmatrix}1\\0\end{bmatrix} = 3/8\begin{bmatrix}2\\1\end{bmatrix} + -1/8\begin{bmatrix}-2\\3\end{bmatrix}$$

The graph of this equation is given in Figure 1.10. ■

Another method for solving linear equations will be discussed in Section 2.1.

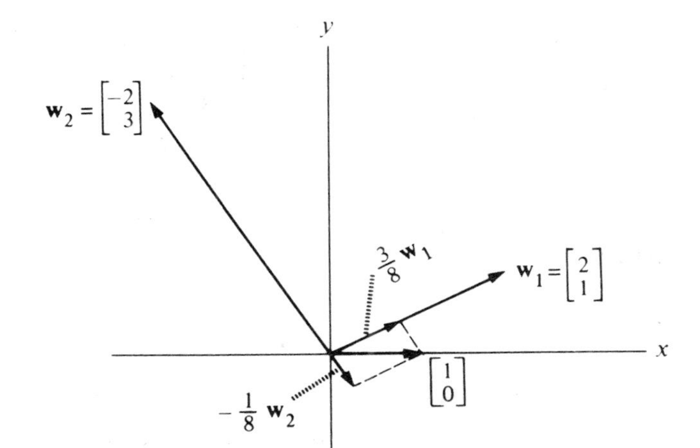

Figure 1.10

Exercise Set 1.5

In Exercises 1–5 express each of the vectors $\begin{bmatrix} 1 \\ 0 \end{bmatrix}$ and $\begin{bmatrix} 0 \\ 1 \end{bmatrix}$ as a linear combination of the vectors $\mathbf{w}_1$, $\mathbf{w}_2$ (whenever possible). Also draw the geometric picture associated with each of these exercises (see Example 1 of this section).

1. $\mathbf{w}_1 = \begin{bmatrix} 1 \\ 0 \end{bmatrix}$, $\mathbf{w}_2 = \begin{bmatrix} 2 \\ -1 \end{bmatrix}$

2. $\mathbf{w}_1 = \begin{bmatrix} -2 \\ 6 \end{bmatrix}$, $\mathbf{w}_2 = \begin{bmatrix} 1 \\ -3 \end{bmatrix}$

3. $\mathbf{w}_1 = \begin{bmatrix} 5 \\ 6 \end{bmatrix}$, $\mathbf{w}_2 = \begin{bmatrix} 0 \\ -3 \end{bmatrix}$

4. $\mathbf{w}_1 = \begin{bmatrix} -2 \\ -4 \end{bmatrix}$, $\mathbf{w}_2 = \begin{bmatrix} 1 \\ 3 \end{bmatrix}$

5. $\mathbf{w}_1 = \begin{bmatrix} -12 \\ -9 \end{bmatrix}$, $\mathbf{w}_2 = \begin{bmatrix} 4 \\ 3 \end{bmatrix}$

6. Let $f: \mathbb{R}^2 \to \mathbb{R}$ be a linear function with $f(\mathbf{w}_1) = 3$ and $f(\mathbf{w}_2) = -1$. For each pair of vectors $\mathbf{w}_1$ and $\mathbf{w}_2$ in Exercises 1–5, find all functions f satisfying this condition.

7. Let $f: \mathbb{R}^2 \to \mathbb{R}$

$$\begin{bmatrix} x \\ y \end{bmatrix} \to 2x + y$$

a. Find two nonzero vectors $\mathbf{v}_1$ and $\mathbf{v}_2$ in $\mathbb{R}^2$ such that $f(\mathbf{v}_1) = 0$ and $f(\mathbf{v}_2) = 1$.

b. Graph $\mathbf{v}_1$ and $\mathbf{v}_2$ and show that these two vectors form a basis for $\mathbb{R}^2$.

c. Let $\mathbf{w} = \begin{bmatrix} 1 \\ 1 \end{bmatrix}$. Find numbers a, b such that $\mathbf{w} = a\mathbf{v}_1 + b\mathbf{v}_2$.

d. Use your result in part c and the fact that f is linear to compute $f(\mathbf{w})$. Compare this to the result you would obtain by computing $f(\mathbf{w})$ using the original definition of f.

★**8.** Let $f: \mathbb{R}^2 \to \mathbb{R}$ be a linear function.

a. Show that there is a nonzero vector $\mathbf{v}_0$ such that $f(\mathbf{v}_0) = 0$. (*Hint*: Start by writing f in the form given in Theorem 1.2).

b. Show that if $\mathbf{v}_0$ is the vector of part a, then $f(c\mathbf{v}_0) = 0$ for every scalar c.

c. Show that if $\mathbf{w}$ is a vector of $\mathbb{R}^2$ that is not parallel to $\mathbf{v}_0$, and if $f(\mathbf{w}) = \mathbf{0}$, then $f(\mathbf{v}) = \mathbf{0}$ for every vector $\mathbf{v}$ of $\mathbb{R}^2$.

★9. Consider the linear function

$$f: \mathbb{R}^2 \to \mathbb{R}$$
$$\begin{bmatrix} x \\ y \end{bmatrix} \to 3x - 2y$$

a. Find the set N of all vectors $\mathbf{v}$ of $\mathbb{R}^2$ such that $f(\mathbf{v}) = 0$.

b. Graph the vectors in N.

c. Find the set S_1 of all vectors $\mathbf{w}$ of $\mathbb{R}^2$ such that $f(\mathbf{w}) = 1$.

d. Graph S_1.

e. Graph the set S_2 of all vectors $\mathbf{u}$ of $\mathbb{R}^2$ such that $f(\mathbf{u}) = 2$.

f. What can you say about the sets N, S_1, S_2, ...?

Applications 1 Linear Algebra, Complex Numbers, and the Roots of $y = x^n - 1$

Although we have only scratched the surface in our study of linear algebra, we can relate many of the ideas discussed to a mathematical problem which, at first glance, seems to have no connection with vectors and linear functions. The problem is to find all roots of equations of the form $0 = x^n - 1$. The Fundamental Theorem of Algebra states that every polynomial of degree n with real (or complex) coefficients has exactly n roots. The difficulty with this theorem is that it does not tell what the roots are or how to find them. In fact, no general methods can be found for solving all polynomials of degree 5 or higher. However, for certain classes of polynomials, it is possible to find *all* the roots. One such class is the set of all polynomials of the form $x^n - 1$.

For $n = 1$ and $n = 2$, the polynomials are $x - 1$ and $x^2 - 1$, and the roots of these two polynomials are $x = 1$ and $x = 1, -1$. For $n = 4$, we factor the equation to obtain

$$x^4 - 1 = (x^2 - 1)(x^2 + 1) = (x - 1)(x + 1)(x - i)(x + i)$$

The roots of this polynomial are

$$x = 1, -1, i, -i \qquad \text{where } i = \sqrt{-1}$$

EXAMPLE 1 The roots of $x^3 - 1$ are

$$x = 1, \quad x = \frac{-1 + \sqrt{3}\,i}{2} \quad \text{and} \quad x = \frac{-1 - \sqrt{3}\,i}{2}$$

We verify that

$$x = \frac{-1 + \sqrt{3}\,i}{2}$$

is a root.

$$\left[\frac{-1+\sqrt{3}\,i}{2}\right]^3 = \left[\frac{-1+\sqrt{3}\,i}{2}\right]\left[\frac{-1+\sqrt{3}\,i}{2}\right]^2$$
$$= \left[\frac{-1+\sqrt{3}\,i}{2}\right]\left[\frac{-2-2\sqrt{3}\,i}{4}\right]$$
$$= \left[\frac{-1+\sqrt{3}\,i}{2}\right]\left[\frac{-1-\sqrt{3}\,i}{2}\right]$$
$$= \frac{1-3i^2}{4}$$
$$= 1$$

(A similar calculation shows that $(-1-\sqrt{3}\,i)/2$ is also a root of $x^3 - 1$.) ■

We have now found all the roots for all equations of the form $0 = x^n - 1$ for $n = 1, 2, 3$, and 4. However, this has given us little insight into how to find the solutions of such equations for $n \geq 5$. To see what is happening, we view complex numbers as vectors.

Just as we plot real numbers as points on a line, we can plot complex numbers as points in the plane. The complex number $a + bi$ is graphed in Figure 1.11 as the point (a, b). As we saw in Section 1.3, each point in the plane corresponds to a vector. So we see that we can associate the vector $\begin{bmatrix} a \\ b \end{bmatrix}$ with the complex number $a + bi$.

Figure 1.11

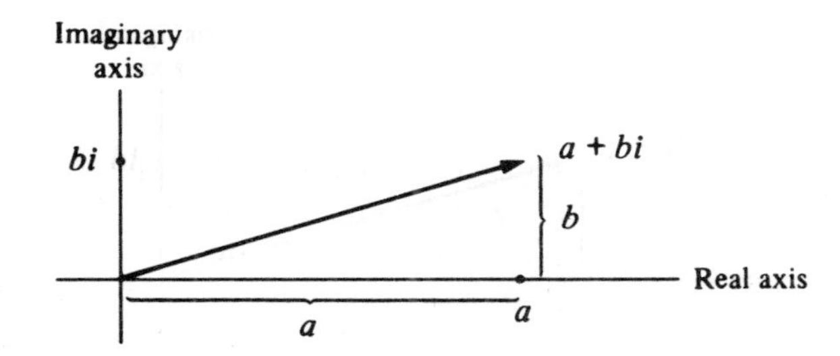

It is important to note that not only is this correspondence between vectors and complex numbers one-to-one (each distinct pair of vectors corresponds to a distinct pair of complex numbers), but also that the two vector operations of addition and multiplication by scalars (real numbers) are preserved. By saying that the operation of addition is preserved, we mean that if we wish to find the vector corresponding to the sum of two complex numbers, we can either add the two numbers and then find the vector corresponding to this sum, or we can find the vectors corresponding to each of the complex numbers and then add these two vectors. Both of these methods give the same result.

$$(a_1 + b_1 i) + (a_2 + b_2 i) = (a_1 + a_2) + i(b_1 + b_2)$$

corresponds to ↓ corresponds to ↓ corresponds to

$$\begin{bmatrix} a_1 \\ b_1 \end{bmatrix} + \begin{bmatrix} a_2 \\ b_2 \end{bmatrix} = \begin{bmatrix} a_1 + a_2 \\ b_1 + b_2 \end{bmatrix}$$

Similarly, scalar multiplication is preserved under this correspondence.

EXAMPLE 2 We illustrate the correspondence between vectors and complex numbers by looking at the sum of two complex numbers and the corresponding sum of vectors, as shown in Figure 1.12

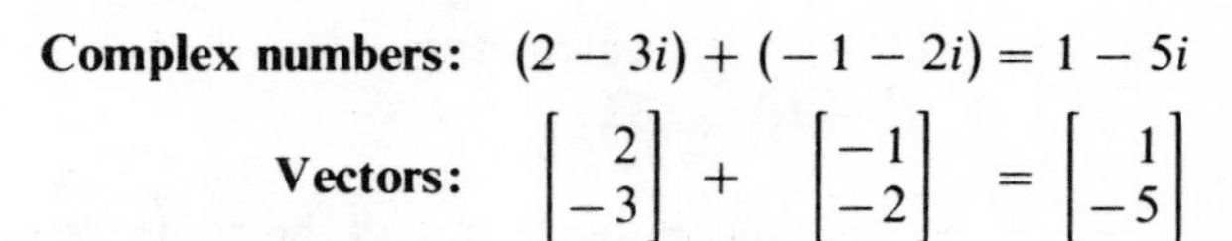

Complex numbers: $(2 - 3i) + (-1 - 2i) = 1 - 5i$

Vectors: $\begin{bmatrix} 2 \\ -3 \end{bmatrix} + \begin{bmatrix} -1 \\ -2 \end{bmatrix} = \begin{bmatrix} 1 \\ -5 \end{bmatrix}$

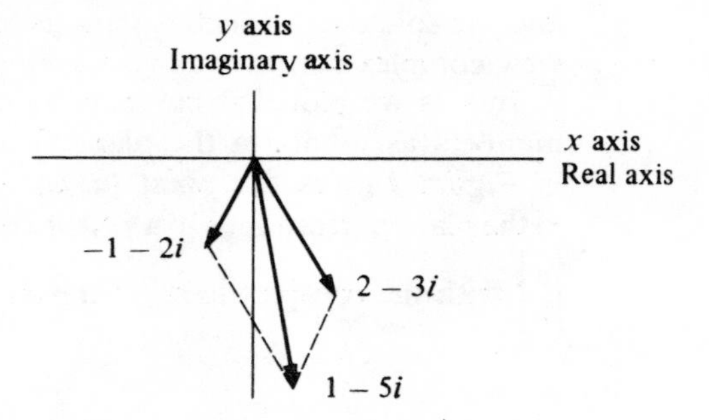

Figure 1.12

■

Vectors in $\mathbb{R}^2$ and complex numbers are essentially the same relative to addition and scalar multiplication. However, since complex numbers can be multiplied together, this gives us a way (which we didn't have until now) of multiplying one vector of $\mathbb{R}^2$ by another vector of $\mathbb{R}^2$.

To begin, we investigate multiplication by i. If we multiply the complex number $a + bi$ by i, we obtain

$$i(a + bi) = ia + i^2 b = -b + ai$$

Using our correspondence between complex numbers and vectors, we can define the function f_i: $\mathbb{R}^2 \to \mathbb{R}^2$, which corresponds to multiplication by i. The function f_i is defined by

$$f_i: \ \mathbb{R}^2 \to \mathbb{R}^2$$

$$\begin{bmatrix} a \\ b \end{bmatrix} \to \begin{bmatrix} -b \\ a \end{bmatrix}$$

Not only is f_i a function, but f_i is a linear function (see Exercise 1 at the end of this section).

We see in Figure 1.13 what f_i does to a typical vector. In the figure, the vector $\begin{bmatrix} a \\ b \end{bmatrix}$ is on the line with slope b/a and the vector $f_i\left(\begin{bmatrix} a \\ b \end{bmatrix}\right) = \begin{bmatrix} -b \\ a \end{bmatrix}$ is on the line with slope $a/(-b)$, which means that these two vectors are perpendicular. Moreover, the two vectors have the same length. We can thus conclude that f_i is the function that rotates each vector by an angle of $\pi/2$ (90°) counterclockwise. This gives us a new interpretation of multiplication by i; it is counterclockwise rotation by $\pi/2$.

Figure 1.13

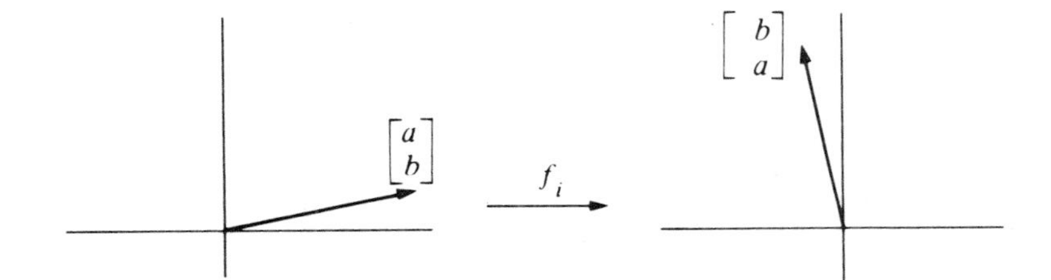

Since multiplication by i^2 is just multiplying by i twice, we see that multiplication by i^2 can be thought of as rotation by $\pi/2$ twice, that is, rotation by π (see Figure 1.14).

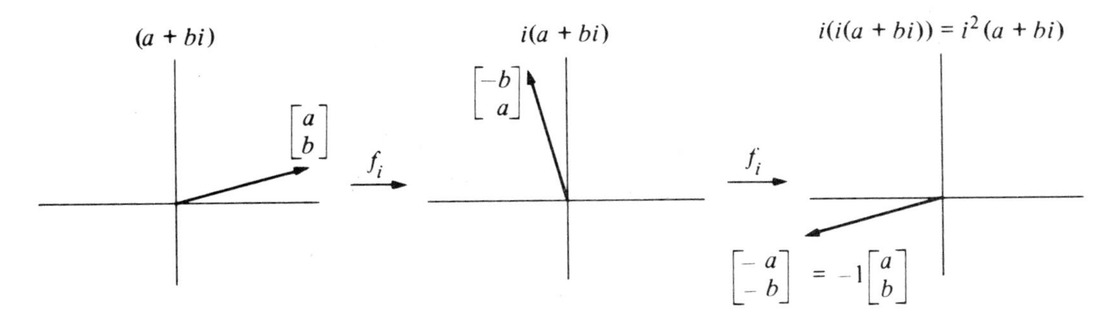

Figure 1.14

In the same way we see (Figure 1.14) that multiplication by i^3 represents a counterclockwise rotation of $3\pi/2$ (270°) and multiplication by $i^4 = 1$ represents rotation by 2π (360°).

EXAMPLE 3 Multiply the complex number $2 + 1i$ by the numbers i, i^2, i^3, and i^4. (This gives a concrete illustration of the geometric interpretation of multiplication by powers of i.)

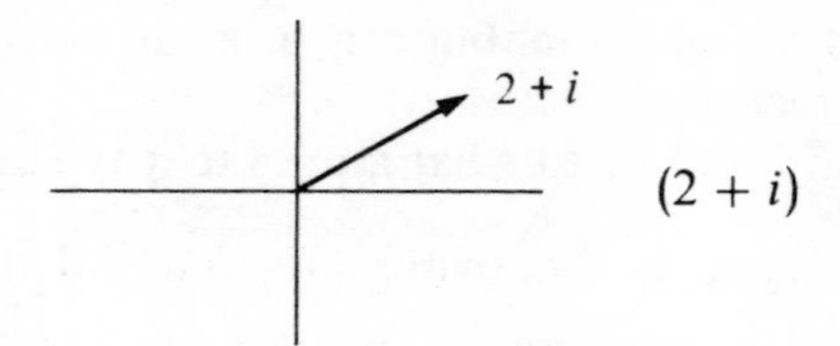

$(2 + i)$

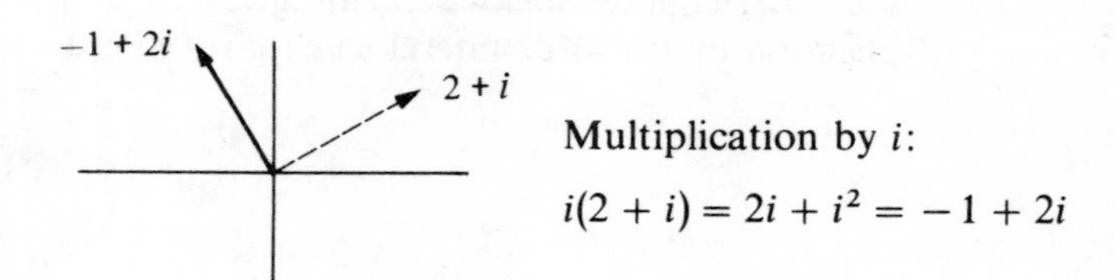

Multiplication by i:

$i(2 + i) = 2i + i^2 = -1 + 2i$

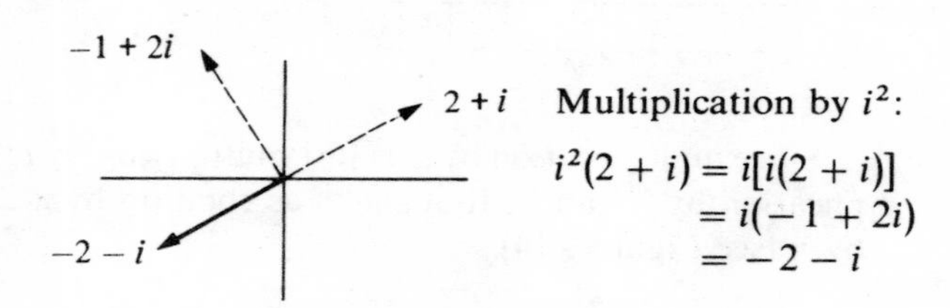

Multiplication by i^2:

$$\begin{aligned} i^2(2 + i) &= i[i(2 + i)] \\ &= i(-1 + 2i) \\ &= -2 - i \end{aligned}$$

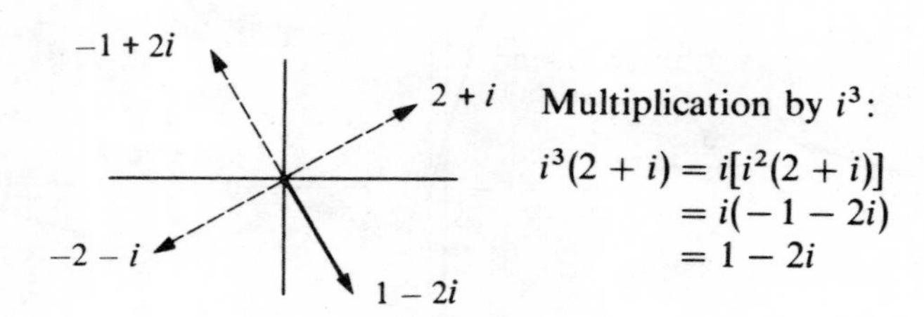

Multiplication by i^3:

$$\begin{aligned} i^3(2 + i) &= i[i^2(2 + i)] \\ &= i(-1 - 2i) \\ &= 1 - 2i \end{aligned}$$

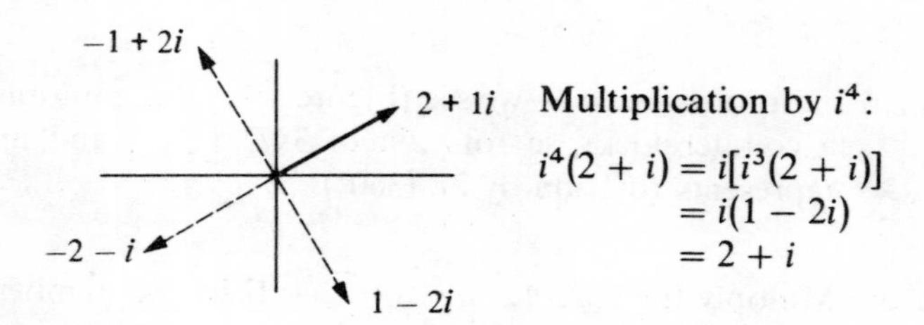

Multiplication by i^4:

$$\begin{aligned} i^4(2 + i) &= i[i^3(2 + i)] \\ &= i(1 - 2i) \\ &= 2 + i \end{aligned}$$

■

Figure 1.15 Geometric interpretation of multiplication by powers of *i*.

Now let us see how this interpretation of multiplication by i as rotation by $\pi/2$ can help us solve our original problem of finding the roots of equations of the form $0 = x^n - 1$. The roots of such an equation are complex numbers z such that $z^n = 1$. In the case in which $n = 4$, we found that i was such a number. Moreover, we have just found that multiplication by i can be thought of as a rotation of 1/4 of the way around.

In order to find complex numbers that satisfy the equation $z^n = 1$, let us see if we can find complex numbers z so that multiplication by z represents a rotation of $1/n$ way around, that is, a rotation of $2\pi/n$. In particular, let us see if this technique will work to find the roots of $y = x^3 - 1$. In this case we want to find a complex number z so that multiplication by z represents a rotation of $2\pi/3$. Although this seems just as hard as solving the original problem, the fact that multiplication by a complex number represents a linear function (see Exercise 2 of this section) allows us to find such a function quite simply. The vectors $\begin{bmatrix}1\\0\end{bmatrix}$ and $\begin{bmatrix}0\\1\end{bmatrix}$ that correspond to the complex numbers $1 + 0i$ and $0 + 1i$ are a basis for $\mathbb{R}^2$. Hence, if we can find a complex number z so that f_z (where f_z is the function that represents multiplication by z) rotates each of these vectors by $2\pi/3$, then, since f_z is linear, it will rotate each vector in $\mathbb{R}^2$ by $2\pi/3$. (See Exercise 7 at the end of this section.)

We begin by finding $f_z\begin{bmatrix}1\\0\end{bmatrix}$. This is the vector of length 1 that is at an angle of $2\pi/3$ from the x axis. It is shown graphically in Figure 1.16.

We have
$$f_z\begin{bmatrix}1\\0\end{bmatrix} = \begin{bmatrix}\cos(2\pi/3)\\ \sin(2\pi/3)\end{bmatrix} = \begin{bmatrix}-1/2\\ \sqrt{3}/2\end{bmatrix}$$

which we translate back into a statement about complex numbers as

$$\begin{aligned} z(1 + 0i) &= \cos(2\pi/3) + i\sin(2\pi/3) \\ &= (-1/2) + i(\sqrt{3}/2) \end{aligned}$$

Figure 1.16

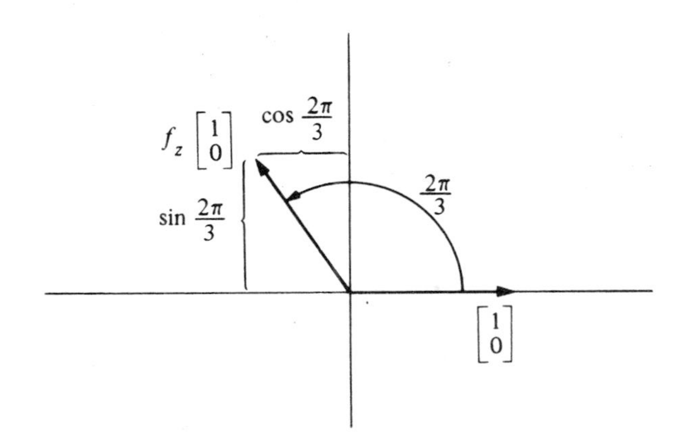

Since $z(1 + 0i) = z$, we have found that if there is a complex number z so that multiplication by z represents a rotation of $2\pi/3$, then

$$z = \frac{-1 + i\sqrt{3}}{2}$$

As we saw above, to see if multiplication by z rotates *every* vector in $\mathbb{R}^2$ by the angle $2\pi/3$, we need only check to see that it rotates the vector corresponding to i by $2\pi/3$, as in Figure 1.17.

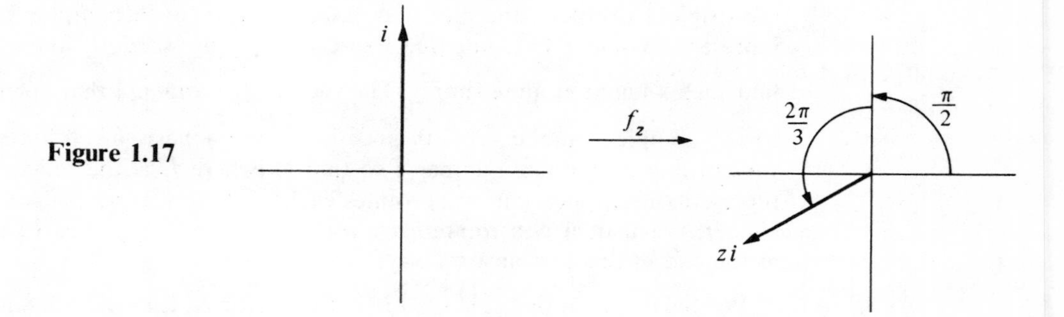

Figure 1.17

It suffices to show that the angle between the vector corresponding to zi and the real axis (x axis) is $(2\pi/3 + \pi/2)$. Since $zi = iz$ and multiplication of z by i corresponds to rotation of z by $\pi/2$, the angle between iz and the real axis is

$$(\text{Angle between } z \text{ and real axis}) + \frac{\pi}{2} = \frac{2\pi}{3} + \frac{\pi}{2}$$

Since multiplication by z rotates the vectors corresponding to 1 and i by $2\pi/3$, multiplication by z rotates each complex number by $2\pi/3$, and since rotation by z three times brings each vector back to its original position, $z^3\alpha = \alpha$ for every complex number α. In particular, $z^3 = 1$, so z is a root of the equation $0 = x^3 - 1$ (as we saw in Example 1).

Furthermore, we note that

$$1 = (z^3)^2 = (z^2)^3$$

which means that z^2 is also a root of the equation $y = x^3 - 1$. Therefore, the three roots of the equation $y = x^3 - 1$ are

$$z^0 = 1, \quad z^1 = (-1/2) + (\sqrt{3}/2)i \qquad \text{and} \qquad z^2 = (-1/2) - (\sqrt{3}/2)i$$

These same ideas can be generalized to give the following theorems.

THEOREM 1.5

> Let $z = \cos\theta + i\sin\theta$. Multiplication by z represents a counterclockwise rotation by the angle θ.

THEOREM 1.6

> The roots of the equation $0 = x^n - 1$ are $1, z, z^2, \ldots, z^{n-1}$, where $z = \cos\theta + i\sin\theta$, and $\theta = 2\pi/n$.

Exercises for Applications 1

1. Show that f_i (as defined in this section) is a linear function; that is, show that f_i satisfies the following two conditions:

a. $f_i\left(\begin{bmatrix} a_1 \\ b_1 \end{bmatrix} + \begin{bmatrix} a_2 \\ b_2 \end{bmatrix}\right) = f_i\left(\begin{bmatrix} a_1 \\ b_1 \end{bmatrix}\right) + f_i\left(\begin{bmatrix} a_2 \\ b_2 \end{bmatrix}\right)$

for all vectors $\begin{bmatrix} a_1 \\ b_1 \end{bmatrix}, \begin{bmatrix} a_2 \\ b_2 \end{bmatrix}$

b. $f_i\left(c\begin{bmatrix} a \\ b \end{bmatrix}\right) = cf_i\left(\begin{bmatrix} a \\ b \end{bmatrix}\right)$

for every vector $\begin{bmatrix} a \\ b \end{bmatrix}$ and every real number c

2. Show that for any complex number z, the function $f_z\colon \mathbb{R}^2 \to \mathbb{R}^2$, which corresponds to multiplication by z, is a linear function.

3. a. Find a complex number z such that multiplication by z corresponds to a rotation by the angle $2\pi/5$.

b. Find a solution (different from $x = 1$) to the equation $x^5 = 1$.

c. Find all roots of the equation $0 = x^5 - 1$.

4. a. Let z be the complex number $z = a + bi$ where a and b have the property that $a^2 + b^2 = 1$. Show that there is an angle θ so that $z = \cos\theta + i\sin\theta$.

b. Show that any complex number z can be written in the form

$$z = r[\cos\theta + i\sin\theta]$$

where r is a real number. This is called the **polar form** of the complex number z.

c. Write the following numbers in polar form:

$$z = 3 + 4i, \quad z = -1/2 + (\sqrt{3}/2)i,$$
$$z = -i, \quad z = 2 + 2i$$

5. Let $z = 2i$. Describe the linear function f_z corresponding to multiplication by z. Also describe the functions f_{z^2}, f_{z^3}, and f_{z^4}.

6. a. Let $z = r(\cos\theta + i\sin\theta)$. Show that $z^n = r^n(\cos n\theta + i\sin n\theta)$.

b. Find all solutions of the equation $x^3 = 8$.

c. Find $(-1 + \sqrt{3}\,i)^6$. (This is not a difficult computation if you use part a).

7. Let $f\colon \mathbb{R}^2 \to \mathbb{R}^2$ be a linear function (see Exercise 1), and suppose that f rotates each of the vectors $\begin{bmatrix} 1 \\ 0 \end{bmatrix}$ and $\begin{bmatrix} 0 \\ 1 \end{bmatrix}$ by an angle θ counterclockwise. Show that for any vector $\begin{bmatrix} a \\ b \end{bmatrix}$, the action of f on $\begin{bmatrix} a \\ b \end{bmatrix}$ is rotation by the angle θ.

Hint: Write $\begin{bmatrix} a \\ b \end{bmatrix} = a\begin{bmatrix} 1 \\ 0 \end{bmatrix} + b\begin{bmatrix} 0 \\ 1 \end{bmatrix}$ and complete Figure 1.18 by finding $f\left(a\begin{bmatrix} 1 \\ 0 \end{bmatrix}\right)$, $f\left(b\begin{bmatrix} 0 \\ 1 \end{bmatrix}\right)$ and $f\begin{bmatrix} a \\ b \end{bmatrix}$.

Figure 1.18

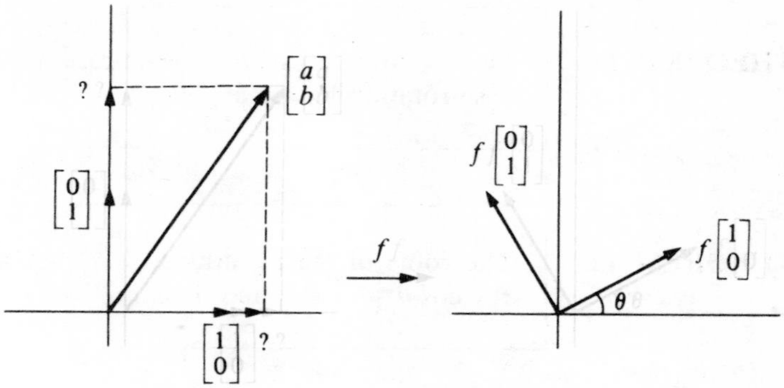

8. Consider the function $f: \mathbb{R}^2 \to \mathbb{R}^2$ defined by

$$f: \begin{bmatrix} a \\ b \end{bmatrix} \to \begin{bmatrix} -b^2 \\ a^2 \end{bmatrix}$$

a. Show that f rotates each of the vectors $\begin{bmatrix} 1 \\ 0 \end{bmatrix}$ and $\begin{bmatrix} 0 \\ 1 \end{bmatrix}$ by the angle $\pi/2$.

b. Evaluate $f\left(\begin{bmatrix} 2 \\ 1 \end{bmatrix}\right)$ and show that f does not rotate the vector $\begin{bmatrix} 2 \\ 1 \end{bmatrix}$ by the angle $\pi/2$.

c. Explain why this function does not violate the result you proved in Exercise 7.

Review Exercises

1. Is the function

$$f: \mathbb{R}^2 \to \mathbb{R}^2$$
$$\begin{bmatrix} x \\ y \end{bmatrix} \to |x| + |y|$$

a linear function? Explain.

2. a. Graph the pair of vectors:

$$\mathbf{v}_1 = \begin{bmatrix} -1 \\ 1 \end{bmatrix} \qquad \mathbf{v}_2 = \begin{bmatrix} 1 \\ -1 \end{bmatrix}$$

b. Is this pair linearly independent? Express $\begin{bmatrix} 1 \\ 0 \end{bmatrix}$ and $\begin{bmatrix} 0 \\ 1 \end{bmatrix}$ as a basis of $\mathbb{R}^2$ of $\mathbf{v}_1$ and $\mathbf{v}_2$ (if possible).

c. Express $\begin{bmatrix} 0 \\ 0 \end{bmatrix}$ in the form $a\mathbf{v}_1 + b\mathbf{v}_2$ where $a \neq 0, b \neq 0$ (if possible).

d. Find all linear functions f with $f(\mathbf{v}_1) = 1$ and $f(\mathbf{v}_2) = 0$.

3. Answer the questions in Exercise 2 for the pair of vectors $\mathbf{v}_1 = \begin{bmatrix} -1 \\ 1 \end{bmatrix}, \mathbf{v}_2 = \begin{bmatrix} 1 \\ 1 \end{bmatrix}$.

4. a. Express each of the vectors $\begin{bmatrix} 1 \\ 0 \end{bmatrix}$ and $\begin{bmatrix} 0 \\ 1 \end{bmatrix}$ as a linear combination of the vectors $\begin{bmatrix} 1 \\ 2 \end{bmatrix}$ and $\begin{bmatrix} 3 \\ -1 \end{bmatrix}$.

b. Suppose $f: \mathbb{R}^2 \to \mathbb{R}$ is a linear function such that $f\begin{bmatrix} 1 \\ 2 \end{bmatrix} = 3$ and $f\begin{bmatrix} 3 \\ -1 \end{bmatrix} = 2$. Find $f\begin{bmatrix} 1 \\ 0 \end{bmatrix}$ and $f\begin{bmatrix} 0 \\ 1 \end{bmatrix}$. Find the general formula for $f\begin{bmatrix} x \\ y \end{bmatrix}$.

5. Let $f: \mathbb{R}^2 \to \mathbb{R}$. What two properties must f satisfy to be linear?

6. Determine which of the following functions are linear:

a. $f\colon\ \mathbb{R}^2 \to \mathbb{R}$

$$\begin{bmatrix} x \\ y \end{bmatrix} \to 0$$

b. $f\colon\ \mathbb{R}^2 \to \mathbb{R}$

$$\begin{bmatrix} x \\ y \end{bmatrix} \to x + y$$

c. $f\colon\ \mathbb{R}^2 \to \mathbb{R}$

$$\begin{bmatrix} x \\ y \end{bmatrix} \to 2x + 1$$

d. $f\colon\ \mathbb{R}^2 \to \mathbb{R}$

$$\begin{bmatrix} x \\ y \end{bmatrix} \to 3$$

e. $f\colon\ \mathbb{R}^2 \to \mathbb{R}$

$$\begin{bmatrix} x \\ y \end{bmatrix} \to \sqrt{2}\,x - 14y$$

7. Determine whether the following statements are true or false. If true, explain why; if false, give a counterexample.

a. If $\mathbf{v}_1$ and $\mathbf{v}_2$ are two vectors that lie on the same line through the origin (in $\mathbb{R}^2$), then $\{\mathbf{v}_1, \mathbf{v}_2\}$ is not a basis for $\mathbb{R}^2$.

b. If $f, g\colon \mathbb{R} \to \mathbb{R}$ are linear functions such that for some number $x_0 \neq 0$ we have $f(x_0) = g(x_0)$, then $f = g$.

c. If $f, g\colon \mathbb{R}^2 \to \mathbb{R}$ are linear functions such that for some vector $\mathbf{v}_0 \neq \begin{bmatrix} 0 \\ 0 \end{bmatrix}$ we have $f(\mathbf{v}_0) = g(\mathbf{v}_0)$, then $f = g$.

8. **a.** Describe all linear functions from $\mathbb{R}$ to $\mathbb{R}$.

b. Describe all linear functions from $\mathbb{R}^2$ to $\mathbb{R}$.

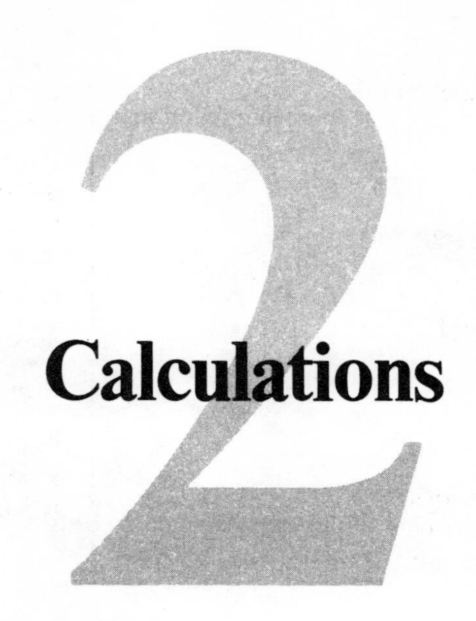

Calculations

The purpose of this chapter is to discuss the elimination method of solving simultaneous linear equations. This is one of the most important computational techniques we will use. A geometric description of why this method works is discussed in Section 2.1, and the computational techniques are explained in Section 2.2.

2.1 The Elimination Method

In Section 1.5 we gave a method for solving systems of two linear equations in two unknowns. This method works well for small systems, but it is inefficient for larger systems. The **elimination method** is a much better way of solving systems of equations. Although this method is probably familiar to you, we will discuss it here to show not only how it works, but why it works. In what follows we will deal with one specific system, but the methods we use will be general and apply to all systems of linear equations. We consider the system

$$2x + 3y = 5$$
$$-x + 2y = 1$$

As in previous sections, we will use geometrical ideas to help us see what is happening.

Each equation in this system represents a line. The points on the line represented by the first equation are those points (a, b) such that $2a + 3b = 5$, and the points on the line represented by the second equation are those

Figure 2.1

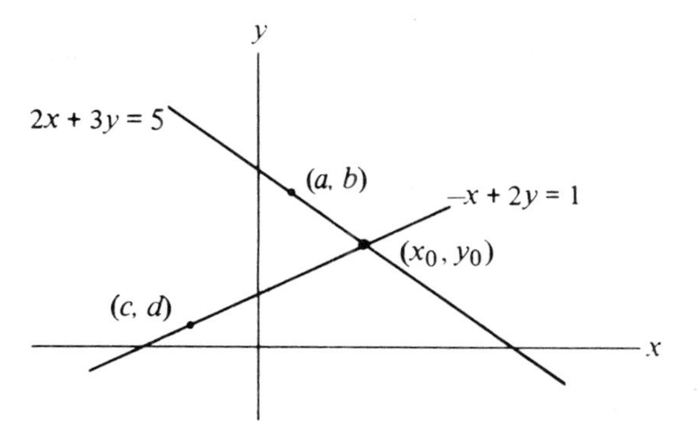

points (c, d) such that $-c + 2d = 1$. The point (x_0, y_0), which is the solution of this system, is on both of these lines—the point of intersection of the lines (see Figure 2.1).

Since we are only interested in the point of intersection of these two lines and not the lines themselves, we replace each line by a new line such that the new lines have the same point of intersection. One of the new lines will be vertical ($x = x_0$), and the other horizontal ($y = y_0$). The point (x_0, y_0) is on both lines.

It is easy to see that, for any numbers c_1 and c_2, the point (x_0, y_0) satisfies the equation

(1) $$c_1(2x + 3y) + c_2(-x + 2y) = 5c_1 + 1c_2$$

since $c_1(2x_0 + 3y_0) = c_1(5)$ and $c_2(-x_0 + 2y_0) = c_2(1)$.

We can rewrite equation (1) in the form

(2) $$(2c_1 - c_2)x + (3c_1 + 2c_2)y = (5c_1 + 1c_2)$$

to see that, for each choice of numbers c_1 and c_2, equation (2) represents a line containing the point (x_0, y_0).

Our goal is to replace one of our original equations with the equation of a horizontal line that contains the point (x_0, y_0). To do this choose c_1 and c_2 so that equation (2) represents a horizontal line, that is, so that the coefficient of x in equation (2) is 0. For example, we can choose $c_1 = 1$ and $c_2 = 2$. Substituting these values in equation (2) we have

$$(0)x + (3 + 4)y = 5 + 2$$

or $$y = 1$$

Thus the horizontal line that contains the point (x_0, y_0) has equation $y = 1$ (therefore, $y_0 = 1$).

Although we could substitute $y = 1$ into one of the original equations and solve for x directly, we will continue this replacement procedure to show how this method works in more complex situations.

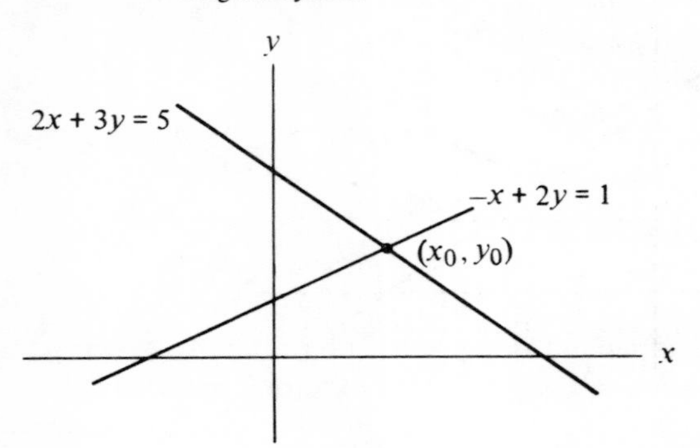

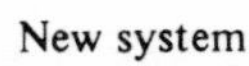

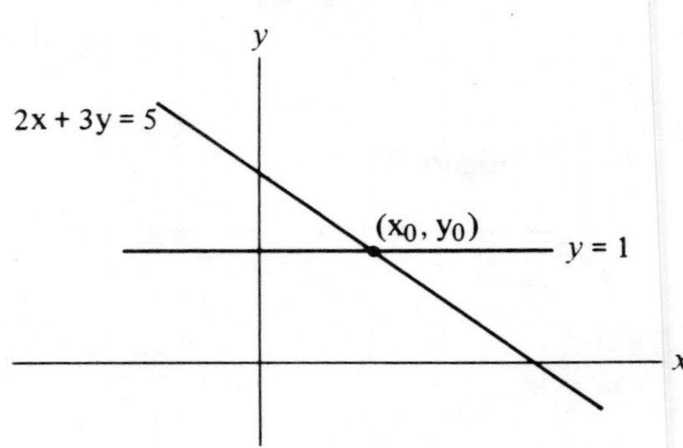

Figure 2.2

We replace the second equation of our original system of equations by the equation $y = 1$ to obtain a new system of equations having the same solution as the original system (see Figure 2.2). Our new system is

$$2x + 3y = 5$$

$$y = 1$$

To solve this system, we use the same procedure as we used to solve the original system, except that we replace the first equation by the equation of a vertical line containing the point (x_0, y_0). The point (x_0, y_0) satisfies the equation

$$c_1(2x + 3y) + c_2(y) = c_1(5) + c_2(1)$$

for any numbers c_1 and c_2. We rewrite the equation as

$$(2c_1)x + (3c_1 + c_2)y = 5c_1 + c_2$$

In this case we want to choose c_1 and c_2 so that the equation above represents a vertical line, that is, so that the coefficient of y is zero. To do this we choose $c_1 = 1$ and $c_2 = -3$. The equation becomes

$$2x + 0y = 2$$

or

$$x = 1$$

So we can replace the second system of equations by a third system

$$x = 1 \qquad y = 1$$

Since the point (x_0, y_0) is the point of intersection of these two lines, $(x_0, y_0) = (1, 1)$. This is shown graphically in Figure 2.3.

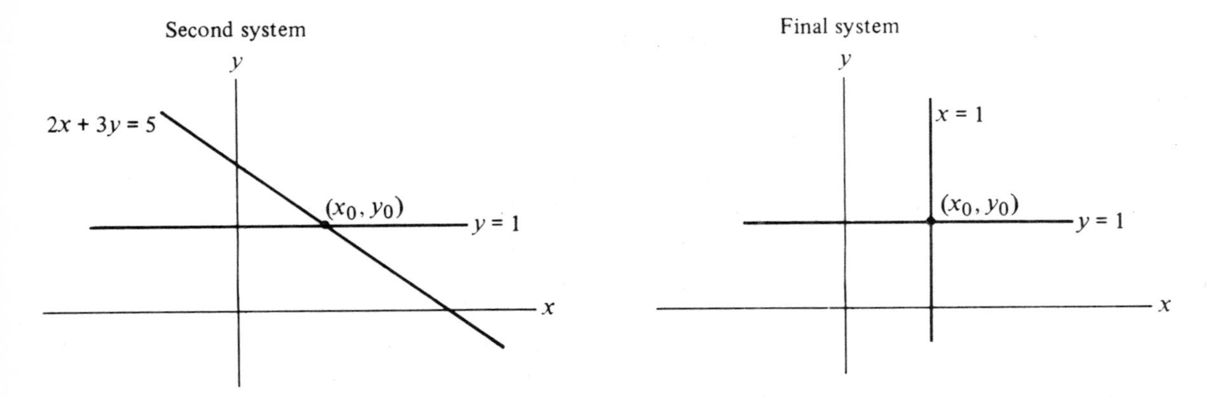

Figure 2.3

Generalizing this procedure, we derive the following theorem.

THEOREM 2.1 If (x_0, y_0) is a solution of a system of linear equations, then (x_0, y_0) is also the solution of the system of equations obtained by replacing one of the equations by the sum of a nonzero multiple of itself and a multiple of the other equation.

The multiples are chosen so that either the coefficient of x or the coefficient of y in the new equation is zero. To see how to do this more succinctly, consider the following example.

EXAMPLE 1 We solve the same system of equations using Theorem 2.1.

$$2x + 3y = 5$$
$$-x + 2y = 1$$

We first solve for y and then solve for x.

$2x + 3y = 5$ (Multiply by 1), $-x + 2y = 1$ (Multiply by 2) — replace 2nd equation $\longrightarrow$ $2x + 3y = 5$, $0x + 7y = 7$ —or→ $2x + 3y = 5$, $y = 1$

$2x + 3y = 5$ (Multiply by 1), $y = 1$ (Multiply by -3) — replace 1st equation $\longrightarrow$ $2x + 0y = 2$, $y = 1$ —or→ $x = 1$, $y = 1$ ■

Exercise Set 2.1

In Exercises 1–3 consider the following systems of linear equations.

a. $\begin{aligned} -2x + 3y &= 6 \\ x + 5y &= 6 \end{aligned}$ **b.** $\begin{aligned} -3x + 4y &= 7 \\ 2x + y &= 3 \end{aligned}$

c. $\begin{aligned} x - 5y &= 9 \\ -x - 2y &= 9 \end{aligned}$ **d.** $\begin{aligned} x - 3y &= 5 \\ -2x + 6y &= -10 \end{aligned}$

e. $\begin{aligned} x - 3y &= 5 \\ -2x + 6y &= 10 \end{aligned}$

1. Graph the lines in each system and determine which systems have only one solution, which systems have no solutions, and which systems have infinitely many solutions.

2. Solve each of the systems of equations using the method of Example 1.

3. Graph the lines corresponding to the equations you got in each step of the calculations you made in Exercise 2 for each of the systems of equations.

4. Consider systems of linear equations of the form

$$\begin{aligned} 2x - y &= 1 \\ -6x + 3y &= k \end{aligned}$$

a. For which values of k does this system of equations have no solutions?

b. For which values of k does this system of equations have more than one solution? How many solutions are there?

c. For which values of k does this system of equations have exactly one solution?

5. Consider systems of equations of the form

$$\begin{aligned} 2x - y &= 0 \\ -6x + my &= 0 \end{aligned}$$

a. For which values of m does this system of equations have no solutions?

b. For which values of m does this system of equations have many solutions? How many solutions does it have?

c. For which values of m does this system have exactly one solution?

2.2 Elimination Method and Matrix Reduction

The elimination method discussed in Section 2.1 for systems of two equations can be applied to systems with three or more equations. In the case of equations with three unknowns, each equation represents a plane in three-dimensional space, and the solution of a system of equations with three unknowns is the point or points of intersection of the planes represented by the equations of the system. The elimination method allows us to replace the original planes by planes parallel to the xy plane, xz plane, and yz plane, giving the z, y, and x coordinates of the point of intersection. For systems with more than three unknowns, the equations represent surfaces called **hyperplanes**, but the elimination method works the same way.

ELIMINATION RULE

> Replace an equation by the sum of a nonzero multiple of itself plus a multiple of another equation.

Note: Along with the elimination rule, there is another operation on systems of linear equations that does not change the solution set of the system. It is changing the order in which the equations are listed. In later chapters we will make use of this operation.

We will use several examples to show how the elimination rule for systems with three unknowns works. First we introduce some notation to make our work easier.

NOTATION

$$\begin{array}{ll} \text{equation } a & (c_1) \\ \text{equation } b & (c_2) \longrightarrow c_1 \cdot (\text{equation } a) + c_2 \cdot (\text{equation } b) \end{array}$$

This notation means that we have replaced equation b of the system by the sum of the products of c_1 times equation a and c_2 times equation b.

For large systems of equations it is useful to follow a rather rigid procedure in eliminating variables so as to avoid confusion. We will always proceed as follows:

1. Begin by eliminating the first variable from all but one of the equations. We generally use the first equation to eliminate the first variable from all the other equations. However, if the first coefficient of the first equation is zero (and sometimes for other reasons), we rearrange the order of the equations so that an equation with a nonzero first coefficient becomes the first equation. This new first equation is called the **pivot equation**.

2. Leave the first equation alone and eliminate the second variable from all but the first two equations. We generally use the second equation to eliminate the second variable from all but the first two equations. However, if the second coefficient of the second equation is zero (and sometimes for other reasons), we rearrange the second equation and those that follow to obtain a system having a second equation (now called the pivot equation) with a nonzero second coefficient.

3. Leave the first two equations alone and eliminate the third variable from all but the first three equations. As before, we generally use the third equation, but in some cases we may have to choose a different pivot equation.

4. Continue in this way until either there are no new variables to eliminate or until all the coefficients of all the remaining equations are zero. At this point the solution of the equations can be easily obtained by **back substitution**, as the following examples illustrate.

EXAMPLE 1 Solve the system

$$\begin{aligned} 2x - y + 3z &= 12 \\ -x + 4y - 2z &= -11 \\ 3x + y + 5z &= 17 \end{aligned}$$

$$\begin{array}{ll} 2x - y + 3z = 12 & (1) \\ -x + 4y - 2z = -11 & (2) \\ 3x + y + 5z = 17 & \end{array} \longrightarrow \begin{array}{ll} 2x - y + 3z = 12 & (-3) \\ 0x + 7y - z = -10 & \\ 3x + y + 5z = 17 & (2) \end{array} \longrightarrow$$

$$\begin{array}{ll} 2x - y + 3z = 12 & \\ 0x + 7y - z = -10 & (-5) \\ 0x + 5y + z = -2 & (7) \end{array} \longrightarrow \begin{array}{l} 2x - y + 3z = 12 \\ 0x + 7y - z = -10 \\ 0x + 0y + 12z = 36 \end{array}$$

From the last equation we have $z = 36/12 = 3$. We use back substitution to solve for x and y. From the second equation and the fact that $z = 3$ we get

$$\begin{aligned} 7y - 3 &= -10 \\ y &= -1 \end{aligned}$$

Putting these values into the first equation gives

$$\begin{aligned} 2x - (-1) + 3(3) &= 12 \\ x &= 1 \end{aligned}$$

Thus, the solution of this system is $x = 1$, $y = -1$, and $z = 3$. We can also write the solution of this system in vector form as

$$\begin{bmatrix} 1 \\ -1 \\ 3 \end{bmatrix} \quad \blacksquare$$

The system of equations in Example 1 had a unique solution. This means that the three planes represented by the equations of this system have a single point in common. In the next example we solve a system in which the three planes represented by the equations have an entire line in common—the solution is not unique.

EXAMPLE 2 We solve the system

$$\begin{aligned} x + 2y + 0z &= 0 \\ -x + y - 3z &= 0 \\ 3x + 4y + 2z &= 0 \end{aligned}$$

$$\begin{aligned} x + 2y + 0z &= 0 \quad (1) \\ -x + y - 3z &= 0 \quad (1) \\ 3x + 4y + 2z &= 0 \end{aligned} \longrightarrow \begin{aligned} x + 2y + 0z &= 0 \quad (-3) \\ 0x + 3y - 3z &= 0 \\ 3x + 4y + 2z &= 0 \quad (1) \end{aligned} \longrightarrow$$

$$\begin{aligned} x + 2y + 0z &= 0 \\ 0x + 3y - 3z &= 0 \quad (2) \\ 0x - 2y + 2z &= 0 \quad (3) \end{aligned} \longrightarrow \begin{aligned} x + 2y + 0z &= 0 \\ 0x + 3y - 3z &= 0 \\ 0x + 0y + 0z &= 0 \end{aligned}$$

The third equation says only that $0 = 0$. From the second equation we have $y = z$. Putting this into the first equation gives $x + 2z + 0z = 0$, which yields $x = -2z$. This means that any point of the form $(-2z, z, z)$ is a solution of this system. For example, if we let $z = 2$ we find that the point $(-4, 2, 2)$ is a solution of all three equations of this system. We write the solutions of the system in vector form as

$$\begin{bmatrix} -2z \\ z \\ z \end{bmatrix} \quad \text{or} \quad z\begin{bmatrix} -2 \\ 1 \\ 1 \end{bmatrix} \quad \blacksquare$$

When solving equations of this sort, it is unnecessary to continually copy the equations. Since it is the coefficients that are important and not the variables, we can rewrite the system without the variables. For example, the system

$$\begin{aligned} x + 2y + 0z &= 0 \\ -x + y - 3z &= 0 \\ 3x + 4y + 2z &= 0 \end{aligned} \quad \text{can be replaced by} \quad \left[\begin{array}{rrr|r} 1 & 2 & 0 & 0 \\ -1 & 1 & -3 & 0 \\ 3 & 4 & 2 & 0 \end{array}\right]$$

DEFINITION 2.1

A rectangular array of numbers with a line between the last two columns (as is written above) is called an **augmented matrix**. A rectangular array of numbers without the line is called a **matrix**. The numbers in a matrix (or augmented matrix) are called the **entries** of the matrix (or augmented matrix).

The rules for solving systems of equations in matrix form are essentially the same as the rules for solving systems written in equation form.

Matrix Row Reduction Rules

Elimination Rule: Replace a row of a matrix by the sum of a nonzero multiple of itself and a multiple of another row.

Rearrangement Rule: Rearrange the rows of the matrix.

Multiplication Rule: Multiply any row by a nonzero number. (This is really a special case of the Elimination Rule in which the multiple of the other row is zero).

EXAMPLE 3 We solve the system of equations given in Example 2 using the matrix form.

$$\begin{aligned} x + 2y + 0z &= 0 \\ -x + y - 3z &= 0 \\ 3x + 4y + 2z &= 0 \end{aligned} \rightarrow \left[\begin{array}{rrr|r} 1 & 2 & 0 & 0 \\ -1 & 1 & -3 & 0 \\ 3 & 4 & 2 & 0 \end{array}\right] \begin{array}{l} (1) \\ (1) \\ \end{array} \rightarrow$$

$$\left[\begin{array}{rrr|r} 1 & 2 & 0 & 0 \\ 0 & 3 & -3 & 0 \\ 3 & 4 & 2 & 0 \end{array}\right] \begin{array}{l} (-3) \\ \\ (1) \end{array} \rightarrow \left[\begin{array}{rrr|r} 1 & 2 & 0 & 0 \\ 0 & 3 & -3 & 0 \\ 0 & -2 & 2 & 0 \end{array}\right] \begin{array}{l} \\ (2) \\ (3) \end{array} \rightarrow \left[\begin{array}{rrr|r} 1 & 2 & 0 & 0 \\ 0 & 3 & -3 & 0 \\ 0 & 0 & 0 & 0 \end{array}\right]$$

which leads to the equations

$$\begin{aligned} x + 2y + 0z &= 0 \\ 0x + 3y - 3z &= 0 \\ 0x + 0y + 0z &= 0 \end{aligned}$$

The solution is found exactly as in Example 2. ■

Some systems of equations represent planes that have no point in common. These systems obviously have no solutions, as the following example shows.

EXAMPLE 4 Solve the following system of equations using the matrix form.

$$\begin{aligned} x + 2y + 0z &= 3 \\ -x + y - 3z &= 0 \\ 3x + 4y + 2z &= 6 \end{aligned}$$

When reducing this matrix, we will write two steps at once to eliminate some unnecessary copying.

$$\left[\begin{array}{rrr|r} 1 & 2 & 0 & 3 \\ -1 & 1 & -3 & 0 \\ 3 & 4 & 2 & 6 \end{array}\right] \begin{array}{l} (1) \quad (-3) \\ (1) \\ \quad (1) \end{array} \rightarrow \left[\begin{array}{rrr|r} 1 & 2 & 0 & 3 \\ 0 & 3 & -3 & 3 \\ 0 & -2 & 2 & -3 \end{array}\right] \begin{array}{l} \\ (2) \\ (3) \end{array} \rightarrow$$

$$\left[\begin{array}{rrr|r} 1 & 2 & 0 & 3 \\ 0 & 3 & -3 & 3 \\ 0 & 0 & 0 & -3 \end{array}\right]$$

The bottom row represents the equation

$$0x + 0y + 0z = -3$$

Since the left-hand side is zero, no equality exists. Therefore, there is no solution for this system. ■

Systems of more than three equations can be solved in the same manner. The solution of such a system is illustrated in the following example. This example also shows how to choose a new pivot equation in the case where a pivot entry is zero.

EXAMPLE 5 Solve the following system of equations.

$$\begin{aligned} x + 2y - z + 3w &= -4 \\ 2x + 4y + 3z - w &= 11 \\ 3x - 2y - 4z - w &= -9 \\ -5x + 2y + z + w &= 8 \end{aligned}$$

$$\left[\begin{array}{rrrr|r} 1 & 2 & -1 & 3 & -4 \\ 2 & 4 & 3 & -1 & 11 \\ 3 & -2 & -4 & -1 & -9 \\ -5 & 2 & 1 & 1 & 8 \end{array}\right] \quad \begin{array}{l} (-2)\ (-3)\ (5) \\ (1)\ \text{row 1} \times (-2) + \text{row 2} \\ (1)\ \text{row 1} \times (-3) + \text{row 3} \\ (1)\ \text{row 1} \times (5) + \text{row 4} \end{array}$$

$$\left[\begin{array}{rrrr|r} 1 & 2 & -1 & 3 & -4 \\ 0 & 0 & 5 & -7 & 19 \\ 0 & -8 & -1 & -10 & 3 \\ 0 & 12 & -4 & 16 & -12 \end{array}\right] \rightleftarrows \left[\begin{array}{rrrr|r} 1 & 2 & -1 & 3 & -4 \\ 0 & -8 & -1 & -10 & 3 \\ 0 & 0 & 5 & -7 & 19 \\ 0 & 12 & -4 & 16 & -12 \end{array}\right] \begin{array}{l} (3) \\ \\ (2) \end{array}$$

$$\left[\begin{array}{rrrr|r} 1 & 2 & -1 & 3 & -4 \\ 0 & -8 & -1 & -10 & 3 \\ 0 & 0 & 5 & -7 & 19 \\ 0 & 0 & -11 & 2 & -15 \end{array}\right] \begin{array}{l} (11) \\ (5) \end{array} \rightarrow \left[\begin{array}{rrrr|r} 1 & 2 & -1 & 3 & -4 \\ 0 & -8 & -1 & -10 & 3 \\ 0 & 0 & 5 & -7 & 19 \\ 0 & 0 & 0 & -67 & 134 \end{array}\right]$$

We use back substitution to solve the equations. From the fourth row we get

$$-67w = 134$$

or

$$w = -2$$

From the third row and the fact that $w = -2$, we get

$$5z - 7(-2) = 19$$

or

$$z = 1$$

From the second row and the facts that $w = -2$ and $z = 1$, we have

$$-8y - 1(1) - 10(-2) = 3$$

or

$$y = 2$$

From the first row and the fact that $w = -2$, $z = 1$, and $y = 2$, we get

$$x + 2(2) - 1(1) + 3(-2) = -4$$

or

$$x = -1$$

The unique solution to this system is $(-1, 2, 1, -2)$. ■

In all the previous examples we have reduced the systems of equations (matrices) only to the point where the value of one unknown was found. We then used this value to solve the remaining equations for the rest of the unknowns. In general this is the most efficient way of solving systems of equations. However, in later sections of the book (see Section 5.3) it will be necessary to reduce systems further. To illustrate this procedure, we solve the system of equations from Example 1 in this way.

EXAMPLE 6 Solve:

$$\begin{aligned} 2x - y + 3z &= 12 \\ -x + 4y - 2z &= -11 \\ 3x + y + 5z &= 17 \end{aligned}$$

As we saw in Example 1 the matrix form is

$$\left[\begin{array}{rrr|r} 2 & -1 & 3 & 12 \\ -1 & 4 & -2 & -11 \\ 3 & 1 & 5 & 17 \end{array}\right] \quad \text{which reduces to} \quad \left[\begin{array}{rrr|r} 2 & -1 & 3 & 12 \\ 0 & 7 & -1 & -10 \\ 0 & 0 & 12 & 36 \end{array}\right]$$

Continuing the reduction

$$\left[\begin{array}{rrr|r} 2 & -1 & 3 & 12 \\ 0 & 7 & -1 & -10 \\ 0 & 0 & 12 & 36 \end{array}\right] \begin{array}{l} \\ \\ (1/12)\rightarrow \end{array} \left[\begin{array}{rrr|r} 2 & -1 & 3 & 12 \\ 0 & 7 & -1 & -10 \\ 0 & 0 & 1 & 3 \end{array}\right] \begin{array}{ll} & (1)\rightarrow \\ (1)\rightarrow & \\ (1) & (-3) \end{array}$$

$$\left[\begin{array}{rrr|r} 2 & -1 & 0 & 3 \\ 0 & 7 & 0 & -7 \\ 0 & 0 & 1 & 3 \end{array}\right] \begin{array}{l} \\ (1/7)\rightarrow \\ \\ \end{array} \left[\begin{array}{rrr|r} 2 & -1 & 0 & 3 \\ 0 & 1 & 0 & -1 \\ 0 & 0 & 1 & 3 \end{array}\right] \begin{array}{l} (1)\rightarrow \\ (1) \\ \\ \end{array}$$

$$\left[\begin{array}{ccc|c} 2 & 0 & 0 & 2 \\ 0 & 1 & 0 & -1 \\ 0 & 0 & 1 & 3 \end{array}\right] (1/2) \longrightarrow \left[\begin{array}{ccc|c} 1 & 0 & 0 & 1 \\ 0 & 1 & 0 & -1 \\ 0 & 0 & 1 & 3 \end{array}\right]$$

So $x = 1$, $y = -1$, and $z = 3$, which is the same as the answer in Example 1. ■

Exercise Set 2.2

Solve the following systems of linear equations:

1. $\begin{aligned} 2x - 3y &= 1 \\ x + 5y &= 0 \end{aligned}$

2. $\begin{aligned} 2x - 3y &= 0 \\ x + 5y &= 1 \end{aligned}$

3. $\begin{aligned} x + 3y - 4z &= 1 \\ 2x - 5y + 2z &= 0 \\ 3x - 2y + 3z &= 0 \end{aligned}$

4. $\begin{aligned} 5x + 2y - z &= 6 \\ 3x - 2y + 4z &= 5 \\ x - 7y + 3z &= -3 \end{aligned}$

5. $\begin{aligned} 3x + 2y - z + 3w &= -8 \\ 2x - 4y + z + w &= 7 \\ -x - 3y + 2z + 3w &= 2 \\ -2x + 2y - z - w &= -5 \end{aligned}$

6. $\begin{aligned} 2x + 3y - z &= 0 \\ 3x - y + 2z &= 0 \end{aligned}$

7. $\begin{aligned} 2x + 3y - z &= 4 \\ 3x - y + 2z &= 4 \end{aligned}$

8. $\begin{aligned} 2x + 3y - z &= 4 \\ -2x - 3y + z &= 4 \end{aligned}$

9. $\begin{aligned} x + 2y &= 1 \\ -x + 3y &= 2 \\ x + y &= 0 \end{aligned}$

10. $\begin{aligned} x + 2y &= 1 \\ -x + 3y &= 2 \\ x + 7y &= 4 \end{aligned}$

11. $\begin{aligned} x + y &= 1 \\ y + z &= 1 \\ x - z &= 2 \end{aligned}$

12. Suppose the matrix for a system of three equations in three unknowns reduces to a matrix of the form

$$\left[\begin{array}{ccc|c} 2 & 3 & -1 & a \\ 0 & -1 & 5 & b \\ 0 & 0 & 0 & c \end{array}\right]$$

a. What condition on the numbers a, b, and c will ensure that the system has more than one solution?

b. What condition on a, b, and c will ensure that the system has no solution?

c. Is there any choice of a, b, c for which this system has exactly one solution?

★**13.** Consider the system of equations

$$\begin{aligned} ax + by &= s \\ cx + dy &= t \end{aligned}$$

a. Solve this system of equations for x and y in terms of a, b, c, d, s, and t.

b. Determine which of the following statements are true and which are false:

i. If $(ad - bc) \neq 0$, then the system has a unique solution (no matter what s and t are).

ii. If this system has more than one solution, then $(ad - bc) = 0$.

iii. If this system has no solution, then $(ad - bc) = 0$.

14. Consider the system of equations

$$\begin{aligned} a_1x + b_1y + c_1z &= 0 \\ a_2x + b_2y + c_2z &= 0 \end{aligned}$$

Explain why it is always possible to find a nontrivial (x, y, z not all zero) solution for such a system of equations. (*Hint:* Look at the geometry.)

Applications 2 Linear Programming

Geometric Ideas

Linear programming was developed in the 1940s as a tool for finding the maximum or minimum value of a linear function subject to certain linear inequalities. It was first used by the Air Force to determine the most efficient and least costly methods of doing various things, such as bidding on contracts, scheduling aircraft maintenance, assigning personnel, and airlift routing (see *Linear Programming and Extensions*, by George B. Dantzig, Princeton University Press, Princeton, N.J., 1963). The main computational tool in linear programming is the simplex algorithm. The simplex algorithm is a modification of the elimination technique discussed in Sections 2.1 and 2.2.

In this section we do not try to give a detailed description of linear programming, but only present the basic ideas of this important subject, using ideas that we have already discussed. The problem we consider is the following: The water in the Great Lakes has become polluted by certain industrial and agricultural chemicals. These chemicals get into the food chains of various species of fish, which are then eaten by people. To protect the public from ingesting harmful amounts of these chemicals, federal and state agencies monitor fish samples taken from various parts of the Great Lakes. Two of the largest and best tasting game fish in the Great Lakes are salmon and lake trout. Unfortunately, these fish also contain the highest percentages of certain toxic chemicals. We want to determine the maximum amount of both of these fish that a person can safely consume. The three major contaminants in these fish are Dieldrin, PCB, and DDT. In Table 2.1 we list the amounts of these contaminants in each of the two species of fish. (The data here are based on the Great Lakes Environmental Contaminants Survey of 1975. There was a wide variance in the data obtained, and the samples were quite small, so that the results of the analysis of this data should not be used as a basis for determining safe levels of fish for people to eat.) The amounts of contaminants listed in Table 2.1 are the amounts of contaminants in the edible portion of the fish.

Table 2.1 Amounts of Chemical Contaminants in Salmon and Lake Trout, and F.D.A. Safe Tolerance Limits

	Salmon (amount per oz)	Lake Trout (amount per oz)	Maximum Daily Intake in oz
Dieldrin	0.07×10^{-6}	0.05×10^{-6}	2.4×10^{-6}
PCB	0.77×10^{-6}	2.58×10^{-6}	40.0×10^{-6}
DDT	0.5×10^{-6}	3.2×10^{-6}	40.0×10^{-6}

To begin to analyze this problem, we must get a better way of representing the data in Table 2.1. We let

x = number of ounces of salmon (edible portion) we wish to eat.

y = number of ounces of lake trout (edible portion) we wish to eat.

Then, using Table 2.1, we find that our portions of salmon and lake trout will be safe if

$$(0.07 \times 10^{-6})x + (0.05 \times 10^{-6})y = \text{amount of Dieldrin} \leq 2.4 \times 10^{-6}$$

$$(0.77 \times 10^{-6})x + (2.58 \times 10^{-6})y = \text{amount of PCB} \leq 40 \times 10^{-6}$$

$$(0.5 \times 10^{-6})x + (3.2 \times 10^{-6})y = \text{amount of DDT} \leq 40 \times 10^{-6}$$

We simplify these equations (by multiplying the first two by 10^8 and the third by 10^7) to obtain

$$7x + 5y \leq 240$$

$$77x + 258y \leq 4000$$

$$5x + 32y \leq 400$$

EXAMPLE 1 Use the inequalities above to determine if a dinner of 6 oz of salmon and 3 oz of lake trout is safe to eat.

We substitute $x = 6$ and $y = 3$ in the left side of each of the inequalities and check to see if these values satisfy all three conditions.

$$7(6) + 5(3) = 57 \qquad \text{which is less than 240}$$

$$77(6) + 258(3) = 1236 \qquad \text{which is less than 4000}$$

$$5(6) + 32(3) = 126 \qquad \text{which is less than 400}$$

This dinner is well within the safe tolerance limits. ■

Although the inequalities above describe the tolerance levels better than Table 2.1, there is still a better way of representing this data. Each of the three inequalities can be graphed. The points that satisfy each of the inequalities form a region of the x, y plane. The intersection of the three regions (which correspond to the three inequalities) is the region consisting of all points (x, y) for which a portion of x ounces of salmon and y ounces of lake trout are safe to eat. The following example illustrates the way to graph an inequality:

EXAMPLE 2 Graph the inequality $7x + 5y \leq 240$

The points in the plane are divided into three classes: those points (x, y) for which $7x + 5y = 240$ (the line that crosses the x axis at $x = 240/7 = 34.3$

and crosses the y axis at $y = 240/5 = 48$) the points for which $7x + 5y < 240$ (the points lying on one side of the line) and the points (x, y) for which $7x + 5y > 240$ (the points lying on the other side of the line) Since

$$7(0) + 5(0) = 0 < 240$$

the point (0, 0) is on the side of the line that corresponds to points where $7x + 5y < 240$. Hence, all points (x, y) on the same side of the line $7x + 5y = 240$ as (0, 0) have the property that $7x + 5y < 240$. All points on the other side of this line have the property that $7x + 5y > 240$ (see Figure 2.4). ■

Figure 2.4

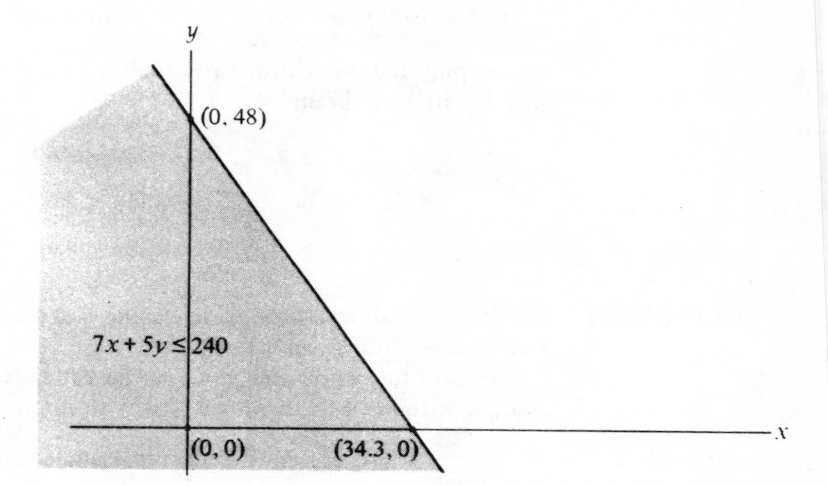

In the same manner we can also graph the regions corresponding to the other two inequalities. There are two more inequalities implicit in our discussion. They are

$$x \geq 0 \quad \text{and} \quad y \geq 0$$

because you can't have a negative amount of fish. Now, any point satisfying all five of these inequalities represents an amount of fish that is safe to eat. The set of all such points corresponds to the region that contains the intersection of the regions corresponding to each of the five inequalities.

This region is called the **region of feasible solutions** and is graphed in Figure 2.5. Any point inside this region satisfies all five inequalities and therefore represents a feasible solution (an amount of fish that is safe to eat) to our problem. Any point outside this region represents a nonfeasible solution (an amount of fish that is not safe to eat). Note that Figure 2.5 shows the coordinates of the vertices (corners) of this region. These points were found by the elimination method (see Section 2.2). These coordinates will be of critical importance in solving this problem.

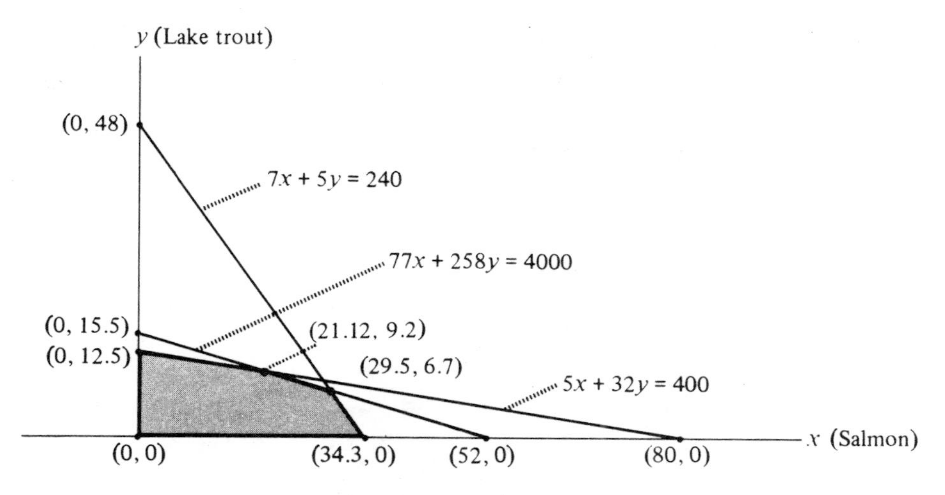

Figure 2.5 Region of feasible solutions.

To maximize the total amount of fish that is safe to eat, we must find a point (x, y) in the region of feasible solutions such that $x + y$ is maximum. To help us see the general picture more clearly, we restate this in more formal terms.

We wish to maximize the function

$$f(x, y) = x + y \qquad \text{for } (x, y) \text{ in the region of feasible solutions}$$

As we saw in Example 1, it was safe to eat 6 oz of salmon and 3 oz of lake trout. This meal would give a total of $6 + 3 = 9$ ounces of fish. Now there are also other amounts of salmon and lake trout that will make a total serving of 9 ounces of fish and be safe to eat. In fact, any point (x, y) such that (x, y) is in the region of feasible solutions and

$$f(x, y) = x + y = 9$$

represents a serving of 9 ounces of fish that is safe to eat. In Figure 2.6 we graph these points.

Figure 2.6

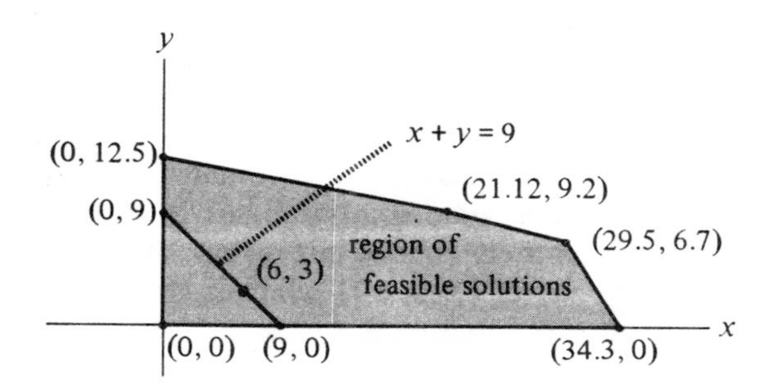

Now let us see if there is a way to safely eat 15 ounces of fish. This means that we must determine if there are any points within the region of feasible solutions for which $x + y = 15$, that is, does the line $x + y = 15$ pass through the region of feasible solutions. As we see in Figure 2.7, it is possible to find a 15-ounce combination of lake trout and salmon that is safe to eat.

Figure 2.7

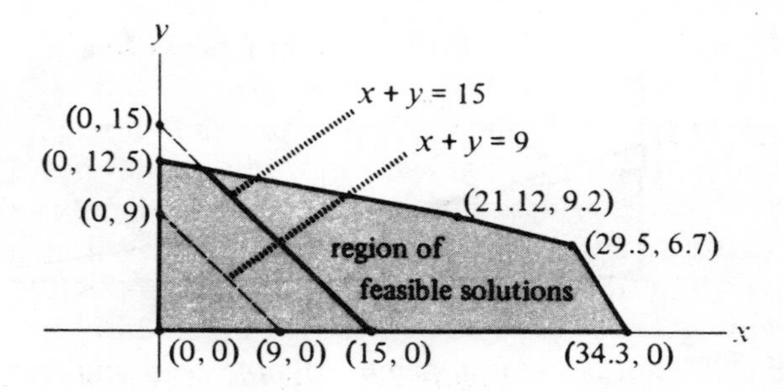

Now let's really get hoggish and see if it is possible to safely eat a 50-ounce serving of these two fish. From Figure 2.8, we see that the line $x + y = 50$ has no points in common with the region of feasible solutions. Hence, there is no combination of salmon and lake trout totaling 50 ounces that is safe to eat.

Figure 2.8

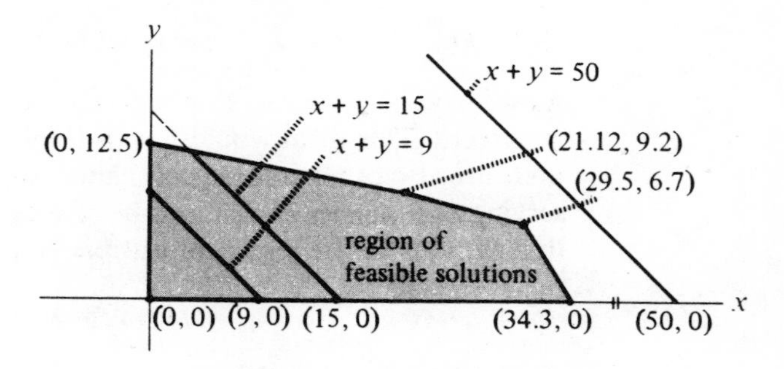

The trial-and-error approach we have been using is quite inefficient, but by looking closely at Figure 2.8 we can get a good idea of what's happening. We are looking at the family of parallel lines of the form $x + y = a$.

We have seen that as we increase the value of a (which represents the total amount of fish in a serving) the lines move up and to the right. Our goal is to maximize a but still stay within the region. Geometrically, this means that we wish to find a so that the line $x + y = a$ crosses the region of feasible solutions, but for any number $b > a$ the line $x + y = b$ lies entirely outside this region. That is, we want the line that goes through the vertex (29.5, 6.7), which is shown in Figure 2.9. Thus

$$a = x + y = 29.5 + 6.7 = 36.2$$

Figure 2.9

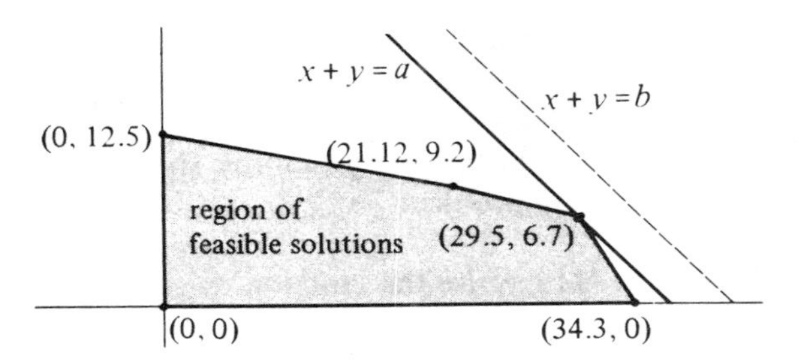

Therefore, the largest serving of lake trout and salmon that is safe to eat contains 29.5 ounces of salmon and 6.7 ounces of lake trout (which gives a total of 36.2 ounces of fish).

The General Linear Programming Problem

In this section we will give a general statement of the procedure we used to solve our two-dimensional problem.

Step 1 Use the information given to obtain a system of inequalities that describes the constraints of the problem. (These were the inequalities that described the amounts of each pollutant that were safe to ingest.)

Step 2 Graph the inequalities obtained in Step 1 to get the region of feasible solutions and label all vertices.

Step 3 Find the function f to be maximized. This function is called the **objective function**. In our example the function was $f(x, y) = x + y$.

Step 4 Find a value a so that the line $f(x, y) = a$ crosses the region of feasible solutions at at least one point, but for any number $b > a$ the line $f(x, y) = b$ does not intersect the region of feasible solutions.

In our example we noticed that all the lines of the form

$$f(x, y) = \text{constant}$$

were parallel and that the one that gave rise to the maximum value of f in the region of feasible solutions crossed the region at a vertex. This is always the case (in some instances the line $f(x, y) = a$ may coincide with a boundary of the region in which case it will contain two vertices). This means that the function $f(x, y)$ is maximum in the region of feasible solutions at a vertex of this region. Therefore, to find the maximum value of f, we need only evaluate $f(x, y)$ at the vertices of the region. The vertex that gives the maximum value of f is the point in the region of feasible solutions where f is maximum. Hence, we can replace Step 4 by Step 4′.

Step 4′ Evaluate f at each vertex of the region of feasible solutions. The vertex at which f is maximum is the point where f is maximum in the region.

The following example shows how Step 4′ can be used in the solution of our problem.

EXAMPLE 3 Maximize the function $f(x, y) = x + y$

subject to the constraints: $x \geq 0$, $y \geq 0$ and

$$7x + 5y \leq 240$$
$$77x + 258y \leq 4000$$
$$5x + 32y \leq 400$$

We list the vertices of the region of feasible solutions and the values of f at each of these vertices:

Vertex	Value of f	
(0, 0)	0	
(0, 12.5)	12.5	
(21.12, 9.2)	30.32	
(29.5, 6.7)	36.2	(Maximum value)
(34.3, 0)	34.3	

Hence, by Step 4′ we see that f attains its maximum value in the region of feasible solutions at the point (29.5, 6.7). ■

This procedure works well for problems involving only two variables. However, in cases involving three variables, the region of feasible solutions becomes a three-dimensional polyhedron and finding the vertices is difficult. In four or more variables things become even more confusing. An example of a four-variable linear programming problem would be to assume that two additional species of fish (perhaps whitefish and perch) were included in our original problem. To solve such a problem, a more sophisticated method is needed. Such a method is the simplex algorithm.

The Simplex Method

The simplex method is an algebraic (as opposed to geometric) method of solving linear programming problems. The main computational tool is row reduction of matrices. Although this method may seem cumbersome when the reductions are carried out by hand, this method is extremely efficient when a computer is used. We will illustrate how the simplex algorithm works by solving the same problem we solved earlier.

We begin by restating the problem. We wish to maximize a certain function called the **objective function** subject to certain inequalities called the **constraint inequalities**. In this case the objective function is $f(x, y) = x + y$ subject to the constraints $x \geq 0$, $y \geq 0$, and

$$7x + 5y \leq 240$$
$$77x + 258y \leq 4000$$
$$5x + 32y \leq 400$$

The first step in the simplex method is to convert the constraint inequalities (other than the nonnegativity constraints $x \geq 0$, $y \geq 0$) into equalities by the addition of slack variables. For example, the set of all points (x, y), where $x \geq 0$ and $y \geq 0$ which satisfy the inequality

$$7x + 5y \leq 240$$

is the same as the set of all points (x, y) satisfying the equation

$$7x + 5y + u = 240 \qquad \text{where } x \geq 0,\ y \geq 0,\ u \geq 0$$

The variable u is called a **slack variable** (it takes up the slack between $7x + 5y$ and 240). Now, for any fixed value of u, the equation

$$7x + 5y + u = 240$$

represents a line. For example, if $u = 0$, then this is the line

$$7x + 5y = 240$$

which is the boundary of the region of feasible solutions. If $u = 30$, the equation of the line is

$$7x + 5y + 30 = 240 \qquad \text{or} \qquad 7x + 5y = 210$$

and if $u = 240$, the equation of the line is

$$7x + 5y + 240 = 240 \qquad \text{or} \qquad 7x + 5y = 0$$

which is just the point (0, 0) since $x \geq 0$, and $y \geq 0$. (Note that u cannot exceed 240, otherwise this would force x or y to be negative). These lines are illustrated in Figure 2.10.

Figure 2.10

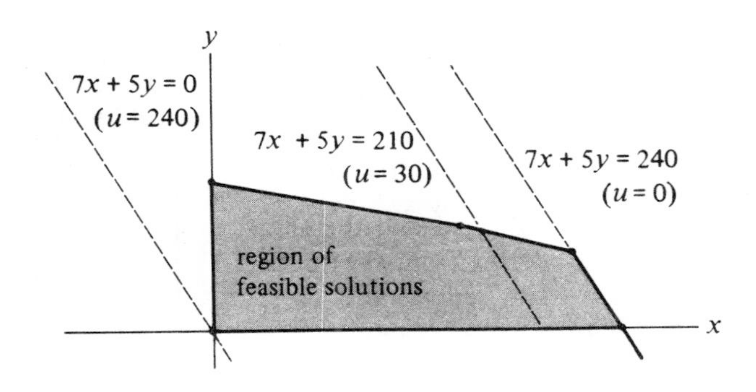

We replace each of the defining inequalities with an equation involving slack variables.

$$
\begin{aligned}
7x + 5y + u &= 240 \\
77x + 258y + v &= 4000 \qquad (1)\\
5x + 32y + w &= 400
\end{aligned}
$$

where x, y, u, v, and w are all nonnegative.

As we saw in the discussion above, each of the boundary lines of the region of feasible solutions corresponds to one of the five variables x, y, u, v, and w being zero. This is illustrated in Figure 2.11. Moreover, each vertex of the region of feasible solutions is the intersection of two of the boundary lines. Hence, each vertex corresponds to two of the variables being zero. In Figure 2.11 the variables that are zero are listed next to the corresponding vertices. It should be noted that setting any pair of vertices equal to zero does not always give a point in the region of feasible solutions. For example, the point corresponding to $w = 0$ and $u = 0$ is the point of intersection of the lines

$$5x + 32y = 400 \qquad \text{and} \qquad 7x + 5y = 240$$

and this point lies outside the region of feasible solutions.

Figure 2.11

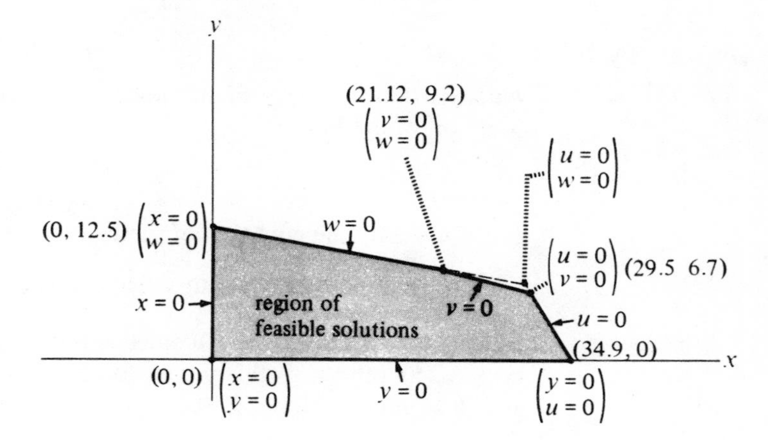

The simplex algorithm provides a way of going from one vertex in the region of feasible solutions to an adjacent vertex in this region in such a way that the objective function increases at each step until the maximum value is attained.

With each vertex of the region of feasible solutions there is an associated matrix called the **simplex tableau**. To obtain the simplex tableau for one vertex from the simplex tableau for an adjacent vertex, a special type of row reduction is used.

The equations used to construct the initial simplex tableau are the equations that describe the region of feasible solutions in terms of x, y, and the slack variables, as well as an equation that describes the objective function. The equation used to describe the objective function of our problem is obtained as follows.

$$\begin{aligned} f(x, y) &= x + y \\ &= x + y + 0u + 0v + 0w \end{aligned}$$

Letting $z = f(x, y)$ we obtain the equation

$$z = x + y + 0u + 0v + 0w$$

or equivalently $$-x - y + 0u + 0v + 0w + z = 0$$

The initial simplex tableau is built from the following equations:

$$\left.\begin{aligned} 7x + 5y + 1u + 0v + 0w + 0z &= 240 \\ 77x + 258y + 0u + 1v + 0w + 0z &= 4000 \\ 5x + 32y + 0u + 0v + 1w + 0z &= 400 \end{aligned}\right\} \text{Region of feasible solutions}$$

$$-1x - 1y + 0u + 0v + 0w + 1z = 0 \quad \} \text{ Objective function}$$

where all the variables are nonnegative.

All the mathematical information about the linear programming problem is contained in these equations. If x, y, u, v, w, and z are nonnegative numbers satisfying these equations, then (x, y) is a point in the region of feasible solutions and z is the value of the objective function at the point (x, y).

The simplex tableau corresponding to this system of equations is

Tableau 1

$$\begin{array}{cccccc|c} x & y & u & v & w & z & \\ 7 & 5 & 1 & 0 & 0 & 0 & 240 \\ 77 & 258 & 0 & 1 & 0 & 0 & 4000 \\ 5 & 32 & 0 & 0 & 1 & 0 & 400 \\ \hline -1 & -1 & 0 & 0 & 0 & 1 & 0 \end{array}$$

This tableau is the one corresponding to the vertex $x = 0$, $y = 0$. By letting the variables x and y be zero, we see that

$$\begin{aligned} 7(0) + 5(0) + 1u + 0v + 0w + 0z &= 240 &\quad \text{or} \quad u &= 240 \\ 77(0) + 258(0) + 0u + 1v + 0w + 0z &= 4000 &\quad \text{or} \quad v &= 4000 \\ 5(0) + 32(0) + 0u + 0v + 1w + 0z &= 400 &\quad \text{or} \quad w &= 400 \\ -1(0) - 1(0) + 0u + 0v + 0w + 1z &= 0 &\quad \text{or} \quad z &= 0 \end{aligned}$$

Now we change to another tableau using the following special row operations. (How these operations are chosen will be described later.)

$$\begin{array}{c} \begin{matrix} x & y & u & v & w & z & \end{matrix} \\ \left[\begin{array}{cccccc|c} 7 & 5 & 1 & 0 & 0 & 0 & 240 \\ 77 & 258 & 0 & 1 & 0 & 0 & 4000 \\ 5 & 32 & 0 & 0 & 1 & 0 & 400 \\ \hline -1 & -1 & 0 & 0 & 0 & 1 & 0 \end{array}\right] \end{array} \quad \begin{matrix} (1) \\ (1) \\ \left(\frac{-5}{32}\right) \quad \left(\frac{-258}{32}\right) \quad \left(\frac{1}{32}\right) \rightarrow \left(\frac{1}{32}\right) \\ (1) \end{matrix}$$

Tableau 2

$$\begin{array}{c} \begin{matrix} x & y & u & v & w & z & \end{matrix} \\ \left[\begin{array}{cccccc|c} 6.2 & 0 & 1 & 0 & -.156 & 0 & 177.5 \\ 36.7 & 0 & 0 & 1 & -8.06 & 0 & 775.0 \\ 0.16 & 1 & 0 & 0 & 0.03 & 0 & 12.5 \\ \hline -0.84 & 0 & 0 & 0 & 0.03 & 1 & 12.5 \end{array}\right] \end{array}$$

Note that the set of all six-tuples (x, y, u, v, w, z) that are solutions of the equations represented by this tableau is the same as the set of all six-tuples satisfying Tableau 1 and hence the original set of equations. This, however, is the tableau associated with the vertex $x = 0$, $w = 0$. Substituting these values in the equations represented by this tableau gives

$$6.2(0) + 0y + 1u + 0v - 1.56(0) + 0z = 177.5$$

$$36.7(0) = 0y + 0u + 1v - 8.06(0) + 0z = 775.0$$

$$0.16(0) + 1y + 0u + 0v + 0.03(0) + 0z = 12.5$$

$$-0.84(0) + 0y + 0u + 0v + 0.03(0) + 1z = 12.5$$

or $$u = 177.5 \quad v = 775.0 \quad y = 12.5 \quad z = 12.5$$

Tableau 2 tells us that at the vertex where $x = 0$ and $w = 0$ the value of the objective function is $z = 12.5$ (see Figure 2.11). However, we can learn even more from this tableau. The bottom row gives an expression for z in terms of x and w (these are the variables we let be zero). The equation represented by the bottom row is

$$-0.84x + 0y + 0u + 0v + 0.03w + 1z = 12.5$$

or $$z = 12.5 + 0.84x - 0.03w$$

From this we see that if we increase the value of x (it is presently zero) by moving to an adjacent vertex, the value of z will increase; however, increasing the value of w will decrease the value of z. Thus, we are not at a maximum since the value of z can still be increased.

We proceed to the next vertex to continue to increase z using the following row operations:

$$\begin{array}{c} \begin{array}{cccccc|c} x & y & u & v & w & z & \\ \end{array} \\ \left[\begin{array}{cccccc|c} 6.2 & 0 & 1 & 0 & -0.156 & 0 & 177.5 \\ 36.7 & 0 & 0 & 1 & -8.06 & 0 & 775.0 \\ 0.16 & 1 & 0 & 0 & 0.03 & 0 & 12.5 \\ \hline -0.84 & 0 & 0 & 0 & 0.03 & 1 & 12.5 \end{array}\right] \end{array} \quad \left(\frac{-6.2}{36.7}\right) \to (1) \quad \left(\frac{1}{36.7}\right) \to \left(\frac{-0.16}{36.7}\right) \to (1) \quad \left(\frac{0.84}{36.7}\right) \to (1)$$

Tableau 3

$$\begin{array}{c} \begin{array}{cccccc|c} x & y & u & v & w & z & \end{array} \\ \left[\begin{array}{cccccc|c} 0 & 0 & 1 & -0.168 & 1.2 & 0 & 46.6 \\ 1 & 0 & 0 & 0.027 & -0.22 & 0 & 21.1 \\ 0 & 1 & 0 & -0.004 & 0.065 & 0 & 9.1 \\ \hline 0 & 0 & 0 & 0.023 & -0.154 & 1 & 30.2 \end{array}\right] \end{array}$$

From this tableau we learn that if $v = 0$ and $w = 0$, then $x = 21.1$, $y = 9.1$, and $u = 46.6$, and the value of the objective function is $z = 30.2$. Moreover, increasing w will increase the value of the objective function (the coefficient of w in the objective (bottom) row of Tableau 3 is negative). Since the value of the objective function can still be increased, we are not at a maximum and we proceed to the next tableau.

$$\begin{array}{c} \begin{array}{cccccc|c} x & y & u & v & w & z & \end{array} \\ \left[\begin{array}{cccccc|c} 0 & 0 & 1 & -0.168 & 1.2 & 0 & 46.6 \\ 1 & 0 & 0 & 0.027 & -0.22 & 0 & 21.1 \\ 0 & 1 & 0 & -0.004 & 0.065 & 0 & 9.1 \\ \hline 0 & 0 & 0 & 0.023 & -0.154 & 1 & 30.2 \end{array}\right] \end{array} \quad \left(\frac{1}{1.2}\right) \to \left(\frac{0.22}{1.2}\right) \to (1) \quad \left(\frac{-0.065}{1.2}\right) \to (1) \quad \left(\frac{0.154}{1.2}\right) \to (1)$$

Tableau 4

$$\begin{array}{c} \begin{array}{cccccc|c} x & y & u & v & w & z & \end{array} \\ \left[\begin{array}{cccccc|c} 0 & 0 & 0.84 & -0.14 & 1 & 0 & 38.8 \\ 1 & 0 & 0.18 & -0.004 & 0 & 0 & 29.7 \\ 0 & 1 & -0.058 & 0.005 & 0 & 0 & 6.4 \\ \hline 0 & 0 & 0.125 & 0.002 & 0 & 1 & 36.0 \end{array}\right] \end{array}$$

From Tableau 4 we see that when $u = 0$ and $v = 0$, then $x = 29.7$, $y = 6.4$, $w = 38.8$, and the value of the objective function is $z = 36.0$. Since solving the bottom equation for z gives

$$z = 36.0 - 0.125u - 0.002w$$

we see that an increase of either u or w will decrease the value of z. Hence, the maximum value for z in the region of feasible solutions is $z = 36.0$ and we have solved the problem.

The function $z = x + y$ attains its maximum value in the region of feasible solutions when $x = 29.7$ and $y = 6.4$; the maximum value of z is 36. (*Note:* The slight discrepancy between this answer and the one obtained earlier is due to the error incurred by rounding off the entries in the various tableaus).

The question that remains to be answered to complete our understanding of the simplex algorithm is how to select the proper row operation to allow us to proceed from one tableau to the next. To begin, let us look at the form of each of the tableaus. There are six columns to the left of the vertical line. The column under z is always the same, having a 1 in the bottom position and 0's in all the other positions (for this reason this column is generally omitted when writing simplex tableaus). Of the remaining five columns to the left of the vertical line, the three columns that we label $\mathbf{e}_1$, $\mathbf{e}_2$, and $\mathbf{e}_3$ always appear in the form

$$\mathbf{e}_1 = \begin{bmatrix} 1 \\ 0 \\ 0 \\ \hline 0 \end{bmatrix} \qquad \mathbf{e}_2 = \begin{bmatrix} 0 \\ 1 \\ 0 \\ \hline 0 \end{bmatrix} \qquad \mathbf{e}_2 = \begin{bmatrix} 0 \\ 0 \\ 1 \\ \hline 0 \end{bmatrix}$$

The way we use the tableau is to let the variables (other than z) associated with the columns that are different from $\mathbf{e}_1$, $\mathbf{e}_2$, and $\mathbf{e}_3$ be zero. Then the values of the remaining variables can be read directly from the tableau.

For example, in Tableau 2 the variables x and w are set equal to zero. The column $\mathbf{e}_1$ is under u, the column $\mathbf{e}_2$ is under v, and the column $\mathbf{e}_3$ is under y. From this tableau we see that when $x = 0$ and $w = 0$, then $y = 12.5$, $u = 177.5$, $v = 775.0$, and $z = 12.5$.

As we go from one tableau to the next, the pattern of columns changes. We leave two of the $\mathbf{e}_i$ columns alone and replace one of the non-$\mathbf{e}_i$ columns, called the **pivot column**, by an $\mathbf{e}_i$ column. Replacing a non-$\mathbf{e}_i$ column by an $\mathbf{e}_i$ column is equivalent to changing a variable that was equal to zero to a non-zero variable.

For example, in going from Tableau 2 to Tableau 3, we changed the variable x from zero to a nonzero value (in this case x became 21.1). The column under x is the pivot column.

Now we want to know which of the two non-$\mathbf{e}_i$ columns to use as the pivot column. We do this by examining the entries in the bottom (objective) row. As in our previous discussion, if one of the entries in the bottom row is negative, then increasing the value of the variable above this entry increases the value of z. If there are two negative entries in the objective row, then increasing either entry increases the value of z, so we choose the entry with the largest absolute value. This gives the largest increase in z for each unit increase in the variable.

Therefore, we see that by looking at a tableau we can determine if it represents a maximal solution and, if not, how to choose a pivot column which will lead to a new tableau representing a larger value of the objective function. We state these ideas more formally.

> **Maximality** A tableau repesents a maximal solution to a linear programming problem if all entries in the bottom row are nonnegative.
>
> **Pivot Column** The pivot column is the column that contains the negative entry of largest absolute value in the bottom row.

In going from one tableau to the next there is a single pivot row. The choice of the pivot row is a little more subtle than the choice of the pivot column. To illustrate what can happen if we choose the wrong pivot row, we operate on Tableau 2. However, instead of using the second row as the pivot row, we use the first row.

$$
\begin{array}{c}
\begin{array}{cccccc|c}
x & y & u & v & w & z & \\
\end{array}\\
\left[\begin{array}{cccccc|c}
6.2 & 0 & 1 & 0 & -0.156 & 0 & 177.5\\
36.7 & 0 & 0 & 1 & -8.06 & 0 & 775.0\\
0.16 & 1 & 0 & 0 & 0.03 & 0 & 12.5\\
\hline
-0.84 & 0 & 0 & 0 & 0.03 & 1 & 12.5
\end{array}\right]
\end{array}
\quad
\left(\frac{1}{6.2}\right)\rightarrow\left(\frac{-36.7}{6.2}\right)(1)\rightarrow\left(\frac{-0.16}{6.2}\right)(1)\rightarrow\left(\frac{0.84}{6.2}\right)(1)\rightarrow
$$

Tableau 3a

$$
\begin{array}{c}
\begin{array}{cccccc|c}
x & y & u & v & w & z & \\
\end{array}\\
\left[\begin{array}{cccccc|c}
1 & 0 & 0.16 & 0 & -0.025 & 0 & 28.6\\
0 & 0 & -5.9 & 1 & -7.14 & 0 & \boxed{-275.7}\\
0 & 1 & -0.025 & 0 & 0.034 & 0 & 7.92\\
\hline
0 & 1 & 0.135 & 0 & 0.009 & 1 & 36.5
\end{array}\right]
\end{array}
$$

From Tableau 3a we see that when $u = 0$ and $w = 0$, then $v = -275.6$. Since $v < 0$, this tableau corresponds to a point that is outside the region of feasible solutions. So, by choosing the wrong pivot row we went outside the region of feasible solutions.

To avoid leaving the region of feasible solutions, we must make sure that all entries in the column to the right of the vertical line are nonnegative.

To see how to do this we must examine the special row operations we have used to go from one tableau to the next.

The object of the row operations is to replace the pivot column by a column containing 1 in the pivot row position and 0's in all the other positions. Moreover, the $\mathbf{e}_i$'s that do not have a 1 in the pivot row position are to be left alone. Below we show what happens to the pivot row and a typical nonpivot row under the row operations.

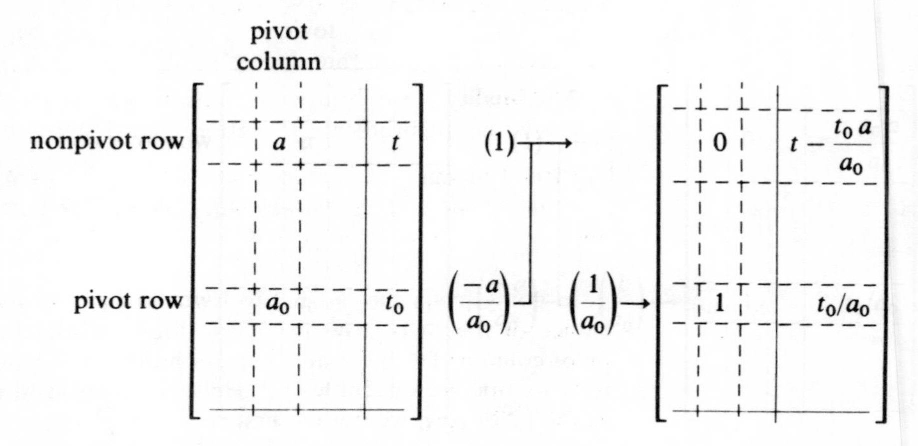

Therefore, the conditions that will ensure that the entries in the right-hand column of the new tableau will be nonnegative are

$$t - (t_0 a/a_0) \geq 0 \quad \text{and} \quad (t_0/a_0) \geq 0$$

that is,

$$(t_0/a_0) \leq (t/a) \quad \text{and} \quad t_0/a_0 \geq 0$$

Therefore, to ensure that all entries of the last column of the new tableau will be nonnegative, the pivot row must have a positive entry in the pivot column and the ratio t_0/a_0 of its last entry to the entry in the pivot column is the minimal such nonnegative ratio. We state this more formally.

Pivot row To find the pivot row examine the ratios of the last entry of each row to the entry in the pivot column in that row. The row with the least positive ratio is the pivot row. (*Note:* If no ratio is positive, then it can be shown that the problem has no solution).

To illustrate these ideas, let us look at Tableau 3. First note that there is a negative entry in the bottom row. Thus the tableau does not represent a maximal solution to the problem. Since -0.154 is the least entry in the bottom row, the column containing -0.154 is the pivot column (this is the column corresponding to w). To find the pivot row, we examine the ratios of the element in the last column to the element in the pivot column:

$$\text{First row:} \quad \frac{46.6}{1.2} = 38.8$$

$$\text{Second row:} \quad \frac{21.1}{-0.22} \quad \text{is negative}$$

$$\text{Third row:} \quad \frac{9.1}{0.065} = 140$$

So the first row is the pivot row since its ratio is the least positive number.

Another example of how the simplex algorithm is applied follows.

EXAMPLE 4 Determine whether the following tableau is maximal, if not proceed to the next tableau.

$$\begin{array}{cccccc|c} x & y & u & v & w & z & \\ \hline 0 & 0 & 1 & 1 & 5 & 0 & 5 \\ 1 & 0 & -2 & 0 & -1 & 0 & 2 \\ 0 & 1 & 3 & 0 & 2 & 0 & 1 \\ \hline 0 & 0 & -2 & 0 & -1 & 1 & 6 \end{array}$$

Since the last row of this tableau contains negative entries (-1 and -2), this tableau does not represent the maximal solution. Since -2 is the least entry in the bottom row, the column under u is the pivot column. The ratios of the entry in the last column to the pivot column are

$$\text{First row:} \quad \frac{5}{1} = 5$$

$$\text{Second row:} \quad \frac{2}{-2} \quad \text{is negative}$$

$$\text{Third row:} \quad \frac{1}{3} = 0.33$$

Since the ratio for the third row is the smallest positive ratio, the third row is the pivot row.

The tableau is reduced as follows:

$$
\begin{array}{c}
\begin{array}{cccccc} x & y & u & v & w & z \end{array} \\
\left[\begin{array}{cccccc|c}
0 & 0 & 1 & 1 & 5 & 0 & 5 \\
1 & 0 & -2 & 0 & -1 & 0 & 2 \\
0 & 1 & \textcircled{3} & 0 & 2 & 0 & 1 \\
\hline
0 & 0 & -2 & 0 & -1 & 1 & 6
\end{array}\right]
\end{array}
\quad
\left(\frac{-1}{3}\right) \to (1) \quad \left(\frac{2}{3}\right) \to (1) \quad \left(\frac{1}{3}\right) \quad \left(\frac{2}{3}\right) \to (1)
$$

$$
\begin{array}{c}
\begin{array}{cccccc} x & y & u & v & w & z \end{array} \\
\left[\begin{array}{cccccc|c}
0 & -0.33 & 0 & 1 & 4.33 & 0 & 4.67 \\
1 & 0.67 & 0 & 0 & 0.33 & 0 & 2.67 \\
0 & 0.33 & 1 & 0 & 0.67 & 0 & 0.33 \\
\hline
0 & 0.67 & 0 & 0 & 0.33 & 1 & 6.67
\end{array}\right]
\end{array}
$$

So we conclude that the value of the objective function has been increased from 6 to 6.67, and since no entry in the bottom row is negative, this is the maximum value of the objective function in the region of feasible solutions. It occurs when $x = 2.67$ and $y \doteq 0$. ■

The work in this section is only the briefest introduction to the subject of linear programming. It was intended to give you a general idea of what kinds of problems can be solved and how the technique of Gaussian elimination (matrix row reduction) can be applied to linear programming. A much more extensive treatment of this subject is contained in the book *Linear Programming and Extensions*, by George Dantzig.

Exercises for Applications 2

1. a. Graph the region of feasible solutions defined by the following inequalities (label all vertices):

$$
\begin{gathered}
x \geq 0, \qquad y \geq 0 \\
x + 3y \leq 30 \\
x + 2y \leq 24 \\
5x + 3y \leq 105
\end{gathered}
$$

b. Use the geometrical method to find the point that maximizes the function

$$f(x, y) = 11x + 27y$$

subject to the constraints in part a.

c. Use the geometrical method to find the point that maximizes the function

$$f(x, y) = 30x + 23y$$

subject to the constraints in part a.

2. Solve problems 1b and 1c using the simplex method.

3. A manufacturing plant has three divisions (call them D_1, D_2, and D_3) and three products (call them P_1, P_2, and P_3). To produce one unit of P_1 requires 6 hours in D_1, 4 hours in D_2, and 2 hours in D_3. One unit of P_2 requires

2 hours in D_1, 3 hours in D_2, and 3 hours in D_3. One unit of P_3 requires 3 hours in D_1 and 3 hours in D_2 with no time in D_3. Each unit of product P_1, P_2, and P_3 can be sold for a profit of \$100, \$110, and \$120 respectively. There is a limit of 150 hours per week available in D_1, 200 hours per week in D_2, and 30 hours per week in D_3. How should production be scheduled to maximize profit?

★**4. a.** Suppose the function g attains its minimum value in a given region at the point (x_0, y_0). Show that the function $f(x, y) = -g(x, y)$ attains its maximum value at the point (x_0, y_0).

b. Show that in order to use the simplex algorithm to solve a linear programming problem in which the objective function is to be minimized over the region of feasible solutions, you can use the simplex algorithm to maximize the negative of the original objective function.

c. Minimize the function $g(x, y) = -x - 2y$ in the region described in Exercise 1a using the simplex method and the result of Exercise 4b.

Review Exercises

1. For each of the following systems of linear equations, use row reduction (Gaussian elimination) to determine all solutions of each system.

a.
$$\begin{aligned} x + y + z &= 1 \\ x - y \phantom{{}+z} &= 7 \\ x + 3y + 2z &= 4 \end{aligned}$$

b.
$$\begin{aligned} x + y + z &= 7 \\ 2x + 3y + z &= 18 \\ 3x + 5y + z &= 29 \end{aligned}$$

c.
$$\begin{aligned} x + y + z &= 1 \\ x + y - z &= 2 \\ x - y - z &= 3 \end{aligned}$$

d.
$$\begin{aligned} x + y + 2z &= 7 \\ 2x + 2y - z &= 4 \\ 3x + 3y + 4z &= 17 \end{aligned}$$

e.
$$\begin{aligned} x + y + z &= -2 \\ 2x + y + 2z &= 3 \\ 3x + 2y - z &= 1 \end{aligned}$$

2. Consider the system of linear equations

$$\begin{aligned} 2x - y &= 6 \\ -4x + 2y &= a \end{aligned}$$

a. Find a number a so that this system has no solutions.

b. Find a number a so that this system has infinitely many solutions.

c. Explain why there is no number a so that this system has exactly one solution.

3. Suppose there are two systems of equations. The augmented matrix associated with the first system is M_1 and the augmented matrix associated with the second system is M_2. Is the following statement true or false? If M_1 and M_2 can be reduced to the same matrix, then the two systems have the same set of solutions. Explain your answer.

4. Solve the following systems of equations:

a.
$$\begin{aligned} 2x - 3y + z + 4w &= 0 \\ -3x + y + 2z - w &= 0 \end{aligned}$$

b.
$$\begin{aligned} 2x - 4y &= 0 \\ x + y &= 0 \\ -x - 2y &= 0 \\ x + 3y &= 0 \end{aligned}$$

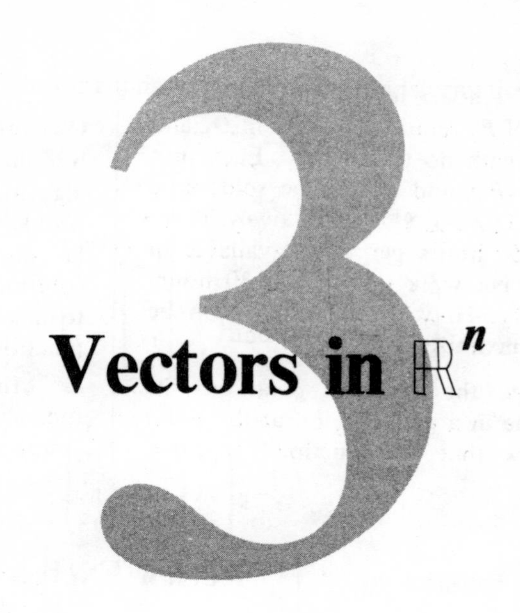

Vectors in $\mathbb{R}^n$

In this chapter we begin the general study of linear algebra by studying linear transformations (linear functions that map vectors to vectors) and by giving a more general definition of a vector. In Section 3.2 you will see that not all vectors are directed line segments; we will use this fact to apply linear algebra to solve problems in differential equations and in calculus in Chapter 9. In Section 3.3, matrices are introduced. Matrices represent linear transformations and provide an important tool for studying linear algebra.

3.1 Linear Transformations on $\mathbb{R}^n$

We introduce vectors with three or more coordinates in the same way as we did for vectors with two coordinates in Chapter 1.

DEFINITION 3.1

An ordered n-tuple of the form $\begin{bmatrix} a_1 \\ a_2 \\ \vdots \\ a_n \end{bmatrix}$

where the a_i's are real numbers, is called a **vector in** $\mathbb{R}^n$. We use the symbol $\mathbb{R}^n$ to denote the set of all vectors having n coordinates (entries).

As in Chapter 1, it is often helpful to think of vectors as line segments starting at the origin of a coordinate system and ending at the point whose coordinates are the entries of the vector. In Chapter 1, we relied heavily on graphs to help us understand important concepts in $\mathbb{R}^1$ and $\mathbb{R}^2$. Graphs will continue to play an important role as we generalize these ideas to $\mathbb{R}^n$ ($n \geq 3$) and introduce new concepts. Of special importance are graphs in $\mathbb{R}^3$.

Throughout this text we will use the three-dimensional coordinate system having the following orientation of the coordinate axes: The positive x axis points out from the page, the positive y axis points to the right, and the positive z axis points up (see Figure 3.1). This is called a right-handed

Figure 3.1

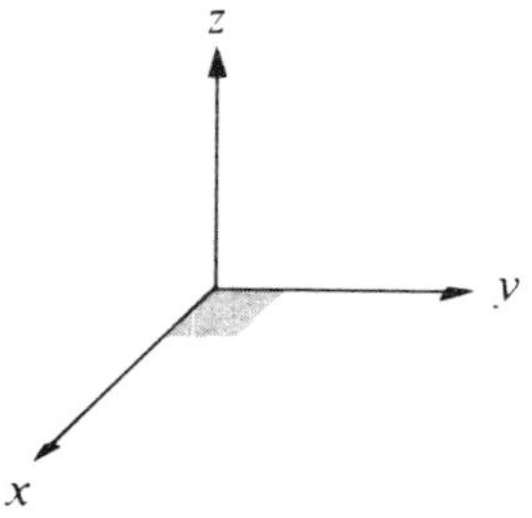

coordinate system because if you curl the fingers of your right hand in the direction from the positive x axis to the positive y axis, your thumb will point in the direction of the positive z axis. To plot a point with coordinates (a, b, c), we begin at the origin and go a units in the x direction, then go b units in the y direction, and then go c units in the z direction. We plot the point (4, 3, 5) in Figure 3.2 to illustrate the procedure. The same procedure is used to plot points with negative coefficients. We illustrate this by plotting the point $(-3, -4, 1)$ in Figure 3.3.

As with vectors in $\mathbb{R}^2$, we graph vectors in $\mathbb{R}^3$ as line segments that begin at the origin and end at the point having the coordinates of the vector.

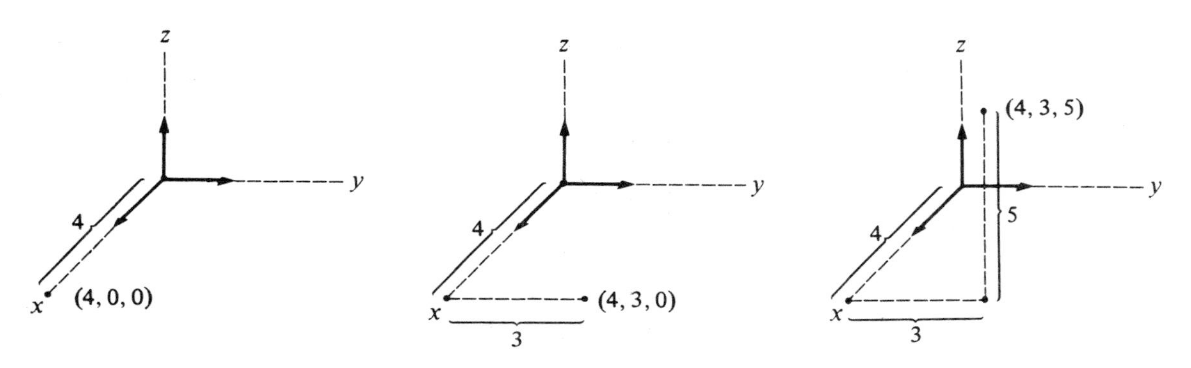

Figure 3.2

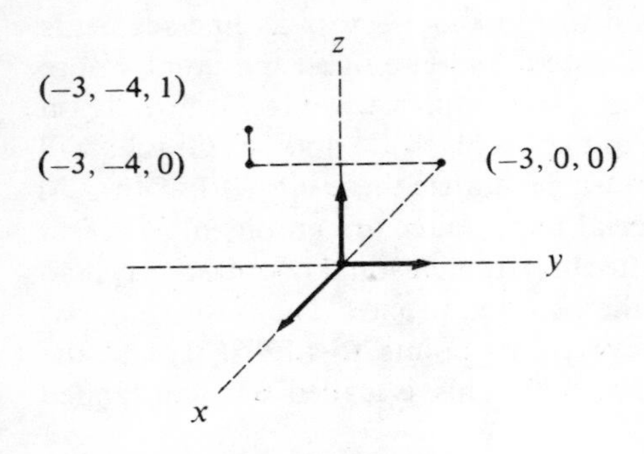

Figure 3.3

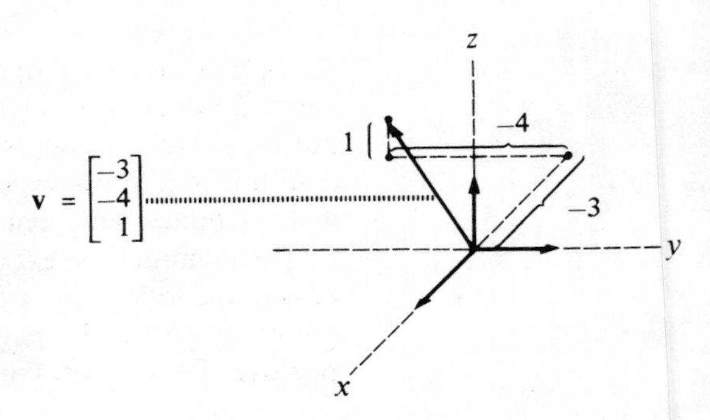

Figure 3.4

The vector

$$\mathbf{v} = \begin{bmatrix} -3 \\ -4 \\ 1 \end{bmatrix}$$

is graphed in Figure 3.4.

With these ideas in mind, we generalize the concepts of addition and scalar multiplication of vectors. These two concepts are defined in $\mathbb{R}^n$ (for $n \geq 3$) as they were in $\mathbb{R}^2$ (see Section 1.3).

DEFINITION 3.2 The **sum of two vectors** is defined by

$$\begin{bmatrix} a_1 \\ a_2 \\ \vdots \\ a_n \end{bmatrix} + \begin{bmatrix} b_1 \\ b_2 \\ \vdots \\ b_n \end{bmatrix} = \begin{bmatrix} a_1 + b_1 \\ a_2 + b_2 \\ \vdots \\ a_n + b_n \end{bmatrix}$$

where a_i and b_i are real numbers.

DEFINITION 3.3 The **product of a scalar** c **and a vector** is defined by

$$c\begin{bmatrix} a_1 \\ a_2 \\ \vdots \\ a_n \end{bmatrix} = \begin{bmatrix} ca_1 \\ ca_2 \\ \vdots \\ ca_n \end{bmatrix}$$

where a_i are real numbers.

Continuing our analogy with Chapter 1, we look at linear functions of vectors in $\mathbb{R}^n$. We will not restrict ourselves only to those functions with range equal to the real numbers, but also will allow the range to be $\mathbb{R}^m$ for some integer m. Such linear functions that map vectors to vectors are called linear transformations.

DEFINITION 3.4

A function $T: \mathbb{R}^n \to \mathbb{R}^m$ is called a **linear transformation** if it satisfies the following two conditions:

1. $T(\mathbf{v}_1 + \mathbf{v}_2) = T(\mathbf{v}_1) + T(\mathbf{v}_2)$ for all vectors $\mathbf{v}_1$ and $\mathbf{v}_2$ in the domain of T.

2. $cT(\mathbf{v}) = T(c\mathbf{v})$ for all scalars c and all vectors $\mathbf{v}$ in the domain of T.

Each of the conditions of Definition 3.4 can be interpreted geometrically. To check Condition 1, compare the diagonal of the parallelogram formed by $T(\mathbf{v}_1)$ and $T(\mathbf{v}_2)$ to the vector $T(\mathbf{v}_1 + \mathbf{v}_2)$ in Figure 3.5. If they are the same, then this condition holds.

Figure 3.5

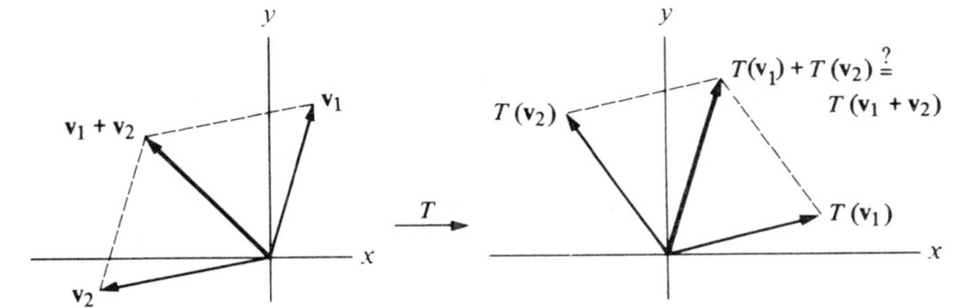

To verify Condition 2, check to see that both $T(\mathbf{v})$ and $T(c\mathbf{v})$ lie on the same line through the origin, and also that $T(c\mathbf{v})$ is c times as long as $T(\mathbf{v})$ (see Figure 3.6).

The following examples should help clarify the definition of linear transformation.

Figure 3.6

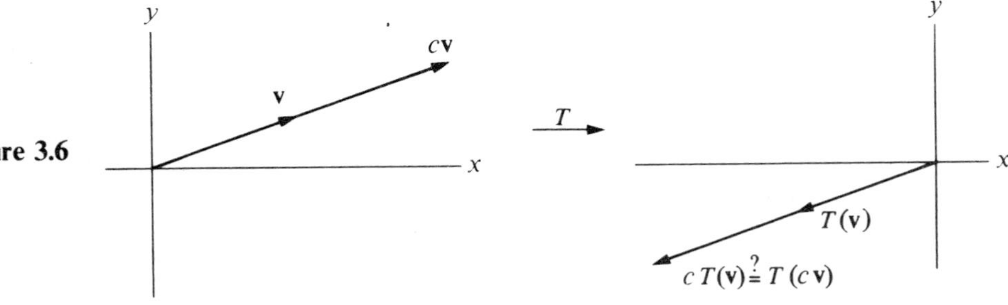

EXAMPLE 1 Show that the following function T is a linear transformation.

$$T: \mathbb{R}^3 \to \mathbb{R}^2$$

$$\begin{bmatrix} x \\ y \\ z \end{bmatrix} \to \begin{bmatrix} x + y \\ y - z \end{bmatrix}$$

Let $\quad \mathbf{v}_1 = \begin{bmatrix} x_1 \\ y_1 \\ z_1 \end{bmatrix} \quad$ and $\quad \mathbf{v}_2 = \begin{bmatrix} x_2 \\ y_2 \\ z_2 \end{bmatrix}$

Then

$$\begin{aligned}
T(\mathbf{v}_1) + T(\mathbf{v}_2) &= T\begin{bmatrix} x_1 \\ y_1 \\ z_1 \end{bmatrix} + T\begin{bmatrix} x_2 \\ y_2 \\ z_2 \end{bmatrix} \\
&= \begin{bmatrix} x_1 + y_1 \\ y_1 - z_1 \end{bmatrix} + \begin{bmatrix} x_2 + y_2 \\ y_2 - z_2 \end{bmatrix} && \text{by definition of } T \\
&= \begin{bmatrix} (x_1 + x_2) + (y_1 + y_2) \\ (y_1 + y_2) - (z_1 + z_2) \end{bmatrix} && \text{by adding the two vectors} \\
&= T\begin{bmatrix} x_1 + x_2 \\ y_1 + y_2 \\ z_1 + z_2 \end{bmatrix} && \text{by definition of } T \\
&= T(\mathbf{v}_1 + \mathbf{v}_2)
\end{aligned}$$

So Condition 1 holds.

Now we check to see if Condition 2 holds. Let c be a scalar and

$$\mathbf{v} = \begin{bmatrix} x \\ y \\ z \end{bmatrix}$$

be an arbitrary vector in $\mathbb{R}^3$. Then

$$\begin{aligned}
T(c\mathbf{v}) &= T\left(c\begin{bmatrix} x \\ y \\ z \end{bmatrix}\right) \\
&= T\begin{bmatrix} cx \\ cy \\ cz \end{bmatrix} && \text{by scalar multiplication} \\
&= \begin{bmatrix} cx + cy \\ cy - cz \end{bmatrix} && \text{by definition of } T \\
&= c\begin{bmatrix} x + y \\ y - z \end{bmatrix} && \text{by scalar multiplication} \\
&= cT(\mathbf{v}) && \text{by definition of } T
\end{aligned}$$

Condition 2 holds, and we have shown that T is a linear transformation.

To clarify the geometric meaning of this transformation, Figure 3.7 shows what T does to the three vectors

$$\begin{bmatrix}1\\0\\0\end{bmatrix}, \begin{bmatrix}0\\1\\0\end{bmatrix}, \begin{bmatrix}0\\0\\1\end{bmatrix}$$

Figure 3.7

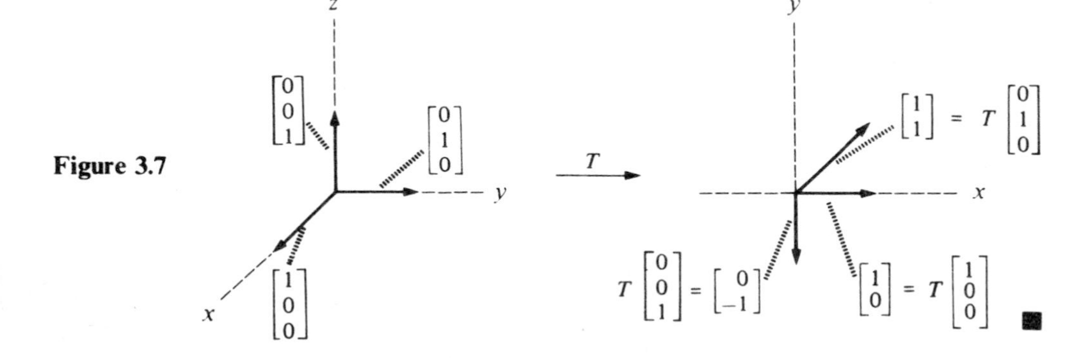

Some of the most important types of linear transformations can be described geometrically as the next example shows.

EXAMPLE 2 Let $T: \mathbb{R}^2 \to \mathbb{R}^2$ be reflection about the x axis, that is, for any vector $\mathbf{v}$, the vector $T(\mathbf{v})$ is the vector that $\mathbf{v}$ would fall on if the paper were folded along the x axis (see Figure 3.8). Show that T is linear.

Figure 3.8

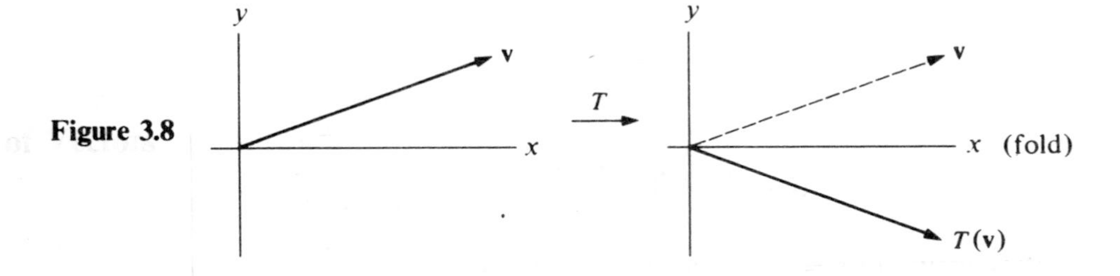

To check Condition 1, we must see if the diagonal of the parallelogram formed by $T(\mathbf{v}_1)$ and $T(\mathbf{v}_2)$ is the same as the image $T(\mathbf{v}_1 + \mathbf{v}_2)$ of the diagonal of the parallelogram formed by $\mathbf{v}_1$ and $\mathbf{v}_2$, for any pair of vectors $\mathbf{v}_1$ and $\mathbf{v}_2$. Since the reflection (fold) T leaves the shape of the parallelogram unchanged, as can be seen in Figure 3.9, Condition 1 holds.

To check Condition 2, we note that T maps the line containing $\mathbf{v}$ and $c\mathbf{v}$

Figure 3.9

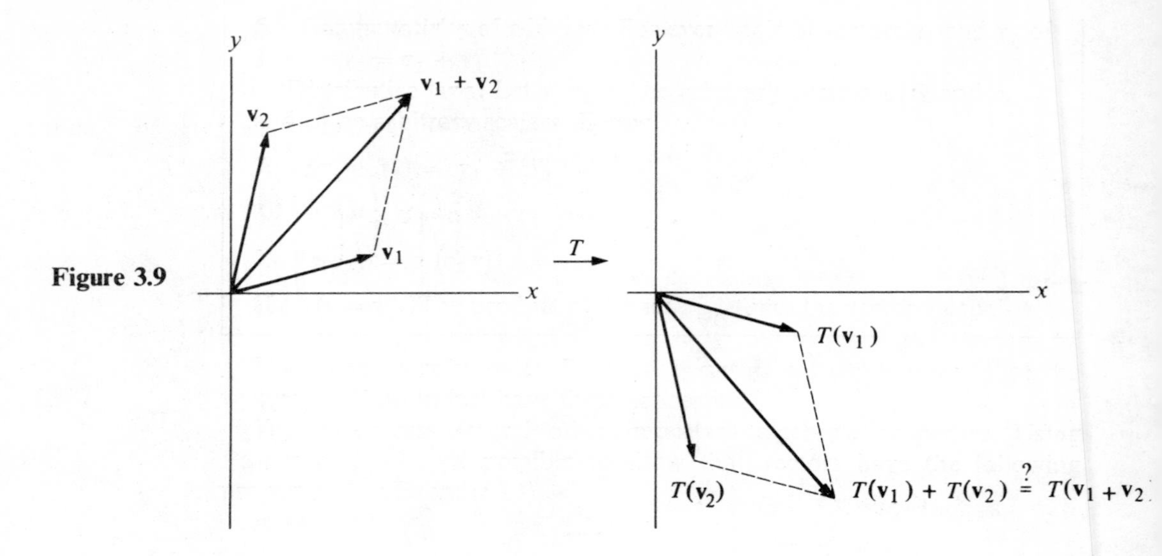

onto the line containing $T(\mathbf{v})$ and $T(c\mathbf{v})$ for every scalar c and any vector $\mathbf{v}$. Moreover, the length of $T(c\mathbf{v})$ is c times as long as $T(\mathbf{v})$, since the reflection T does not change the lengths of vectors. So, Condition 2 also holds and T is a linear transformation. ■

EXAMPLE 3 Let $T: \mathbb{R}^2 \to \mathbb{R}^2$ be the function that shifts the tip of each vector one unit to the right (see Figure 3.10). Show that T is not a linear transformation.

Figure 3.10

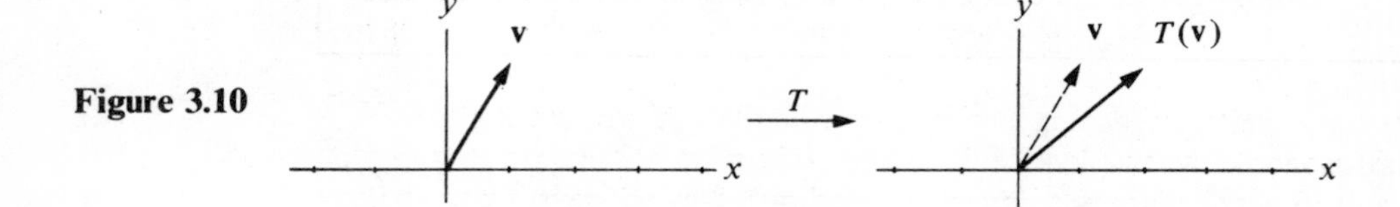

To show that T is not a linear transformation, it is only necessary to show that one of the two conditions of Definition 3.4 does not hold. However, to help you get a better idea of what these two conditions mean, we will examine each of the conditions and show that, in fact, neither one holds.

We begin by showing that Condition 1 does not hold for the vectors $\mathbf{v}_1 = \begin{bmatrix} 1 \\ 0 \end{bmatrix}$ and $\mathbf{v}_2 = \begin{bmatrix} 0 \\ 1 \end{bmatrix}$. (*Note:* To show that Condition 1 holds, it is necessary to check *all* pairs of vectors $\mathbf{v}_1$ and $\mathbf{v}_2$, but to show that it does not hold, it is only necessary to find two vectors for which Condition 1 is false.) It is apparent from Figure 3.11, that since

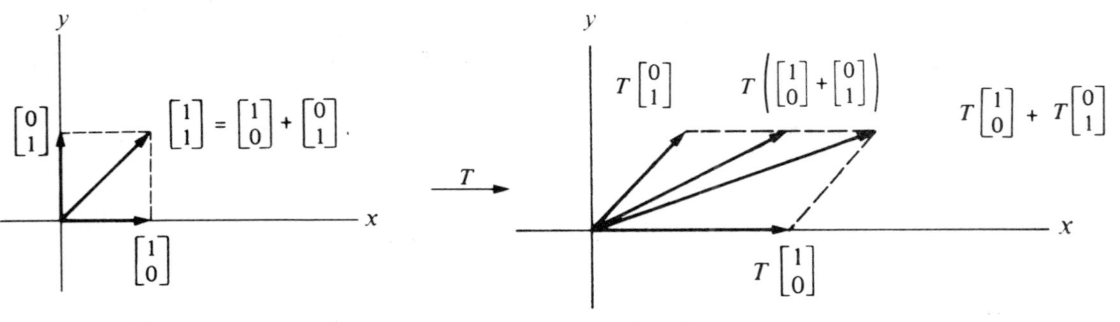

Figure 3.11

$$T(\mathbf{v}_1 + \mathbf{v}_2) = T\left(\begin{bmatrix}1\\0\end{bmatrix} + \begin{bmatrix}0\\1\end{bmatrix}\right) = T\begin{bmatrix}1\\1\end{bmatrix} = \begin{bmatrix}2\\1\end{bmatrix}$$

is not equal to $\quad T(\mathbf{v}_1) + T(\mathbf{v}_2) = \begin{bmatrix}2\\0\end{bmatrix} + \begin{bmatrix}1\\1\end{bmatrix} = \begin{bmatrix}3\\1\end{bmatrix}$

Condition 1 does not hold.

To show that Condition 2 does not hold, let $c = 2$ and $\mathbf{v} = \begin{bmatrix}0\\1\end{bmatrix}$. Then, as can be seen in Figure 3.12

$$T(c\mathbf{v}) = T\left(2\begin{bmatrix}0\\1\end{bmatrix}\right) = T\begin{bmatrix}0\\2\end{bmatrix} = \begin{bmatrix}1\\2\end{bmatrix}$$

which is not equal to $\quad cT(\mathbf{v}) = 2T\begin{bmatrix}0\\1\end{bmatrix} = 2\begin{bmatrix}1\\1\end{bmatrix} = \begin{bmatrix}2\\2\end{bmatrix}$

showing that our transformation does not satisfy Condition 2 either. ■

Figure 3.12

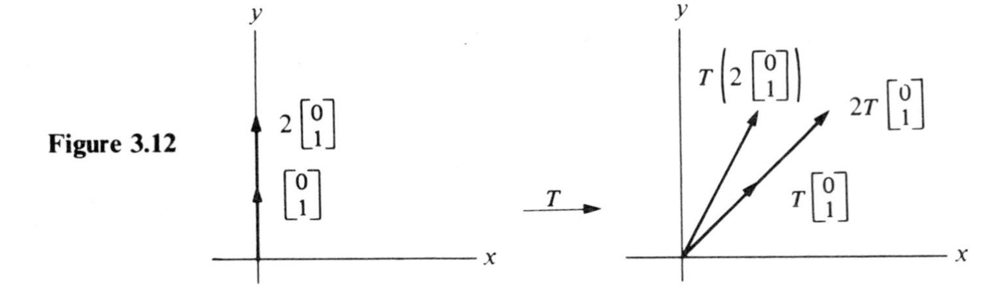

We conclude this section by mentioning two very important linear transformations. One of them is the transformation

$$T: \mathbb{R}^n \to \mathbb{R}^m$$
$$\mathbf{v} \to \mathbf{0}$$

where $\mathbf{0} = \begin{bmatrix} 0 \\ \vdots \\ 0 \end{bmatrix}$ is called the **zero vector.**

T maps every vector of $\mathbb{R}^n$ to the zero vector of $\mathbb{R}^m$. It is called the **zero transformation**.

The other important linear transformation is

$$T: \mathbb{R}^n \to \mathbb{R}^n$$
$$\mathbf{v} \to \mathbf{v}$$

which maps every vector of $\mathbb{R}^n$ to itself. It is called the **identity transformation**.

Exercise Set 3.1

1. Graph the following points, vectors, and subsets of $\mathbb{R}^3$:

a. The points $(2, 1, 0)$, $(3, 1, 2)$, $(1, -1, 2)$, $(-2, 3, -2)$, $(-1, -2, -3)$.

b. The vectors

$$\begin{bmatrix} 2 \\ 1 \\ 0 \end{bmatrix}, \begin{bmatrix} -3 \\ 1 \\ 2 \end{bmatrix}, \begin{bmatrix} 0 \\ 0 \\ 0 \end{bmatrix}, \begin{bmatrix} -2 \\ -3 \\ 2 \end{bmatrix}$$

c. The set of all points of the form $(t, 0, t)$ where t is a real number.

d. The set of all points of the form $(t, s, 1)$ where t and s are real numbers.

e. The set of all points of the form $(t, 2t, 0)$ where t is a real number.

2. Let

$$\mathbf{v} = \begin{bmatrix} 1 \\ -2 \\ 5 \end{bmatrix} \quad \text{and} \quad \mathbf{u} = \begin{bmatrix} -2 \\ 1 \\ 7 \end{bmatrix}.$$

a. Find $\mathbf{u} + \mathbf{v}$. **b.** Find $2\mathbf{v} - 3\mathbf{u}$.

c. Find a vector $\mathbf{x}$ so that $\mathbf{v} + \mathbf{x} = \mathbf{u}$.

d. Find scalars c_1 and c_2 so that

$$c_1\mathbf{v} + c_2\mathbf{u} = \begin{bmatrix} 4 \\ -5 \\ 3 \end{bmatrix}.$$

3. Let

$$\mathbf{v} = \begin{bmatrix} 1 \\ 2 \\ 3 \\ -1 \end{bmatrix} \quad \text{and} \quad \mathbf{u} = \begin{bmatrix} -2 \\ 4 \\ 6 \\ 1 \end{bmatrix}$$

a. Find $\mathbf{v} + \mathbf{u}$. **b.** Find $6\mathbf{v} - 3\mathbf{u}$.

c. Find $\mathbf{x}$ so that $\mathbf{v} + \mathbf{x} = \mathbf{u}$.

In Exercises 4–11, determine whether the function is linear.

4. $T: \mathbb{R}^2 \to \mathbb{R}^2$

$$\begin{bmatrix} x \\ y \end{bmatrix} \to \begin{bmatrix} y \\ x \end{bmatrix}$$

5. T: $\mathbb{R}^3 \to \mathbb{R}^2$

$$\begin{bmatrix} x \\ y \\ z \end{bmatrix} \to \begin{bmatrix} x + y + z \\ x \end{bmatrix}$$

6. T: $\mathbb{R}^3 \to \mathbb{R}^2$

$$\begin{bmatrix} x \\ y \\ z \end{bmatrix} \to \begin{bmatrix} 0 \\ 0 \end{bmatrix}$$

7. T: $\mathbb{R}^3 \to \mathbb{R}^2$

$$\begin{bmatrix} x \\ y \\ z \end{bmatrix} \to \begin{bmatrix} 1 \\ 1 \end{bmatrix}$$

8. T: $\mathbb{R}^2 \to \mathbb{R}^2$

$$\begin{bmatrix} x \\ y \end{bmatrix} \to \begin{bmatrix} x \\ y - 1 \end{bmatrix}$$

9. T: $\mathbb{R}^2 \to \mathbb{R}^2$ is reflection about the line $y = x$.

10. T: $\mathbb{R}^2 \to \mathbb{R}^2$ is counterclockwise rotation of 90°.

11. T: $\mathbb{R}^2 \to \mathbb{R}^3$

$$\begin{bmatrix} x \\ y \end{bmatrix} \to \begin{bmatrix} 0 \\ x \\ y \end{bmatrix}$$

12. Give a geometric description of the set of all points of $\mathbb{R}^3$ which are the tips of vectors of the form:

a. $\mathbf{v} = a\begin{bmatrix} 1 \\ 0 \\ 0 \end{bmatrix} + b\begin{bmatrix} 0 \\ 1 \\ 0 \end{bmatrix}$ where a and b are real numbers

b. $\mathbf{v} = a\begin{bmatrix} 1 \\ 2 \\ 0 \end{bmatrix} + b\begin{bmatrix} -1 \\ 3 \\ 0 \end{bmatrix}$ where a and b are real numbers

★ **13.** Let T: $\mathbb{R}^2 \to \mathbb{R}^2$ be a function. Show that T is linear if and only if

$$T\begin{bmatrix} x_1 \\ x_2 \end{bmatrix} = \begin{bmatrix} a_{11}x_1 + a_{12}x_2 \\ a_{21}x_1 + a_{22}x_2 \end{bmatrix}$$ where a_{11}, a_{12}, a_{21}, a_{22} are real numbers

14. Let

T_1: $\mathbb{R}^2 \to \mathbb{R}^2$ and T_2: $\mathbb{R}^2 \to \mathbb{R}^2$

$$\begin{bmatrix} x \\ y \end{bmatrix} \to \begin{bmatrix} -y \\ x \end{bmatrix} \qquad \begin{bmatrix} x \\ y \end{bmatrix} \to \begin{bmatrix} x \\ -y \end{bmatrix}$$

Is $T_1\left(T_2\begin{bmatrix} x \\ y \end{bmatrix}\right) = T_2\left(T_1\begin{bmatrix} x \\ y \end{bmatrix}\right)$?

3.2. Vector Spaces and Subspaces

In this section we will look at vectors from a more abstract point of view. To begin, we list a number of the most important properties of vectors in $\mathbb{R}^n$. For now, we will interpret V to mean $\mathbb{R}^n$, and we will also understand the word *scalar* to mean a real number.

Properties of Vectors

1. **Additive closure:** For every pair of vectors $\mathbf{v}_1$ and $\mathbf{v}_2$ in V, their sum $\mathbf{v}_1 + \mathbf{v}_2$ is a vector in V.

2. **Closure under scalar multiplication:** For any vector $\mathbf{v}$ in V and any scalar c, the product $c\mathbf{v}$ is a vector in V.

3. There is a vector $\mathbf{0}$ in V such that $\mathbf{v} + \mathbf{0} = \mathbf{v}$ for every vector $\mathbf{v}$ in V. (In $\mathbb{R}^n$, $\mathbf{0}$ is the vector all of whose entries are 0.)

4. **Existence of additive inverses:** For each vector $\mathbf{v}$ in V, there is a vector $-\mathbf{v}$ in V such that $\mathbf{v} + (-\mathbf{v}) = \mathbf{0}$.

5. **Associativity of addition:** For any vectors $\mathbf{v}_1$, $\mathbf{v}_2$, $\mathbf{v}_3$ of V, $\mathbf{v}_1 + (\mathbf{v}_2 + \mathbf{v}_3) = (\mathbf{v}_1 + \mathbf{v}_2) + \mathbf{v}_3$.

6. **Commutativity of addition:** For every pair of vectors $\mathbf{v}_1$ and $\mathbf{v}_2$ of V, $\mathbf{v}_1 + \mathbf{v}_2 = \mathbf{v}_2 + \mathbf{v}_1$.

Distributive laws: Let $\mathbf{v}$, $\mathbf{v}_1$, $\mathbf{v}_2$ be arbitrary vectors of V and c, c_1, and c_2 be arbitrary scalars. Then

7. $c(\mathbf{v}_1 + \mathbf{v}_2) = c\mathbf{v}_1 + c\mathbf{v}_2$
8. $(c_1 + c_2)\mathbf{v} = c_1\mathbf{v} + c_2\mathbf{v}$
9. $(c_1c_2)\mathbf{v} = c_1(c_2\mathbf{v})$
10. $1\mathbf{v} = \mathbf{v}$ (The product of the scalar 1 times the vector $\mathbf{v}$ equals $\mathbf{v}$.)

The reader is referred to Exercise 1 in this section to show that the vectors in $\mathbb{R}^n$ do in fact have these properties.

Vectors possess several other important algebraic properties. Using Properties 1–10 it is possible to show that vectors have the following properties (see Exercise 13).

a. $0\mathbf{v} = \mathbf{0}$
b. $c\mathbf{0} = \mathbf{0}$
c. $c\mathbf{v} = \mathbf{0}$ if and only if $c = 0$ or $\mathbf{v} = \mathbf{0}$.
d. $-1\mathbf{v} = -\mathbf{v}$

where $\mathbf{v}$ is an arbitrary vector in V and c is an arbitrary scalar.

It turns out that the vectors of $\mathbb{R}^n$ are not the only mathematical objects that have Properties 1–10. In fact, there are many other important mathematical systems with these properties.

DEFINITION 3.5 A set of objects that has addition and scalar multiplication operations defined and satisfies Properties 1–10 is called a **vector space** over the real numbers. The objects of a vector space are called **vectors**.

Vector spaces can be defined over different number systems (rational or complex numbers, for example); but in this book we will concentrate on vector spaces over the real numbers. Although the main focus of this book will be on the vector spaces $\mathbb{R}^n$, we will often apply the knowledge gained from their study to gain insight into the structure of other vector spaces. We will give two examples of vector spaces different from $\mathbb{R}^n$ later in this section.

For most of this text, the most important types of vector spaces we will discuss are subspaces. These are subsets of $\mathbb{R}^n$ (or other vector spaces), which are themselves vector spaces.

DEFINITION 3.6 A subset of a vector space which is itself a vector space (under the same addition and scalar multiplication as the larger vector space) is called a **subspace**.

To check all the properties of a vector space can be quite tedious; however, by looking at these properties of vectors carefully, we can eliminate a lot of work when looking for subspaces.

Properties 5–10 are called **inherited properties**; that is, since they are valid for all vectors of V, they are also valid for all vectors of subsets of V. For example, if V is a vector space, then all vectors of V commute under addition (Property 6), and so all vectors of any subset of V must also commute. The fact that the rest of Properties 5–10 are inherited is left as an exercise.

Property 3 is true if Property 2 holds, since if $\mathbf{v}$ is in a subset S, then $0\mathbf{v} = \mathbf{0}$ is also in S. Property 4 follows from Property 2, since if $\mathbf{v}$ is in S and S has Property 2, then $-1(\mathbf{v}) = -\mathbf{v}$ is in S. Therefore, to show that a nonempty subset S of V is a subspace, it is only necessary to check Properties 1 and 2. However, since Property 3 is so easily verified, it is a good idea to make sure that S contains $\mathbf{0}$ before trying to verify Properties 1 and 2.

THEOREM 3.1

A nonempty subset S of a vector space V is a subspace if and only if both of the following properties hold:

1. For every pair of vectors $\mathbf{v}_1$ and $\mathbf{v}_2$ of S, their sum $\mathbf{v}_1 + \mathbf{v}_2$ is in S.

2. For every scalar c and every vector $\mathbf{v}$ of S, their product $c\mathbf{v}$ is in S.

From Theorem 3.1 we see that both V and the set $\{\mathbf{0}\}$ are subspaces of every vector space V. These two subspaces are called the **trivial subspaces** of V.

EXAMPLE 1 Show that the set S of all vectors of the form $\begin{bmatrix} x \\ 2x \end{bmatrix}$, for x a real number, is a subspace of $\mathbb{R}^2$. $\left(Note\text{: } \mathbf{0} = \begin{bmatrix} 0 \\ 2 \cdot 0 \end{bmatrix} \text{ is in } S.\right)$

First, we check Property 1 of Theorem 3.1. Let

$$\mathbf{v}_1 = \begin{bmatrix} x_1 \\ 2x_1 \end{bmatrix} \quad \text{and} \quad \mathbf{v}_2 = \begin{bmatrix} x_2 \\ 2x_2 \end{bmatrix}$$

be arbitrary vectors of S. Then

$$\mathbf{v}_1 + \mathbf{v}_2 = \begin{bmatrix} x_1 \\ 2x_1 \end{bmatrix} + \begin{bmatrix} x_2 \\ 2x_2 \end{bmatrix} = \begin{bmatrix} x_1 + x_2 \\ 2x_1 + 2x_2 \end{bmatrix} = \begin{bmatrix} (x_1 + x_2) \\ 2(x_1 + x_2) \end{bmatrix}$$

So $\mathbf{v}_1 + \mathbf{v}_2$ is a vector whose second coordinate is twice its first, and is thus an element of S.

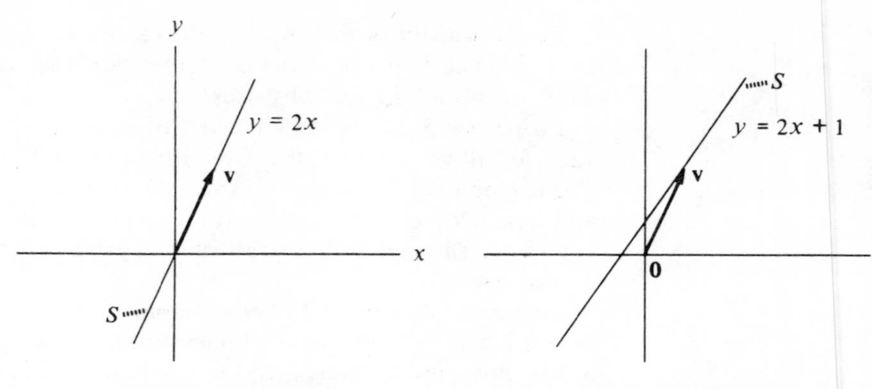

Figure 3.13 **Figure 3.14**

To check Property 2, if c is a scalar and $\mathbf{v} = \begin{bmatrix} x \\ 2x \end{bmatrix}$ is an element of S, then

$$c\mathbf{v} = c\begin{bmatrix} x \\ 2x \end{bmatrix} = \begin{bmatrix} cx \\ 2cx \end{bmatrix}$$

So $c\mathbf{v}$ is an element of S.

It should be noted that this subspace S consists of all vectors $\mathbf{v}$ that have their endpoints on the line $y = 2x$ (see Figure 3.13.) ■

EXAMPLE 2 Show that the set of all vectors $\mathbf{v}$ of $\mathbb{R}^2$ that have their endpoints on the line $y = 2x + 1$ is not a subspace of $\mathbb{R}^2$.

As is obvious from Figure 3.14 and Definition 3.6, the set of all vectors $\mathbf{v}$ of $\mathbb{R}^2$ that have their endpoints on the line $y = 2x + 1$ is not a subspace of $\mathbb{R}^2$ because the vector $\mathbf{0}$ is not a member of this set. ■

EXAMPLE 3 The subset S of all vectors of $\mathbb{R}^2$ of length less than or equal to 1 is not a subspace of $\mathbb{R}^2$.

By letting $\mathbf{v}_1 = \begin{bmatrix} 1 \\ 0 \end{bmatrix}$ and $\mathbf{v}_2 = \begin{bmatrix} 0 \\ 1 \end{bmatrix}$, we see that Property 1 of Theorem 3.1 is violated since

$$\mathbf{v}_1 + \mathbf{v}_2 = \begin{bmatrix} 1 \\ 0 \end{bmatrix} + \begin{bmatrix} 0 \\ 1 \end{bmatrix} = \begin{bmatrix} 1 \\ 1 \end{bmatrix}$$

which has length $\sqrt{2}$, which is greater than 1, and hence is not in S. So S is not a subspace of $\mathbb{R}^2$ (see Figure 3.15). ■

We conclude this section with two examples of vector spaces that are different from $\mathbb{R}^n$ and are not subspaces of $\mathbb{R}^n$.

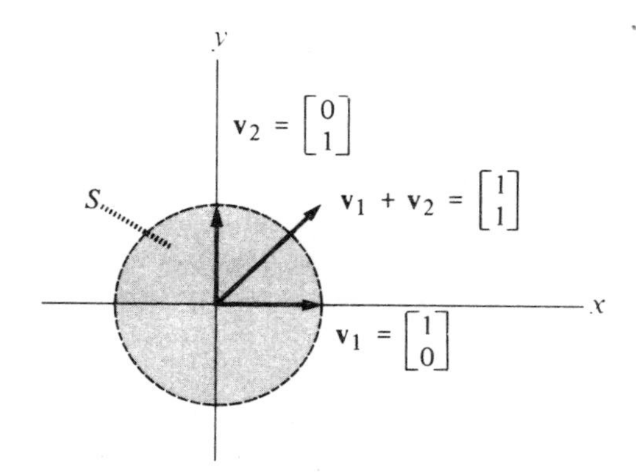

Figure 3.15

EXAMPLE 4 Let V be the set of all polynomials of one variable with real coefficients; that is, an element $p(x)$ of V is a polynomial of the form

$$p(x) = a_n x^n + a_{n-1} x^{n-1} + \cdots + a_1 x + a_0$$

The sum of two typical "vectors" in V might be

$$(x^4 - 3x^3 + x + 1) + (x^5 - 6x^2 + 2) = (x^5 + x^4 - 3x^3 - 6x^2 + x + 3)$$

A typical scalar multiplication might be

$$6(x^5 - 3x^3 + x + 5) = (6x^5 - 18x^3 + 6x + 30)$$

It is a simple exercise to verify that V satisfies the properties of Definition 3.5 of a vector space (see Exercise 2).

What is more interesting about this example is the fact that there is a very important linear transformation associated with this vector space of polynomials. (Definition 3.4 of a linear transformation given in Section 3.1 is valid for all vector spaces, not only $\mathbb{R}^n$.) The linear transformation is the differential operator D

$$D: V \to V$$
$$p(x) \to p'(x)$$

where $p(x)$ is a polynomial of V and $p'(x)$ is its derivative; for example, $D(x^2 + 3x - 2) = 2x + 3$. We show that D is a linear transformation by checking the two conditions of Definition 3.4.

We check Condition 1, for any two vectors $\mathbf{v}_1 = p_1(x)$ and $\mathbf{v}_2 = p_2(x)$

$$\begin{aligned} D(\mathbf{v}_1 + \mathbf{v}_2) &= D(p_1(x) + p_2(x)) \\ &= (p_1(x) + p_2(x))' && \text{by definition of } D \\ &= p_1'(x) + p_2'(x) && \text{property of derivatives} \\ &= D(\mathbf{v}_1) + D(\mathbf{v}_2) && \text{by definition of } D \end{aligned}$$

and Condition 2, for any vector $\mathbf{v}$ and any scalar c

$$\begin{aligned} D(c\mathbf{v}) &= D(cp(x)) = [cp(x)]' \\ &= cp'(x) = cD(\mathbf{v}) \end{aligned}$$

So D is a linear transformation. Applications of linear algebra to vector spaces like this one and to the differential operator D play an important role in the study of differential equations. ■

EXAMPLE 5 (See Applications 1.) Let V be the set of all complex numbers, that is, all numbers of the form $a + bi$ where a and b are real numbers and $i = \sqrt{-1}$. The sum of a typical pair of vectors (complex numbers) is

$$(2 + 3i) + (4 - 5i) = 6 - 2i$$

and the product of a typical scalar times a typical vector is

$$-4(2 - 3i) = -8 + 12i$$

The verification that the complex numbers are indeed a vector space is left to the reader.

Since any vector (complex number) $\mathbf{v} = a + bi$ can be expressed

$$\begin{aligned} \mathbf{v} &= a + bi \\ &= a(1 + 0i) + b(0 + 1i) \end{aligned}$$

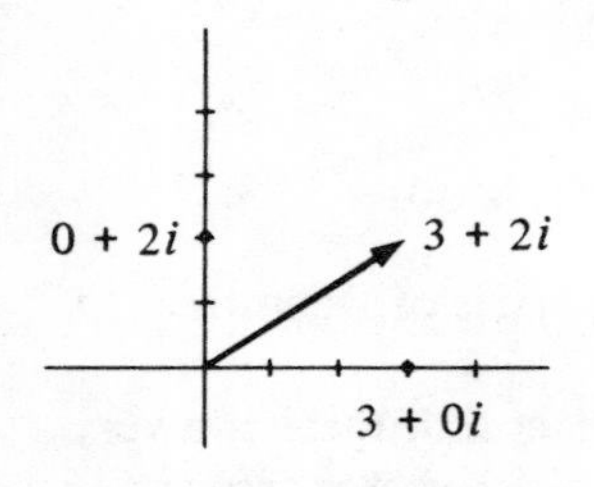

Figure 3.16

we see that the vectors $(1 + 0i)$ and $(0 + 1i)$ play the same role in the vector space of the complex numbers as the vectors $\begin{bmatrix} 1 \\ 0 \end{bmatrix}$ and $\begin{bmatrix} 0 \\ 1 \end{bmatrix}$ play in $\mathbb{R}^2$. In fact, in Figure 3.16 we graph the complex numbers just as we graph vectors in $\mathbb{R}^2$.

An interesting linear transformation on this vector space is multiplication by i.

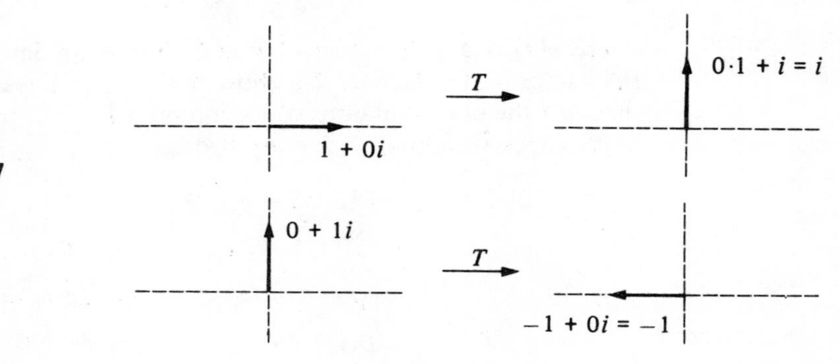

Figure 3.17

$$T: V \to V$$
$$(a + bi) \to i(a + bi) = -b + ai \qquad \text{since } i^2 = (\sqrt{-1})^2 = -1$$

To verify that T is a linear transformation, see Exercise 3.

Note that $T(1 + 0i) = i$ and $T(0 + 1i) = -1$. As can be seen in Figure 3.17, T rotates each of these vectors 90° counterclockwise. It is easily seen in Figure 3.18 that T rotates any complex number (vector) 90° counterclockwise.

Figure 3.18

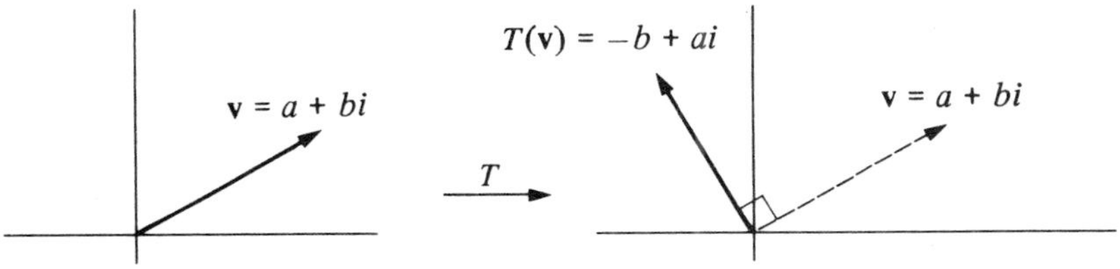

v = $a + bi$ $\xrightarrow{T}$ T(v) = $-b + ai$, v = $a + bi$

■

Exercise Set 3.2

1. Show that $\mathbb{R}^2$ is a vector space.

2. Show that the set of all polynomials (see Example 4) is a vector space.

3. Show that the mapping T defined in Example 5 is a linear transformation.

4. For each of the sets of vectors described below, determine which are subspaces of $\mathbb{R}^2$ and graph the set of endpoints of the vectors in each set.

a. The set of all vectors of the form $\begin{bmatrix} x \\ x \end{bmatrix}$.

b. The set of all vectors of $\mathbb{R}^2$ of length less than or equal to 2.

c. The set of all vectors with endpoints on the line $y = -x + 1$.

d. The set of all vectors with endpoints on the line $2x - 3y = 0$.

e. The set of all vectors with endpoints on the curve $y = x^2$.

★**5.** Let V be the set of all polynomials of one variable with real coefficients (as in Example 4) and the function I be defined by

$$I: \ V \to \mathbb{R}$$
$$p(x) \to \int_0^1 p(x)\,dx$$

Show that I is a linear transformation.

★**6.** Show that the set $\mathbb{R}_3[x]$ of all polynomials of degree at most 3 along with the zero polynomial is a subspace of the vector space of all polynomials (with real coefficients).

★**7.** Let $\mathbb{R}[x]$ denote the vector space of all polynomials with real coefficients. Show that each of the following functions is a linear transformation.

a. $X: \mathbb{R}[x] \to \mathbb{R}[x]$
$p(x) \to xp(x)$

b. $T: \mathbb{R}[x] \to \mathbb{R}$
$p(x) \to p(0)$

★**8.** **a.** Describe the set N of all polynomials $p(x)$ such that the transformation T, defined in Exercise 7, maps $p(x)$ to 0. Is this set a subspace of $\mathbb{R}[x]$?

b. Is the set of polynomials mapped by T to 1 a subspace of $\mathbb{R}[x]$?

★**9.** Suppose U_1 and U_2 are subspaces of a vector

space V. Show that the set of all vectors belonging to both U_1 and U_2 (the intersection of U_1 and U_2) is a subspace of V.

★ **10.** Give an example to show that the union of two subspaces of a vector space is not necessarily a subspace.

11. Determine whether the following sets with the given addition and scalar multiplication are vector spaces:

a. The set of all positive real numbers with addition defined by

$$x + y = xy$$

and scalar multiplication defined by

$$cx = x^c.$$

b. The set of all ordered pairs of real numbers with addition defined by

$$(a_1, b_1) + (a_2, b_2) = (a_1 + a_2 - 1, b_1 + b_2 - 1)$$

and scalar multiplication defined by

$$c(a, b) = (ca, cb)$$

12. Determine which of the following subsets of $\mathbb{R}^3$ are subspaces of $\mathbb{R}^3$:

a. The set of all vectors of the form

$$\begin{bmatrix} t \\ 2t \\ 0 \end{bmatrix} \quad \text{where } t \text{ is a real number.}$$

b. The set of all vectors of the form

$$\begin{bmatrix} t \\ 2t \\ 1 \end{bmatrix} \quad \text{where } t \text{ is a real number.}$$

c. The set of all vectors of the form

$$\begin{bmatrix} t \\ s \\ t+s \end{bmatrix} \quad \text{where } t \text{ and } s \text{ are real numbers.}$$

13. Using only Properties 1–10 of vectors, prove

a. $0\mathbf{v} = \mathbf{0}$ [*Hint*: $(1 + 0)\mathbf{v} = ?$]

b. $c\mathbf{0} = \mathbf{0}$ [*Hint*: $c\mathbf{0} = c(0\mathbf{v}) = (c0)\mathbf{v} = ?$]

c. $c\mathbf{v} = \mathbf{0}$ if and only if $c = \mathbf{0}$ or $\mathbf{v} = \mathbf{0}$. [*Hint*: Show that if $\mathbf{v} \neq \mathbf{0}$ and $c \neq 0$, then $c\mathbf{v} \neq \mathbf{0}$. To do this, write $\mathbf{v} = (1/c)(c\mathbf{v})$ and use part b].

d. $-1\mathbf{v} = -\mathbf{v}$ [*Hint*: $\mathbf{0} = [1 + (-1)]\mathbf{v} = ?$]

14. Show that the set of all ordered n-tuples of real numbers with addition defined by

$$(a_1, a_2, \ldots, a_n) + (b_1, b_2, \ldots, b_n) = (a_1 + b_1, a_2 + b_2, \ldots, a_n + b_n)$$

and scalar multiplication defined by

$$c(a_1, a_2, \ldots, a_n) = (ca_1, ca_2, \ldots, ca_n)$$

is a vector space. Such vectors are called **row vectors**.

3.3 Matrices and Linear Transformations

We pick up an idea discussed in Chapter 1. In Section 1.4, we saw that for a linear function $f: \mathbb{R}^2 \to \mathbb{R}$, the value of $f(\mathbf{v})$ for any vector $\mathbf{v}$ in $\mathbb{R}^2$ could be found if we knew the values of $f\left(\begin{bmatrix} 1 \\ 0 \end{bmatrix}\right)$ and $f\left(\begin{bmatrix} 0 \\ 1 \end{bmatrix}\right)$. The reasons for this are

1. $\begin{bmatrix} x \\ y \end{bmatrix} = x\begin{bmatrix} 1 \\ 0 \end{bmatrix} + y\begin{bmatrix} 0 \\ 1 \end{bmatrix}$

2. f is a linear function, so $\quad f\left(\begin{bmatrix} x \\ y \end{bmatrix}\right) = xf\left(\begin{bmatrix} 1 \\ 0 \end{bmatrix}\right) + yf\left(\begin{bmatrix} 0 \\ 1 \end{bmatrix}\right)$

The same ideas can be used for linear transformations $T: \mathbb{R}^n \to \mathbb{R}^m$.

1. For any vector $\mathbf{v}$ in $\mathbb{R}^n$ where

$$\mathbf{v} = \begin{bmatrix} x_1 \\ x_2 \\ \vdots \\ x_n \end{bmatrix} \quad \text{then} \quad \mathbf{v} = x_1 \begin{bmatrix} 1 \\ 0 \\ 0 \\ \vdots \\ 0 \end{bmatrix} + x_2 \begin{bmatrix} 0 \\ 1 \\ 0 \\ \vdots \\ 0 \end{bmatrix} + \cdots + x_n \begin{bmatrix} 0 \\ 0 \\ 0 \\ \vdots \\ 1 \end{bmatrix}$$

2. T is linear, so

$$T(\mathbf{v}) = T\left(x_1 \begin{bmatrix} 1 \\ 0 \\ 0 \\ \vdots \\ 0 \end{bmatrix} + x_2 \begin{bmatrix} 0 \\ 1 \\ 0 \\ \vdots \\ 0 \end{bmatrix} + \cdots + x_n \begin{bmatrix} 0 \\ 0 \\ 0 \\ \vdots \\ 1 \end{bmatrix} \right)$$

$$= T\left(x_1 \begin{bmatrix} 1 \\ 0 \\ 0 \\ \vdots \\ 0 \end{bmatrix} \right) + T\left(x_2 \begin{bmatrix} 0 \\ 1 \\ 0 \\ \vdots \\ 0 \end{bmatrix} \right) + \cdots + T\left(x_n \begin{bmatrix} 0 \\ 0 \\ 0 \\ \vdots \\ 1 \end{bmatrix} \right)$$

Condition 1 of Definition 3.4

$$= x_1 T \begin{bmatrix} 1 \\ 0 \\ 0 \\ \vdots \\ 0 \end{bmatrix} + x_2 T \begin{bmatrix} 0 \\ 1 \\ 0 \\ \vdots \\ 0 \end{bmatrix} + \cdots + x_n T \begin{bmatrix} 0 \\ 0 \\ 0 \\ \vdots \\ 1 \end{bmatrix}$$

Condition 2 of Definition 3.4

So we have proved Theorem 3.2.

THEOREM 3.2

Let $T: \mathbb{R}^n \to \mathbb{R}^m$ be a linear transformation. Then T is completely determined if the vectors

$$T \begin{bmatrix} 1 \\ 0 \\ 0 \\ \vdots \\ 0 \end{bmatrix}, \quad T \begin{bmatrix} 0 \\ 1 \\ 0 \\ \vdots \\ 0 \end{bmatrix}, \quad \ldots, \quad T \begin{bmatrix} 0 \\ 0 \\ \vdots \\ 0 \\ 1 \end{bmatrix}$$

are known. Specifically, if $\mathbf{v} = \begin{bmatrix} x_1 \\ x_2 \\ \vdots \\ x_n \end{bmatrix}$

then $$T(\mathbf{v}) = x_1 T \begin{bmatrix} 1 \\ 0 \\ 0 \\ \vdots \\ 0 \end{bmatrix} + x_2 T \begin{bmatrix} 0 \\ 1 \\ 0 \\ \vdots \\ 0 \end{bmatrix} + \cdots + x_n T \begin{bmatrix} 0 \\ 0 \\ \vdots \\ 0 \\ 1 \end{bmatrix}$$

A compact way to store this important information about a particular linear transformation T is to use a matrix whose columns are the vectors

$$T\begin{bmatrix}1\\0\\0\\\vdots\\0\end{bmatrix},\quad T\begin{bmatrix}0\\1\\0\\\vdots\\0\end{bmatrix},\quad \ldots,\quad T\begin{bmatrix}0\\0\\0\\\vdots\\1\end{bmatrix}$$

EXAMPLE 1 We will find the matrix associated with the linear transformation T.

$$T\colon \mathbb{R}^3 \to \mathbb{R}^2$$

$$\begin{bmatrix}x_1\\x_2\\x_3\end{bmatrix} \to \begin{bmatrix}x_1 + x_2\\x_2 - x_3\end{bmatrix}$$

Since $T\begin{bmatrix}1\\0\\0\end{bmatrix} = \begin{bmatrix}1+0\\0-0\end{bmatrix} = \begin{bmatrix}1\\0\end{bmatrix}$,

$$T\begin{bmatrix}0\\1\\0\end{bmatrix} = \begin{bmatrix}0+1\\1-0\end{bmatrix} = \begin{bmatrix}1\\1\end{bmatrix}, \text{ and } T\begin{bmatrix}0\\0\\1\end{bmatrix} = \begin{bmatrix}0+0\\0-1\end{bmatrix} = \begin{bmatrix}0\\-1\end{bmatrix}$$

the matrix associated with T is

$$M = \begin{bmatrix}1 & 1 & 0\\0 & 1 & -1\end{bmatrix} \quad \blacksquare$$

To make the matrix a useful tool we must define how it operates on vectors. Since a matrix M represents a linear transformation T, we want M to operate on vectors the same way T does. Since the columns of M are the vectors

$$T\begin{bmatrix}1\\0\\0\\\vdots\\0\end{bmatrix},\quad T\begin{bmatrix}0\\1\\0\\\vdots\\0\end{bmatrix},\quad \ldots,\quad T\begin{bmatrix}0\\0\\\vdots\\0\\1\end{bmatrix}$$

we can use Theorem 3.2 to define how M operates on a vector. If

$$\mathbf{v} = \begin{bmatrix}x_1\\x_2\\\vdots\\x_n\end{bmatrix}$$

then we define $M\mathbf{v}$ by

$$M\mathbf{v} = x_1 \text{ (1st column of } M) + x_2 \text{ (2nd column of } M) + \cdots + x_m \text{ (}m\text{th column of } M)$$

In this way we have made $M\mathbf{v} = T\mathbf{v}$ for every vector $\mathbf{v}$.

EXAMPLE 2 Let T and M be as in Example 1. We illustrate the way in which M operates on an arbitrary vector.

$$M\begin{bmatrix} x_1 \\ x_2 \\ x_3 \end{bmatrix} = \begin{bmatrix} 1 & 1 & 0 \\ 0 & 1 & -1 \end{bmatrix}\begin{bmatrix} x_1 \\ x_2 \\ x_3 \end{bmatrix} = x_1\begin{bmatrix} 1 \\ 0 \end{bmatrix} + x_2\begin{bmatrix} 1 \\ 1 \end{bmatrix} + x_3\begin{bmatrix} 0 \\ -1 \end{bmatrix}$$

$$= \begin{bmatrix} 1x_1 + 1x_2 + 0x_3 \\ 0x_1 + 1x_2 - 1x_3 \end{bmatrix} = \begin{bmatrix} x_1 + x_2 \\ x_2 - x_3 \end{bmatrix} \qquad ■$$

DEFINITION 3.7

> The operation of a matrix M on a vector $\mathbf{v}$ is called **multiplication** of a vector $\mathbf{v}$ by M. The resulting $M\mathbf{v}$ is called the **product** of M and $\mathbf{v}$.

Note: If M is a matrix with n rows and m columns, then M operates on vectors with m entries to give vectors with n entries.

$$n\left\{\overbrace{\begin{bmatrix} * & * & \cdots & * \\ * & * & \cdots & * \\ \vdots & \vdots & & \vdots \\ * & * & \cdots & * \end{bmatrix}}^{m}\right.\begin{bmatrix} * \\ * \\ \vdots \\ * \\ * \\ * \end{bmatrix} = \left.\begin{bmatrix} * \\ * \\ \vdots \\ * \end{bmatrix}\right\}n$$

Thus, the transformation represented by M has domain $\mathbb{R}^m$ and range $\mathbb{R}^n$—$M\colon \mathbb{R}^m \to \mathbb{R}^n$. Such a matrix is called an $\boldsymbol{n \times m}$ **matrix**.

The following definition gives a concise way of finding the product of a matrix and a vector.

DEFINITION 3.8

> Let $$M = \begin{bmatrix} a_{11} & a_{12} & \cdots & a_{1n} \\ a_{21} & a_{22} & \cdots & a_{2n} \\ \vdots & \vdots & & \vdots \\ a_{m1} & a_{m2} & \cdots & a_{mn} \end{bmatrix} \quad \text{and} \quad \mathbf{v} = \begin{bmatrix} x_1 \\ x_2 \\ \vdots \\ x_n \end{bmatrix}$$

Then

$$M\mathbf{v} = \begin{bmatrix} a_{11} & a_{12} & \cdots & a_{1n} \\ a_{21} & a_{22} & \cdots & a_{2n} \\ \vdots & \vdots & & \vdots \\ a_{m1} & a_{m2} & \cdots & a_{mn} \end{bmatrix} \begin{bmatrix} x_1 \\ x_2 \\ \vdots \\ x_n \end{bmatrix} = \begin{bmatrix} a_{11}x_1 + a_{12}x_2 + \cdots + a_{1n}x_n \\ a_{21}x_1 + a_{22}x_2 + \cdots + a_{2n}x_n \\ \vdots \\ a_{m1}x_1 + a_{m2}x_2 + \cdots + a_{mn}x_n \end{bmatrix}$$

This rather formal definition may seem formidable at first reading, but matrix multiplication is not difficult.

It helps to use two hands when multiplying matrices. Start with the index finger of your left hand pointing at the first entry in the first row of the matrix, and the index finger of your right hand pointing at the first entry of the vector. Multiply these numbers—$a_{11}x_1$. Then move your left hand to the second entry of the first row of the matrix and your right hand down to the second entry of the vector. Multiply these two entries—$a_{12}x_2$—and add this to the previous product—$a_{11}x_1 + a_{12}x_2$. Then move your left hand to the third entry in the first row and move your right hand down to the third entry of your vector. Multiply these two entries—$a_{13}x_3$—and add this to the previous sum of products—$a_{11}x_1 + a_{12}x_2 + a_{13}x_3$. Continue in this way until you have traversed the entire first row of the matrix. The final sum of products is the first entry in the vector $M\mathbf{v}$. To get the second entry of $M\mathbf{v}$, move your left hand down to the second row and begin the procedure all over again. Continue in this way until all entries have been computed.

EXAMPLE 3 This example is a detailed example of multiplying a matrix times a vector. The computation is shown step by step. The circled entries are where you should be pointing at each stage. This example is only an illustration of the process of matrix multiplication. You should develop enough facility with matrix multiplication so that you can multiply a matrix times a vector in only one step. We find the product of

the matrix $\begin{bmatrix} 1 & 2 & 3 \\ -1 & 1 & 1 \\ 2 & 3 & 2 \end{bmatrix}$ and the vector $\begin{bmatrix} 5 \\ 3 \\ -2 \end{bmatrix}$

$$\begin{bmatrix} \textcircled{1} & 2 & 3 \\ -1 & 1 & 1 \\ 2 & 3 & 2 \end{bmatrix} \begin{bmatrix} \textcircled{5} \\ 3 \\ -2 \end{bmatrix} \rightarrow \begin{bmatrix} 5 + \\ \\ \\ \end{bmatrix}$$

$$\begin{bmatrix} 1 & \textcircled{2} & 3 \\ -1 & 1 & 1 \\ 2 & 3 & 2 \end{bmatrix} \begin{bmatrix} 5 \\ \textcircled{3} \\ -2 \end{bmatrix} \rightarrow \begin{bmatrix} 5 + 6 \\ \\ \\ \end{bmatrix}$$

$$\begin{bmatrix} 1 & 2 & \textcircled{3} \\ -1 & 1 & 1 \\ 2 & 3 & 2 \end{bmatrix} \begin{bmatrix} 5 \\ 3 \\ \textcircled{-2} \end{bmatrix} \rightarrow \begin{bmatrix} 5 + 6 - 6 \\ \\ \\ \end{bmatrix} = \begin{bmatrix} 5 \\ \\ \\ \end{bmatrix}$$

So we have found the first entry of $M\mathbf{v}$.

$$\begin{bmatrix} 1 & 2 & 3 \\ \textcircled{-1} & 1 & 1 \\ 2 & 3 & 2 \end{bmatrix}\begin{bmatrix} \textcircled{5} \\ 3 \\ -2 \end{bmatrix} \to \begin{bmatrix} 5 \\ -5+ \\ \end{bmatrix}$$

$$\begin{bmatrix} 1 & 2 & 3 \\ -1 & \textcircled{1} & 1 \\ 2 & 3 & 2 \end{bmatrix}\begin{bmatrix} 5 \\ \textcircled{3} \\ -2 \end{bmatrix} \to \begin{bmatrix} 5 \\ -5+3+ \\ \end{bmatrix}$$

$$\begin{bmatrix} 1 & 2 & 3 \\ -1 & 1 & \textcircled{1} \\ 2 & 3 & 2 \end{bmatrix}\begin{bmatrix} 5 \\ 3 \\ \textcircled{-2} \end{bmatrix} \to \begin{bmatrix} 5 \\ -5+3-2 \\ \end{bmatrix} = \begin{bmatrix} 5 \\ -4 \\ \end{bmatrix}$$

Now we have found the first two entries of $M\mathbf{v}$.

$$\begin{bmatrix} 1 & 2 & 3 \\ -1 & 1 & 1 \\ \textcircled{2} & 3 & 2 \end{bmatrix}\begin{bmatrix} \textcircled{5} \\ 3 \\ -2 \end{bmatrix} \to \begin{bmatrix} 5 \\ -4 \\ 10+ \end{bmatrix}$$

$$\begin{bmatrix} 1 & 2 & 3 \\ -1 & 1 & 1 \\ 2 & \textcircled{3} & 2 \end{bmatrix}\begin{bmatrix} 5 \\ \textcircled{3} \\ -2 \end{bmatrix} \to \begin{bmatrix} 5 \\ -4 \\ 10+9 \end{bmatrix}$$

$$\begin{bmatrix} 1 & 2 & 3 \\ -1 & 1 & 1 \\ 2 & 3 & \textcircled{2} \end{bmatrix}\begin{bmatrix} 5 \\ 3 \\ \textcircled{-2} \end{bmatrix} \to \begin{bmatrix} 5 \\ -4 \\ 10+9-4 \end{bmatrix} = \begin{bmatrix} 5 \\ -4 \\ 15 \end{bmatrix}$$

So we conclude
$$\begin{bmatrix} 1 & 2 & 3 \\ -2 & 1 & 1 \\ 2 & 3 & 2 \end{bmatrix}\begin{bmatrix} 5 \\ 3 \\ -2 \end{bmatrix} = \begin{bmatrix} 5 \\ -4 \\ 15 \end{bmatrix} \qquad \blacksquare$$

Numerous exercises on matrix multiplication are included at the end of this section to help you gain proficiency in matrix multiplication. You should practice matrix multiplication until you can multiply a matrix times a vector in one step as shown in the example below.

EXAMPLE 4

$$\begin{bmatrix} 1 & 2 & -3 & 4 \\ -1 & 2 & 1 & 1 \\ 3 & -2 & 2 & 4 \end{bmatrix}\begin{bmatrix} 1 \\ 2 \\ -2 \\ 3 \end{bmatrix} = \begin{bmatrix} 23 \\ 4 \\ 7 \end{bmatrix} \qquad \blacksquare$$

It should be noted that every linear transformation can be represented by a matrix, and conversely, that every matrix represents a linear transformation.

EXAMPLE 5 **a.** Find the matrix that represents the linear transformation $T\colon \mathbb{R}^2 \to \mathbb{R}^2$, which is counterclockwise rotation by 90°.

Since $T\begin{bmatrix}1\\0\end{bmatrix} = \begin{bmatrix}0\\1\end{bmatrix}$ and $T\begin{bmatrix}0\\1\end{bmatrix} = \begin{bmatrix}-1\\0\end{bmatrix}$

the matrix for T is $M = \begin{bmatrix}0 & -1\\1 & 0\end{bmatrix}$

b. Use the matrix M to find $T\begin{bmatrix}-3\\7\end{bmatrix}$.

$$T\begin{bmatrix}-3\\7\end{bmatrix} = M\begin{bmatrix}-3\\7\end{bmatrix} = \begin{bmatrix}0 & -1\\1 & 0\end{bmatrix}\begin{bmatrix}-3\\7\end{bmatrix} = \begin{bmatrix}-7\\-3\end{bmatrix} \quad \blacksquare$$

Similarly, we can prove the following theorem.

THEOREM 3.3 Let $T: \mathbb{R}^2 \to \mathbb{R}^2$ be counterclockwise rotation by an angle θ. Then the matrix that represents T is

$$\begin{bmatrix}\cos\theta & -\sin\theta\\ \sin\theta & \cos\theta\end{bmatrix}$$

EXAMPLE 6 Find the matrix that represents rotation by 30° in $\mathbb{R}^2$.

We use Theorem 3.3. Since $\cos 30° = \sqrt{3}/2$ and $\sin 30° = \frac{1}{2}$, the matrix representing rotation by 30° counterclockwise is

$$M = \begin{bmatrix}\cos 30° & -\sin 30°\\ \sin 30° & \cos 30°\end{bmatrix} = \begin{bmatrix}\sqrt{3}/2 & -1/2\\ 1/2 & \sqrt{3}/2\end{bmatrix} \quad \blacksquare$$

Relationships between matrices and linear transformations can be used to learn more about the systems of linear equations discussed in Chapter 2. Let us reconsider the system of equations we solved in Example 6 in Section 2.2. The system is

$$\begin{aligned} 2x - y + 3z &= 12\\ -x + 4y - 2z &= -11\\ 3x + y + 5z &= 17 \end{aligned}$$

This can be thought of as equating two vectors in $\mathbb{R}^3$, namely

$$\begin{bmatrix}2x - y + 3z\\ -x + 4y - 2z\\ 3x + y + 5z\end{bmatrix} = \begin{bmatrix}12\\-11\\17\end{bmatrix}$$

The vector on the left can be written as the product of a matrix and a vector to give

$$\begin{bmatrix} 2 & -1 & 3 \\ -1 & 4 & -2 \\ 3 & 1 & 5 \end{bmatrix}\begin{bmatrix} x \\ y \\ z \end{bmatrix} = \begin{bmatrix} 12 \\ -11 \\ 17 \end{bmatrix}$$

In this form the system of equations can be viewed differently. The matrix

$$M = \begin{bmatrix} 2 & -1 & 3 \\ -1 & 4 & -2 \\ 3 & 1 & 5 \end{bmatrix}$$

represents a linear transformation, call it T, from $\mathbb{R}^3$ to $\mathbb{R}^3$. The solution of this equation is a vector

$$\mathbf{v} = \begin{bmatrix} x \\ y \\ z \end{bmatrix} \qquad \text{such that} \qquad T\mathbf{v} = \begin{bmatrix} 12 \\ -11 \\ 17 \end{bmatrix}$$

That is,

$$\begin{bmatrix} x \\ y \\ z \end{bmatrix}$$

is the vector in the domain of T mapped onto the vector

$$\begin{bmatrix} 12 \\ -11 \\ 17 \end{bmatrix} \qquad \text{(See Figure 3.19.)}$$

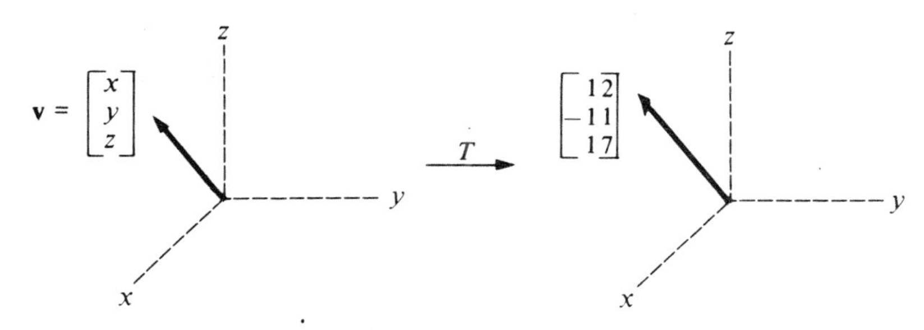

Figure 3.19

The following is another example of how we can use the concepts of matrices and linear transformations to help understand more about systems of linear equations.

EXAMPLE 7 Solve the system

$$\begin{aligned} 2x - y &= 3 \\ -6x + 3y &= -9 \end{aligned}$$

a. We solve this system as we did in Section 2.2.

$$\left[\begin{array}{rr|r} 2 & -1 & 3 \\ -6 & 3 & -9 \end{array}\right] \begin{array}{l} (3) \\ (1) \end{array} \longrightarrow \left[\begin{array}{rr|r} 2 & -1 & 3 \\ 0 & 0 & 0 \end{array}\right]$$

So $2x - y = 3$ or $y = 2x - 3$.

b. We write the equation in matrix form

$$\begin{bmatrix} 2 & -1 \\ -6 & 3 \end{bmatrix}\begin{bmatrix} x \\ y \end{bmatrix} = \begin{bmatrix} 3 \\ -9 \end{bmatrix}$$

c. We combine these results. The solutions of the equation

$$\begin{bmatrix} 2 & -1 \\ -6 & 3 \end{bmatrix}\begin{bmatrix} x \\ y \end{bmatrix} = \begin{bmatrix} 3 \\ -9 \end{bmatrix}$$

are all vectors of the form $\begin{bmatrix} x \\ 2x - 3 \end{bmatrix}$

that is, all vectors with tip on the line $y = 2x - 3$. The graphs in Figure 3.20 illustrate that each vector with endpoint on the line $y = 2x - 3$ is mapped by T onto the vector $\begin{bmatrix} 3 \\ -9 \end{bmatrix}$. Moreover, these are the only vectors that T maps onto $\begin{bmatrix} 3 \\ -9 \end{bmatrix}$, where T is the linear transformation represented by the matrix

$$M = \begin{bmatrix} 2 & -1 \\ -6 & 3 \end{bmatrix}$$

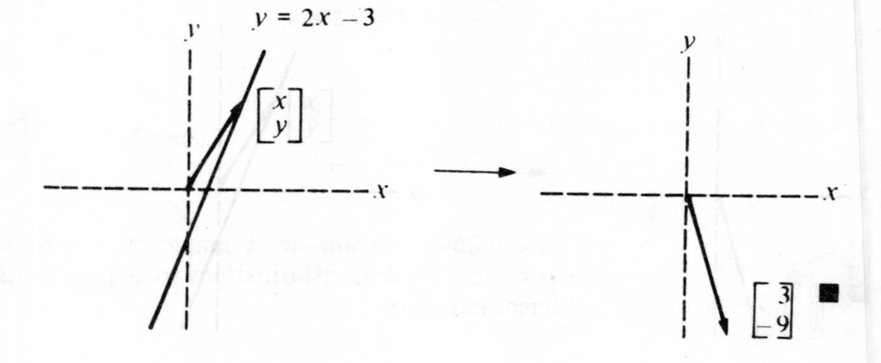

Figure 3.20 ■

Sometimes we are given a matrix and wish to find the linear transformation that it represents. This is particularly useful when the linear transformation has a simple geometric description as in the next example.

EXAMPLE 8 Describe the linear transformation represented by the matrix

$$M = \begin{bmatrix} 0 & 1 \\ 1 & 0 \end{bmatrix}$$

Since M has 2 rows and 2 columns, $T\colon \mathbb{R}^2 \to \mathbb{R}^2$. Moreover,

$$M\begin{bmatrix} x \\ y \end{bmatrix} = \begin{bmatrix} 0 & 1 \\ 1 & 0 \end{bmatrix}\begin{bmatrix} x \\ y \end{bmatrix} = \begin{bmatrix} y \\ x \end{bmatrix}$$

So

$$T\begin{bmatrix} x \\ y \end{bmatrix} = \begin{bmatrix} y \\ x \end{bmatrix}$$

Reflection about the line $y = x$ is a linear transformation that has the same action on each vector of $\mathbb{R}^2$ as M. Thus T is the linear transformation represented by M—two functions that act the same are equal (see Example 6 in Section 1.1). ■

Two matrices are of special interest. If

$$T_0\colon \mathbb{R}^n \to \mathbb{R}^m$$
$$\mathbf{v} \to \mathbf{0}$$

is the zero linear transformation, then the matrix that represents T_0 is called the **zero matrix**. The zero matrix is the $m \times n$ matrix in which every entry is zero.

The other matrix of note is the **identity matrix**. It represents the identity transformation

$$I\colon \mathbb{R}^n \to \mathbb{R}^n$$
$$\mathbf{v} \to \mathbf{v}$$

The identity matrix has the form

$$\begin{bmatrix} 1 & 0 & \cdots & 0 & 0 \\ 0 & 1 & 0 & \cdots & 0 \\ \vdots & & \ddots & & \vdots \\ 0 & & \cdots & 0 & 1 \end{bmatrix}$$

The following theorem summarizes the important connection between matrices and linear transformations. It can be proved using Theorem 3.2 and the definition of matrix multiplication.

THEOREM 3.4 Let T: $\mathbb{R}^n \to \mathbb{R}^m$ be a linear transformation and let M be the matrix whose columns are the vectors

$$T\begin{bmatrix}1\\0\\0\\\vdots\\0\end{bmatrix}, \quad T\begin{bmatrix}0\\1\\0\\\vdots\\0\end{bmatrix}, \quad \ldots, \quad T\begin{bmatrix}0\\\vdots\\0\\0\\1\end{bmatrix}$$

Then, for any vector $\mathbf{v}$ of $\mathbb{R}^n$, $\quad T\mathbf{v} = M\mathbf{v}$

In the following example we illustrate how the concept of multiplication of a vector by a matrix can be extended to answer a rather complicated question. The material in this example will be covered in detail in Sections 7.1 and 7.2.

EXAMPLE 9† Let $\quad T_1$: $\mathbb{R}^2 \to \mathbb{R}^2$ be reflection about the line $y = x$
T_2: $\mathbb{R}^2 \to \mathbb{R}^2$ be reflection about the x axis.

We determine the net effect of applying the linear transformation T_2 to an arbitrary vector and then applying the linear transformation T_1 to the resulting vector. Figure 3.21 illustrates this.

We begin by finding the matrices M_1 and M_2, which represent T_1 and T_2, respectively.

$$M_1 = \begin{bmatrix}0 & 1\\1 & 0\end{bmatrix} \quad \text{and} \quad M_2 = \begin{bmatrix}1 & 0\\0 & -1\end{bmatrix}$$

We first reflect $\mathbf{v}$ about the x axis. This gives the vector $T_2(\mathbf{v})$. Then we reflect this vector about the line $y = x$ to obtain the vector $T_1(T_2(\mathbf{v}))$. To determine the net effect of these two transformations, we use matrices M_1 and M_2. Since the image of a vector under a linear transformation is the product of the matrix representing that linear transformation and the vector, we have

$$\begin{aligned}T_1(T_2(\mathbf{v})) &= M_1(M_2(\mathbf{v}))\\ &= \begin{bmatrix}0 & 1\\1 & 0\end{bmatrix}\left(\begin{bmatrix}1 & 0\\0 & -1\end{bmatrix}\begin{bmatrix}x\\y\end{bmatrix}\right) \qquad \text{where } \mathbf{v} = \begin{bmatrix}x\\y\end{bmatrix}\end{aligned}$$

† This example may be omitted by those instructors not wishing to introduce the concept of matrix multiplication at this time. Those instructors wishing to discuss matrix multiplication in greater depth may cover Sections 7.1 and 7.2 after this example with no loss of continuity.

$$= \begin{bmatrix} 0 & 1 \\ 1 & 0 \end{bmatrix} \begin{bmatrix} x \\ -y \end{bmatrix}$$

$$= \begin{bmatrix} -y \\ x \end{bmatrix}$$

Hence, the net effect of first reflecting a vector about the x axis and then reflecting the resulting vector about the line $y = x$ is to map the vector $\mathbf{v} = \begin{bmatrix} x \\ y \end{bmatrix}$ to the vector $\begin{bmatrix} -y \\ x \end{bmatrix}$, that is to rotate the vector $\mathbf{v}$ counterclockwise by 90° (see Example 5).

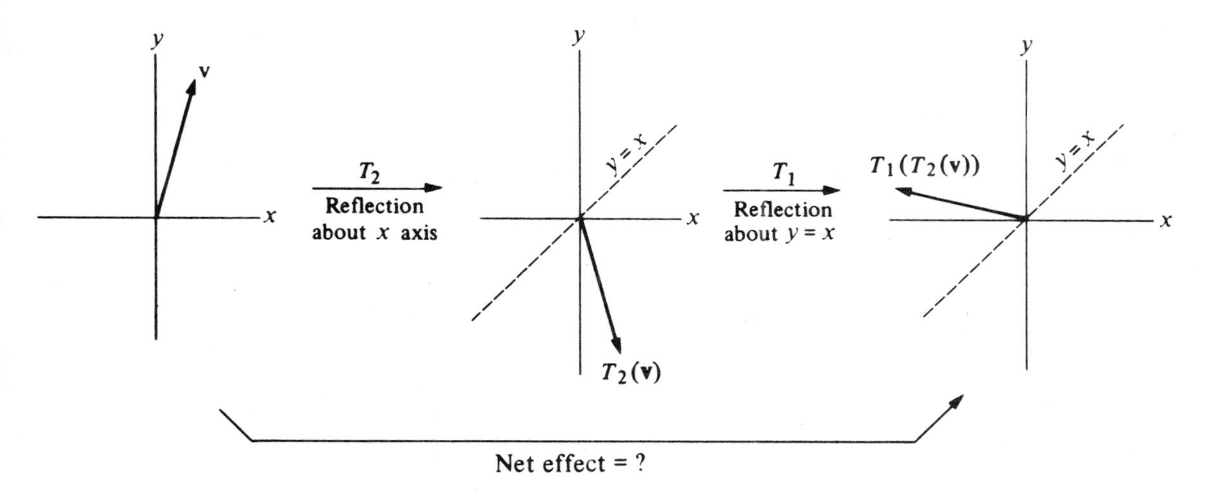

Figure 3.21 ■

Example 9 is a special case of the general result that the net effect of applying two reflections (in $\mathbb{R}^2$) is a rotation. (This result is discussed in Exercise 10 in Section 8.2.)

Exercise Set 3.3

1. Write the matrix that represents each of the following linear transformations:

a. $T: \mathbb{R}^2 \to \mathbb{R}^2$ is reflection about the line $y = x$.

b. $T: \mathbb{R}^2 \to \mathbb{R}^2$ is reflection about the y axis.

c. $T: \mathbb{R}^2 \to \mathbb{R}^2$ is counterclockwise rotation by 45°.

d. $T: \mathbb{R}^2 \to \mathbb{R}^2$ is the linear transformation that stretches each vector by a factor of 2.

e. $T: \mathbb{R}^3 \to \mathbb{R}^3$

$$\begin{bmatrix} x \\ y \\ z \end{bmatrix} \to \begin{bmatrix} x + y - z \\ 2x \\ x - 4z \end{bmatrix}$$

f. $T: \mathbb{R}^4 \to \mathbb{R}$

$$\begin{bmatrix} x \\ y \\ z \\ w \end{bmatrix} \to (x + y - 2z + 3w)$$

g. $T: \mathbb{R}^2 \to \mathbb{R}^4$

$$\begin{bmatrix} x \\ y \end{bmatrix} \to \begin{bmatrix} 2x - y \\ 0 \\ -x - y \\ y \end{bmatrix}$$

2. Find the following products:

a. $\begin{bmatrix} 2 & -3 \\ 4 & 1 \end{bmatrix}\begin{bmatrix} 2 \\ -4 \end{bmatrix}$

b. $\begin{bmatrix} 6 & 1 & 7 & -2 \\ -3 & 4 & 1 & 0 \end{bmatrix}\begin{bmatrix} 1 \\ -1 \\ 2 \\ 3 \end{bmatrix}$

c. $\begin{bmatrix} 1 & 2 & 4 \\ -1 & -3 & 1 \\ 2 & 4 & 3 \end{bmatrix}\begin{bmatrix} -2 \\ -1 \\ 6 \end{bmatrix}$

d. $\begin{bmatrix} -1 & 2 \\ 3 & 1 \\ 4 & 1 \\ 2 & 3 \end{bmatrix}\begin{bmatrix} -3 \\ 4 \end{bmatrix}$

e. $\begin{bmatrix} 1 & 5 & 7 & -1 \\ -2 & -4 & 0 & -3 \\ 2 & 6 & -2 & -2 \\ 3 & 1 & 4 & 5 \end{bmatrix}\begin{bmatrix} 6 \\ -1 \\ 7 \\ 4 \end{bmatrix}$

f. $\begin{bmatrix} 1 & 9 & 16 & 15 & 12 \\ 2 & 12 & 24 & 24 & 13 \\ -3 & -40 & -9 & 27 & -2 \\ 6 & 35 & -12 & 19 & 19 \\ 8 & 2 & 14 & 8 & 11 \end{bmatrix}\begin{bmatrix} 0 \\ 0 \\ 0 \\ 1 \\ 0 \end{bmatrix}$

g. $\begin{bmatrix} 1 & 3 & 2 & -3 & 5 \end{bmatrix}\begin{bmatrix} 1 \\ 1 \\ 0 \\ 3 \\ -2 \end{bmatrix}$

h. $\begin{bmatrix} 1 & -3 \\ -2 & 0 \\ 7 & 1 \\ 6 & -3 \\ -5 & 2 \end{bmatrix}\begin{bmatrix} 4 \\ 4 \end{bmatrix}$

i. $\begin{bmatrix} 2 & -3 & 4 & -7 & -1 \\ 1 & 0 & 0 & 0 & 0 \\ 6 & 3 & 8 & -5 & -2 \\ 4 & -1 & 0 & 0 & -2 \\ 1 & 1 & 0 & 1 & 3 \\ 3 & 1 & 2 & -1 & -1 \end{bmatrix}\begin{bmatrix} 2 \\ 5 \\ 4 \\ 1 \\ 0 \end{bmatrix}$

3. For each of the following equations, solve for the vector $\mathbf{v} = \begin{bmatrix} x \\ y \end{bmatrix}$ and graph the solution set as was done in Example 7. Let

$$M = \begin{bmatrix} -1 & 3 \\ 4 & -12 \end{bmatrix}$$

a. $M\mathbf{v} = \begin{bmatrix} -1 \\ 4 \end{bmatrix}$ **b.** $M\mathbf{v} = \begin{bmatrix} 3 \\ -12 \end{bmatrix}$

c. $M\mathbf{v} = \begin{bmatrix} 2 \\ -8 \end{bmatrix}$ **d.** $M\mathbf{v} = \begin{bmatrix} 0 \\ 0 \end{bmatrix}$

e. $M\mathbf{v} = \begin{bmatrix} 1 \\ 0 \end{bmatrix}$

4. From your answers to Exercise 3, can you make any conjectures about the action of the linear transformation represented by M?

5. Give a geometric description for each of the linear transformations represented by the following matrices.

a. $\begin{bmatrix} 3 & 0 \\ 0 & 3 \end{bmatrix}$ **b.** $\begin{bmatrix} -1 & 0 \\ 0 & -1 \end{bmatrix}$

c. $\begin{bmatrix} 0 & 0 \\ 0 & 1 \end{bmatrix}$ **d.** $\begin{bmatrix} 0 & -1 \\ 1 & 0 \end{bmatrix}$

e. $\begin{bmatrix} 0 & 1 \\ -1 & 0 \end{bmatrix}$ **f.** $\begin{bmatrix} \sqrt{2} & -\sqrt{2} \\ \sqrt{2} & \sqrt{2} \end{bmatrix}$

6. Find the matrix that represents each of the following linear transformations.

a. $T: \mathbb{R}^3 \to \mathbb{R}^3$, $\mathbf{v} \to \mathbf{0}$ **b.** $T: \mathbb{R}^3 \to \mathbb{R}^3$, $\mathbf{v} \to \mathbf{v}$

c. What are the names of the transformations in parts **a** and **b**?

★**7.** Let $\mathbb{R}_3[x]$ be the vector space consisting of all polynomials of degree at most 3 with real coefficients as well as the zero polynomial.

a. Show that every polynomial p of $\mathbb{R}_3[x]$ can be written uniquely as a linear combination of the polynomials

$$p_0(x) = 1, \quad p_1(x) = x, \quad p_2(x) = x^2, \quad \text{and} \quad p_3(x) = x^3$$

b. Let $D: \mathbb{R}_3[x] \to \mathbb{R}_3[x]$

$$p(x) \to p'(x)$$

be the differentiation operator. Find $D(p_0(x))$, $D(p_1(x))$, $D(p_2(x))$, and $D(p_3(x))$.

c. In part **a** you showed that if

$$p(x) = c_0 + c_1 x + c_2 x^2 + c_3 x^3$$

then the numbers c_0, c_1, c_2, and c_3 completely determine the polynomial $p(x)$. Let us represent the polynomial $p(x)$ by the vector having entries c_0, c_1, c_2, c_3

$$p(x) \text{ corresponds to } \begin{bmatrix} c_0 \\ c_1 \\ c_2 \\ c_3 \end{bmatrix}$$

Similarly, the derivative of $p(x)$ will correspond to a vector in $\mathbb{R}^4$. Find this vector.

d. The differentiation operator D induces a linear transformation $\tilde{D}$

$$\tilde{D}: \mathbb{R}^4 \to \mathbb{R}^4$$

$$\begin{pmatrix} \text{vector} \\ \text{corresponding} \\ \text{to } p(x) \end{pmatrix} \to \begin{pmatrix} \text{vector} \\ \text{corresponding} \\ \text{to } p'(x) \end{pmatrix}$$

Find the matrix representing $\tilde{D}$.

8. Let $T: \mathbb{R}^m \to \mathbb{R}^n$ be a linear transformation. Define the functions T_j for $j = 1, 2, \ldots, m$ as follows:

$$T_j: \mathbb{R}^m \to \mathbb{R}$$

$$\mathbf{v} \to j\text{th coordinate of } T(\mathbf{v})$$

For each of the linear transformations listed below, find the matrix that represents T_2.

a. $T: \mathbb{R}^2 \to \mathbb{R}^2$ is counterclockwise rotation by 90°.

b. $T: \mathbb{R}^3 \to \mathbb{R}^3$

$$\mathbf{v} \to 2\mathbf{v}$$

c. T is the linear transformation represented by the matrix

$$M = \begin{bmatrix} 1 & -3 & 2 \\ 2 & 1 & 0 \\ 4 & -7 & -1 \end{bmatrix}$$

★ **9.** Let $T: \mathbb{R}^m \to \mathbb{R}^n$ be a linear transformation and let T_j be defined as in Exercise 8.

a. Show that for any vector $\mathbf{v}$ in $\mathbb{R}^m$

$$T(\mathbf{v}) = T_1(\mathbf{v})\begin{bmatrix} 1 \\ 0 \\ 0 \\ \vdots \\ 0 \end{bmatrix} + T_2(\mathbf{v})\begin{bmatrix} 0 \\ 1 \\ 0 \\ \vdots \\ 0 \end{bmatrix} + \cdots + T_n(\mathbf{v})\begin{bmatrix} 0 \\ 0 \\ \vdots \\ 0 \\ 1 \end{bmatrix}$$

b. Suppose the matrix that represents T is M. Describe the matrix that represents T_j.

c. Show that the following statement is true: The transformation $T: \mathbb{R}^m \to \mathbb{R}^n$ is linear if and only if each of the transformations T_j is linear for $j = 1, 2, \ldots, m$.

10. Prove Theorem 3.3.

11. (Optional) Find the net effect of performing the operations indicated (in the order given) to an arbitrary vector $\mathbf{v}$ of $\mathbb{R}^2$.

a. First rotate counterclockwise by 90°, then reflect about the line $y = x$.

b. First reflect about the line $y = x$, then reflect about the x axis (compare your answer to Example 9).

c. First reflect about the x axis, then reflect about the y axis.

Applications 3 4 9 23 8 4 24 21 4 8 13 14 8 20 25 22 17 10 14 10 18 24 14 1 12

In this section we study ways of writing and deciphering secret messages. We will even learn how to read the title of this section. Technically, this section deals with ciphers, not codes. Ciphering is a technique of making secret messages, whereas coding is a way of including redundant information in a message so that errors that can distort the message during transmission can be corrected. For example, when the pilot of a plane identifies himself as "One Niner Charlie Able Baker," he is making his identity (which is 19CAB) unambiguously clear by adding redundant information. (Coding theory is the topic of the applications section of Chapter 5.)

Not all ciphers are used by secret agents and spies in international intrigues. In fact, there are many ciphers that most of us encounter every day. For example, the number on a driver's license, a social security number, a credit card number, plant identification numbers on canned food, and automobile serial numbers are all ciphers—they contain secret information.

To begin our study of ciphers we translate the alphabet into numerical information. This is the simplest cipher. It consists of just numbering the letters of the alphabet.

Cipher 1

A	B	C	D	E	F	G	H	I	J	K	L	M
0	1	2	3	4	5	6	7	8	9	10	11	12

N	O	P	Q	R	S	T	U	V	W	X	Y	Z
13	14	15	16	17	18	19	20	21	22	23	24	25

Note: Starting with zero will make things somewhat simpler later on. Using this cipher, we could encipher the message

M A T H I S E N J O Y A B L E

as 12 0 19 7 8 18 4 13 9 14 24 0 1 11 4

Cipher 1 is extremely easy to encipher as well as to decipher. To make things a bit more difficult, we could assign the numbers 0–25 to the letters of the alphabet in a different order. One way to do this is to shift the numbering. The cipher wheel shown in Figure 3.22 is a simple device for doing this. We will call ciphers made this way **additive ciphers.**

The cipher wheel consists of two circular discs. The inner disc can be rotated so that 26 different ciphers can be formed (one for each of the 26 different positions of the inner disc). For example, using the cipher shown on the wheel in Figure 3.22 (we will call it Cipher 2), the message MATH IS ENJOYABLE is enciphered as

17 5 24 12 13 23 9 18 14 19 3 5 6 16 9

Now let us look more closely at the difference between the two ways we have enciphered the message "MATH IS ENJOYABLE." The ciphered message was

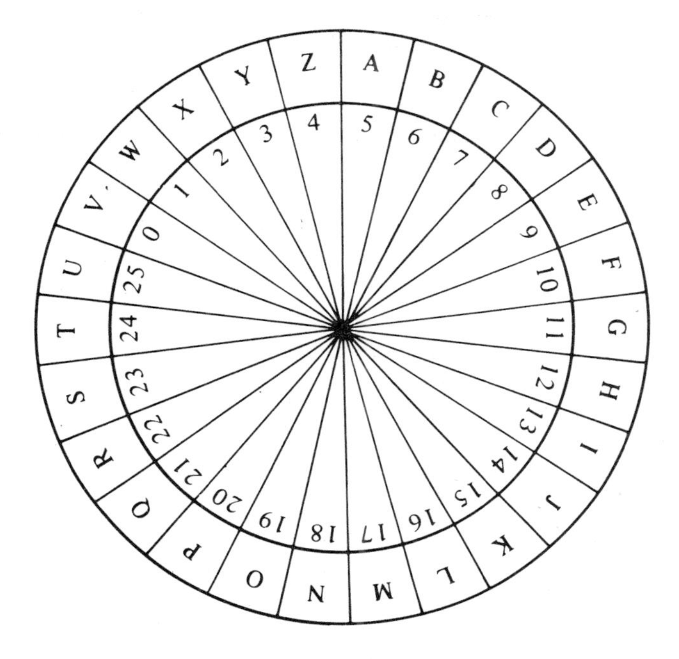

Figure 3.22

Cipher wheel.

12 0 19 7 8 18 4 13 9 14 24 0 1 11 4 (Cipher 1)

17 5 24 12 13 23 9 18 14 19 3 5 6 16 9 (Cipher 2)

It is easily seen that for all the numbers except 24, the way to obtain Cipher 2 from Cipher 1 is to add 5 to each number. Let us look at the number 24 more closely. Adding 5 to 24 would give 29, which would not be reasonable since all the numbers in our cipher are between 0 and 25. However, if we look at the inner disc of the cipher wheel and go 5 spaces beyond 24, we come to the number 3. So we see that Cipher 2 is obtained from Cipher 1 by adding 5 to each number in this special sense.

This type of arithmetic is called **clock arithmetic**. We think of the inner disc of the cipher wheel as a 26-hour clock. When we add the numbers 24 and 5, we can think of this as representing the time 5 hours past 24 on the 26-hour clock. These ideas can easily be extended to show that the time (on the 26-hour clock represented by any number n) is just the remainder when n is divided by 26. For example, since

$$26\overline{)29}^{\;1R3}$$

the time represented by the number 29 is 3. The mathematical notation for this is

$$29 \equiv 3 \qquad (\text{mod } 26)$$

which is read "29 is **congruent to** 3 **modulo** 26."

We can use this congruence notation to express addition. For example

$$5 + 24 \equiv 3 \qquad (\text{mod } 26)$$

We can also multiply modulo 26. For example,

$$3 \times 15 = 45 \equiv 19 \qquad (\text{mod } 26)$$

Multiplication modulo 26 gives another way to encipher messages. We can multiply each number in Cipher 1 by a fixed number modulo 26. We will call this type of cipher a **multiplicative cipher.**

EXAMPLE 1 We encipher the message MATH IS ENJOYABLE using the multiple 3.

M	A	T	H	I	S	E	N	J	O	Y	A	B	L	E	
12	0	19	7	8	18	4	13	9	14	24	0	1	11	4	Cipher 1
36	0	57	21	24	54	12	39	27	42	72	0	3	33	12	multiply by 3
10	0	5	21	24	2	12	13	1	16	20	0	3	7	12	reduce mod 26 ■

Now the problem of deciphering a multiplicative cipher is somewhat more complicated. We cannot simply divide each number by 3, because when we reduced modulo 26, we got numbers that are not divisible by 3. We must find some other way of undoing multiplication by 3 modulo 26. To do this we notice that

$$3 \times 9 = 27 \equiv 1 \qquad (\text{mod } 26)$$

This means that if n is any number, then

$$\begin{aligned} n &\equiv n \times 1 && (\text{mod } 26) \\ &\equiv n \times (3 \times 9) && (\text{mod } 26) \\ &\equiv (n \times 3) \times 9 && (\text{mod } 26) \end{aligned}$$

Therefore, if we start with the coded message $(n \times 3)$, multiply by 9, and reduce modulo 26, we will get the original message.

EXAMPLE 2 We decipher the message we enciphered in Example 1.

10	0	5	21	24	2	12	13	1	16	20	0	3	7	12	Cipher
90	0	45	189	216	18	108	117	9	144	180	0	27	63	108	multiply by 9
12	0	19	7	8	18	4	13	9	14	24	0	1	11	4	reduce mod 26
M	A	T	H	I	S	E	N	J	O	Y	A	B	L	E	decipher Cipher 1

■

Since multiplication by 9 modulo 26 undoes multiplication by 3, the number 9 is said to be the **inverse of 3 modulo 26**. In the same way we can find inverses of other numbers modulo 26. For example, 7 is the inverse of 15 modulo 26 since

$$15 \times 7 = 105 = (4 \times 26) + 1$$

which implies that

$$15 \times 7 \equiv 1 \qquad (\text{mod } 26)$$

Hence, if a code is enciphered by multiplication by 15 and reducing modulo 26, it can be deciphered by multiplying by 7 and reducing modulo 26 (See Exercises 2c and 3c).

Caution is required when making up multiplicative ciphers. Certain numbers should not be chosen as multipliers. The problem that occurs if the wrong number is chosen is illustrated in the next example.

EXAMPLE 3 We look at what happens to certain letters if we choose 2 as the multiplier in a multiplicative cipher.

$$\begin{aligned} A &\to 0 \xrightarrow{\text{(mult by 2)}} 0 \xrightarrow{\text{(reduce mod 26)}} 0 \\ N &\to 13 \xrightarrow{\text{(mult by 2)}} 26 \xrightarrow{\text{(reduce mod 26)}} 0 \\ B &\to 1 \xrightarrow{\text{(mult by 2)}} 2 \xrightarrow{\text{(reduce mod 26)}} 2 \\ O &\to 14 \xrightarrow{\text{(mult by 2)}} 28 \xrightarrow{\text{(reduce mod 26)}} 2 \end{aligned}$$

and so on.

The problem with such a cipher is obvious. If we have a zero in the enciphered message, there is no way of determining whether it should be deciphered as an A or an N. Similarly, the number 2 can be deciphered as either B or O. This difficulty can be avoided if we choose the multiplier m for the cipher correctly. ■

As we saw in Example 3, it is important for a multiplicative cipher to have the following property: *Two distinct letters should not be enciphered as the same number*. This means that if a and b are different numbers modulo 26, then

$$am \not\equiv bm \pmod{26} \qquad \text{or equivalently} \qquad (a-b)m \not\equiv 0 \pmod{26}$$

which means that the number $(a - b)m$ should not be divisible by 26. Now neither $(a - b)$ nor m is divisible by 26. If 26 divides $(a - b)$ then $a \equiv b$ (mod 26), and if 26 divides m, then $m = 0$ (mod 26). Therefore, if 26 divides the product $(a - b)m$, either m is divisible by 2 (and $(a - b)$ is divisible by 13), or m is divisible by 13 (and $(a - b)$ is divisible by 2). To prevent two different letters from being enciphered as the same letter we must choose m so that it is neither divisible by 2 nor by 13. Conversely, if m is divisible by either 2 or 13, then it is always possible to find two distinct letters that are enciphered as the same number.

There are two problems with the two types of ciphers that we have discussed so far. The first problem is that there are relatively few different ciphers (26 different additive ciphers from the cipher wheel and 12 multiplicative ciphers). The second problem is that the frequency of the occurrence of letters is preserved. Using frequency data for the occurrence of letters in the English language would give someone trying to break the cipher a good clue as to how the message was enciphered. Even if a message is enciphered by first using the cipher wheel and then multiplying, the frequencies would still be preserved in the resulting cipher.

To get a better cipher, it would be desirable to correct both of these deficiencies; that is, to have a vast collection of ciphers in which the frequency of occurrence of letters was different from the frequency of occurrence of letters in the English language. Such a collection of ciphers can be found by using matrices.

One way to change the frequency of occurrence of individual letters is to deal with pairs (or triples) of letters, and encipher pairs (or triples) of letters instead of individual letters. Since there are $26 \times 26 = 676$ pairs of letters, looking for frequency of pairs of letters is considerably more difficult than looking for the frequency of individual letters (the number of triples is $26^3 = 17{,}576$). The following procedure will produce such a cipher:

ENCIPHERING PROCEDURE

Step 1 Divide the message into pairs of letters (if there are an odd number of letters add the letter X to the end of the message).

Step 2 Convert the pairs of letters into vectors using Cipher 1.

Step 3 Apply a nonsingular matrix M (we will define the term *nonsingular* soon) to each of the vectors obtained in Step 2.

Step 4 Reduce the vectors modulo 26.

Step 5 Write the message.

The following example should help clarify this procedure.

EXAMPLE 4 We encipher the message PIG using the matrix method of enciphering. The matrix we choose is

$$M = \begin{bmatrix} 2 & 3 \\ 1 & 2 \end{bmatrix}$$

Step 1 Form pairs.

$$\text{PIG} \longrightarrow \text{PI GX}$$

Step 2 Get vectors.

$$\begin{bmatrix} \text{P} \\ \text{I} \end{bmatrix} \begin{bmatrix} \text{G} \\ \text{X} \end{bmatrix} \xrightarrow{\text{cipher 1}} \begin{bmatrix} 15 \\ 8 \end{bmatrix} \begin{bmatrix} 6 \\ 23 \end{bmatrix}$$

Step 3 Multiply by M.

$$\begin{bmatrix} 2 & 3 \\ 1 & 2 \end{bmatrix} \begin{bmatrix} 15 \\ 8 \end{bmatrix} = \begin{bmatrix} 54 \\ 31 \end{bmatrix}$$

$$\begin{bmatrix} 2 & 3 \\ 1 & 2 \end{bmatrix} \begin{bmatrix} 6 \\ 23 \end{bmatrix} = \begin{bmatrix} 81 \\ 52 \end{bmatrix}$$

Step 4 Reduce mod 26.

$$\begin{bmatrix} 54 \\ 31 \end{bmatrix} \equiv \begin{bmatrix} 2 \\ 5 \end{bmatrix} \pmod{26}$$

$$\begin{bmatrix} 81 \\ 52 \end{bmatrix} \equiv \begin{bmatrix} 3 \\ 0 \end{bmatrix} \pmod{26}$$

Step 5 Write enciphered message.

2 5 3 0 ■

In the following example we streamline this procedure a little and encipher a longer message.

EXAMPLE 5 Encipher the message MATH IS ENJOYABLE using the matrix M used in Example 4.

Step 1 Form pairs of letters and add X.

$$\begin{bmatrix} M \\ A \end{bmatrix} \begin{bmatrix} T \\ H \end{bmatrix} \begin{bmatrix} I \\ S \end{bmatrix} \begin{bmatrix} E \\ N \end{bmatrix} \begin{bmatrix} J \\ O \end{bmatrix} \begin{bmatrix} Y \\ A \end{bmatrix} \begin{bmatrix} B \\ L \end{bmatrix} \begin{bmatrix} E \\ X \end{bmatrix}$$

Steps 2 and 3 Convert to vectors and multiply by M.

$$\begin{bmatrix} 2 & 3 \\ 1 & 2 \end{bmatrix} : \begin{bmatrix} 12 \\ 0 \end{bmatrix} \begin{bmatrix} 19 \\ 7 \end{bmatrix} \begin{bmatrix} 8 \\ 18 \end{bmatrix} \begin{bmatrix} 4 \\ 13 \end{bmatrix} \begin{bmatrix} 9 \\ 14 \end{bmatrix} \begin{bmatrix} 24 \\ 0 \end{bmatrix} \begin{bmatrix} 1 \\ 11 \end{bmatrix} \begin{bmatrix} 4 \\ 23 \end{bmatrix}$$

$$\rightarrow \begin{bmatrix} 24 \\ 12 \end{bmatrix} \begin{bmatrix} 59 \\ 33 \end{bmatrix} \begin{bmatrix} 70 \\ 44 \end{bmatrix} \begin{bmatrix} 47 \\ 30 \end{bmatrix} \begin{bmatrix} 60 \\ 37 \end{bmatrix} \begin{bmatrix} 48 \\ 24 \end{bmatrix} \begin{bmatrix} 35 \\ 23 \end{bmatrix} \begin{bmatrix} 77 \\ 50 \end{bmatrix}$$

Step 4 Reduce mod 26.

$$\equiv \begin{bmatrix} 24 \\ 12 \end{bmatrix} \begin{bmatrix} 7 \\ 7 \end{bmatrix} \begin{bmatrix} 18 \\ 18 \end{bmatrix} \begin{bmatrix} 21 \\ 4 \end{bmatrix} \begin{bmatrix} 8 \\ 11 \end{bmatrix} \begin{bmatrix} 22 \\ 24 \end{bmatrix} \begin{bmatrix} 9 \\ 23 \end{bmatrix} \begin{bmatrix} 25 \\ 24 \end{bmatrix}$$

Step 5 Write the enciphered message.

24 12 7 7 18 18 21 4 8 11 22 24 9 23 25 24 ■

This will make an excellent cipher provided we can find a simple procedure for deciphering it. In the next example we will see how to decipher a single pair of numbers.

EXAMPLE 6 We decipher the pair of numbers 2,5. (These are the first two numbers in the message we enciphered in Example 4).

Let the pair of numbers x,y be the numbers that correspond under Cipher 1 to the two letters we enciphered as 2,5. This means that

$$M \begin{bmatrix} x \\ y \end{bmatrix} = \begin{bmatrix} 2 & 3 \\ 1 & 2 \end{bmatrix} \begin{bmatrix} x \\ y \end{bmatrix} \equiv \begin{bmatrix} 2 \\ 5 \end{bmatrix} \pmod{26}$$

or equivalently

$$2x + 3y \equiv 2 \pmod{26}$$
$$1x + 2y \equiv 5 \pmod{26}$$

To solve this system of equations, we proceed as in Chapter 2, by using the elimination method and then reduce modulo 26.

$$\left[\begin{array}{cc|c} 2 & 3 & 2 \\ 1 & 2 & 5 \end{array}\right] \begin{array}{c}(1) \\ (-2)\end{array} \to \left[\begin{array}{cc|c} 2 & 3 & 2 \\ 0 & -1 & -8 \end{array}\right] \begin{array}{c}(1) \\ (3)\end{array} \to \left[\begin{array}{cc|c} 2 & 0 & -22 \\ 0 & -1 & -8 \end{array}\right] \to \left[\begin{array}{cc|c} 1 & 0 & -11 \\ 0 & 1 & 8 \end{array}\right]$$

Since $-11 \equiv -11 + 26 \equiv 15 \pmod{26}$, reducing the entries in this matrix modulo 26 yields

$$\left[\begin{array}{cc|c} 1 & 0 & 15 \\ 0 & 1 & 8 \end{array}\right]$$

So $x = 15$ and $y = 8$. Since P is the letter corresponding to the number 15 in Cipher 1 and I is the letter corresponding to 8, the pair 2,5 is deciphered as P I. ■

Although the method of Example 6 is rather cumbersome and is not suitable for deciphering long messages, it can be generalized to obtain a very efficient method for deciphering such ciphers. To begin, let us assume that the enciphered message contains the pair of numbers m,n (in the example, the pair m,n is the pair 2,5). We must find the pair of numbers x,y such that

$$M\begin{bmatrix} x \\ y \end{bmatrix} = \begin{bmatrix} m \\ n \end{bmatrix} \pmod{26}$$

In this case we must solve the equation

$$\left[\begin{array}{cc|c} 2 & 3 & x \\ 1 & 2 & y \end{array}\right] = \begin{bmatrix} m \\ n \end{bmatrix} \pmod{26}$$

To do this we row reduce the matrix

$$\left[\begin{array}{cc|c} 2 & 3 & m \\ 1 & 2 & n \end{array}\right]$$

to obtain

$$\left[\begin{array}{cc|c} 2 & 3 & m \\ 1 & 2 & n \end{array}\right] \begin{array}{c}(1) \\ (-2)\end{array} \to \left[\begin{array}{cc|c} 2 & 3 & m \\ 0 & -1 & m - 2n \end{array}\right] \begin{array}{c}(1) \\ (3)\end{array} \to$$
$$\left[\begin{array}{cc|c} 2 & 0 & 4m - 6n \\ 0 & -1 & m - 2n \end{array}\right] \to \left[\begin{array}{cc|c} 1 & 0 & 2m - 3n \\ 0 & 1 & -m + 2n \end{array}\right]$$

So

$$\begin{bmatrix} x \\ y \end{bmatrix} = \begin{bmatrix} 2m - 3n \\ -m + 2n \end{bmatrix} = \begin{bmatrix} 2 & -3 \\ -1 & 2 \end{bmatrix}\begin{bmatrix} m \\ n \end{bmatrix}$$

Therefore, to find the vector $\begin{bmatrix} x \\ y \end{bmatrix}$, we simply multiply the vector $\begin{bmatrix} m \\ n \end{bmatrix}$ by the matrix

$$N = \begin{bmatrix} 2 & -3 \\ -1 & 2 \end{bmatrix}$$

and then reduce modulo 26. (The matrix N is called the **inverse** of M; inverses of matrices will be discussed in detail in Chapter 7).

We summarize this discussion by writing the method for deciphering messages enciphered using matrix multiplication on pairs of letters.

DECIPHERING PROCEDURE

> **Step 1** Write each pair of numbers as a vector.
>
> **Step 2** Apply the inverse matrix N to each vector.
>
> **Step 3** Reduce the vectors obtained in Step 2 modulo 26.
>
> **Step 4** Convert the coefficients of the vectors obtained in Step 3 into letters using Cipher 1.
>
> **Step 5** Write the message.

We illustrate this method of deciphering messages in the following example.

EXAMPLE 7 Decipher the title of this section. This cipher was made using the matrix M we defined in Example 4. The title is:

4 9 23 8 4 24 21 4 8 13 14 8 20 25 22 17 10 14 10 18 24 1 12

Steps 1 and 2 Convert to vectors and multiply by N.

$$\begin{bmatrix} 2 & -3 \\ -1 & 2 \end{bmatrix} : \begin{bmatrix} 4 \\ 9 \end{bmatrix} \begin{bmatrix} 23 \\ 8 \end{bmatrix} \begin{bmatrix} 4 \\ 24 \end{bmatrix} \begin{bmatrix} 21 \\ 4 \end{bmatrix} \begin{bmatrix} 8 \\ 13 \end{bmatrix} \begin{bmatrix} 14 \\ 8 \end{bmatrix} \begin{bmatrix} 20 \\ 25 \end{bmatrix} \begin{bmatrix} 22 \\ 17 \end{bmatrix} \begin{bmatrix} 10 \\ 14 \end{bmatrix} \begin{bmatrix} 10 \\ 18 \end{bmatrix} \begin{bmatrix} 24 \\ 14 \end{bmatrix} \begin{bmatrix} 1 \\ 12 \end{bmatrix}$$

which equals

$$\begin{bmatrix} -19 \\ 14 \end{bmatrix} \begin{bmatrix} 22 \\ -7 \end{bmatrix} \begin{bmatrix} -64 \\ 44 \end{bmatrix} \begin{bmatrix} 30 \\ -13 \end{bmatrix} \cdot \begin{bmatrix} -23 \\ 18 \end{bmatrix} \begin{bmatrix} 4 \\ 2 \end{bmatrix} \begin{bmatrix} -35 \\ 30 \end{bmatrix} \begin{bmatrix} -7 \\ 12 \end{bmatrix} \begin{bmatrix} -22 \\ 18 \end{bmatrix} \begin{bmatrix} -34 \\ 26 \end{bmatrix} \begin{bmatrix} 6 \\ 4 \end{bmatrix} \begin{bmatrix} -34 \\ 13 \end{bmatrix}$$

Step 3 Reduce mod 26.

$$\begin{bmatrix} 7 \\ 14 \end{bmatrix} \begin{bmatrix} 22 \\ 19 \end{bmatrix} \begin{bmatrix} 14 \\ 18 \end{bmatrix} \begin{bmatrix} 4 \\ 13 \end{bmatrix} \begin{bmatrix} 3 \\ 18 \end{bmatrix} \begin{bmatrix} 4 \\ 2 \end{bmatrix} \begin{bmatrix} 17 \\ 4 \end{bmatrix} \begin{bmatrix} 19 \\ 12 \end{bmatrix} \begin{bmatrix} 4 \\ 18 \end{bmatrix} \begin{bmatrix} 18 \\ 0 \end{bmatrix} \begin{bmatrix} 6 \\ 4 \end{bmatrix} \begin{bmatrix} 18 \\ 23 \end{bmatrix}$$

Step 4 Convert to letters using Cipher 1.

$$\begin{bmatrix} \text{H} \\ \text{O} \end{bmatrix} \begin{bmatrix} \text{W} \\ \text{T} \end{bmatrix} \begin{bmatrix} \text{O} \\ \text{S} \end{bmatrix} \begin{bmatrix} \text{E} \\ \text{N} \end{bmatrix} \begin{bmatrix} \text{D} \\ \text{S} \end{bmatrix} \begin{bmatrix} \text{E} \\ \text{C} \end{bmatrix} \begin{bmatrix} \text{R} \\ \text{E} \end{bmatrix} \begin{bmatrix} \text{T} \\ \text{M} \end{bmatrix} \begin{bmatrix} \text{E} \\ \text{S} \end{bmatrix} \begin{bmatrix} \text{S} \\ \text{A} \end{bmatrix} \begin{bmatrix} \text{G} \\ \text{E} \end{bmatrix} \begin{bmatrix} \text{S} \\ \text{X} \end{bmatrix}$$

Step 5 Write the message.

HOW TO SEND SECRET MESSAGES X ■

Now we discuss which types of matrices can be used to make matrix multiplication ciphers. Just as we could not use any number between 0 and 25 as the multiplier in a multiplication cipher, we also cannot use any 2×2 matrix as the multiplier in a matrix cipher. The reason is the same in both cases. In the multiplication cipher we had to avoid the situation in which two different letters were enciphered as the same number. In the matrix cipher we must avoid the situation in which two different pairs of letters are enciphered as the same pair of numbers. The matrices that have the property that they do not map any pair of distinct vectors onto the same vector are called **nonsingular matrices**. A matrix that has the property that it does not map any pair of distinct vectors onto vectors that are equivalent modulo 26 is called **nonsingular modulo** 26. Theorem 3.4 describes a simple test to determine if a 2×2 matrix is nonsingular modulo 26. See Exercise 9.

THEOREM 3.4

> The matrix $$M = \begin{bmatrix} a & b \\ c & d \end{bmatrix}$$
> is **nonsingular mod 26** if and only if the quantity $(ad - bc)$ is not divisible by either 2 or 13. The quantity $(ad - bc)$ is called the determinant of M.

(The theory of determinants is discussed in Chapter 8.)

EXAMPLE 8 The matrix

$$M = \begin{bmatrix} 1 & 3 \\ -1 & 1 \end{bmatrix} \quad \text{has determinant} \quad [(1)(1) - (3)(-1)] = 4$$

Since 4 is divisible by 2, M is singular and cannot be used to form a matrix cipher.

Note that

$$\begin{bmatrix} 1 & 3 \\ -1 & 1 \end{bmatrix}\begin{bmatrix} 0 \\ 0 \end{bmatrix} = \begin{bmatrix} 0 \\ 0 \end{bmatrix} \quad \text{and} \quad \begin{bmatrix} 1 & 3 \\ -1 & 1 \end{bmatrix}\begin{bmatrix} 13 \\ 13 \end{bmatrix} = \begin{bmatrix} 52 \\ 0 \end{bmatrix} \equiv \begin{bmatrix} 0 \\ 0 \end{bmatrix}$$

This means that if M were used as the multiplier matrix in a matrix cipher, then the pairs A,A and I,I would both be enciphered as 0 0. ■

If a matrix M is nonsingular mod 26, then, to decipher the matrix code with multiplier M, it is necessary to invert M modulo 26. The next example illustrates the method of finding the inverse of a matrix modulo 26.

EXAMPLE 9 We invert the matrix

$$M = \begin{bmatrix} 4 & 5 \\ 1 & 2 \end{bmatrix} \pmod{26}$$

As we saw above, this involves reducing the matrix

$$\left[\begin{array}{cc|c} 4 & 5 & m \\ 1 & 2 & n \end{array}\right]$$

$$\left[\begin{array}{cc|c} 4 & 5 & m \\ 1 & 2 & n \end{array}\right] \begin{array}{l} (1) \\ (-4) \end{array} \longrightarrow \left[\begin{array}{cc|c} 4 & 5 & m \\ 0 & -3 & m-4n \end{array}\right] \begin{array}{l} (3) \\ (5) \end{array} \longrightarrow$$

$$\left[\begin{array}{cc|c} 12 & 0 & 8m-20n \\ 0 & -3 & m-4n \end{array}\right] \begin{array}{l} (1/4) \longrightarrow \\ (-1) \longrightarrow \end{array} \left[\begin{array}{cc|c} 3 & 0 & 2m-5n \\ 0 & 3 & -m+4n \end{array}\right]$$

To finish the reduction, we cannot simply divide by 3 since we are dealing with numbers modulo 26. However, we can use the same technique as before when we wanted to divide by 3—that is, we can multiply by 9, since $3 \times 9 \equiv 1 \pmod{26}$.

$$\left[\begin{array}{cc|c} 3 & 0 & 2m-5n \\ 0 & 3 & -m+4n \end{array}\right] \begin{array}{l} (9) \longrightarrow \\ (9) \longrightarrow \end{array} \left[\begin{array}{cc|c} 27 & 0 & 18m-45n \\ 0 & 27 & -9m+36m \end{array}\right]$$

which is congruent to

$$\left[\begin{array}{cc|cc} 1 & 0 & 18 & 7 \\ 0 & 1 & 17 & 10 \end{array}\right] \pmod{26}$$

So the inverse (mod 26) of the matrix M is

$$N = \begin{bmatrix} 18 & 7 \\ 17 & 10 \end{bmatrix} \quad ■$$

There is one question about matrix ciphers that we have not yet answered. How many 2×2 matrix ciphers are there? The answer is: There are

$$(13^2 - 1)(13^2 - 13)(2^2 - 1)(2^2 - 2) = 157{,}248$$

The number of different 3×3 matrix ciphers is

$$(13^3 - 1)(13^3 - 13)(13^3 - 13^2)(2^3 - 1)(2^3 - 2)(2^3 - 2^2) \cong 1.6 \times 10^{12}$$

Therefore, we can conclude that matrix ciphers are excellent ciphers. If the multiplier matrix is known, messages can easily be enciphered and deciphered (especially with the aid of an electronic computer). If, however, the multiplier matrix is not known, it is almost impossible to break this cipher because frequency considerations are of little value, and also because there are so many different ciphers of this type.

Exercises for Applications 3

1. For each of the equations below find a number x between 0 and 25 that satisfies the equation.

a. $5 + x \equiv 0 \pmod{26}$

b. $263 \equiv x \pmod{26}$

c. $6x \equiv 4 \pmod{26}$ Find two solutions.

d. $5x \equiv 1 \pmod{26}$

e. $5x \equiv 6 \pmod{26}$

f. $7(x + 2) \equiv 11 \pmod{26}$

2. Encipher the message

TO BE OR NO TT OB E

a. Using Cipher 1.

b. Using the additive cipher by adding 5.

c. Using the multiplicative cipher with multiplier 7.

d. Using the matrix cipher with matrix

$$M = \begin{bmatrix} 4 & 5 \\ 1 & 2 \end{bmatrix}$$

3. Decipher each of the messages you enciphered in Problem 2.

4. Decipher the message

11 14 21 4 8 18 6 17 0 13 3

using Cipher 1.

5. The message

21 24 5 14 18 2 16 1 10 23 13

was enciphered using the additive cipher by adding 10. Decipher this message.

6. The message

3 18 1 20 14 12 4 7 0 13 15

is a multiplicative cipher with multiplier 5. Decipher this message.

7. Consider the matrices

$$M_1 = \begin{bmatrix} 1 & 3 \\ 0 & 7 \end{bmatrix} \qquad M_2 = \begin{bmatrix} 2 & 3 \\ 2 & 1 \end{bmatrix}$$

$$M_3 = \begin{bmatrix} 1 & 2 \\ 3 & 11 \end{bmatrix} \qquad M_4 = \begin{bmatrix} 3 & 5 \\ 6 & 10 \end{bmatrix}$$

a. Find the determinant of each of these matrices.

b. Find the inverses modulo 26 of those matrices that are nonsingular modulo 26.

c. For each of the matrices that are singular modulo 26, find two different vectors (with coefficients between 0 and 25) that are mapped by the matrix to the same vector (see Example 8).

8. Decipher the message

14 20 12 16 9 3 18 16 21 25 5 22 19 0 8 5

It is a matrix cipher with matrix $M = \begin{bmatrix} 1 & 3 \\ 0 & 7 \end{bmatrix}$.

9. Let M be the matrix

$$M = \begin{bmatrix} a & b \\ c & d \end{bmatrix}$$

a. Show how to reduce the matrix

$$\left[\begin{array}{cc|c} a & b & m \\ c & d & n \end{array}\right]$$

to obtain

$$\left[\begin{array}{cc|cc} ad-bc & 0 & dm & -bn \\ 0 & ad-bc & -cm & an \end{array}\right]$$

b. Show that the matrix M is invertible modulo 26 if and only if the number $(ad - bc)$ has an inverse modulo 26.

c. Give a formula for the inverse modulo 26 of the matrix M if it is invertible (express the inverse of $(ad - bc)$ as $(ad - bc)^{-1}$).

Review Exercises

1. Define the following terms:

a. Vector space

b. Linear transformation

c. Linear combination

2. For each of the linear transformations described below, find the matrix that represents that transformation:

a. $T: \mathbb{R}^2 \to \mathbb{R}^2$ is reflection about the x axis.

b. $T: \mathbb{R}^2 \to \mathbb{R}^2$ is counterclockwise rotation by an angle θ (assume $0 \le \theta \le \pi/2$).

c. $T: \mathbb{R}^3 \to \mathbb{R}^2$ is the linear transformation such that

$$T\begin{bmatrix}1\\0\\0\end{bmatrix} = \begin{bmatrix}3\\4\end{bmatrix}, \quad T\begin{bmatrix}0\\1\\0\end{bmatrix} = \begin{bmatrix}1\\1\end{bmatrix}, \quad T\begin{bmatrix}0\\0\\1\end{bmatrix} = \begin{bmatrix}0\\1\end{bmatrix}$$

d. $T: \mathbb{R}^3 \to \mathbb{R}$ is the linear transformation such that

$$T\begin{bmatrix}x\\y\\z\end{bmatrix} = 3x - y + 2z$$

3. a. Find all vectors $\mathbf{v}$ such that $M\mathbf{v} = \mathbf{0}$ where

$$M = \begin{bmatrix} 1 & -1 & -1 \\ 3 & 2 & 7 \\ -2 & 5 & 8 \end{bmatrix}$$

b. Is the set of vectors you found in part **a** a vector space? Explain.

4. Determine which of the following sets S is a vector space (subspace):

a. S is the set of all vectors of $\mathbb{R}^2$ that lie in either the first or third quadrant.

b. S is the set of all vectors of $\mathbb{R}^2$ that have their tips on the line $y = x + 1$.

c. S is the set of all vectors of $\mathbb{R}^3$ that lie in the yz plane.

d. S is the set of all vectors of $\mathbb{R}^3$ that lie within the sphere of radius 1 centered at the origin.

5. Compute the following products (when possible):

a. $\begin{bmatrix} 2 & -3 & 1 \\ 3 & 1 & 1 \\ 2 & -3 & 7 \end{bmatrix}\begin{bmatrix}-2\\1\\3\end{bmatrix}$

b. $\begin{bmatrix} 2 & 3 \\ 3 & -1 \\ 1 & 1 \end{bmatrix}\begin{bmatrix}-1\\1\end{bmatrix}$

c. $\begin{bmatrix} 1 & -2 & 3 & 0 \end{bmatrix}\begin{bmatrix}2\\3\\-1\\4\end{bmatrix}$

d. $\begin{bmatrix} 2 & 3 \\ 3 & -1 \\ 1 & 1 \end{bmatrix}\begin{bmatrix}-1\\1\\0\end{bmatrix}$

6. Describe the linear transformation represented by each of the following matrices:

a. $\begin{bmatrix} 1 & 0 & 0 \\ 0 & 1 & 0 \\ 0 & 0 & 1 \end{bmatrix}$ **b.** $\begin{bmatrix} 1 & 0 & 0 \\ 0 & 1 & 0 \\ 0 & 0 & 0 \end{bmatrix}$

c. $\begin{bmatrix} 0 & 1 \\ -1 & 0 \end{bmatrix}$ **d.** $\begin{bmatrix} 0 & -1 \\ -1 & 0 \end{bmatrix}$

e. $\begin{bmatrix} 3 & 0 & 0 \\ 0 & 3 & 0 \\ 0 & 0 & 3 \end{bmatrix}$

Vector Spaces, Subspaces, Basis, and Dimension

We begin this chapter by finding the vector equations of lines and planes and by finding that some lines and some planes are subspaces. Then we return to the concept of basis, *which was introduced in Chapter 1, and give a formal definition of this term. We show that any two bases of a vector space have the same number of elements—this number is called the* dimension *of the vector space. The concepts of basis and dimension are extremely important because they allow us to gain an overview of many mathematical situations.*

4.1 Lines and Planes in $\mathbb{R}^n$

Just as we associate a point with the vector ending at that point, we can also associate curves (and surfaces) in $\mathbb{R}^n$ with the set of vectors whose endpoints are the points of the curve (surface). One of the simplest curves is the straight line. We begin our discussion with the equation of the straight line in $\mathbb{R}^2$. The line l with slope m and y intercept b has equation $y = mx + b$. This equation means that the line l consists of all points of the form $(x, mx + b)$. Therefore, a typical vector with its endpoint on the line is of the form

$$\mathbf{v} = \begin{bmatrix} x \\ mx + b \end{bmatrix} \qquad \text{or} \qquad \mathbf{v} = x\begin{bmatrix} 1 \\ m \end{bmatrix} + \begin{bmatrix} 0 \\ b \end{bmatrix}$$

We denote a general vector on this line by $\mathbf{v}(x)$ and write the vector equation of the line as

$$\mathbf{v}(x) = x\begin{bmatrix}1\\m\end{bmatrix} + \begin{bmatrix}0\\b\end{bmatrix}$$

The vector $\begin{bmatrix}1\\m\end{bmatrix}$ gives the direction of the line—the line is parallel to the vector $\begin{bmatrix}1\\m\end{bmatrix}$. The vector $\begin{bmatrix}0\\b\end{bmatrix}$ is a vector that ends on the line. Both vectors are shown in Figure 4.1.

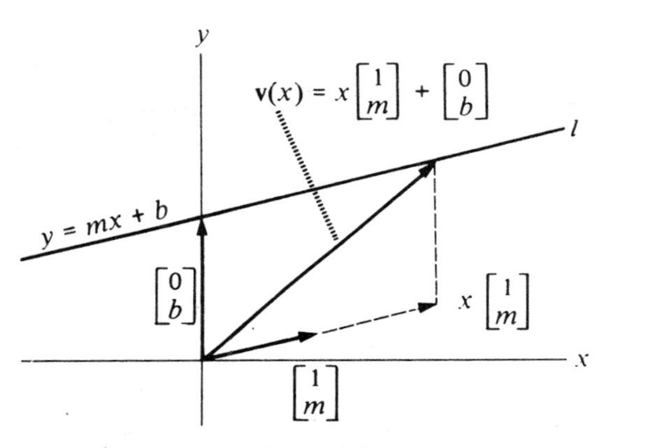

Figure 4.1

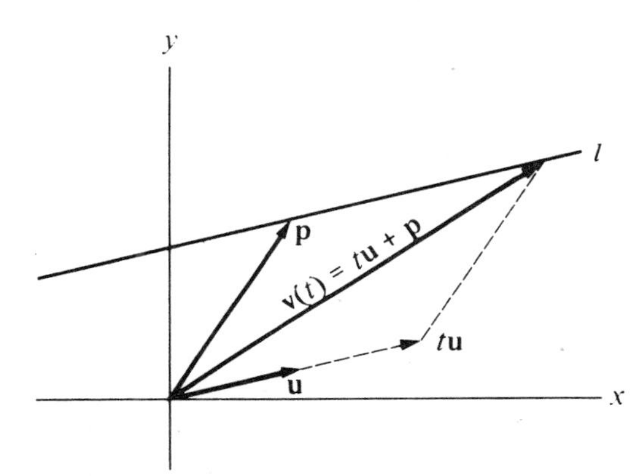

Figure 4.2

The vector equation of a line is not unique. In general, the line l parallel to the vector $\mathbf{u}$ and passing through the tip of the vector $\mathbf{p}$ has equation

$$\mathbf{v}(t) = t\mathbf{u} + \mathbf{p}$$

as shown in Figure 4.2. It is convenient to think of the vector equation of a line as describing a route for getting to any point on the line from the origin. You begin by traveling along $\mathbf{p}$ to reach the line and then travel in the direction parallel to the vector $\mathbf{u}$. The scalar t tells you how far you should go in this direction; for example, if $t = 2$, you travel 2 lengths of $\mathbf{u}$ from the tip of $\mathbf{p}$, or if $t = -1$, you go 1 length of $\mathbf{u}$ in the opposite direction. These situations are illustrated in Figure 4.3.

Figure 4.3

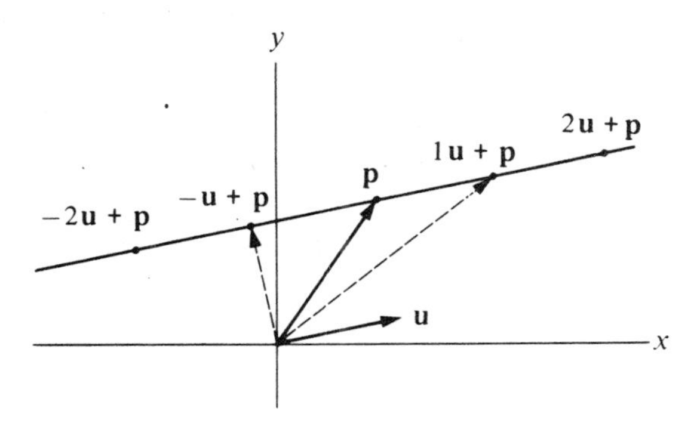

EXAMPLE 1 We graph the set of vectors

$$\mathbf{v}(t) = t\begin{bmatrix}1\\1\end{bmatrix} + \begin{bmatrix}1\\3\end{bmatrix}$$

The graph is shown in Figure 4.4.

Figure 4.4

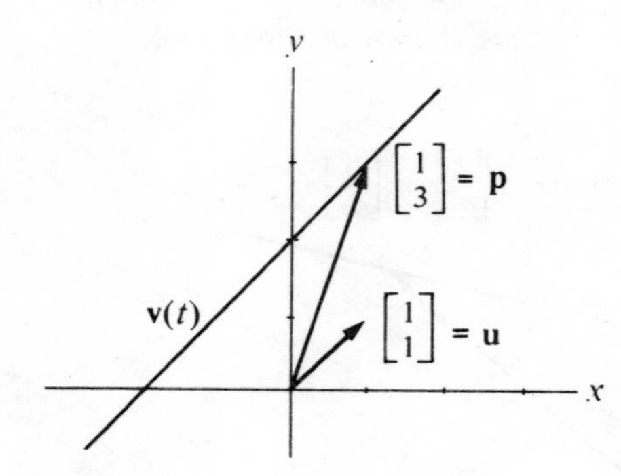

■

EXAMPLE 2 Write the vector equation of the line with slope 3 and y intercept -1.

Since the ordinary equation of this line is $y = 3x - 1$, the points of this line are all of the form $(t, 3t - 1)$. Therefore, a typical vector whose tip lies on the line is

$$\mathbf{v}(t) = \begin{bmatrix}t\\3t-1\end{bmatrix} = t\begin{bmatrix}1\\3\end{bmatrix} + \begin{bmatrix}0\\-1\end{bmatrix}$$

which is the vector equation of this line. ■

The formula that worked in $\mathbb{R}^2$ also gives the equation of a line in $\mathbb{R}^n$ for $n \geq 3$.

> **Vector Equation of a Line:** The line in $\mathbb{R}^n$ that is parallel to the vector $\mathbf{u}$ and contains the point at the tip of the vector $\mathbf{p}$ has vector equation
>
> $$\mathbf{v}(t) = t\mathbf{u} + \mathbf{p}$$

EXAMPLE 3 Find the equation of the line in $\mathbb{R}^3$ that is parallel to the vector

$$\begin{bmatrix}1\\-1\\2\end{bmatrix}$$

and passes through the tip of the vector

$$\begin{bmatrix} -2 \\ 3 \\ 2 \end{bmatrix}$$

In this case

$$\mathbf{u} = \begin{bmatrix} 1 \\ -1 \\ 2 \end{bmatrix} \quad \text{and} \quad \mathbf{p} = \begin{bmatrix} -2 \\ 3 \\ 2 \end{bmatrix}$$

So the equation of this line is

$$\mathbf{v}(t) = t\begin{bmatrix} 1 \\ -1 \\ 2 \end{bmatrix} + \begin{bmatrix} -2 \\ 3 \\ 2 \end{bmatrix}$$

The graph is shown in Figure 4.5.

Figure 4.5

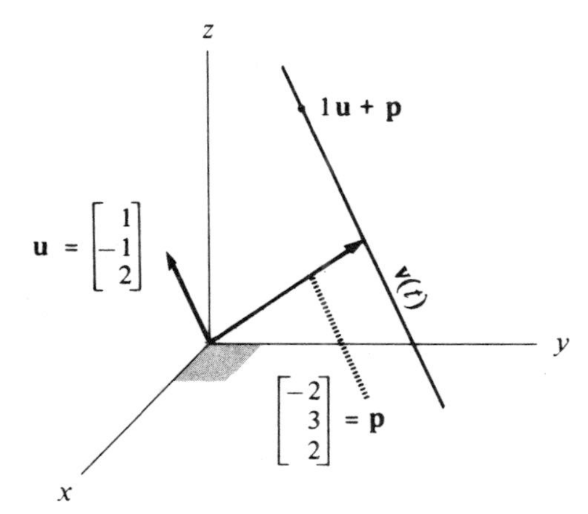

■

The next example illustrates how to find the equation of a line through two given points.

EXAMPLE 4 Find the equation of the line in $\mathbb{R}^3$ that contains the points (1, 2, 2) and (−1, 3, 4).

We begin by looking for a vector **u** parallel to this line. For convenience, we let

$$\mathbf{p}_1 = \begin{bmatrix} 1 \\ 2 \\ 2 \end{bmatrix} \quad \text{and} \quad \mathbf{p}_2 = \begin{bmatrix} -1 \\ 3 \\ 4 \end{bmatrix}$$

Using the parallelogram rule (see Section 1.3), we see that the vector $\mathbf{p}_2 - \mathbf{p}_1$ is parallel to the line joining $\mathbf{p}_1$ and $\mathbf{p}_2$. So we let $\mathbf{u} = \mathbf{p}_2 - \mathbf{p}_1$. Thus

$$\mathbf{u} = \mathbf{p}_2 - \mathbf{p}_1 = \begin{bmatrix} -1 \\ 3 \\ 4 \end{bmatrix} - \begin{bmatrix} 1 \\ 2 \\ 2 \end{bmatrix}$$

$$= \begin{bmatrix} -2 \\ 1 \\ 2 \end{bmatrix}$$

For the point $\mathbf{p}$ on the line we can choose either of the points $\mathbf{p}_1$ or $\mathbf{p}_2$. We let $\mathbf{p} = \mathbf{p}_1$. So the equation of the line is

$$\mathbf{v}(t) = t\mathbf{u} + \mathbf{p} = t(\mathbf{p}_2 - \mathbf{p}_1) + \mathbf{p}_1 = t\begin{bmatrix} -2 \\ 1 \\ 2 \end{bmatrix} + \begin{bmatrix} 1 \\ 2 \\ 2 \end{bmatrix}$$

The graph is shown in Figure 4.6.

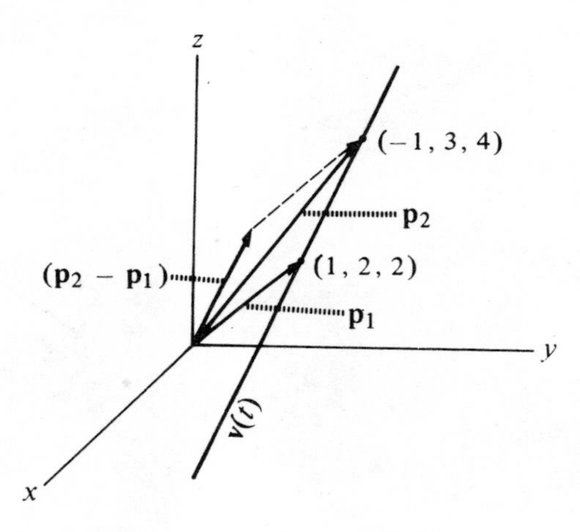

Figure 4.6

■

A given line may have many different vector equations. It is often necessary to determine when two vector equations represent the same line. This can be done in several ways, but the most straightforward method is this. Suppose we wish to find out if the equations $\mathbf{v}_1(t) = t\mathbf{u}_1 + \mathbf{p}_1$ and $\mathbf{v}_2(s) = s\mathbf{u}_2 + \mathbf{p}_2$ represent the same line. We first check to see if the lines they represent are parallel. To do this we must see if the direction vectors $\mathbf{u}_1$ and $\mathbf{u}_2$ are parallel (are scalar multiples of one another). If the two equations represent parallel lines, then we proceed to the next step. Since two parallel lines either have no points in common or are the same line, we need only see if some point of one line is on the other line. Since $\mathbf{p}_1$ is on the first line, it is sufficient to see if $\mathbf{p}_1$ is also on the second line. Thus we get the following theorem.

THEOREM 4.1 Let $\mathbf{v}_1(t) = t\mathbf{u}_1 + \mathbf{p}_1$ and $\mathbf{v}_2(s) = s\mathbf{u}_2 + \mathbf{p}_2$ be vector equations in $\mathbb{R}^n$. Then the lines represented by the equations are the same if and only if $\mathbf{u}_1$ is a scalar multiple of $\mathbf{u}_2$, and there is a number s so that $\mathbf{p}_1 = s\mathbf{u}_2 + \mathbf{p}_2$ (in other words, $\mathbf{p}_1$ is on the line represented by $\mathbf{v}_2(s) = s\mathbf{u}_2 + \mathbf{p}_2$).

Figure 4.7 is a graphical interpretation of this theorem.

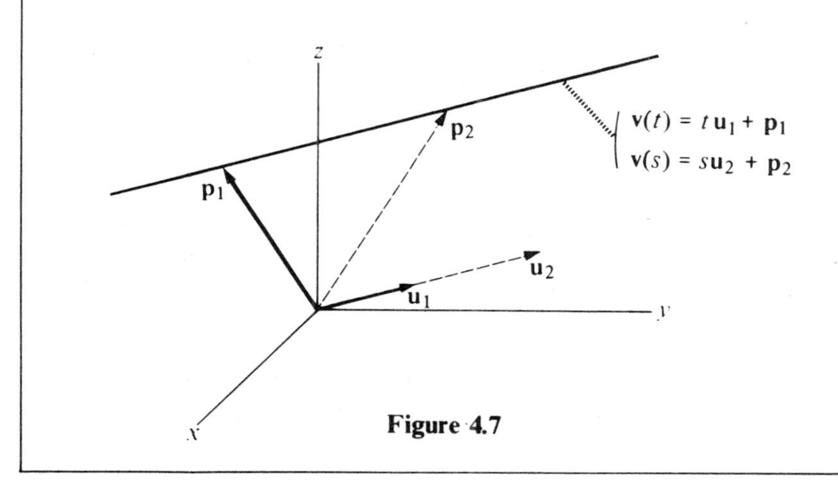

Figure 4.7

EXAMPLE 5 Determine if the following two equations represent the same line:

$$\mathbf{v}_1(t) = t\begin{bmatrix} -1 \\ 2 \\ -3 \end{bmatrix} + \begin{bmatrix} 2 \\ -4 \\ 5 \end{bmatrix} \quad \text{and} \quad \mathbf{v}_2(s) = s\begin{bmatrix} 2 \\ -4 \\ 6 \end{bmatrix} + \begin{bmatrix} 3 \\ -6 \\ 8 \end{bmatrix}$$

In this example,

$$\mathbf{u}_1 = \begin{bmatrix} -1 \\ 2 \\ -3 \end{bmatrix} \quad \text{and} \quad \mathbf{u}_2 = \begin{bmatrix} 2 \\ -4 \\ 6 \end{bmatrix}$$

so we see that $\mathbf{u}_2 = -2\mathbf{u}_1$. Therefore, the lines represented by these equations are parallel.

Also

$$\mathbf{p}_1 = \begin{bmatrix} 2 \\ -4 \\ 5 \end{bmatrix}$$

To see if the tip of this vector lies on the line represented by the second equation, we must see if there is a number s so that

$$\begin{bmatrix} 2 \\ -4 \\ 5 \end{bmatrix} = s \begin{bmatrix} 2 \\ -4 \\ 6 \end{bmatrix} + \begin{bmatrix} 3 \\ -6 \\ 8 \end{bmatrix}$$

Equating coefficients, we see that if there is such an s, it must satisfy each of the following three equations:

$$\begin{aligned} 2 &= 2s + 3 \\ -4 &= -4s - 6 \\ 5 &= 6s + 8 \end{aligned}$$

The number $s = -\frac{1}{2}$ is a solution to each of these equations. Thus we have shown that these two equations represent the same line. ■

Now we discuss the vector equation of a plane in $\mathbb{R}^n$, for $n \geq 3$. The first thing we must do is discuss what we mean by the word *plane*. There is one plane we understand fairly well, namely $\mathbb{R}^2$. To describe planes in $\mathbb{R}^n$ ($n \geq 3$), we make use of the description of points in $\mathbb{R}^2$ given in Sections 1.4 and 1.5, especially Theorem 1.4. From this theorem we see that if $\mathbf{w}_1$ and $\mathbf{w}_2$ are a pair of nonparallel vectors of $\mathbb{R}^2$, then each vector of $\mathbb{R}^2$ can be expressed as a linear combination of the vectors $\mathbf{w}_1$ and $\mathbf{w}_2$, in other words, $\mathbb{R}^2$ is the set of all linear combinations of the vectors $\mathbf{w}_1$ and $\mathbf{w}_2$. We generalize this description of the plane $\mathbb{R}^2$ to give the vector equations of the planes of $\mathbb{R}^n$ that contain the origin.

DEFINITION 4.1 Let $\mathbf{w}_1$ and $\mathbf{w}_2$ be two nonparallel vectors of $\mathbb{R}^n$. The set of all vectors of the form $t_1\mathbf{w}_1 + t_2\mathbf{w}_2$, where t_1 and t_2 are arbitrary scalars, is called the **plane spanned by $\mathbf{w}_1$ and $\mathbf{w}_2$**.

Vector Equation of the Plane Spanned by $\mathbf{w}_1$ and $\mathbf{w}_2$: The vector equation of the plane spanned by two nonparallel vectors $\mathbf{w}_1$ and $\mathbf{w}_2$ is

$$\mathbf{v}(t_1, t_2) = t_1\mathbf{w}_1 + t_2\mathbf{w}_2 \qquad \text{where } t_1 \text{ and } t_2 \text{ are scalars.}$$

You should note that the planes discussed so far all contain the origin ($\mathbf{0} = \mathbf{v}(0, 0)$). We will discuss planes that do not contain the origin later in this section.

EXAMPLE 6 Find the vector equation of the plane containing the origin and the points $(1, -1, 3)$ and $(2, 2, 4)$.

In this case, the vectors $\mathbf{w}_1$ and $\mathbf{w}_2$ are

$$\mathbf{w}_1 = \begin{bmatrix} 1 \\ -1 \\ 3 \end{bmatrix} \quad \text{and} \quad \mathbf{w}_2 = \begin{bmatrix} 2 \\ 2 \\ 4 \end{bmatrix}$$

The equation of this plane is

$$\mathbf{v}(t_1, t_2) = t_1 \begin{bmatrix} 1 \\ -1 \\ 3 \end{bmatrix} + t_2 \begin{bmatrix} 2 \\ 2 \\ 4 \end{bmatrix} \quad \blacksquare$$

Now we find the vector equations for planes in $\mathbb{R}^n$ that do not contain the origin. Each plane in $\mathbb{R}^n$ belongs to a family of parallel planes. One plane of each family contains the origin and the other planes are all parallel to this one. The vector equations of the planes in each family are related to one another in the same way that vector equations of parallel lines are related. Let P be a plane parallel to the plane containing the origin that is spanned by $\mathbf{w}_1$ and $\mathbf{w}_2$, and let $\mathbf{p}$ be a vector which has its tip on P. The points of P can be thought of as being obtained by placing the plane spanned by $\mathbf{w}_1$ and $\mathbf{w}_2$ on top of the vector $\mathbf{p}$ (making sure not to tilt the plane). The vectors $\mathbf{v}$ of the plane P are gotten by adding $\mathbf{p}$ to the vectors of the plane spanned by $\mathbf{w}_1$ and $\mathbf{w}_2$. Summarizing, we have

Vector Equation of a Plane: Let P be the plane in $\mathbb{R}^n$ ($n \geq 3$) that is parallel to the plane spanned by $\mathbf{w}_1$ and $\mathbf{w}_2$ and contains the point at the tip of the vector $\mathbf{p}$. Then the equation of P is

$$\mathbf{v}(t_1, t_2) = (t_1\mathbf{w}_1 + t_2\mathbf{w}_2) + \mathbf{p} \qquad \text{where } t_1 \text{ and } t_2 \text{ are scalars}$$

The geometrical meaning of this equation is shown in Figure 4.8.

Figure 4.8

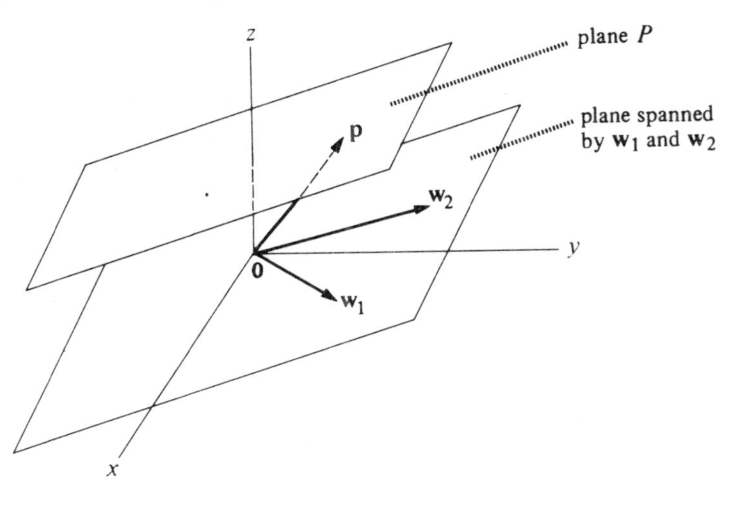

EXAMPLE 7 Find the vector equation of the plane parallel to the plane spanned by

$$\begin{bmatrix} 1 \\ -1 \\ 3 \end{bmatrix} \quad \text{and} \quad \begin{bmatrix} 2 \\ 2 \\ 4 \end{bmatrix}$$

and containing the point (3, 1, 1).

The equation is

$$\mathbf{v}(t_1, t_2) = \left(t_1 \begin{bmatrix} 1 \\ -1 \\ 3 \end{bmatrix} + t_2 \begin{bmatrix} 2 \\ 2 \\ 4 \end{bmatrix} \right) + \begin{bmatrix} 3 \\ 1 \\ 1 \end{bmatrix} \quad \blacksquare$$

EXAMPLE 8 Find the vector equation of the plane P containing the points (1, −2, 3), (−1, 4, 1), (−3, 2, 2).

Since we are given three points of the plane, the choice of $\mathbf{p}$ is no problem. What we must find are the vectors $\mathbf{w}_1$ and $\mathbf{w}_2$. For convenience, let

$$\mathbf{p}_1 = \begin{bmatrix} 1 \\ -2 \\ 3 \end{bmatrix} \quad \mathbf{p}_2 = \begin{bmatrix} -1 \\ 4 \\ 1 \end{bmatrix} \quad \mathbf{p}_3 = \begin{bmatrix} -3 \\ 2 \\ 2 \end{bmatrix}$$

Now the vector $\mathbf{p}_2 - \mathbf{p}_1$ is parallel to the line joining the tip of $\mathbf{p}_1$ to the tip of $\mathbf{p}_2$. Since this line is in the plane P, the vector $\mathbf{p}_2 - \mathbf{p}_1$ is parallel to the plane P. In the same way, we see that the vector $\mathbf{p}_3 - \mathbf{p}_1$ is also parallel to the plane P. So the plane (containing the origin) spanned by the vectors $\mathbf{p}_2 - \mathbf{p}_1$ and $\mathbf{p}_3 - \mathbf{p}_1$ is the plane containing the origin and parallel to P. We choose $\mathbf{w}_1 = \mathbf{p}_2 - \mathbf{p}_1$ and $\mathbf{w}_2 = \mathbf{p}_3 - \mathbf{p}_1$ and obtain the following equation of the plane P:

$$\begin{aligned} \mathbf{v}(t_1, t_2) &= (t_1 \mathbf{w}_1 + t_2 \mathbf{w}_2) + \mathbf{p} \\ &= [t_1(\mathbf{p}_2 - \mathbf{p}_1) + t_2(\mathbf{p}_3 - \mathbf{p}_1)] + \mathbf{p}_1 \\ &= \left[t_1 \left(\begin{bmatrix} -1 \\ 4 \\ 1 \end{bmatrix} - \begin{bmatrix} 1 \\ -2 \\ 3 \end{bmatrix} \right) + t_2 \left(\begin{bmatrix} -3 \\ 2 \\ 2 \end{bmatrix} - \begin{bmatrix} 1 \\ -2 \\ 3 \end{bmatrix} \right) \right] + \begin{bmatrix} 1 \\ -2 \\ 3 \end{bmatrix} \\ &= \left(t_1 \begin{bmatrix} -2 \\ 6 \\ -2 \end{bmatrix} + t_2 \begin{bmatrix} -4 \\ 4 \\ -1 \end{bmatrix} \right) + \begin{bmatrix} 1 \\ -2 \\ 3 \end{bmatrix} \end{aligned}$$

Note that we could have chosen any one of the vectors $\mathbf{p}_1$, $\mathbf{p}_2$, or $\mathbf{p}_3$ as $\mathbf{p}$ in the equation above. Moreover, we could use the difference of any two different pairs of $\mathbf{p}_i$'s as $\mathbf{w}_1$ and $\mathbf{w}_2$. Figure 4.9 illustrates this example.

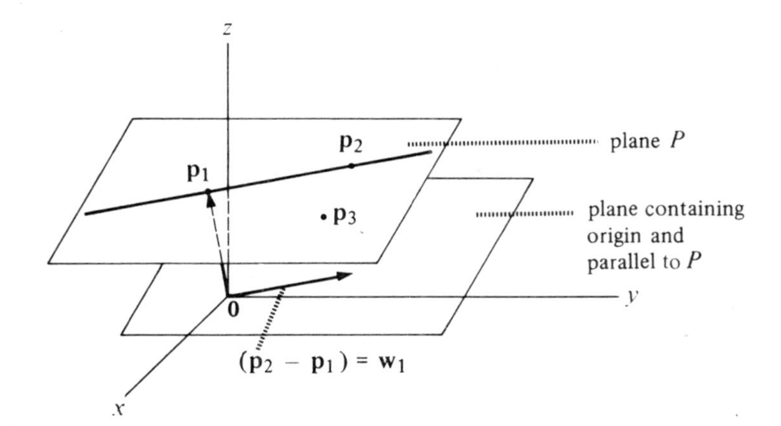

Figure 4.9

■

Exercise Set 4.1

1. Graph the vectors in each set below on the same set of coordinate axes.

a. $\begin{bmatrix}1\\2\end{bmatrix}$, $\begin{bmatrix}1\\2\end{bmatrix} + \begin{bmatrix}1\\0\end{bmatrix}$, $\begin{bmatrix}1\\2\end{bmatrix} + 2\begin{bmatrix}1\\0\end{bmatrix}$, $\begin{bmatrix}1\\2\end{bmatrix} + 3\begin{bmatrix}1\\0\end{bmatrix}$, $\begin{bmatrix}1\\2\end{bmatrix} - \begin{bmatrix}1\\0\end{bmatrix}$, $\begin{bmatrix}1\\2\end{bmatrix} - 2\begin{bmatrix}1\\0\end{bmatrix}$

b. $\mathbf{v}(t) = \begin{bmatrix}1\\2\end{bmatrix} + t\begin{bmatrix}1\\0\end{bmatrix}$.

2. For each vector equation below, find the equation of the form $y = mx + b$ that represents the same line. Graph this line.

a. $\mathbf{v}(t) = t\begin{bmatrix}1\\3\end{bmatrix} + \begin{bmatrix}0\\2\end{bmatrix}$

b. $\mathbf{v}(t) = t\begin{bmatrix}-1\\3\end{bmatrix} + \begin{bmatrix}2\\1\end{bmatrix}$

c. $\mathbf{v}(t) = t\begin{bmatrix}3\\4\end{bmatrix} - \begin{bmatrix}1\\1\end{bmatrix}$

3. For each of the following pairs of points find a vector equation of the line containing them.

a. (1, 1) and (3, −2)

b. (1, 3) and (−2, 4)

c. (1, −1, 1) and (2, 0, 2)

d. (1, 3, 2) and (−2, −3, 4)

4. Find the vector equation of each of the following planes.

a. The plane containing the point (0, 2, 3) and parallel to the plane spanned by the vectors

$$\begin{bmatrix}1\\-1\\2\end{bmatrix} \quad \text{and} \quad \begin{bmatrix}2\\3\\4\end{bmatrix}$$

b. The plane containing the origin and parallel to the plane

$$\mathbf{v}(t_1, t_2) = t_1\begin{bmatrix}1\\3\\2\end{bmatrix} + t_2\begin{bmatrix}2\\2\\4\end{bmatrix} + \begin{bmatrix}1\\5\\2\end{bmatrix}$$

c. The plane containing the points (1, −2, 3), (2, 5, 7), and (2, 2, 3).

d. The plane containing the points (2, 4, 5), (7, 1, 2), and (3, 3, 3).

5. Determine whether the two equations in each part represent the same line or plane.

a. $\mathbf{v}_1(t) = t\begin{bmatrix}1\\3\end{bmatrix} + \begin{bmatrix}2\\4\end{bmatrix}$

$\mathbf{v}_2(s) = s\begin{bmatrix}-1\\-3\end{bmatrix} + \begin{bmatrix}3\\7\end{bmatrix}$

b. $\mathbf{v}_1(t) = t\begin{bmatrix}1\\-2\\3\end{bmatrix} + \begin{bmatrix}0\\3\\-4\end{bmatrix}$

$\mathbf{v}_2(s) = s\begin{bmatrix}-2\\4\\-6\end{bmatrix} + \begin{bmatrix}-1\\5\\-7\end{bmatrix}$

c. $\mathbf{v}_1(t) = t\begin{bmatrix}-1\\2\\4\end{bmatrix} + \begin{bmatrix}0\\5\\2\end{bmatrix}$

$\mathbf{v}_2(s) = s\begin{bmatrix}3\\-6\\-12\end{bmatrix} + \begin{bmatrix}-2\\9\\9\end{bmatrix}$

d. $\mathbf{v}_1(t_1, t_2) = t_1\begin{bmatrix}3\\-2\\6\end{bmatrix} + t_2\begin{bmatrix}-1\\-1\\2\end{bmatrix}$

$\mathbf{v}_2(s_1, s_2) = s_1\begin{bmatrix}2\\-3\\6\end{bmatrix} + s_2\begin{bmatrix}5\\0\\0\end{bmatrix}$

★**6. a.** Let $\mathbf{p}, \mathbf{p}_1, \mathbf{p}_2$ be vectors in $\mathbb{R}^n$ and $\mathbf{p}_1, \mathbf{p}_2$ linearly independent. Prove the following statement:
There are real numbers a_1, a_2 such that $a_1\mathbf{p}_1 + a_2\mathbf{p}_2 = \mathbf{p}$ if and only if the matrix

$$\left[\begin{array}{cc|c} \mathbf{p}_1 & \mathbf{p}_2 & \mathbf{p} \end{array}\right]$$

can be reduced to the matrix

$$\left[\begin{array}{cc|c} 1 & 0 & a_1 \\ 0 & 1 & a_2 \\ 0 & 0 & 0 \\ \vdots & \vdots & \vdots \\ 0 & 0 & 0 \end{array}\right]$$

b. Show that the vector $\mathbf{p}$ is in the plane $\mathbf{v}(t_1, t_2) = t_1\mathbf{p}_1 + t_2\mathbf{p}_2$ if and only if the matrix

$$\left[\begin{array}{cc|c} \mathbf{p}_1 & \mathbf{p}_2 & \mathbf{p} \end{array}\right] \xrightarrow{\text{reduces to}} \left[\begin{array}{cc|c} 1 & 0 & a_1 \\ 0 & 1 & a_2 \\ 0 & 0 & 0 \\ 0 & 0 & 0 \end{array}\right]$$

for some real numbers a_1, a_2.

c. Use this technique to determine if the vector

$$\begin{bmatrix}2\\-4\\6\end{bmatrix}$$

is in the plane

$$\mathbf{v}(t_1, t_2) = t_1\begin{bmatrix}3\\-2\\4\end{bmatrix} + t_2\begin{bmatrix}-1\\-2\\2\end{bmatrix}$$

d. Show that the two planes

$$\mathbf{v}(t_1, t_2) = t_1\mathbf{p}_1 + t_2\mathbf{p}_2 + \mathbf{p}$$

and

$$\mathbf{w}(s_1, s_2) = s_1\mathbf{q}_1 + s_2\mathbf{q}_2 + \mathbf{q}$$

are parallel if and only if the matrix

$$\left[\begin{array}{cc|cc} \mathbf{p}_1 & \mathbf{p}_2 & \mathbf{q}_1 & \mathbf{q}_2 \end{array}\right] \xrightarrow{\text{reduces to}} \left[\begin{array}{cc|cc} 1 & 0 & a_1 & b_1 \\ 0 & 1 & a_2 & b_2 \\ 0 & 0 & 0 & 0 \\ \vdots & \vdots & \vdots & \vdots \\ 0 & 0 & 0 & 0 \end{array}\right]$$

7. Show that the following two equations represent the same plane (see Exercise 6).

$$\mathbf{v}_1(t_1, t_2) = t_1\begin{bmatrix}3\\-2\\4\end{bmatrix} + t_2\begin{bmatrix}-1\\-2\\2\end{bmatrix} + \begin{bmatrix}3\\3\\2\end{bmatrix}$$

$$\mathbf{v}_2(s_1, s_2) = s_1\begin{bmatrix}2\\-4\\6\end{bmatrix} + s_2\begin{bmatrix}0\\-8\\10\end{bmatrix} + \begin{bmatrix}-1\\3\\0\end{bmatrix}$$

8. Determine the points of intersection of the line

$$\mathbf{v}(t) = t\begin{bmatrix}1\\-1\\2\end{bmatrix} + \begin{bmatrix}2\\0\\3\end{bmatrix}$$

and the plane

$$\mathbf{w}(t_1, t_2) = t_1\begin{bmatrix}1\\3\\0\end{bmatrix} + t_2\begin{bmatrix}0\\3\\2\end{bmatrix} + \begin{bmatrix}2\\1\\4\end{bmatrix}$$

9. Let P be a plane in $\mathbb{R}^3$. Show that there is exactly one plane parallel to P that is also a subspace of $\mathbb{R}^3$.

10. Consider the set P of all vectors of the form

$$\mathbf{v} = \begin{bmatrix} x \\ y \\ z \end{bmatrix}$$

where x, y, and z satisfy the equation

$$ax + by + cz = 0$$

for some fixed numbers a, b, c. Show that the tips of the vectors of P form a plane that passes through the origin. Find the vector equation of this plane.

11. Suppose that (x, y, z) is a point on the plane with the vector equation

$$\mathbf{v}(t_1, t_2) = t_1 \begin{bmatrix} -3 \\ 2 \\ 0 \end{bmatrix} + t_2 \begin{bmatrix} -2 \\ 2 \\ 2 \end{bmatrix}$$

Find numbers a, b, c so that the coordinates x, y, z satisfy an equation of the form

$$ax + by + cz = 0$$

(*Hint:* Find the expression for a typical vector in this plane and then express the relationship between the x, y, and z coordinates of this vector.)

4.2 Pictures of Subspaces, Spanning Sets, and Cosets

In this section we investigate generalizations of the ideas of line and plane that we discussed in Section 4.1. This will allow us to apply our geometrical understanding of $\mathbb{R}^2$ and $\mathbb{R}^3$ to higher dimensional vector spaces. We begin by discussing the subspaces of $\mathbb{R}^2$.

EXAMPLE 1 We will find all possible types of subspaces of $\mathbb{R}^2$.

There are two trivial subspaces of every vector space—the subspace containing the zero vector alone, and the subspace that is the entire vector space. Let us now look for the nontrivial subspaces of $\mathbb{R}^2$.

Suppose W is a nontrivial subspace of $\mathbb{R}^2$. This means that W contains a nonzero vector $\mathbf{w}_1$. Since W is a subspace, all scalar multiples of $\mathbf{w}_1$ also lie in W (by Property 2 of Theorem 3.1 in Section 3.2)—the line through $\mathbf{w}_1$ (containing the origin) is contained in W. If these are the only vectors of W then we are done (it is easy to see that the set of all scalar multiples of a single vector is a subspace). Let us suppose that there is a vector $\mathbf{w}_2$ of W, that is not a scalar multiple of $\mathbf{w}_1$. As before, we see that all scalar multiples of $\mathbf{w}_2$ must lie in W (Property 2 of Theorem 3.1); but in addition we can use Property 1 of the theorem to see that sums of scalar multiples of $\mathbf{w}_1$ and scalar multiples of $\mathbf{w}_2$ also must lie in W. That is, W must contain all linear combinations of $\mathbf{w}_1$ and $\mathbf{w}_2$. However, since $\mathbf{w}_1$ and $\mathbf{w}_2$ are not parallel, the set of all linear combinations of $\mathbf{w}_1$ and $\mathbf{w}_2$ is all of $\mathbb{R}^2$ (Theorem 1.4). Thus, there is only one type of nontrivial subspace of $\mathbb{R}^2$—a line through the origin. ■

The technique used for finding subspaces of $\mathbb{R}^2$ in this example can be applied to many other situations. In the two theorems and definition that follow, we elucidate important aspects of this technique.

THEOREM 4.2 If W is a subspace of a vector space V and $\mathbf{w}_1, \mathbf{w}_2, \ldots, \mathbf{w}_k$ are vectors in W, then all linear combinations of the vectors $\mathbf{w}_1, \mathbf{w}_2, \ldots, \mathbf{w}_k$ (all vectors of the form $\mathbf{v} = a_1\mathbf{w}_1 + a_2\mathbf{w}_2 + \cdots + a_k\mathbf{w}_k$ for scalars $a_1, a_2, \ldots, a_k$) are also elements of W.

The proof of this theorem follows directly from Theorem 3.1 (Section 3.2). From Property 2, we see that all scalar multiples $a_i\mathbf{w}_i$ of the vectors $\mathbf{w}_i$ ($i = 1, 2, \ldots, k$) are elements of W; and then by Property 1, we see that all sums of the form $a_1\mathbf{w}_1 + a_2\mathbf{w}_2 + \cdots + a_k\mathbf{w}_k$ are also elements of W.

THEOREM 4.3 Let $\mathbf{w}_1, \mathbf{w}_2, \ldots, \mathbf{w}_k$ be vectors of a vector space W. Then the set of all linear combinations $a_1\mathbf{w}_1 + a_2\mathbf{w}_2 + \cdots + a_k\mathbf{w}_k$ (for scalars $a_1, a_2, \ldots, a_k$) is a subspace of W.

The proof of this theorem follows directly from Theorem 3.1 (see Exercise 7).

Theorem 4.3 leads us to the following very important definition.

DEFINITION 4.2 If W is the vector space formed by all linear combinations of the vectors $\mathbf{w}_1, \mathbf{w}_2, \ldots, \mathbf{w}_k$, then we say that W is the subspace **spanned** by the vectors $\mathbf{w}_1, \mathbf{w}_2, \ldots, \mathbf{w}_k$. This subspace is denoted by $\langle \mathbf{w}_1, \mathbf{w}_2, \ldots, \mathbf{w}_k \rangle$.

EXAMPLE 2 The vector space

$$\left\langle \begin{bmatrix} 1 \\ 1 \\ 1 \end{bmatrix}, \begin{bmatrix} 1 \\ 1 \\ 4 \end{bmatrix} \right\rangle$$

consists of all vectors of the form

$$\mathbf{v} = a_1 \begin{bmatrix} 1 \\ 1 \\ 1 \end{bmatrix} + a_2 \begin{bmatrix} 1 \\ 1 \\ 4 \end{bmatrix}$$

and is, therefore, a plane.

The graph is shown in Figure 4.10.

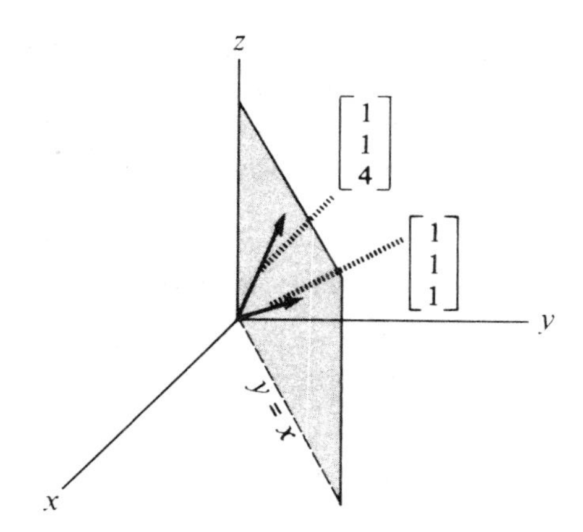

Figure 4.10

■

EXAMPLE 3 Find the subspace spanned by

$$\mathbf{w}_1 = \begin{bmatrix} 1 \\ -1 \\ 2 \end{bmatrix} \quad \text{and} \quad \mathbf{w}_2 = \begin{bmatrix} -3 \\ 3 \\ -6 \end{bmatrix}$$

A typical vector of this vector space is

$$\mathbf{v} = a_1 \begin{bmatrix} 1 \\ -1 \\ 2 \end{bmatrix} + a_2 \begin{bmatrix} -3 \\ 3 \\ -6 \end{bmatrix}$$

Since

$$\begin{bmatrix} -3 \\ 3 \\ -6 \end{bmatrix} = -3 \begin{bmatrix} 1 \\ -1 \\ 2 \end{bmatrix}$$

we see that

$$\mathbf{v} = a_1 \begin{bmatrix} 1 \\ -1 \\ 2 \end{bmatrix} + a_2 \begin{bmatrix} -3 \\ 3 \\ -6 \end{bmatrix}$$

$$= a_1 \begin{bmatrix} 1 \\ -1 \\ 2 \end{bmatrix} - 3a_2 \begin{bmatrix} 1 \\ -1 \\ 2 \end{bmatrix}$$

$$= (a_1 - 3a_2) \begin{bmatrix} 1 \\ -1 \\ 2 \end{bmatrix}$$

So the vector space spanned by these two vectors consists of the scalar multiples of one vector—a line through the origin. In fact,

$$W = \left\langle \begin{bmatrix} 1 \\ -1 \\ 2 \end{bmatrix} \right\rangle \quad ■$$

The process used to find subspaces of $\mathbb{R}^2$ can be used to find the types of subspaces of other vector spaces as well. We use Theorems 4.2 and 4.3 to streamline this procedure. Taken together, the theorems say that you can build up a subspace W $(\neq \langle \mathbf{0} \rangle)$ in the following way:

1. Begin with a nonzero vector $\mathbf{w}_1$ of W. If $W = \langle \mathbf{w}_1 \rangle$ (all scalar multiples of $\mathbf{w}_1$), we are done—W is a line.

2. If $\langle \mathbf{w}_1 \rangle \neq W$, then choose a vector $\mathbf{w}_2$ in W that is not in $\langle \mathbf{w}_1 \rangle$. If the space $\langle \mathbf{w}_1, \mathbf{w}_2 \rangle = W$, we are done—$W$ is a plane.

3. If $\langle \mathbf{w}_1, \mathbf{w}_2 \rangle \neq W$, then choose $\mathbf{w}_3$ in W but not in $\langle \mathbf{w}_1, \mathbf{w}_2 \rangle$. If $\langle \mathbf{w}_1, \mathbf{w}_2, \mathbf{w}_3 \rangle = W$, we are done. If not, continue this procedure.

We apply this method to finding all types of subspaces of $\mathbb{R}^3$.

EXAMPLE 4 We build up the subspaces of $\mathbb{R}^3$ one vector at a time.

We begin with the trivial subspace $\langle \mathbf{0} \rangle$. Next, let us consider a nontrivial subspace W that contains a nonzero vector $\mathbf{w}_1$. If $\langle \mathbf{w}_1 \rangle = W$, then W is a line through the origin. If $\langle \mathbf{w}_1 \rangle \neq W$, then there is a vector $\mathbf{w}_2$ of W that is not in the space $\langle \mathbf{w}_1 \rangle$. If $W = \langle \mathbf{w}_1, \mathbf{w}_2 \rangle$, then W is a plane containing the origin. If not, then there is a vector $\mathbf{w}_3$ of W, and $\mathbf{w}_3$ is not in the plane spanned by $\mathbf{w}_1$ and $\mathbf{w}_2$. We will now show that $\langle \mathbf{w}_1, \mathbf{w}_2, \mathbf{w}_3 \rangle = \mathbb{R}^3$.

What we must show is that any vector $\mathbf{v}$ of $\mathbb{R}^3$ can be written as a linear combination of the vectors $\mathbf{w}_1$, $\mathbf{w}_2$, and $\mathbf{w}_3$. So we must find scalars a_1, a_2, a_3 so that

$$\mathbf{v} = a_1\mathbf{w}_1 + a_2\mathbf{w}_2 + a_3\mathbf{w}_3$$

or

$$\mathbf{v} = (a_1\mathbf{w}_1 + a_2\mathbf{w}_2) + a_3\mathbf{w}_3$$

We interpret the last equation as meaning that we want to express $\mathbf{v}$ as the sum of a vector in the plane $\langle \mathbf{w}_1, \mathbf{w}_2 \rangle$ (spanned by $\mathbf{w}_1$ and $\mathbf{w}_2$) and a scalar multiple of $\mathbf{w}_3$. If $\mathbf{v}$ is in the plane spanned by $\mathbf{w}_1$ and $\mathbf{w}_2$, this is no problem, so we will assume that $\mathbf{v}$ is not in $\langle \mathbf{w}_1, \mathbf{w}_2 \rangle$. We begin (see Figure 4.11) by constructing the plane P that is parallel to $\langle \mathbf{w}_1, \mathbf{w}_2 \rangle$ and passes through the tip of the vector $\mathbf{v}$. Since $\mathbf{w}_3$ is not in the plane $\langle \mathbf{w}_1, \mathbf{w}_2 \rangle$, the line containing $\mathbf{w}_3$ intersects the plane P. Let a_3 be the scalar so that $a_3\mathbf{w}_3$ is the vector ending at the point where this line intersects P. Then

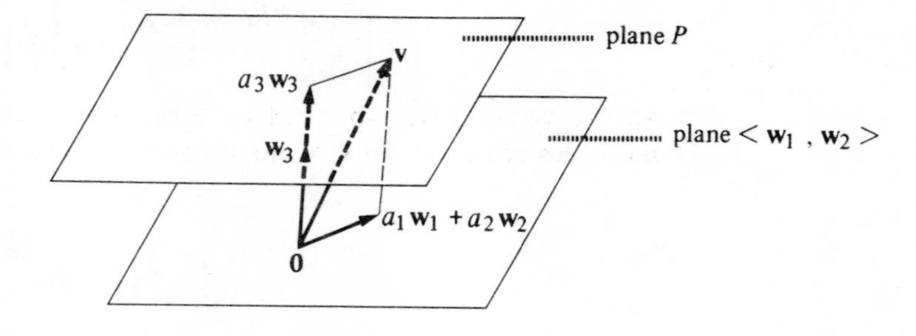

Figure 4.11

$a_3\mathbf{w}_3$ and a vector in the plane $\langle \mathbf{w}_1, \mathbf{w}_2 \rangle$ form two sides of a parallelogram having $\mathbf{v}$ as its diagonal. So $\mathbf{v}$ is the sum of a vector in $\langle \mathbf{w}_1, \mathbf{w}_2 \rangle$ and $a_3\mathbf{w}_3$. If $a_1\mathbf{w}_1 + a_2\mathbf{w}_2$ is the vector in $\langle \mathbf{w}_1, \mathbf{w}_2 \rangle$, then we have $\mathbf{v} = (a_1\mathbf{w}_1 + a_2\mathbf{w}_2) + a_3\mathbf{w}_3$. Thus $\mathbb{R}^3 = \langle \mathbf{w}_1, \mathbf{w}_2, \mathbf{w}_3 \rangle$.

We conclude that there are four types of subspaces of $\mathbb{R}^3$. They are $\langle \mathbf{0} \rangle$, lines through the origin, planes containing the origin, and $\mathbb{R}^3$ itself. ■

As we have shown, lines and planes containing the origin are subspaces. The question remains as to what can be said about other lines and planes. As we saw in Section 4.1, lines and planes fall into families of parallel lines and planes. In each of these families only one line or plane is a subspace—the one containing the origin.

By analogy, we can also group higher dimensional subsets of $\mathbb{R}^n$ into parallel families. If W is a subspace of $\mathbb{R}^n$ and $\mathbf{p}$ is a vector of $\mathbb{R}^n$, then the set of all vectors of the form

$$\mathbf{v} = \mathbf{p} + \mathbf{w} \qquad \text{where } \mathbf{w} \text{ is an arbitrary vector of } W$$

can be considered as the subset of $\mathbb{R}^n$ that is parallel to W and contains $\mathbf{p}$. Such subsets play an important role in linear algebra and are given a special name.

DEFINITION 4.3

> Let W be a subspace of a vector space V and $\mathbf{p}$ a vector of V. The set of all elements of the form $\mathbf{p} + \mathbf{w}$ where $\mathbf{w}$ can be any element of W is called the **coset** of W containing $\mathbf{p}$. It is denoted by $\mathbf{p} + W$.

EXAMPLE 5 Let W be the set of all vectors of $\mathbb{R}^3$ that lie in the xy plane. Describe the coset

$$\begin{bmatrix} 0 \\ 0 \\ 1 \end{bmatrix} + W$$

W consists of all vectors of the form

$$\begin{bmatrix} x \\ y \\ 0 \end{bmatrix}$$

So the coset

$$\begin{bmatrix} 0 \\ 0 \\ 1 \end{bmatrix} + W$$

consists of all vectors of the form

$$\begin{bmatrix} 0 \\ 0 \\ 1 \end{bmatrix} + \begin{bmatrix} x \\ y \\ 0 \end{bmatrix}.$$

or all vectors of the form
$$\begin{bmatrix} x \\ y \\ 1 \end{bmatrix}$$

This is just the plane $z = 1$. The solution is shown graphically in Figure 4.12.

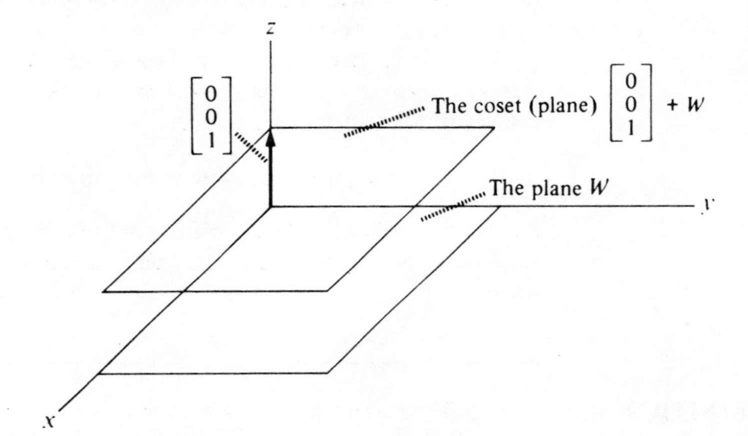

Figure 4.12

■

Two important properties of families of parallel lines (or parallel planes) are held by all cosets. Let W be a subspace of the vector space V. Then every point of V lies in some coset of W. (If W is a line, this translates as: Every point of V is on some line of the family of lines parallel to W). Moreover, no two different cosets have any points in common. (If W is a

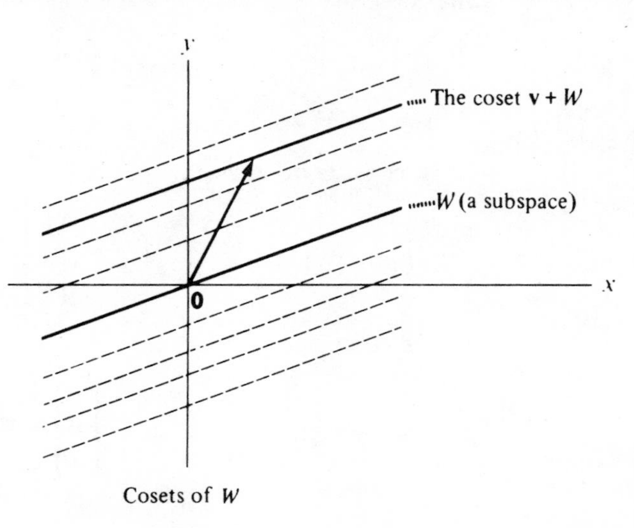

Cosets of W

Figure 4.13

line, this translates as: Distinct parallel lines have no points in common). See Figure 4.13 for an illustration.

The examples of the concepts of span, coset, and how subspaces look geometrically were only discussed in $\mathbb{R}^2$ and $\mathbb{R}^3$ in this section. This was done so that you could see how these concepts correspond to the concrete geometrical ideas developed in Section 4.1. These concepts provide the link between the spaces we can see ($\mathbb{R}^1$, $\mathbb{R}^2$, and $\mathbb{R}^3$) and the ones we can't ($\mathbb{R}^n$ for $n \geq 4$).

Exercise Set 4.2

1. Let $W = \left\langle \begin{bmatrix} 2 \\ 0 \\ 4 \end{bmatrix}, \begin{bmatrix} 1 \\ 1 \\ 1 \end{bmatrix} \right\rangle$

a. Find two different spanning sets for W.

b. Find a spanning set for W containing three (different) vectors.

c. Why is there no spanning set for W containing only one vector?

2. For each pair of spanning sets, determine if they span the same vector space or not.

a. $\left\{ \begin{bmatrix} 1 \\ 1 \end{bmatrix} \right\}$ and $\left\{ \begin{bmatrix} 1 \\ 0 \end{bmatrix}, \begin{bmatrix} 0 \\ 1 \end{bmatrix} \right\}$

b. $\left\{ \begin{bmatrix} -1 \\ 2 \end{bmatrix}, \begin{bmatrix} 3 \\ 7 \end{bmatrix} \right\}$ and $\left\{ \begin{bmatrix} 1 \\ 0 \end{bmatrix}, \begin{bmatrix} 0 \\ 1 \end{bmatrix} \right\}$

c. $\left\{ \begin{bmatrix} 3 \\ 3 \end{bmatrix}, \begin{bmatrix} 2 \\ 5 \end{bmatrix}, \begin{bmatrix} -3 \\ 6 \end{bmatrix} \right\}$ and $\left\{ \begin{bmatrix} -1 \\ 2 \end{bmatrix}, \begin{bmatrix} 3 \\ 7 \end{bmatrix} \right\}$

d. $\left\{ \begin{bmatrix} 1 \\ 1 \\ -2 \end{bmatrix}, \begin{bmatrix} 3 \\ 7 \\ 3 \end{bmatrix} \right\}$ and $\left\{ \begin{bmatrix} 4 \\ 8 \\ 1 \end{bmatrix}, \begin{bmatrix} 2 \\ 6 \\ 5 \end{bmatrix} \right\}$

e. $\left\{ \begin{bmatrix} 1 \\ 1 \\ -2 \end{bmatrix}, \begin{bmatrix} 3 \\ 7 \\ 3 \end{bmatrix} \right\}$ and $\left\{ \begin{bmatrix} 4 \\ 8 \\ 1 \end{bmatrix}, \begin{bmatrix} 2 \\ 6 \\ 5 \end{bmatrix}, \begin{bmatrix} 1 \\ 1 \\ 1 \end{bmatrix} \right\}$

f. $\left\{ \begin{bmatrix} 1 \\ 1 \\ -2 \end{bmatrix}, \begin{bmatrix} 3 \\ 7 \\ 3 \end{bmatrix} \right\}$ and $\left\{ \begin{bmatrix} 4 \\ 8 \\ 1 \end{bmatrix}, \begin{bmatrix} 2 \\ 6 \\ 5 \end{bmatrix}, \begin{bmatrix} 1 \\ 1 \\ -2 \end{bmatrix} \right\}$

3. Let $W = \left\langle \begin{bmatrix} -1 \\ 2 \end{bmatrix} \right\rangle$

Graph the following cosets of W on the same set of coordinate axes:

W, $\begin{bmatrix} -2 \\ 4 \end{bmatrix} + W$, $\begin{bmatrix} 1 \\ 0 \end{bmatrix} + W$, $\begin{bmatrix} 0 \\ 1 \end{bmatrix} + W$, $\begin{bmatrix} 0 \\ 2 \end{bmatrix} + W$, $\begin{bmatrix} 1 \\ 1 \end{bmatrix} + W$

4. Let

$$W = \left\langle \begin{bmatrix} 1 \\ 0 \\ 0 \end{bmatrix}, \begin{bmatrix} 0 \\ 1 \\ 0 \end{bmatrix} \right\rangle$$

Sketch the following surfaces on the same graph:

W, $\begin{bmatrix} 3 \\ -2 \\ 0 \end{bmatrix} + W$, $\begin{bmatrix} 0 \\ 0 \\ 1 \end{bmatrix} + W$, $\begin{bmatrix} 0 \\ 0 \\ 2 \end{bmatrix} + W$, $\begin{bmatrix} 3 \\ 5 \\ 2 \end{bmatrix} + W$

5. Which of the following vectors are in the coset

$$\begin{bmatrix} 1 \\ -1 \\ 2 \end{bmatrix} + \left\langle \begin{bmatrix} 2 \\ 1 \\ 1 \end{bmatrix}, \begin{bmatrix} -1 \\ 2 \\ 3 \end{bmatrix} \right\rangle ?$$

a. $\begin{bmatrix} 1 \\ -1 \\ 2 \end{bmatrix}$ **b.** $\begin{bmatrix} 2 \\ 1 \\ 1 \end{bmatrix}$ **c.** $\begin{bmatrix} 0 \\ 1 \\ 5 \end{bmatrix}$ **d.** $\begin{bmatrix} 1 \\ 3 \\ 4 \end{bmatrix}$

6. Consider the coset $\begin{bmatrix} 1 \\ -2 \end{bmatrix} + \left\langle \begin{bmatrix} 1 \\ 1 \end{bmatrix} \right\rangle$

a. Find vectors $\mathbf{v}_1$ and $\mathbf{v}_2$ of this coset so that $\mathbf{v}_1 + \mathbf{v}_2$ is not in this coset. Graph the coset, $\mathbf{v}_1$, $\mathbf{v}_2$, and $\mathbf{v}_1 + \mathbf{v}_2$.

b. Find a vector $\mathbf{v}$ of this coset and a scalar c so that $c\mathbf{v}$ is not in this coset. Graph the coset, $\mathbf{v}$, and $c\mathbf{v}$.

c. Is this coset a subspace?

7. Let $\mathbf{w}_1, \mathbf{w}_2, \mathbf{w}_3$ be vectors in $\mathbb{R}^4$.

a. Show that

$$(a_1\mathbf{w}_1 + a_2\mathbf{w}_2 + a_3\mathbf{w}_3) + (b_1\mathbf{w}_1 + b_2\mathbf{w}_2 + b_3\mathbf{w}_3)$$

is a linear combination of $\mathbf{w}_1, \mathbf{w}_2, \mathbf{w}_3$ for any scalars $a_1, a_2, a_3, b_1, b_2, b_3$.

b. Show that

$$c(a_1\mathbf{w}_1 + a_2\mathbf{w}_2 + a_3\mathbf{w}_3)$$

is a linear combination of $\mathbf{w}_1, \mathbf{w}_2, \mathbf{w}_3$ for any scalars c, a_1, a_2, a_3.

c. Show that the set of all linear combinations of $\mathbf{w}_1, \mathbf{w}_2, \mathbf{w}_3$ is a subspace of $\mathbb{R}^4$.

★ **8.** Consider the vector space $\mathbb{R}_3[x]$ of all polynomials of degree at most 3 with real coefficients (and the polynomial $p(x) = 0$).

a. Show that $\mathbb{R}_3[x] = \langle 1, x, x^2, x^3 \rangle$.

b. Show that $\mathbb{R}_3[x] = \langle 1, (x-1), (x-1)^2, (x-1)^3 \rangle$.

★ **9.** Let $p_1(x), p_2(x)$ be polynomials in $\mathbb{R}_3[x]$. If $p_1(x) = p_2(x)$ for $x = 1, 2, 3, 4$, show that $p_1(x) = p_2(x)$ for *all* values of x.

★ **10.** Consider the following polynomials (called the **Lagrange interpolation polynomials**):

$$L_1(x) = \frac{(x-2)(x-3)(x-4)}{(1-2)(1-3)(1-4)}$$

$$L_2(x) = \frac{(x-1)(x-3)(x-4)}{(2-1)(2-3)(2-4)}$$

$$L_3(x) = \frac{(x-1)(x-2)(x-4)}{(3-1)(3-2)(3-4)}$$

$$L_4(x) = \frac{(x-1)(x-2)(x-3)}{(4-1)(4-2)(4-3)}$$

a. Evaluate each of these polynomials at $x = 1, 2, 3, 4$.

b. Show that if $p(x)$ is an arbitrary polynomial in $\mathbb{R}_3[x]$, then

$$p(x) = p(1)L_1(x) + p(2)L_2(x) + p(3)L_3(x) + p(4)L_4(x)$$

(*Hint:* Use Exercise 9).

c. Show that the polynomials $L_1(x), L_2(x), L_3(x), L_4(x)$ span $\mathbb{R}_3[x]$.

4.3 Linear Independence and Basis

One of the things that came up several times in Section 4.2 was that the lists of vectors that spanned a subspace sometimes contained redundant vectors. One case where this occurred was in Example 3. Here the subspace W, as initially defined, was spanned by two vectors; but after a little investigation, we saw that one of the two vectors was redundant and could be eliminated.

In general, when giving a set of vectors that spans a vector space, it is best to give as few vectors as possible—that is, to eliminate redundant vectors. A set of vectors with no redundancies is called a **linearly independent set of vectors**, and a linearly independent set of vectors that spans a vector space is called a **basis** of that vector space. In this section we will study these concepts.

To begin, we must give a technical meaning to the term *redundant vector*. Let W be the space spanned by the vectors $\mathbf{w}_1, \mathbf{w}_2, \ldots, \mathbf{w}_k$. We say that the vector $\mathbf{w}_i$ of this spanning set is **redundant** if each vector $\mathbf{w}$ of W can be written as a linear combination of the vectors $\mathbf{w}_1, \ldots, \mathbf{w}_{i-1}, \mathbf{w}_{i+1}, \ldots, \mathbf{w}_k$ (all the vectors except $\mathbf{w}_i$). The following example illustrates this.

EXAMPLE 1 Let

$$\mathbf{w}_1 = \begin{bmatrix} -1 \\ 1 \\ 2 \\ 0 \end{bmatrix} \quad \mathbf{w}_2 = \begin{bmatrix} 2 \\ 2 \\ -1 \\ 3 \end{bmatrix} \quad \mathbf{w}_3 = \begin{bmatrix} 0 \\ 4 \\ 3 \\ 3 \end{bmatrix}$$

and $W = \langle \mathbf{w}_1, \mathbf{w}_2, \mathbf{w}_3 \rangle$. We will show that the vector

$$\mathbf{w} = \mathbf{w}_1 + \mathbf{w}_2 + \mathbf{w}_3$$

can also be written as a linear combination of the vectors $\mathbf{w}_1$ and $\mathbf{w}_2$ alone.

$$\mathbf{w} = \mathbf{w}_1 + \mathbf{w}_2 + \mathbf{w}_3 = \begin{bmatrix} -1 \\ 1 \\ 2 \\ 0 \end{bmatrix} + \begin{bmatrix} 2 \\ 2 \\ -1 \\ 3 \end{bmatrix} + \begin{bmatrix} 0 \\ 4 \\ 3 \\ 3 \end{bmatrix} = \begin{bmatrix} 1 \\ 7 \\ 4 \\ 6 \end{bmatrix}$$

This can also be expressed in terms of $\mathbf{w}_1$ and $\mathbf{w}_2$ as

$$\mathbf{w} = 3\mathbf{w}_1 + 2\mathbf{w}_2 = 3\begin{bmatrix} -1 \\ 1 \\ 2 \\ 0 \end{bmatrix} + 2\begin{bmatrix} 2 \\ 2 \\ -1 \\ 3 \end{bmatrix} = \begin{bmatrix} 1 \\ 7 \\ 4 \\ 6 \end{bmatrix}$$

So this particular vector can be expressed as a linear combination of $\mathbf{w}_1$ and $\mathbf{w}_2$.

Now we will show that every vector of W can be expressed as a linear combination of only $\mathbf{w}_1$ and $\mathbf{w}_2$—that is, $\mathbf{w}_3$ is redundant.

Note that $\mathbf{w}_3$ is a linear combination of the two vectors $\mathbf{w}_1$ and $\mathbf{w}_2$

$$\mathbf{w}_3 = 2\mathbf{w}_1 + \mathbf{w}_2$$

Using this fact, it is easy to see that any vector $\mathbf{u}$ of W is a linear combination of $\mathbf{w}_1$ and $\mathbf{w}_2$. We note that if

$$\mathbf{u} = a_1\mathbf{w}_1 + a_2\mathbf{w}_2 + a_3\mathbf{w}_3$$

then

$$\begin{aligned} \mathbf{u} &= a_1\mathbf{w}_1 + a_2\mathbf{w}_2 + a_3(2\mathbf{w}_1 + \mathbf{w}_2) \\ &= (a_1 + 2a_3)\mathbf{w}_1 + (a_2 + a_3)\mathbf{w}_2 \end{aligned}$$

So $\mathbf{u}$ is a linear combination of $\mathbf{w}_1$ and $\mathbf{w}_2$ alone, and we have shown that $\mathbf{w}_3$ is redundant which implies that $W = \langle \mathbf{w}_1, \mathbf{w}_2 \rangle$. ■

Now we do the same thing in general. If $\mathbf{w}_i$ is a redundant vector in the spanning set $\{\mathbf{w}_1, \mathbf{w}_2, \ldots, \mathbf{w}_k\}$, then this means that every vector of W can be written as a linear combination of the vectors $\mathbf{w}_1, \ldots, \mathbf{w}_{i-1}, \mathbf{w}_{i+1}, \ldots, \mathbf{w}_k$ (all the vectors except $\mathbf{w}_i$). So, in particular, $\mathbf{w}_i$ itself is a linear combination of the other vectors of the spanning set.

Conversely, if we have a vector of a spanning set that is a linear combination of the other vectors, then this element is redundant. This follows because any expression involving this vector can be replaced by an expression involving only the other elements of the spanning set. To do this, simply replace the vector by the linear combination of the other vectors that equals it (see Example 1).

We can conclude that a vector of a spanning set is redundant if and only if it is a linear combination of the other vectors of the spanning set.

EXAMPLE 2 For the subspace $W = \left\langle \begin{bmatrix} 1 \\ 2 \\ -3 \\ 1 \end{bmatrix}, \begin{bmatrix} 2 \\ -2 \\ 0 \\ 4 \end{bmatrix}, \begin{bmatrix} -1 \\ 1 \\ 0 \\ -2 \end{bmatrix} \right\rangle$

we eliminate redundant vectors to obtain a spanning set containing no redundant vectors.

We begin by seeing if the first vector is redundant. We look for scalars a_2 and a_3 so that

$$\begin{bmatrix} 1 \\ 2 \\ -3 \\ 1 \end{bmatrix} = a_2 \begin{bmatrix} 2 \\ -2 \\ 0 \\ 4 \end{bmatrix} + a_3 \begin{bmatrix} -1 \\ 1 \\ 0 \\ -2 \end{bmatrix}$$

This is equivalent to the system

$$\begin{aligned} 1 &= 2a_2 - a_3 \\ 2 &= -2a_2 + a_3 \\ -3 &= 0a_2 + 0a_3 \\ 1 &= 4a_2 - 2a_3 \end{aligned}$$

By looking at the third equation, it is obvious that this system has no solution. So the first vector is not redundant.

To see if the second vector is redundant, we must see if there are scalars a_1 and a_3 such that

$$\begin{bmatrix} 2 \\ -2 \\ 0 \\ 4 \end{bmatrix} = a_1 \begin{bmatrix} 1 \\ 2 \\ -3 \\ 1 \end{bmatrix} + a_3 \begin{bmatrix} -1 \\ 1 \\ 0 \\ -2 \end{bmatrix}$$

This is equivalent to the system

$$\begin{aligned} 2 &= a_1 - a_3 \\ -2 &= 2a_1 + a_3 \\ 0 &= -3a_1 + 0a_3 \\ 4 &= a_1 - 2a_3 \end{aligned}$$

From the third equation, we get $a_1 = 0$. Using this value in the first equation of this system gives $a_3 = -2$. So

$$\begin{bmatrix} 2 \\ -2 \\ 0 \\ 4 \end{bmatrix} = -2 \begin{bmatrix} -1 \\ 1 \\ 0 \\ -2 \end{bmatrix}$$

which means that the second vector is redundant and may be eliminated.

So $$W = \left\langle \begin{bmatrix} 1 \\ 2 \\ -3 \\ 1 \end{bmatrix}, \begin{bmatrix} -1 \\ 1 \\ 0 \\ -2 \end{bmatrix} \right\rangle$$

It is easy to see that neither of these vectors is a linear combination of the other (scalar multiple), and so we are done. ■

The procedure used in Example 2 is cumbersome. There is a more efficient method of finding redundant vectors. We will describe it for a subspace W spanned by three vectors to keep the notation from getting out of hand; but you should have no difficulty applying these ideas to subspaces spanned by other numbers of vectors.

Suppose $W = \langle \mathbf{w}_1, \mathbf{w}_2, \mathbf{w}_3 \rangle$:

If $\mathbf{w}_1$ is redundant, then $\mathbf{w}_1 = a_2\mathbf{w}_2 + a_3\mathbf{w}_3$, or equivalently,

$$\mathbf{0} = -1\mathbf{w}_1 + a_2\mathbf{w}_2 + a_3\mathbf{w}_3 \qquad \text{for some scalars } a_2, a_3$$

If $\mathbf{w}_2$ is redundant, then $\mathbf{w}_2 = b_1\mathbf{w}_1 + b_3\mathbf{w}_3$, or equivalently,

$$\mathbf{0} = b_1\mathbf{w}_1 + (-1)\mathbf{w}_2 + b_3\mathbf{w}_3 \qquad \text{for some scalars } b_1, b_3$$

If $\mathbf{w}_3$ is redundant, then $\mathbf{w}_3 = c_1\mathbf{w}_1 + c_2\mathbf{w}_2$, or equivalently,

$$\mathbf{0} = c_1\mathbf{w}_1 + c_2\mathbf{w}_2 + (-1)\mathbf{w}_3 \qquad \text{for some scalars } c_1, c_2$$

From this we can conclude that if one of the vectors $\mathbf{w}_1$, $\mathbf{w}_2$, or $\mathbf{w}_3$ is redundant, then there are scalars x_1, x_2, and x_3, not all of which are zero (one is -1), that make the following equation hold:

$$\mathbf{0} = x_1\mathbf{w}_1 + x_2\mathbf{w}_2 + x_3\mathbf{w}_3$$

Thus, instead of trying to solve for each of the $\mathbf{w}_i$'s in terms of the others, we need only solve one vector equation. For if we find a solution to the equation

$$\mathbf{0} = x_1\mathbf{w}_1 + x_2\mathbf{w}_2 + x_3\mathbf{w}_3$$

in which some x_i is not zero, then we can express $\mathbf{w}_i$ as a linear combination of the other vectors, that is, $\mathbf{w}_i$ is redundant.

EXAMPLE 3 Express one of the vectors $\mathbf{w}_1$, $\mathbf{w}_2$, or $\mathbf{w}_3$ as a linear combination of the others, where

$$\mathbf{w}_1 = \begin{bmatrix} 2 \\ -1 \\ 3 \end{bmatrix} \quad \mathbf{w}_2 = \begin{bmatrix} 3 \\ -4 \\ 5 \end{bmatrix} \quad \mathbf{w}_3 = \begin{bmatrix} 1 \\ 2 \\ 1 \end{bmatrix}$$

From the previous discussion we see that we must solve the equation

$$\mathbf{0} = x_1 \begin{bmatrix} 2 \\ -1 \\ 3 \end{bmatrix} + x_2 \begin{bmatrix} 3 \\ -4 \\ 5 \end{bmatrix} + x_3 \begin{bmatrix} 1 \\ 2 \\ 1 \end{bmatrix}$$

which gives the system of equations

$$\begin{aligned} 0 &= 2x_1 + 3x_2 + x_3 \\ 0 &= -x_1 - 4x_2 + 2x_3 \\ 0 &= 3x_1 + 5x_2 + x_3 \end{aligned}$$

Putting this in matrix form and reducing we get (see Section 2.2)

$$\left[\begin{array}{rrr|r} 2 & 3 & 1 & 0 \\ -1 & -4 & 2 & 0 \\ 3 & 5 & 1 & 0 \end{array}\right] \xrightarrow{\text{reduces to}} \left[\begin{array}{rrr|r} 2 & 3 & 1 & 0 \\ 0 & -5 & 5 & 0 \\ 0 & 0 & 0 & 0 \end{array}\right]$$

Solving this equation, we see that $x_2 = x_3$ and $x_1 = -2x_3$, and so

$$\begin{aligned} \mathbf{0} &= x_1\mathbf{w}_1 + x_2\mathbf{w}_2 + x_3\mathbf{w}_3 \\ &= (-2x_3)\mathbf{w}_1 + x_3\mathbf{w}_2 + x_3\mathbf{w}_3 \\ &= x_3(-2\mathbf{w}_1 + \mathbf{w}_2 + \mathbf{w}_3) \qquad \text{for any number } x_3 \end{aligned}$$

If we let $x_3 = 1$, this equation becomes

$$\mathbf{0} = -2\mathbf{w}_1 + \mathbf{w}_2 + \mathbf{w}_3$$

and we can solve this for any one of the vectors $\mathbf{w}_1$, $\mathbf{w}_2$, or $\mathbf{w}_3$ (since none of these vectors has a coefficient of zero in this equation). We solve for $\mathbf{w}_2$ to get

$$\mathbf{w}_2 = 2\mathbf{w}_1 - \mathbf{w}_3 \qquad \blacksquare$$

For larger sets of vectors, a more formal method for eliminating redundant vectors is needed. We now show how to obtain such a method. The problem we want to solve is this: If we are given a set of vectors

$$S = \{\mathbf{w}_1, \mathbf{w}_2, \ldots, \mathbf{w}_n\}$$

we wish to eliminate redundant vectors to obtain a subset B of S such that every vector of S is a linear combination of the vectors of B. In Example 3 we found that the set B could be chosen as $B = \{\mathbf{w}_1, \mathbf{w}_3\}$. As we have shown above, the way to find redundant vectors is to obtain nontrivial solutions to the equation

$$x_1\mathbf{w}_1 + x_2\mathbf{w}_2 + \cdots + x_n\mathbf{w}_n = \mathbf{0}$$

If we let

$$M = [\mathbf{w}_1 \quad \mathbf{w}_2 \quad \cdots \quad \mathbf{w}_n]$$

be the matrix with columns $\mathbf{w}_1, \mathbf{w}_2, \ldots, \mathbf{w}_n$, then this equation can be written (see Section 3.3)

$$M\mathbf{x} = \mathbf{0} \qquad \text{where } \mathbf{x} = \begin{bmatrix} x_1 \\ \vdots \\ x_n \end{bmatrix}$$

To solve this equation, we reduce the augmented matrix

$$\left[\begin{array}{c|c} M & \begin{matrix} 0 \\ \vdots \\ 0 \end{matrix} \end{array}\right] \xrightarrow{\text{reduces to}} \left[\begin{array}{c|c} P & \begin{matrix} 0 \\ \vdots \\ 0 \end{matrix} \end{array}\right]$$

The matrix P has the property that the solution set of the equation $P\mathbf{x} = \mathbf{0}$ is the same as the solution set of the equation $M\mathbf{x} = \mathbf{0}$ (see Section 2.2). Combining these ideas, we see that if we can find scalars $t_1, t_2, \ldots, t_n$ such that

$$t_1\mathbf{p}_1 + t_2\mathbf{p}_2 + \cdots + t_n\mathbf{p}_n = \mathbf{0} \qquad \text{where } \mathbf{p}_1, \mathbf{p}_2, \ldots, \mathbf{p}_n \text{ are the columns of } P$$

then

$$\mathbf{t} = \begin{bmatrix} t_1 \\ \vdots \\ t_n \end{bmatrix}$$

is a solution of the equation $P\mathbf{x} = \mathbf{0}$. Hence, $\mathbf{t}$ is a solution of the equation $M\mathbf{x} = \mathbf{0}$, which means that

$$t_1\mathbf{w}_1 + t_2\mathbf{w}_2 + \cdots + t_n\mathbf{w}_n = \mathbf{0}$$

We recall that $\mathbf{w}_i$ is a redundant vector if and only if $t_i \neq 0$. This means that the vector $\mathbf{w}_i$ is a redundant vector of the set $\{\mathbf{w}_1, \mathbf{w}_2, \ldots, \mathbf{w}_n\}$ if and only if the corresponding column vector $\mathbf{p}_i$ is a redundant vector in the set $\{\mathbf{p}_1, \mathbf{p}_2, \ldots, \mathbf{p}_n\}$. Therefore, we have reduced the problem of finding redundant vectors in the set $\{\mathbf{w}_1, \mathbf{w}_2, \ldots, \mathbf{w}_n\}$ to the problem of finding redundant vectors in the set $\{\mathbf{p}_1, \mathbf{p}_2, \ldots, \mathbf{p}_n\}$. As the following example shows, it is easy to find which of the column vectors of a reduced matrix P are redundant.

EXAMPLE 4 Suppose $W = \langle \mathbf{w}_1, \mathbf{w}_2, \mathbf{w}_3, \mathbf{w}_4 \rangle$, where

$$\mathbf{w}_1 = \begin{bmatrix} 1 \\ 1 \\ -1 \\ 2 \end{bmatrix} \quad \mathbf{w}_2 = \begin{bmatrix} 2 \\ -1 \\ 3 \\ 1 \end{bmatrix} \quad \mathbf{w}_3 = \begin{bmatrix} 3 \\ 0 \\ 2 \\ 3 \end{bmatrix} \quad \mathbf{w}_4 = \begin{bmatrix} 0 \\ -3 \\ 5 \\ -3 \end{bmatrix}$$

To eliminate redundant vectors we begin by using row reduction.

$$\left[\begin{array}{c|c} & 0 \\ M & 0 \\ & 0 \\ & 0 \end{array}\right] = \left[\begin{array}{rrrr|r} 1 & 2 & 3 & 0 & 0 \\ 1 & -1 & 0 & -3 & 0 \\ -1 & 3 & 2 & 5 & 0 \\ 2 & 1 & 3 & -3 & 0 \end{array}\right] \xrightarrow{\text{reduces to}} \left[\begin{array}{rrrr|r} 1 & 2 & 3 & 0 & 0 \\ 0 & 3 & 3 & 3 & 0 \\ 0 & 0 & 0 & 0 & 0 \\ 0 & 0 & 0 & 0 & 0 \end{array}\right] = \left[\begin{array}{c|c} & 0 \\ P & 0 \\ & 0 \\ & 0 \end{array}\right]$$

Looking at the columns of the reduced matrix P, it is easy to see that the third and fourth columns of P are each linear combinations of the first two columns of P. Specifically

$$\begin{bmatrix} 3 \\ 3 \\ 0 \\ 0 \end{bmatrix} = 1\begin{bmatrix} 1 \\ 0 \\ 0 \\ 0 \end{bmatrix} + 1\begin{bmatrix} 2 \\ 3 \\ 0 \\ 0 \end{bmatrix} \quad \text{and} \quad \begin{bmatrix} 0 \\ 3 \\ 0 \\ 0 \end{bmatrix} = -2\begin{bmatrix} 1 \\ 0 \\ 0 \\ 0 \end{bmatrix} + 1\begin{bmatrix} 2 \\ 3 \\ 0 \\ 0 \end{bmatrix}$$

Therefore we see that

$$\mathbf{w}_3 = 1\mathbf{w}_1 + 1\mathbf{w}_2 \quad \text{and} \quad \mathbf{w}_4 = -2\mathbf{w}_1 + 1\mathbf{w}_2$$

Hence $\mathbf{w}_3$ and $\mathbf{w}_4$ are redundant and

$$W = \langle \mathbf{w}_1, \mathbf{w}_2 \rangle \quad \blacksquare$$

Theorem 4.4 summarizes the techniques illustrated by this example.

THEOREM 4.4 Suppose $S = \{\mathbf{w}_1, \mathbf{w}_2, \ldots, \mathbf{w}_n\}$ is a spanning set for the vector space W. To find a subset B of S that also spans W, proceed as follows:

Step 1 Reduce the matrix:

$$[M \mid \mathbf{0}] = \left[\begin{array}{c|c} & 0 \\ \mathbf{w}_1 \mathbf{w}_2 \ldots \mathbf{w}_n & \vdots \\ & 0 \end{array}\right] \xrightarrow{\text{reduces to}} \left[\begin{array}{c|c} & 0 \\ \mathbf{p}_1 \mathbf{p}_2 \cdots \mathbf{p}_n & \vdots \\ & 0 \end{array}\right] = [P \mid \mathbf{0}]$$

Step 2 Eliminate redundant vectors from the set $\{\mathbf{p}_1, \mathbf{p}_2, \ldots, \mathbf{p}_n\}$ to obtain a subset Q of the columns of P such that each $\mathbf{p}_i$ can be expressed as a linear combination of elements of Q.

Step 3 The subset B that spans W is the subset of W that corresponds to Q.

EXAMPLE 5 Use Theorem 4.4 to find a smaller spanning set for W where

$$W = \left\langle \mathbf{w}_1 = \begin{bmatrix} 1 \\ 1 \\ 2 \end{bmatrix}, \quad \mathbf{w}_2 = \begin{bmatrix} -1 \\ 1 \\ 1 \end{bmatrix}, \quad \mathbf{w}_3 = \begin{bmatrix} 0 \\ 1 \\ 1 \end{bmatrix}, \quad \mathbf{w}_4 = \begin{bmatrix} 2 \\ 2 \\ 0 \end{bmatrix}, \quad \mathbf{w}_5 = \begin{bmatrix} 3 \\ 1 \\ 2 \end{bmatrix} \right\rangle$$

Step 1

$$\left[\begin{array}{c|c} & 0 \\ M & 0 \\ & 0 \end{array}\right] = \left[\begin{array}{ccccc|c} 1 & -1 & 0 & 2 & 3 & 0 \\ 1 & 1 & 1 & 2 & 1 & 0 \\ 2 & 1 & 1 & 0 & 2 & 0 \end{array}\right] \xrightarrow{\text{reduces to}} \left[\begin{array}{ccccc|c} 1 & -1 & 0 & 2 & 3 & 0 \\ 0 & -2 & -1 & 0 & 2 & 0 \\ 0 & 0 & -1 & -8 & -2 & 0 \end{array}\right] = \left[\begin{array}{c|c} & 0 \\ P & 0 \\ & 0 \end{array}\right]$$

Step 2 We express each of the last two columns of P as a linear combination of the first three columns of P. Since the matrix P is reduced, this is relatively easy. To explicitly show how this is done, we will express $\begin{bmatrix} 2 \\ 0 \\ -8 \end{bmatrix}$ as a linear combination of the first three columns of P. We write

$$\begin{bmatrix} 2 \\ 0 \\ -8 \end{bmatrix} = t_1 \begin{bmatrix} 1 \\ 0 \\ 0 \end{bmatrix} + t_2 \begin{bmatrix} -1 \\ -2 \\ 0 \end{bmatrix} + t_3 \begin{bmatrix} 0 \\ -1 \\ -1 \end{bmatrix}$$

Looking at the bottom row, we see that

$$-8 = t_1(0) + t_2(0) + t_3(-1)$$

Therefore, $t_3 = 8$. We substitute this value for t_3 in the original equation and examine the second row to obtain

$$0 = t_1(0) + t_2(-2) + 8(-1)$$

From this we see that $t_2 = -4$. We substitute the values for t_2 and t_3 into the original equation and look at the top row to obtain

$$2 = t_1(1) + (-4)(-1) + (8)(0)$$

From this we see that $t_1 = -2$. Hence, we have shown that

$$\begin{bmatrix} 2 \\ 0 \\ -8 \end{bmatrix} = -2 \begin{bmatrix} 1 \\ 0 \\ 0 \end{bmatrix} - 4 \begin{bmatrix} -1 \\ -2 \\ 0 \end{bmatrix} + 8 \begin{bmatrix} 0 \\ -1 \\ -1 \end{bmatrix}$$

In exactly the same way, we can show that

$$\begin{bmatrix} 3 \\ 2 \\ -2 \end{bmatrix} = 1 \begin{bmatrix} 1 \\ 0 \\ 0 \end{bmatrix} - 2 \begin{bmatrix} -1 \\ -2 \\ 0 \end{bmatrix} + 2 \begin{bmatrix} 0 \\ -1 \\ -1 \end{bmatrix}$$

Thus the subset Q is $\{\mathbf{p}_1, \mathbf{p}_2, \mathbf{p}_3\}$.

Step 3 The set B is $B = \{\mathbf{w}_1, \mathbf{w}_2, \mathbf{w}_3\}$. Hence

$$W = \left\langle \mathbf{w}_1 = \begin{bmatrix} 1 \\ 1 \\ 2 \end{bmatrix}, \quad \mathbf{w}_2 = \begin{bmatrix} -1 \\ 1 \\ 1 \end{bmatrix}, \quad \mathbf{w}_3 = \begin{bmatrix} 0 \\ 1 \\ 1 \end{bmatrix} \right\rangle \quad \blacksquare$$

As a consequence of Theorem 4.4 we see that whenever a spanning set contains no redundant vectors, then the only solution to the equation

$$\mathbf{0} = x_1\mathbf{w}_1 + x_2\mathbf{w}_2 + \cdots + x_k\mathbf{w}_k \qquad \text{where the } \mathbf{w}_i \text{ are the vectors of the spanning set and the } x_i \text{ are scalars}$$

is $x_1 = 0, x_2 = 0, \ldots, x_k = 0$. Such a set of vectors is said to be linearly independent.

DEFINITION 4.4

A set $\mathbf{w}_1, \mathbf{w}_2, \ldots, \mathbf{w}_k$ of vectors is said to be **linearly independent** if the only solution to the equation

$$\mathbf{0} = x_1\mathbf{w}_1 + x_2\mathbf{w}_2 + \cdots + x_k\mathbf{w}_k$$

is $x_1 = 0, x_2 = 0, \ldots, x_k = 0$.

DEFINITION 4.5

If W is a vector space and the vectors $\mathbf{w}_1, \mathbf{w}_2, \ldots, \mathbf{w}_k$ span W and are also linearly independent, then the set $\{\mathbf{w}_1, \mathbf{w}_2, \ldots, \mathbf{w}_k\}$ is called a **basis** for W.

(*Note*: Just as there were many spanning sets for a particular vector space, there are also many different bases for any given vector space.)

The concept of basis is of great importance in the study of vector spaces. It is an idea that focuses our attention on the essence of a particular space.

EXAMPLE 6 Find a basis for the vector space $W = \langle \mathbf{w}_1, \mathbf{w}_2, \mathbf{w}_3, \mathbf{w}_4 \rangle$ discussed in Example 5.

In Example 5 we showed (using Theorem 4.4) that

$$W = \langle \mathbf{w}_1, \mathbf{w}_2, \mathbf{w}_3 \rangle$$

We will show that $\{\mathbf{w}_1, \mathbf{w}_2, \mathbf{w}_3\}$ is a basis for W. To do this, we must show that the only solution of the equation

$$x_1\mathbf{w}_1 + x_2\mathbf{w}_2 + x_3\mathbf{w}_3 = \mathbf{0}$$

is $x_1 = x_2 = x_3 = 0$. We begin by using the same matrix reduction we used in Example 5 to obtain

$$\left[\begin{array}{ccc|c} 1 & -1 & 0 & 0 \\ 1 & 1 & 1 & 0 \\ 2 & 1 & 1 & 0 \end{array}\right] \xrightarrow{\text{reduces to}} \left[\begin{array}{ccc|c} 1 & -1 & 0 & 0 \\ 0 & -2 & -1 & 0 \\ 0 & 0 & -1 & 0 \end{array}\right]$$

As before, we see that $x_1\mathbf{w}_1 + x_2\mathbf{w}_2 + x_3\mathbf{w}_3 = \mathbf{0}$ if and only if

$$x_1\begin{bmatrix} 1 \\ 0 \\ 0 \end{bmatrix} + x_2\begin{bmatrix} -1 \\ -2 \\ 0 \end{bmatrix} + x_3\begin{bmatrix} 0 \\ -1 \\ -1 \end{bmatrix} = \begin{bmatrix} 0 \\ 0 \\ 0 \end{bmatrix}$$

From this equation it is obvious that

$$x_1 = x_2 = x_3 = 0$$

Hence $\mathbf{w}_1$, $\mathbf{w}_2$, and $\mathbf{w}_3$ are linearly independent, and $\{\mathbf{w}_1, \mathbf{w}_2, \mathbf{w}_3\}$ is a basis for W. ■

The following theorem summarizes the most important properties of a basis of a vector space.

THEOREM 4.5

If $\{\mathbf{w}_1, \mathbf{w}_2, \ldots, \mathbf{w}_k\}$ is a basis for the vector space W, then

1. Every vector $\mathbf{w}$ of W can be expressed as a linear combination of the vectors $\mathbf{w}_1, \mathbf{w}_2, \ldots, \mathbf{w}_k$, and

2. This expression for $\mathbf{w}$ is unique; that is, if there are two expressions for $\mathbf{w}$

$$\mathbf{w} = a_1\mathbf{w}_1 + a_2\mathbf{w}_2 + \cdots + a_k\mathbf{w}_k$$

and

$$\mathbf{w} = b_1\mathbf{w}_1 + b_2\mathbf{w}_2 + \cdots + b_k\mathbf{w}_k$$

then $a_1 = b_1$, $a_2 = b_2$, $\ldots$, $a_k = b_k$.

Statement 1 just says that a basis is a spanning set, and Statement 2 follows since, if

$$a_1\mathbf{w}_1 + a_2\mathbf{w}_2 + \cdots + a_k\mathbf{w}_k = b_1\mathbf{w}_1 + b_2\mathbf{w}_2 + \cdots + b_k\mathbf{w}_k$$

then

$$\mathbf{0} = (a_1 - b_1)\mathbf{w}_1 + (a_2 - b_2)\mathbf{w}_2 + \cdots + (a_k - b_k)\mathbf{w}_k$$

Since the vectors $\mathbf{w}_1, \mathbf{w}_2, \ldots, \mathbf{w}_k$ are linearly independent, we see that this means that $a_1 - b_1 = 0, a_2 - b_2 = 0, \ldots, a_k - b_k = 0$, or $a_1 = b_1, a_2 = b_2, \ldots, a_k = b_k$.

As we have seen and will continue to see, the concept of basis for a vector space will play a key role in our study of linear algebra. One question that we have not yet answered is whether every vector space (other than $\langle \mathbf{0} \rangle$) has a basis. This is discussed in the following theorem.

THEOREM 4.6 If $V \neq \langle \mathbf{0} \rangle$ is a vector space, then there exists a basis for V.

Exercise Set 4.3

In Exercises 1–5 find a basis for each of the vector spaces listed.

1. $W = \left\langle \begin{bmatrix} 1 \\ 1 \end{bmatrix}, \begin{bmatrix} 0 \\ 2 \end{bmatrix}, \begin{bmatrix} -3 \\ 2 \end{bmatrix} \right\rangle$

2. $W = \left\langle \begin{bmatrix} 1 \\ -1 \\ 2 \end{bmatrix}, \begin{bmatrix} 3 \\ 3 \\ -2 \end{bmatrix}, \begin{bmatrix} 4 \\ 1 \\ 3 \end{bmatrix} \right\rangle$

3. $W = \left\langle \begin{bmatrix} 1 \\ 1 \\ -1 \\ 2 \end{bmatrix}, \begin{bmatrix} 3 \\ 3 \\ -2 \\ 0 \end{bmatrix}, \begin{bmatrix} 1 \\ 1 \\ 0 \\ -4 \end{bmatrix}, \begin{bmatrix} 1 \\ 1 \\ 0 \\ 3 \end{bmatrix} \right\rangle$

4. $W = \left\langle \begin{bmatrix} 1 \\ 2 \\ -3 \\ 2 \end{bmatrix}, \begin{bmatrix} 2 \\ 4 \\ 1 \\ 0 \end{bmatrix}, \begin{bmatrix} 3 \\ 6 \\ -2 \\ 2 \end{bmatrix}, \begin{bmatrix} 5 \\ 10 \\ 0 \\ -2 \end{bmatrix} \right\rangle$

5. $W = \left\langle \begin{bmatrix} 1 \\ 2 \\ -3 \\ 2 \end{bmatrix}, \begin{bmatrix} 2 \\ 4 \\ 1 \\ 0 \end{bmatrix}, \begin{bmatrix} 3 \\ 6 \\ -2 \\ 2 \end{bmatrix}, \begin{bmatrix} 5 \\ 10 \\ 0 \\ -5 \end{bmatrix}, \begin{bmatrix} 1 \\ 1 \\ 0 \\ 1 \end{bmatrix} \right\rangle$

6. For each of the following sets of vectors, determine if the vectors all lie on the same line, all lie in the same plane (containing the origin), or span a three-dimensional space.

a. $\left\{ \begin{bmatrix} -1 \\ 2 \\ 3 \end{bmatrix}, \begin{bmatrix} 3 \\ -2 \\ 5 \end{bmatrix}, \begin{bmatrix} 1 \\ 2 \\ 11 \end{bmatrix} \right\}$

b. $\left\{ \begin{bmatrix} -1 \\ 2 \\ 3 \end{bmatrix}, \begin{bmatrix} 3 \\ -2 \\ 5 \end{bmatrix}, \begin{bmatrix} 1 \\ 2 \\ 10 \end{bmatrix} \right\}$

c. $\left\{ \begin{bmatrix} -1 \\ 2 \\ 5 \end{bmatrix}, \begin{bmatrix} 3 \\ -6 \\ -15 \end{bmatrix}, \begin{bmatrix} 1 \\ -2 \\ -5 \end{bmatrix} \right\}$

d. $\left\{ \begin{bmatrix} -1 \\ 2 \\ 5 \end{bmatrix}, \begin{bmatrix} 3 \\ -6 \\ -15 \end{bmatrix}, \begin{bmatrix} 1 \\ -2 \\ 5 \end{bmatrix} \right\}$

7. Find a basis for each of the following subspaces of $\mathbb{R}^3$.

a. The line through the origin containing the point $(1, -1, 2)$.

b. The plane containing the origin and the points $(2, 0, 3)$ and $(-1, 0, 3)$.

c. $\mathbb{R}^3$ itself.

8. If two vectors are linearly independent, they cannot lie on the same line through the origin. Explain this statement.

9. Suppose that $\mathbf{w}_1, \mathbf{w}_2, \mathbf{w}_3$ are linearly independent vectors of $\mathbb{R}^3$. Show that these three vectors cannot all lie in the same plane containing the origin. (*Hint:* Look at the discussion of subspaces of $\mathbb{R}^3$ in Section 4.2).

10. Show that a basis for the vector space

$$V = \left\langle \begin{bmatrix} 1 \\ 3 \end{bmatrix}, \begin{bmatrix} 2 \\ -4 \end{bmatrix} \right\rangle$$

is

$$\left\{ \begin{bmatrix} 1 \\ 0 \end{bmatrix}, \begin{bmatrix} 0 \\ 1 \end{bmatrix} \right\}$$

★ **11.** Two functions f and g are said to be linearly independent on the interval $[a, b]$ if $c_1 = c_2 = 0$ are the only real numbers such that

$$c_1 f(x) + c_2 g(x) = 0 \qquad \text{for all } x \text{ in } [a, b]$$

a. Show that the functions $f(x) = x^2$ and $g(x) = 1/(x - 2)$ are linearly independent on the interval $[0, 1]$.

b. Show that the functions $f(x) = \sin x$ and $g(x) = \cos x$ are linearly independent on the interval $[0, 2\pi]$.

c. Show that the functions

$$f_1(x) = 1 \qquad f_2(x) = \cos 2x \qquad f_3(x) = \cos^2 x$$

are linearly dependent on the interval $[0, 2\pi]$.

★ **12.** Show that the Lagrange interpolation polynomials (see Exercise 10 in Section 4.2) form a basis for $\mathbb{R}_3[x]$.

13. Let W be a subspace of the vector space V, and let $\{\mathbf{b}_1, \mathbf{b}_2, \ldots, \mathbf{b}_k\}$ be a basis for W. If $\mathbf{b}$ is a vector of V that is not in W, show that

a. The set $\{\mathbf{b}_1, \mathbf{b}_2, \ldots, \mathbf{b}_k; \mathbf{b}\}$ is linearly independent.

b. W is a proper subset of (contained in, but not equal to) the subspace $\langle \mathbf{b}_1, \mathbf{b}_2, \ldots, \mathbf{b}_k; \mathbf{b} \rangle$.

14. Let $\mathbb{R}[x]$ be the set of all polynomials in x with real coefficients. Find a basis for $\mathbb{R}[x]$ that contains infinitely many vectors (polynomials). (See Example 4 in Section 3.2).

15. **a.** Show that every basis for $\mathbb{R}^2$ contains exactly two vectors.

b. Show that every basis for $\mathbb{R}^3$ contains exactly three vectors.

4.4 Dimension

We continue our discussion of the concept of basis for a vector space. For any vector space (other than $\langle \mathbf{0} \rangle$) there can be many different bases. There is one thing that all these bases have in common:

THEOREM 4.7

If V is a vector space, then any two bases for V contain the same number of vectors.

DEFINITION 4.6

The number of vectors in a basis for a vector space V is called the **dimension** of V and is denoted $\mathbf{dim}(V)$.

EXAMPLE 1 $\dim(\mathbb{R}^3) = 3$, since a basis for $\mathbb{R}^3$ is

$$\left\{ \begin{bmatrix} 1 \\ 0 \\ 0 \end{bmatrix}, \begin{bmatrix} 0 \\ 1 \\ 0 \end{bmatrix}, \begin{bmatrix} 0 \\ 0 \\ 1 \end{bmatrix} \right\}$$

which contains three vectors. Moreover, we can see that neither of the sets

$$\left\{\begin{bmatrix} 2 \\ 6 \\ -1 \end{bmatrix}, \begin{bmatrix} 0 \\ 1 \\ 4 \end{bmatrix}\right\} \quad \text{nor} \quad \left\{\begin{bmatrix} -1 \\ 3 \\ 2 \end{bmatrix}, \begin{bmatrix} 2 \\ -4 \\ 5 \end{bmatrix}, \begin{bmatrix} 0 \\ -1 \\ -3 \end{bmatrix}, \begin{bmatrix} 2 \\ 2 \\ -3 \end{bmatrix}\right\}$$

is a basis for $\mathbb{R}^3$ since neither of these sets contains exactly three vectors. ■

In the same way, we can see that the vectors

$$\mathbf{e}_1 = \begin{bmatrix} 1 \\ 0 \\ 0 \\ \vdots \\ 0 \end{bmatrix}, \quad \mathbf{e}_2 = \begin{bmatrix} 0 \\ 1 \\ 0 \\ \vdots \\ 0 \end{bmatrix}, \quad \ldots, \quad \mathbf{e}_n = \begin{bmatrix} 0 \\ 0 \\ \vdots \\ 0 \\ 1 \end{bmatrix}$$

form a basis for $\mathbb{R}^n$ (for any n). This basis is called the **standard basis for** $\mathbb{R}^n$.

Another important theorem concerning dimension is:

THEOREM 4.8

If V is a vector space of dimension n, then any subset of linearly independent vectors of V contains no more than n vectors. In particular, if W is a subspace of V, then $\dim(W)$ is less than or equal to n. Moreover, if W is a subspace of V and $\dim(V) = \dim(W)$, then $V = W$.

The discussion of the proofs of Theorems 4.7 and 4.8 will be postponed until the end of this section. First we will show how these ideas can be applied.

Using the concept of dimension we can complete our discussion of the types of subspaces that a vector space can have. As we saw in Section 4.2, there are three types of subspaces of $\mathbb{R}^2$—the subspace $\langle \mathbf{0} \rangle$, the lines containing the origin, and $\mathbb{R}^2$ itself. These types of subspaces are completely determined by their dimension—the subspace $\langle \mathbf{0} \rangle$ has dimension 0 (no basis vectors), the lines containing the origin are subspaces of dimension 1, and $\mathbb{R}^2$ has dimension 2. In $\mathbb{R}^3$ the same holds true; the type of a subspace is determined by its dimension. The subspace $\langle \mathbf{0} \rangle$ has dimension 0, the lines through the origin have dimension 1, the planes containing the origin have dimension 2, and $\mathbb{R}^3$ has dimension 3. Compare this method with the lengthy discussion of the subspaces of $\mathbb{R}^3$ in Section 4.2. Similarly, we use dimension to classify the different types of subspaces of $\mathbb{R}^n$ ($n > 3$). In these higher dimensional vector spaces, we cannot actually see these subspaces as we can in $\mathbb{R}^2$ and $\mathbb{R}^3$, but we can get a feeling for what they are like by comparing them to the lower dimensional vector spaces we can see.

EXAMPLE 2 Find all types of subspaces of $\mathbb{R}^5$.

If W is a subspace of $\mathbb{R}^5$, then $\dim(W) \leq 5$ (Theorem 4.8). So the types of subspaces of $\mathbb{R}^5$ are:

Dimension 0—the subspace $\langle \mathbf{0} \rangle$

Dimension 1—lines through the origin

Dimension 2—planes containing the origin

Dimension 3—copies of $\mathbb{R}^3$ which contain the origin

Dimension 4—copies of $\mathbb{R}^4$ which contain the origin

Dimension 5—$\mathbb{R}^5$ itself ■

A very useful application of Theorem 4.7 is the following theorem.

THEOREM 4.9

> Let V be a vector space of dimension n.
> **1.** If $\{\mathbf{v}_1, \mathbf{v}_2, \ldots, \mathbf{v}_n\}$ is a set of n linearly independent vectors of V, then these vectors span V, and $\{\mathbf{v}_1, \mathbf{v}_2, \ldots, \mathbf{v}_n\}$ is a basis for V.
> **2.** If $\{\mathbf{v}_1, \mathbf{v}_2, \ldots, \mathbf{v}_n\}$ is a set of n vectors spanning V, then these vectors are linearly independent and $\{\mathbf{v}_1, \mathbf{v}_2, \ldots, \mathbf{v}_n\}$ is a basis for V.

For a discussion of the proof of this theorem, see Exercises 2 and 3 of this section.

Theorem 4.9 is useful in the following situation. Suppose you have n vectors of a vector space of dimension n, and you want to determine if these vectors are a basis for this vector space. What Theorem 4.9 says is that it is not necessary to prove that the vectors of this set are both linearly independent and also span V; checking either one of these properties is sufficient.

EXAMPLE 3 Show that

$$\left\{\mathbf{v}_1 = \begin{bmatrix} 1 \\ -3 \end{bmatrix}, \quad \mathbf{v}_2 = \begin{bmatrix} \sqrt{2} \\ 4 \end{bmatrix}\right\}$$

is a basis for $\mathbb{R}^2$.

Since $\mathbf{v}_1$ is not a scalar multiple of $\mathbf{v}_2$, these two vectors are linearly independent. So, by Statement 1 of Theorem 4.9, we see that $\{\mathbf{v}_1, \mathbf{v}_2\}$ is a basis for $\mathbb{R}^2$. ■

* The remainder of this section involves proofs of Theorems 4.7 and 4.8. This discussion is more abstract than most of the other material contained in this book, but it is optional since no other topics in this book will be based on it. However, it contains some fundamental ideas of linear algebra that may be of interest. We begin by discussing the following lemma.

LEMMA 4.1 Let V be a vector space that has a basis containing n vectors. Then any subset of V containing more than n vectors is linearly dependent.

To see why this lemma is true, we look at the case $n = 3$.

Let V be a vector space with basis $\{\mathbf{v}_1, \mathbf{v}_2, \mathbf{v}_3\}$, and let $\mathbf{w}_1, \mathbf{w}_2, \ldots, \mathbf{w}_k$ be k linearly independent vectors of V. We will show that $k \leq 3$. We know that V contains sets containing one, two, or three linearly independent vectors (for example, the sets $\{\mathbf{v}_1\}$, $\{\mathbf{v}_1, \mathbf{v}_2\}$, and $\{\mathbf{v}_1, \mathbf{v}_2, \mathbf{v}_3\}$). So we will assume that $k \geq 3$ and then show that k can be no larger than 3. The method of proof involves replacing each of the vectors $\mathbf{v}_1$, $\mathbf{v}_2$, and $\mathbf{v}_3$ by the vectors $\mathbf{w}_1$, $\mathbf{w}_2$, and $\mathbf{w}_3$. This replacement is done one vector at a time. After each replacement, we will show that the new set still spans V. After the third replacement, we will show that V is spanned by $\mathbf{w}_1$, $\mathbf{w}_2$, and $\mathbf{w}_3$, so that if there were a $\mathbf{w}_4$, it would have to be a linear combination of the vectors $\mathbf{w}_1$, $\mathbf{w}_2$, and $\mathbf{w}_3$; but this can't happen, since the $\mathbf{w}_i$ are linearly independent. So there can be at most three $\mathbf{w}_i$.

We begin the replacement procedure:

Step 1 Since $\{\mathbf{v}_1, \mathbf{v}_2, \mathbf{v}_3\}$ is a basis for V, the vectors $\mathbf{v}_1$, $\mathbf{v}_2$, and $\mathbf{v}_3$ span V and $\mathbf{w}_1$ is a linear combination of these three vectors. Therefore, we can find scalars a_1, a_2, and a_3 such that

$$\mathbf{w}_1 = a_1\mathbf{v}_1 + a_2\mathbf{v}_2 + a_3\mathbf{v}_3$$

Since $\mathbf{w}_1 \neq \mathbf{0}$, at least one of the a_i is also not 0. For simplicity, let us assume that $a_1 \neq 0$. By this assumption, we can solve the above equation for $\mathbf{v}_1$ to get

$$\mathbf{v}_1 = -\frac{a_2}{a_1}\mathbf{v}_2 - \frac{a_3}{a_1}\mathbf{v}_3 + \frac{1}{a_1}\mathbf{w}_1$$

So $\mathbf{v}_1$ is a linear combination of $\mathbf{v}_2$, $\mathbf{v}_3$, and $\mathbf{w}_1$. Therefore, the space spanned by $\mathbf{v}_1$, $\mathbf{v}_2$, and $\mathbf{v}_3$ is also spanned by $\mathbf{v}_2$, $\mathbf{v}_3$, and $\mathbf{w}_1$. Thus $V = \langle \mathbf{v}_2, \mathbf{v}_3, \mathbf{w}_1 \rangle$.

Step 2 Since $V = \langle \mathbf{v}_2, \mathbf{v}_3, \mathbf{w}_1 \rangle$, and $\mathbf{w}_2$ is in V, $\mathbf{w}_2$ is a linear combination of $\mathbf{v}_2$, $\mathbf{v}_3$, and $\mathbf{w}_1$. Therefore, we can find scalars b_1, b_2, and b_3 so that

$$\mathbf{w}_2 = b_2\mathbf{v}_2 + b_3\mathbf{v}_3 + b_1\mathbf{w}_1$$

Since $\mathbf{w}_1$ and $\mathbf{w}_2$ are linearly independent, at least one of the scalars b_2 or b_3 is not zero (otherwise $\mathbf{w}_2$ would be a scalar multiple of $\mathbf{w}_1$). Let us assume that $b_2 \neq 0$. We can solve the above equation for b_2 to get

$$\mathbf{v}_2 = -\frac{b_3}{b_2}\mathbf{v}_3 - \frac{b_1}{b_2}\mathbf{w}_1 + \frac{1}{b_2}\mathbf{w}_2$$

So we conclude that V is spanned by $\mathbf{v}_3$, $\mathbf{w}_1$, $\mathbf{w}_2$.

Step 3 Since $V = \langle \mathbf{v}_3, \mathbf{w}_1, \mathbf{w}_2 \rangle$, and $\mathbf{w}_3$ is in V, we can find scalars c_1, c_2, and c_3 so that

$$\mathbf{w}_3 = c_3\mathbf{v}_3 + c_1\mathbf{w}_1 + c_2\mathbf{w}_2$$

Now since $\mathbf{w}_1$, $\mathbf{w}_2$, and $\mathbf{w}_3$ are linearly independent, $\mathbf{w}_3$ is not a linear combination of $\mathbf{w}_2$ and $\mathbf{w}_3$ alone. Therefore, c_3 is not zero. Thus, we can solve the above equation for $\mathbf{v}_3$ to get

$$\mathbf{v}_3 = -\frac{c_1}{c_3}\mathbf{w}_1 - \frac{c_2}{c_3}\mathbf{w}_2 + \frac{1}{c_3}\mathbf{w}_3$$

So we conclude that $V = \langle \mathbf{w}_1, \mathbf{w}_2, \mathbf{w}_3 \rangle$.

Therefore there can be no other $\mathbf{w}_i$. If there were another $\mathbf{w}_i$, it would be a linear combination of $\mathbf{w}_1$, $\mathbf{w}_2$, $\mathbf{w}_3$ (since the $\mathbf{w}_i$ lies in V). This cannot happen, since the $\mathbf{w}_i$'s were assumed to be linearly independent. Thus we have shown that $k \leq 3$, completing the proof for this case. The general result is proved the same way.

The proof of Theorem 4.7 follows easily from Lemma 4.1. Suppose a vector space V has two different bases $\{\mathbf{v}_1, \mathbf{v}_2, \ldots, \mathbf{v}_n\}$ and $\{\mathbf{w}_1, \mathbf{w}_2, \ldots, \mathbf{w}_k\}$. Since the set $\{\mathbf{v}_1, \mathbf{v}_2, \ldots, \mathbf{v}_n\}$ is a basis for V and $\{\mathbf{w}_1, \mathbf{w}_2, \ldots, \mathbf{w}_k\}$ is a linearly independent set of vectors (since it forms a basis for V), we can apply Lemma 4.1 to obtain $k \leq n$. On the other hand, using the fact that $\{\mathbf{w}_1, \mathbf{w}_2, \ldots, \mathbf{w}_k\}$ is a basis for V and that $\{\mathbf{v}_1, \mathbf{v}_2, \ldots, \mathbf{v}_n\}$ is a linearly independent set of vectors of V, we get $n \leq k$. Hence, since $n \geq k$ and $n \leq k$, we have $n = k$.

Theorem 4.8 follows directly from Lemma 4.1 and Theorem 4.7.

Exercise Set 4.4

1. Use Theorem 4.9 to help determine which of the following sets are bases for $\mathbb{R}^3$:

a. $\left\{\begin{bmatrix}1\\0\\-1\end{bmatrix}, \begin{bmatrix}2\\0\\-3\end{bmatrix}, \begin{bmatrix}0\\2\\4\end{bmatrix}\right\}$

b. $\left\{\begin{bmatrix}1\\1\\1\end{bmatrix}, \begin{bmatrix}2\\2\\2\end{bmatrix}, \begin{bmatrix}1\\-1\\2\end{bmatrix}\right\}$

c. $\left\{\begin{bmatrix}1\\2\\-4\end{bmatrix}, \begin{bmatrix}-3\\2\\4\end{bmatrix}, \begin{bmatrix}-3\\10\\4\end{bmatrix}, \begin{bmatrix}0\\0\\0\end{bmatrix}\right\}$

d. $\left\{\begin{bmatrix}1\\1\\1\end{bmatrix}, \begin{bmatrix}3\\-1\\1\end{bmatrix}, \begin{bmatrix}4\\1\\5\end{bmatrix}, \begin{bmatrix}-2\\3\\1\end{bmatrix}\right\}$

e. $\left\{\begin{bmatrix}1\\0\\1\end{bmatrix}, \begin{bmatrix}0\\1\\0\end{bmatrix}\right\}$

★**2. a.** Let $\mathbf{v}_1$ and $\mathbf{v}_2$ be linearly independent vectors of $\mathbb{R}^2$. Use Lemma 4.1 to show that if $\mathbf{v}$ is any vector of $\mathbb{R}^2$, then the set $\{\mathbf{v}_1, \mathbf{v}_2, \mathbf{v}\}$ is linearly dependent. From this show that $\mathbf{v}_1$ and $\mathbf{v}_2$ span $\mathbb{R}^2$, and hence that $\{\mathbf{v}_1, \mathbf{v}_2\}$ is a basis for $\mathbb{R}^2$.

b. Let $\mathbf{v}_1$, $\mathbf{v}_2$, and $\mathbf{v}_3$ be linearly independent vectors of $\mathbb{R}^3$. Using Lemma 4.1, show that $\{\mathbf{v}_1, \mathbf{v}_2, \mathbf{v}_3\}$ is a basis for $\mathbb{R}^3$.

★**3.** Suppose $\mathbf{v}_1$, $\mathbf{v}_2$, and $\mathbf{v}_3$ span $\mathbb{R}^3$. Using

Theorem 4.7, show that $\{\mathbf{v}_1, \mathbf{v}_2, \mathbf{v}_3\}$ is a basis for $\mathbb{R}^3$. (*Hint:* If $\mathbf{v}_1$, $\mathbf{v}_2$, and $\mathbf{v}_3$ are not linearly independent, what does this say about the dimension of the space they span?)

4. List the types of subspaces of $\mathbb{R}^4$.

5. Let

$$A = \begin{bmatrix} 1 & 2 & -1 \\ 0 & 1 & 3 \\ -1 & 0 & 1 \end{bmatrix} \quad \text{and} \quad \mathbf{v}_0 = \begin{bmatrix} 1 \\ 0 \\ 0 \end{bmatrix}$$

a. Compute $\mathbf{v}_1 = A\mathbf{v}_0$, $\mathbf{v}_2 = A\mathbf{v}_1$, $\mathbf{v}_3 = A\mathbf{v}_2$.

b. Is $\{\mathbf{v}_0, \mathbf{v}_1, \mathbf{v}_2\}$ a basis for $\mathbb{R}^3$?

c. Express $\mathbf{v}_3$ as a linear combination of $\mathbf{v}_0, \mathbf{v}_1, \mathbf{v}_2$.

6. Let $T: \mathbb{R}^3 \to \mathbb{R}^3$ be a linear transformation, and $\{\mathbf{b}_1, \mathbf{b}_2, \mathbf{b}_3\}$ be a basis for $\mathbb{R}^3$. Is it always true that $\{T(\mathbf{b}_1), T(\mathbf{b}_2), T(\mathbf{b}_3)\}$ is also a basis for $\mathbb{R}^3$?

7. Let $T: \mathbb{R}^3 \to \mathbb{R}^3$ be a linear transformation and $\{\mathbf{v}_1, \mathbf{v}_2, \mathbf{v}_3\}$ be three vectors of $\mathbb{R}^3$. If $\{T(\mathbf{v}_1), T(\mathbf{v}_2), T(\mathbf{v}_3)\}$ is a basis for $\mathbb{R}^3$, show that $\{\mathbf{v}_1, \mathbf{v}_2, \mathbf{v}_3\}$ must have been a basis for $\mathbb{R}^3$.

8. Let W_1 and W_2 be 3-dimensional subspaces of $\mathbb{R}^5$. Show that there is a nontrivial vector $\mathbf{v}$ that lies in both W_1 and W_2. (*Hint:* Pick a basis for each of the subspaces and find a dependency relation between the six vectors of these two bases. Use this dependency relationship to find a vector that is in both of the subspaces.)

★**9.** Let $W_1 = \langle \mathbf{w}_1, \mathbf{u}_1 \rangle$ and $W_2 = \langle \mathbf{w}_2, \mathbf{u}_2 \rangle$ be two-dimensional subspaces (planes) in $\mathbb{R}^4$ that intersect in a line. Show that the subspace $W = \langle \mathbf{w}_1, \mathbf{u}_1, \mathbf{w}_2, \mathbf{u}_2 \rangle$ is of dimension 3. (*Hint:* Let $\mathbf{w}$ be a nonzero vector on the line of intersection of W_1 and W_2. Find a basis for W that contains $\mathbf{w}$.)

★**10.** Let $W_1 = \langle \mathbf{w}_1, \mathbf{u}_1 \rangle$ and $W_2 = \langle \mathbf{w}_2, \mathbf{u}_2 \rangle$ be two-dimensional subspaces of $\mathbb{R}^4$. Show that if $W = \langle \mathbf{w}_1, \mathbf{u}_1, \mathbf{w}_2, \mathbf{u}_2 \rangle$, then

$$\dim(W) = \dim(W_1) + \dim(W_2) - \dim(\text{intersection of } W_1 \text{ and } W_2)$$

(*Hint:* Use Exercise 9).

11. Suppose $\{\mathbf{u}, \mathbf{v}, \mathbf{w}\}$ is a basis for the vector space V. Determine which of the following sets are also bases for V.

a. $\{\mathbf{u}, \mathbf{u} + \mathbf{v}, \mathbf{u} + \mathbf{v} + \mathbf{w}\}$

b. $\{\mathbf{u} + \mathbf{v}, \mathbf{v} + \mathbf{w}, \mathbf{w} + \mathbf{u}\}$

c. $\{\mathbf{u} - \mathbf{v}, \mathbf{u} + \mathbf{w}, \mathbf{v} + \mathbf{w}, \mathbf{u} + 2\mathbf{v}\}$

d. $\{\mathbf{u} + \mathbf{v} + \mathbf{w}, \mathbf{u} - \mathbf{v} - \mathbf{w}\}$

e. $\{2\mathbf{u} - \mathbf{v} + \mathbf{w}, \mathbf{u} - 3\mathbf{v} + 2\mathbf{w}, 3\mathbf{u} - 2\mathbf{v} + \mathbf{w}\}$

Review Exercises

1. Define each of the following:

a. Subspace of $\mathbb{R}^n$

b. Linear independence

c. Basis

d. Dimension

e. Spanning set

f. Coset

2. a. Reduce the matrix

$$\begin{bmatrix} 1 & 2 & 1 & 0 \\ 1 & 3 & 2 & 1 \\ 0 & 1 & 2 & 2 \end{bmatrix}$$

b. Does

$$\left\langle \begin{bmatrix} 1 \\ 1 \\ 0 \end{bmatrix}, \begin{bmatrix} 2 \\ 3 \\ 1 \end{bmatrix} \right\rangle = \left\langle \begin{bmatrix} 1 \\ 2 \\ 2 \end{bmatrix}, \begin{bmatrix} 0 \\ 1 \\ 2 \end{bmatrix} \right\rangle ?$$

Why or why not?

3. **a.** Find the vector equation of the line determined by the points (1, 3, 2) and (−2, −3, 4).

b. Express the equation

$$\mathbf{v}(t) = t\begin{bmatrix} 1 \\ -2 \end{bmatrix} + \begin{bmatrix} 3 \\ 4 \end{bmatrix}$$

in the form $y = mx + b$.

4. Which of the following sets span $\mathbb{R}^3$; which are linearly independent; and which are bases for $\mathbb{R}^3$?

a. $\left\{\begin{bmatrix} 1 \\ 0 \\ 1 \end{bmatrix}, \begin{bmatrix} 2 \\ 0 \\ 3 \end{bmatrix}, \begin{bmatrix} -4 \\ 0 \\ 5 \end{bmatrix}, \begin{bmatrix} 17 \\ 0 \\ -2 \end{bmatrix}\right\}$

b. $\left\{\begin{bmatrix} 1 \\ 0 \\ 1 \end{bmatrix}, \begin{bmatrix} 17 \\ 0 \\ -2 \end{bmatrix}\right\}$

c. $\left\{\begin{bmatrix} 1 \\ 0 \\ 1 \end{bmatrix}, \begin{bmatrix} 17 \\ 0 \\ -2 \end{bmatrix}, \begin{bmatrix} 16 \\ 1 \\ -3 \end{bmatrix}\right\}$

d. $\left\{\begin{bmatrix} 1 \\ 0 \\ 2 \end{bmatrix}, \begin{bmatrix} 0 \\ 1 \\ 2 \end{bmatrix}, \begin{bmatrix} 2 \\ -1 \\ 2 \end{bmatrix}\right\}$

5. Find a basis for each of the subspaces listed in Exercise 4.

6. Graph the following equations:

a. $\mathbf{v}(t) = t\begin{bmatrix} -1 \\ 2 \end{bmatrix} + \begin{bmatrix} 3 \\ 1 \end{bmatrix}$

b. $\mathbf{v}(t_1,t_2) = t_1\begin{bmatrix} 1 \\ -1 \\ 1 \end{bmatrix} + t_2\begin{bmatrix} 1 \\ 0 \\ 0 \end{bmatrix} + \begin{bmatrix} 2 \\ 0 \\ 0 \end{bmatrix}$

7. Let W be the subspace of $\mathbb{R}^3$ with spanning set

$$\left\{\begin{bmatrix} 1 \\ -1 \\ 2 \end{bmatrix}, \begin{bmatrix} -2 \\ 2 \\ -6 \end{bmatrix}\right\}$$

a. Graph W.

b. Graph the coset $W + \begin{bmatrix} 0 \\ 0 \\ 3 \end{bmatrix}$

8. Consider the following pair of lines in $\mathbb{R}^3$:

$$\mathbf{v}_1(t) = t\begin{bmatrix} 2 \\ 1 \\ 1 \end{bmatrix} \quad \text{and} \quad \mathbf{v}_2(s) = s\begin{bmatrix} -1 \\ -1 \\ 2 \end{bmatrix} + \begin{bmatrix} 1 \\ 0 \\ 0 \end{bmatrix}$$

a. Do these lines have a point in common?

b. Are these lines parallel?

c. Are these the same line?

9. Do Exercise 8 with the lines

$$\mathbf{v}_1(t) = t\begin{bmatrix} 2 \\ 1 \\ 1 \end{bmatrix} + \begin{bmatrix} 1 \\ 3 \\ 2 \end{bmatrix}$$

and

$$\mathbf{v}_2(s) = s\begin{bmatrix} -4 \\ -2 \\ -2 \end{bmatrix} + \begin{bmatrix} 1 \\ 3 \\ 2 \end{bmatrix}$$

10. Do Exercise 8 with the lines

$$\mathbf{v}_1(t) = t\begin{bmatrix} 2 \\ 1 \\ 1 \end{bmatrix} + \begin{bmatrix} 1 \\ 3 \\ 2 \end{bmatrix}$$

and

$$\mathbf{v}_2(s) = s\begin{bmatrix} -4 \\ -2 \\ -2 \end{bmatrix} + \begin{bmatrix} 2 \\ 1 \\ 1 \end{bmatrix}$$

5 Subspaces Related to Linear Transformations

To this point we have studied vector spaces and linear transformations separately. In this chapter we begin to see how linear transformations and the vector spaces on which they operate are interrelated. The important concepts of null space and image space are introduced, and the fundamental relationship between the dimensions of these two subspaces and the dimension of the domain of a linear transformation is discussed.

5.1 Image Space, Null Space, and Systems of Equations

When solving systems of linear equations (see Section 2.2), students generally are puzzled by three situations:

1. Certain systems of equations have no solution.
2. Certain systems of equations have more than one solution.
3. The same matrix can be the coefficient matrix for a system that has no solutions as well as a system having more than one solution.

In this section we will explain these mysteries. We will begin by examining systems that have the coefficient matrix

$$M = \begin{bmatrix} 1 & 2 & 0 \\ -1 & 1 & -3 \\ 3 & 4 & 2 \end{bmatrix}$$ (This is the coefficient matrix of Examples 3 and 4 in Section 2.2.)

In Example 7 of Section 3.3 we saw that finding a solution to a system of equations such as

$$\begin{aligned} 1x_1 + 2x_2 + 0x_3 &= a_1 \\ -1x_1 + 1x_2 - 3x_3 &= a_2 \\ 3x_1 + 4x_2 + 2x_3 &= a_3 \end{aligned}$$

can be viewed in terms of matrices as trying to find all vectors $\mathbf{v}$ satisfying the equation

$$M\mathbf{v} = \mathbf{w} \qquad \text{where } M = \begin{bmatrix} 1 & 2 & 0 \\ -1 & 1 & -3 \\ 3 & 4 & 2 \end{bmatrix} \qquad \text{and } \mathbf{w} = \begin{bmatrix} a_1 \\ a_2 \\ a_3 \end{bmatrix}$$

Instead of attacking this problem head on as we did in Section 2.2, we will study it from a more sophisticated point of view by using many of the ideas we have developed in Chapters 3 and 4. We shift our focus from looking at the solutions of individual equations to considering the collection of all the vectors that are the images under M of all the vectors in the domain of M. If $\mathbf{w}$ is in this collection, then

$$\mathbf{w} = M\mathbf{v} = \begin{bmatrix} 1 & 2 & 0 \\ -1 & 1 & -3 \\ 3 & 4 & 2 \end{bmatrix}\begin{bmatrix} x_1 \\ x_2 \\ x_3 \end{bmatrix} = x_1\begin{bmatrix} 1 \\ -1 \\ 3 \end{bmatrix} + x_2\begin{bmatrix} 2 \\ 1 \\ 4 \end{bmatrix} + x_3\begin{bmatrix} 0 \\ -3 \\ 2 \end{bmatrix}$$

Thus we see that the vectors of this collection are linear combinations of the vectors that form the columns of M. Moreover, any vector $\mathbf{w}$ that is a linear combination of the columns of M is the image of a vector $\mathbf{v}$. (The coefficients of $\mathbf{v}$ are the coefficients of the column vectors in the expression for $\mathbf{v}$ as a linear combination of the columns of M.)

Let us call the set of all the linear combinations of the column vectors of M the set I. From Theorem 4.3 we see that I is a subspace. We now find a basis for I. To do this we proceed as in Section 4.3.

We can easily see that a basis for I is the set

$$\left\{\begin{bmatrix} 1 \\ -1 \\ 3 \end{bmatrix}, \begin{bmatrix} 2 \\ 1 \\ 4 \end{bmatrix}\right\}$$

Therefore, I is a two-dimensional subspace (a plane) in $\mathbb{R}^3$ (see Figure 5.1). What this means is that if $\mathbf{w}$ is a vector in the plane I, then $M\mathbf{v} = \mathbf{w}$ has a

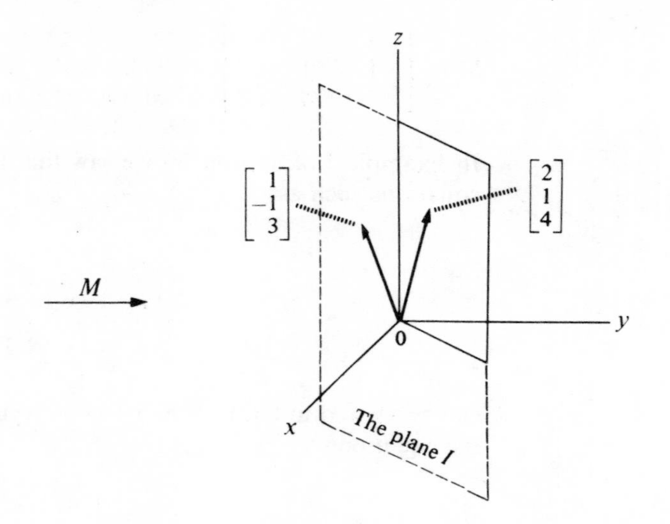

Figure 5.1

solution. On the other hand, if the vector **u** of $\mathbb{R}^3$ is not in I, then $M\mathbf{v} = \mathbf{u}$ can have no solutions because all vectors of the form $M\mathbf{v}$ must be elements of the subspace I.

EXAMPLE 1 The vector

$$\mathbf{w} = 3\begin{bmatrix} 1 \\ -1 \\ 3 \end{bmatrix} + 2\begin{bmatrix} 2 \\ 1 \\ 4 \end{bmatrix} = \begin{bmatrix} 7 \\ -1 \\ 17 \end{bmatrix}$$

is a linear combination of the basis vectors for I. Furthermore,

$$\mathbf{w} = 3\begin{bmatrix} 1 \\ -1 \\ 3 \end{bmatrix} + 2\begin{bmatrix} 2 \\ 1 \\ 4 \end{bmatrix} + 0\begin{bmatrix} 0 \\ -3 \\ 2 \end{bmatrix} = \begin{bmatrix} 1 & 2 & 0 \\ -1 & 1 & -3 \\ 3 & 4 & 2 \end{bmatrix}\begin{bmatrix} 3 \\ 2 \\ 0 \end{bmatrix}$$

So the vector

$$\begin{bmatrix} 3 \\ 2 \\ 0 \end{bmatrix}$$

is a solution of the equation

$$M\mathbf{v} = \begin{bmatrix} 1 & 2 & 0 \\ -1 & 1 & -3 \\ 3 & 4 & 2 \end{bmatrix}\mathbf{v} = \begin{bmatrix} 7 \\ -1 \\ 17 \end{bmatrix} = \mathbf{w} \quad \blacksquare$$

DEFINITION 5.1 Let M be a matrix. The subspace I spanned by the column vectors of M is called the **image space** of M.

So we can solve the first puzzle. The equation $M\mathbf{v} = \mathbf{w}$ has a solution if and only if $\mathbf{w}$ is in the image space of M.

Now we consider the second puzzle—when can a system have multiple solutions? We begin with the special case of equations of the form

$$M\mathbf{v} = \mathbf{0}$$

In our example this is

$$\begin{bmatrix} 1 & 2 & 0 \\ -1 & 1 & -3 \\ 3 & 4 & 2 \end{bmatrix} \begin{bmatrix} x_1 \\ x_2 \\ x_3 \end{bmatrix} = \begin{bmatrix} 0 \\ 0 \\ 0 \end{bmatrix}$$

Let N be the set of all vectors satisfying this equation. As we saw in Example 3 in Section 2.2, N is the set of all multiples of the vector

$$\begin{bmatrix} -2 \\ 1 \\ 1 \end{bmatrix}$$

Thus N is a one-dimensional subspace (a line) of $\mathbb{R}^3$.

Now we study the general case, to see when the equation $M\mathbf{v} = \mathbf{w}$ can have multiple solutions. We note that the equation $M\mathbf{v} = \mathbf{0}$ is special in the sense that we know that there is always a solution to such an equation, namely $\mathbf{v} = \mathbf{0}$. If $\mathbf{w}$ is a vector in the image space I, then $M\mathbf{v} = \mathbf{w}$ has a solution. Let $\mathbf{v}_0$ be such a solution. We want to examine the relationship among the various solutions of this equation. Suppose $\mathbf{v}$ is another solution. Then

$$M\mathbf{v} = \mathbf{w} = M\mathbf{v}_0$$

or

$$M\mathbf{v} - M\mathbf{v}_0 = \mathbf{0}$$

This is equivalent to the equation

$$M(\mathbf{v} - \mathbf{v}_0) = \mathbf{0}$$

since M represents a linear transformation. From the last equation we see that $\mathbf{v} - \mathbf{v}_0$ is in the set N. Let $\mathbf{v} - \mathbf{v}_0 = \mathbf{n}$. Then we see that

$$\mathbf{v} = \mathbf{v}_0 + \mathbf{n}$$

Moreover, whenever $\mathbf{m}$ is any vector of N we have

$$M(\mathbf{v}_0 + \mathbf{m}) = M\mathbf{v}_0 + M\mathbf{m} = \mathbf{w} + \mathbf{0} = \mathbf{w}$$

So the set of all vectors $\mathbf{v}$ for which $M\mathbf{v} = \mathbf{w}$ is the set of all vectors of the form $\mathbf{v}_0 + \mathbf{m}$ where $\mathbf{m}$ is some vector of N—that is, $\mathbf{v}$ is in the coset $\mathbf{v}_0 + N$ (see Figure 5.2).

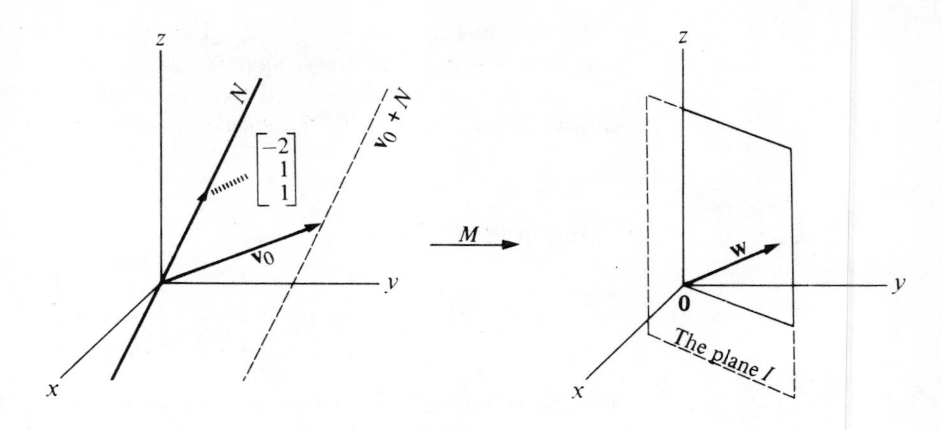

Figure 5.2

DEFINITION 5.2 Let M be a matrix. The set N of all vectors $\mathbf{n}$ such that $M\mathbf{n} = \mathbf{0}$ is called the **null space** or **kernel** of M. (*Note:* As the term indicates, the null space of a matrix is a subspace of the domain of the linear transformation represented by the matrix. See Exercise 14 at the end of this section.)

Summarizing our findings, we have:

THEOREM 5.1 If M is a matrix, then

1. The equation $M\mathbf{v} = \mathbf{w}$ has a solution if and only if $\mathbf{w}$ is in the image space I of M (I is the space spanned by the columns of M).

2. If the equation $M\mathbf{v} = \mathbf{w}$ has a solution $\mathbf{v} = \mathbf{v}_0$, then all other solutions of this equation are the vectors of the coset $\mathbf{v}_0 + N$, where N is the null space of M (N is the set of all vectors $\mathbf{n}$ such that $M\mathbf{n} = \mathbf{0}$). Geometrically, this says that the set of vector solutions to an equation of the form $M\mathbf{v} = \mathbf{w}$ is a set of vectors (coset) parallel to N.

EXAMPLE 2 We consider the matrix

$$M = \begin{bmatrix} -1 & 2 \\ 2 & -4 \end{bmatrix}$$

a. We find a basis for I, the image space of M. I is spanned by the two vectors

$$\begin{bmatrix} -1 \\ 2 \end{bmatrix} \quad \text{and} \quad \begin{bmatrix} 2 \\ -4 \end{bmatrix}$$

Since

$$\begin{bmatrix} 2 \\ -4 \end{bmatrix} = -2\begin{bmatrix} -1 \\ 2 \end{bmatrix}$$

a basis for I is $\left\{\begin{bmatrix} -1 \\ 2 \end{bmatrix}\right\}$, and the image space I is the line containing $\begin{bmatrix} -1 \\ 2 \end{bmatrix}$.

b. We find a basis for N, the null space of M. To solve the equation

$$\mathbf{0} = M\mathbf{n} = \begin{bmatrix} -1 & 2 \\ 2 & -4 \end{bmatrix}\begin{bmatrix} x_1 \\ x_2 \end{bmatrix}$$

we reduce the matrix

$$\left[\begin{array}{rr|r} -1 & 2 & 0 \\ 2 & -4 & 0 \end{array}\right]$$

to get

$$\left[\begin{array}{rr|r} -1 & 2 & 0 \\ 0 & 0 & 0 \end{array}\right]$$

From this we see (using the methods of Section 2.2) that N is the set of all vectors of the form

$$x_2\begin{bmatrix} 2 \\ 1 \end{bmatrix}$$

Hence, N is the line containing the vector $\begin{bmatrix} 2 \\ 1 \end{bmatrix}$, and a basis for N is $\left\{\begin{bmatrix} 2 \\ 1 \end{bmatrix}\right\}$ (see Figure 5.3).

Figure 5.3

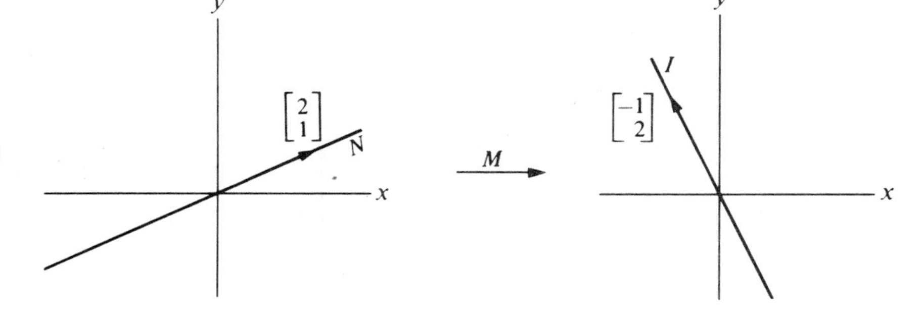

c. We find all solutions to the equation $M\mathbf{v} = \begin{bmatrix} -3 \\ 6 \end{bmatrix}$. The matrix

$$\left[\begin{array}{rr|r} -1 & 2 & -3 \\ 2 & -4 & 6 \end{array}\right] \xrightarrow{\text{reduces to}} \left[\begin{array}{rr|r} -1 & 2 & -3 \\ 0 & 0 & 0 \end{array}\right]$$

So $-x_1 + 2x_2 = -3$, giving the solutions (see Figure 5.4)

$$\begin{bmatrix} 3 + 2x_2 \\ x_2 \end{bmatrix} = \begin{bmatrix} 3 \\ 0 \end{bmatrix} + x_2 \begin{bmatrix} 2 \\ 1 \end{bmatrix} = \begin{bmatrix} 3 \\ 0 \end{bmatrix} + N$$

Figure 5.4

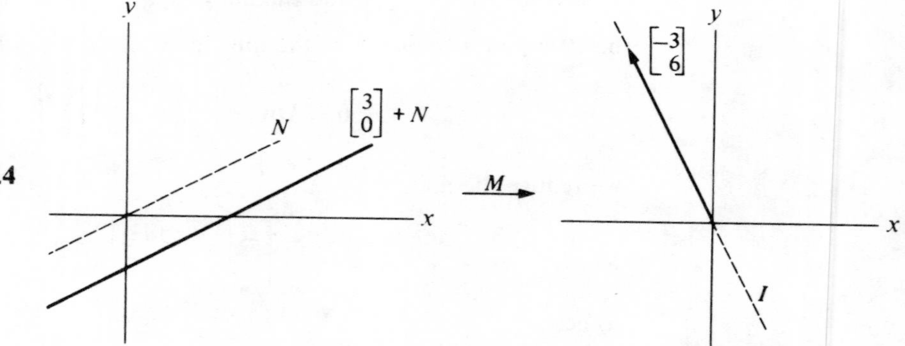

This illustrates the fact that each of the cosets (in this case, lines parallel to N) is mapped by M onto a single vector of the image space of M. ■

EXAMPLE 3 Let

$$M = \begin{bmatrix} 1 & 2 & 3 & 1 \\ -1 & 1 & 0 & 1 \\ 2 & 1 & 1 & -1 \end{bmatrix}$$

a. We see that $M: \mathbb{R}^4 \to \mathbb{R}^3$. To find a basis for the image space of M, we reduce the matrix

$$\left[\begin{array}{cccc|c} 1 & 2 & 3 & 1 & 0 \\ -1 & 1 & 0 & 1 & 0 \\ 2 & 1 & 1 & -1 & 0 \end{array}\right]$$

to get

$$\left[\begin{array}{cccc|c} 1 & 2 & 3 & 1 & 0 \\ 0 & 3 & 3 & 2 & 0 \\ 0 & 0 & -2 & -1 & 0 \end{array}\right]$$

from which we see that

$$\begin{bmatrix} 1 \\ 2 \\ -1 \end{bmatrix} = -\frac{5}{6}\begin{bmatrix} 1 \\ 0 \\ 0 \end{bmatrix} + \frac{1}{6}\begin{bmatrix} 2 \\ 3 \\ 0 \end{bmatrix} + \frac{1}{2}\begin{bmatrix} 3 \\ 3 \\ -2 \end{bmatrix}$$

Therefore (see Section 4.3),

$$\begin{bmatrix} 1 \\ 1 \\ -1 \end{bmatrix} = -\frac{5}{6}\begin{bmatrix} 1 \\ -1 \\ 2 \end{bmatrix} + \frac{1}{6}\begin{bmatrix} 2 \\ 1 \\ 1 \end{bmatrix} + \frac{1}{2}\begin{bmatrix} 3 \\ 0 \\ 1 \end{bmatrix}$$

Moreover, since the first three columns of the reduced matrix are linearly independent, the first three columns of M are linearly independent. Hence a basis for I (the image space of M) is

$$\left\{ \begin{bmatrix} 1 \\ -1 \\ 2 \end{bmatrix}, \begin{bmatrix} 2 \\ 1 \\ 1 \end{bmatrix}, \begin{bmatrix} 3 \\ 0 \\ 1 \end{bmatrix} \right\}$$

Since I is a three-dimensional subspace of $\mathbb{R}^3$, $I = \mathbb{R}^3$.

b. The null space N of M is found by solving the equation $M\mathbf{n} = \mathbf{0}$. This requires reducing the same matrix as in part **a** and leads to the solutions

$$x_4 = x_4 \qquad x_3 = -\frac{1}{2}x_4 \qquad x_2 = -\frac{1}{6}x_4 \qquad x_1 = \frac{5}{6}x_4$$

So the null space of M is all vectors of the form

$$\begin{bmatrix} (5/6)x_4 \\ -(1/6)x_4 \\ -(1/2)x_4 \\ x_4 \end{bmatrix} = \frac{x_4}{6}\begin{bmatrix} 5 \\ -1 \\ -3 \\ 6 \end{bmatrix}$$

which is the one-dimensional subspace of $\mathbb{R}^4$ with basis

$$\left\{ \begin{bmatrix} 5 \\ -1 \\ -3 \\ 6 \end{bmatrix} \right\}$$

c. Now we find all solutions $\mathbf{v} = \begin{bmatrix} x_1 \\ x_2 \\ x_3 \\ x_4 \end{bmatrix}$ to the equation

$$M\mathbf{v} = \begin{bmatrix} 1 & 2 & 3 & 1 \\ -1 & 1 & 0 & 1 \\ 2 & 1 & 1 & -1 \end{bmatrix}\begin{bmatrix} x_1 \\ x_2 \\ x_3 \\ x_4 \end{bmatrix} = \begin{bmatrix} 6 \\ 1 \\ 2 \end{bmatrix}$$

To solve the equation $M\mathbf{v} = \begin{bmatrix} 6 \\ 1 \\ 2 \end{bmatrix}$ we could forge ahead and reduce this system, but this is the difficult way, since we have already calculated the null space of M. Our theory tells us that if we can find a particular solution

$\mathbf{v}_0$ to this equation, then all the other solutions are just the elements of the coset $\mathbf{v}_0 + N$. Moreover, we know that the first three columns of M are a basis for I, so that if

$$\begin{bmatrix} 6 \\ 1 \\ 2 \end{bmatrix}$$

is in I, it is a linear combination of the first three column vectors of M. To find $\mathbf{v}_0$, we need only reduce the matrix

$$\left[\begin{array}{rrr|r} 1 & 2 & 3 & 6 \\ -1 & 1 & 0 & 1 \\ 2 & 1 & 1 & 2 \end{array}\right]$$

This matrix reduces to

$$\left[\begin{array}{rrr|r} 1 & 2 & 3 & 6 \\ 0 & 3 & 3 & 7 \\ 0 & 0 & -2 & -3 \end{array}\right]$$

which gives the solution

$$x_1 = -\frac{1}{6} \qquad x_2 = \frac{5}{6}, \qquad x_3 = \frac{3}{2}$$

Thus

$$\begin{bmatrix} 6 \\ 1 \\ 2 \end{bmatrix} = -\frac{1}{6}\begin{bmatrix} 1 \\ -1 \\ 2 \end{bmatrix} + \frac{5}{6}\begin{bmatrix} 2 \\ 1 \\ 1 \end{bmatrix} + \frac{3}{2}\begin{bmatrix} 3 \\ 0 \\ 1 \end{bmatrix}$$

and we have

$$\begin{aligned} \begin{bmatrix} 6 \\ 1 \\ 2 \end{bmatrix} &= -\frac{1}{6}\begin{bmatrix} 1 \\ -1 \\ 2 \end{bmatrix} + \frac{5}{6}\begin{bmatrix} 2 \\ 1 \\ 1 \end{bmatrix} + \frac{3}{2}\begin{bmatrix} 3 \\ 0 \\ 1 \end{bmatrix} + 0\begin{bmatrix} 1 \\ 1 \\ -1 \end{bmatrix} \\ &= \begin{bmatrix} 1 & 2 & 3 & 1 \\ -1 & 1 & 0 & 1 \\ 2 & 1 & 1 & -1 \end{bmatrix}\begin{bmatrix} -(1/6) \\ (5/6) \\ (3/2) \\ 0 \end{bmatrix} \end{aligned}$$

So the set of all solutions of this equation is the coset $\mathbf{v}_0 + N$ that consists of all vectors of the form

$$\begin{bmatrix} -(1/6) \\ (5/6) \\ (3/2) \\ 0 \end{bmatrix} + x\begin{bmatrix} 5 \\ -1 \\ -3 \\ 6 \end{bmatrix} \qquad \text{where } x \text{ is an arbitrary scalar}$$

■

Exercise Set 5.1

For each of the systems of equations listed in Exercises 1–4, do the following:

a. Write the systems as a matrix equation in the form $M\mathbf{v} = \mathbf{w}$.

b. Find the null space and image space of M.

c. Solve the system of equations.

d. Draw graphs of the domain and range of M (see Example 2), showing the null space of M, the image space of M, the vector $\mathbf{w}$, and the set of all solutions $\mathbf{v}$ of each system.

1. $\begin{aligned} 2x - y &= -1 \\ 3x + 2y &= 9 \end{aligned}$ **2.** $\begin{aligned} 3x + 2y &= -4 \\ -6x - 4y &= 8 \end{aligned}$

3. $\begin{aligned} x + 2y + 3z &= 11 \\ -x + 3y + 2z &= 14 \end{aligned}$ **4.** $\begin{aligned} x + y &= 3 \\ x - y &= 1 \\ -x + 2y &= 1 \end{aligned}$

For each of the matrices in Exercises 5–9, find a basis for the null space and a basis for the image space.

5. $M = \begin{bmatrix} 1 & -1 \\ 2 & -2 \end{bmatrix}$ **6.** $M = \begin{bmatrix} 1 & 3 \\ 2 & -1 \end{bmatrix}$

7. $M = \begin{bmatrix} 1 & 2 & -1 \\ 2 & -1 & 3 \\ 5 & -1 & 7 \end{bmatrix}$

8. $M = \begin{bmatrix} 1 & 2 & -3 & 4 \\ -3 & 1 & 2 & -1 \\ -7 & 0 & 7 & -6 \end{bmatrix}$

9. $M = \begin{bmatrix} 1 & 2 \\ -2 & 3 \\ 3 & 1 \\ 4 & -2 \end{bmatrix}$

10. For each of the matrices M given in Exercises 5–9 above, do the following:

a. Find a vector $\mathbf{w}_1$ such that the equation $M\mathbf{v} = \mathbf{w}_1$ has no solution, or explain why there is no such vector.

b. Find a vector $\mathbf{w}_2$ such that the equation $M\mathbf{v} = \mathbf{w}_2$ has more than one solution, or explain why there is no such vector $\mathbf{w}_2$.

11. Find a basis for the null space and the image space of each of the following matrices:

a. $\begin{bmatrix} 1 & 1 & -2 & 5 & 0 & -3 \\ 3 & 3 & -6 & 2 & 1 & 1 \\ 2 & 2 & 4 & 3 & -2 & 3 \end{bmatrix}$

b. $\begin{bmatrix} -1 & 3 & -5 \\ 1 & 0 & 4 \\ 3 & 1 & -2 \\ -9 & 3 & 1 \\ 0 & 1 & -4 \\ 1 & 1 & 1 \end{bmatrix}$

12. If the equation $M\mathbf{v} = \mathbf{w}_0$ has a unique solution $\mathbf{v} = \mathbf{v}_0$, then show that there is no equation of the form $M\mathbf{v} = \mathbf{w}_1$ (same matrix M) having more than one solution.

13. Let $M: \mathbb{R}^2 \to \mathbb{R}^3$ be a matrix. Show that there is a vector $\mathbf{w}_0$ of $\mathbb{R}^3$ such that the equation $M\mathbf{v} = \mathbf{w}_0$ has no solution.

14. Let M be a matrix. Prove the following statements:

a. If $\mathbf{v}_1$ and $\mathbf{v}_2$ are vectors in the domain of M so that $M\mathbf{v}_1 = M\mathbf{v}_2 = \mathbf{0}$, then $M(\mathbf{v}_1 + \mathbf{v}_2) = \mathbf{0}$.

b. If $M\mathbf{v} = \mathbf{0}$, then $M(c\mathbf{v}) = \mathbf{0}$ for any scalar c.

c. The null space of M is a subspace of the domain of M.

5.2 Rank and Nullity

In Section 5.1 the concepts of image space and null space were defined for matrices. Since every matrix represents a linear transformation, we can give analogous definitions of these concepts for linear transformations.

DEFINITION 5.3

Let $T: V \to W$ be a linear transformation.

1. The **null space** (or **kernel**) of T is the subspace of V consisting of all vectors $\mathbf{n}$ such that $T(\mathbf{n}) = \mathbf{0}$.

2. The **image space** of T is the subspace of W consisting of all vectors $\mathbf{w}$ for which there is a vector $\mathbf{v}$ of V such that $T(\mathbf{v}) = \mathbf{w}$.

(*Note:* If M is the matrix associated with T, then (a) The null space of M = the null space of T and (b) The image space of M = the image space of T.)

The next two examples show how the null space and image space of a linear transformation can be found using Definition 5.3. In these examples, we discuss two important types of linear transformations, projections and reflections. These will be discussed in detail in Chapter 6. For now, we will only need the intuitive definitions of the projection of a vector $\mathbf{v}$ on a line l, and the reflection of a vector $\mathbf{v}$ about a line l. The **projection** of a vector $\mathbf{v}$ onto a line l is the vector that is the shadow of $\mathbf{v}$ on l when a light is placed above (or below) the line so that its rays are perpendicular to l. The **reflection** of $\mathbf{v}$ about the line l is the vector that would be under $\mathbf{v}$ if the paper were folded along the line l. Both of these situations are shown in Figure 5.5.

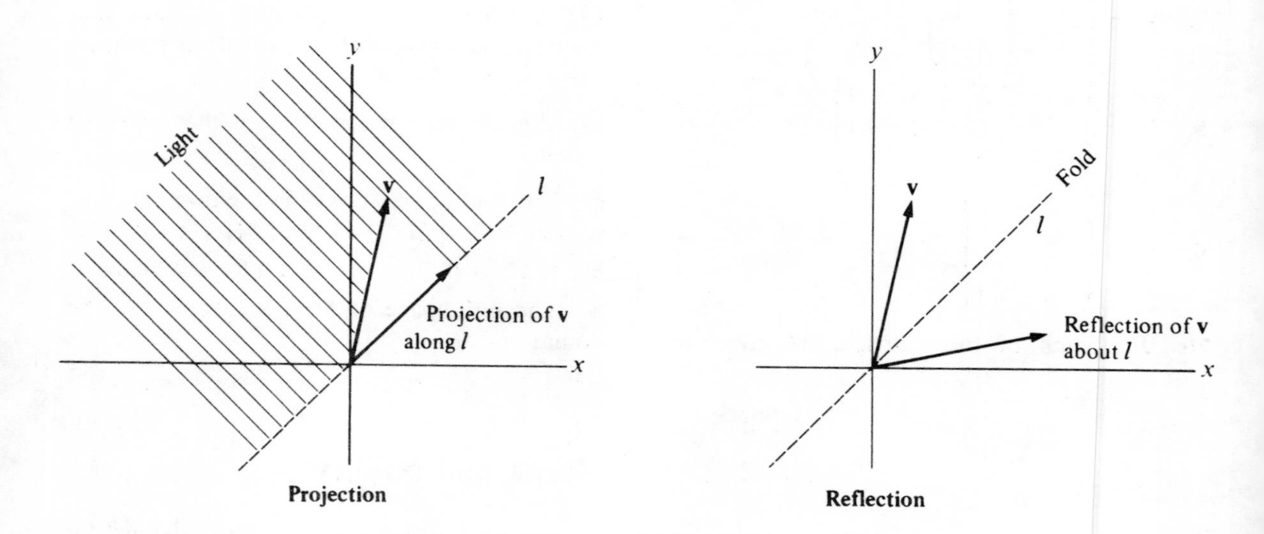

Figure 5.5

Figure 5.6

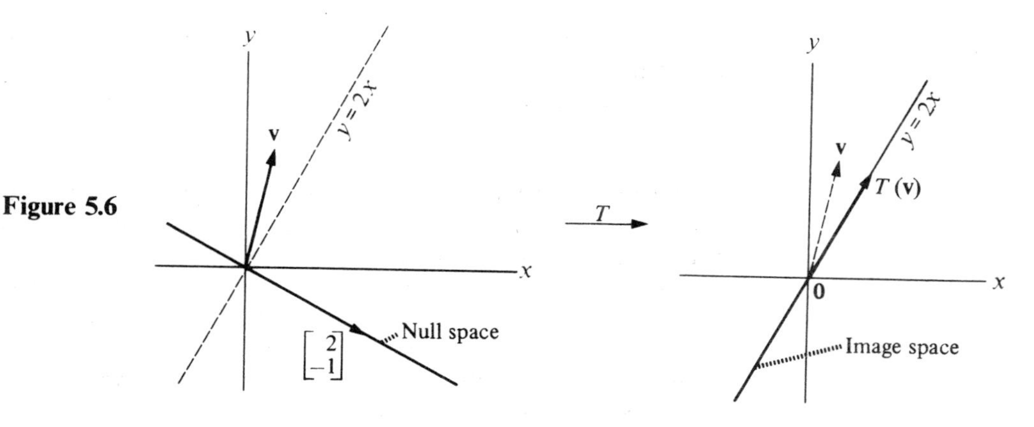

EXAMPLE 1 Let T: $\mathbb{R}^2 \to \mathbb{R}^2$ be projection onto the line $y = 2x$. We find the null space and image space of T. Refer to Figure 5.6.

To find the null space of T, we ask which vectors are projected onto $\mathbf{0}$. These are all vectors lying on the line through the origin perpendicular to the line $y = 2x$. All such vectors are of the form $\begin{bmatrix} 2x \\ -x \end{bmatrix}$. Thus a basis for the null space is $\left\{\begin{bmatrix} 2 \\ -1 \end{bmatrix}\right\}$.

Now we find the image space of T. Since T takes each vector of $\mathbb{R}^2$ onto the line $y = 2x$, we see that the image space consists of all vectors lying along this line—all vectors of the form $\left\{\begin{bmatrix} x \\ 2x \end{bmatrix}\right\}$. So a basis of the image space is $\left\{\begin{bmatrix} 1 \\ 2 \end{bmatrix}\right\}$. ■

EXAMPLE 2 Let T: $\mathbb{R}^2 \to \mathbb{R}^2$ be reflection about the line $y = 2x$. We find the null space and image space of T. Refer to Figure 5.7.

Figure 5.7

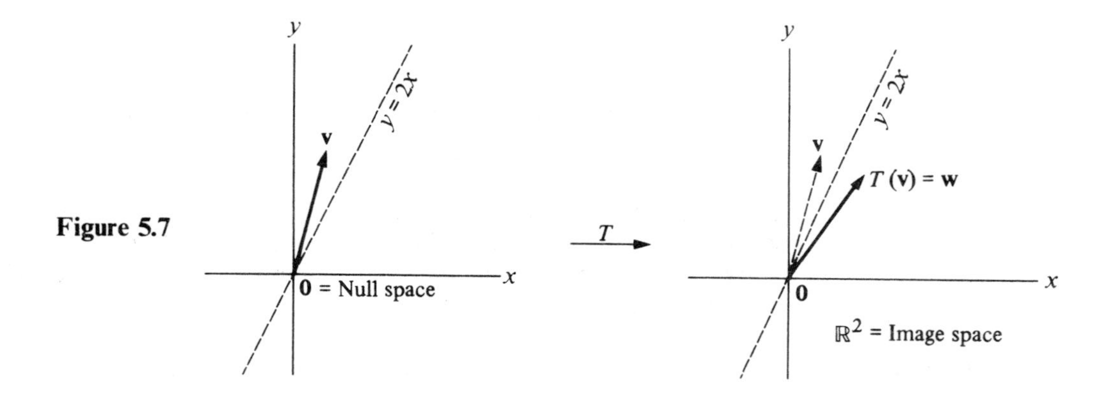

The null space of T is $\langle \mathbf{0} \rangle$, since $\mathbf{0}$ is the only vector whose reflection about the line $y = 2x$ is $\mathbf{0}$.

The image space of T is all of $\mathbb{R}^2$, since for any vector $\mathbf{w}$ of $\mathbb{R}^2$, there is a vector $\mathbf{v}$ of V that is reflected onto $\mathbf{w}$. ■

We now begin a study of the relationships between null spaces and image spaces. In many instances a great deal of information about a linear transformation or matrix can be obtained if one knows the dimension of its null space and image space. Up until now, however, the only way that we could find the dimensions for these spaces was to actually find bases for each of these spaces and then see how many elements each basis contained. In the remainder of this section and in Section 5.3 we will discuss various relationships between the dimensions of these spaces. This will simplify the process of finding them as well as give a deeper understanding of the relationships between linear transformations and the vector spaces on which they operate.

DEFINITION 5.4

1. The dimension of the null space of a linear transformation (or matrix) is called the **nullity** of the linear transformation (or matrix).

2. The dimension of the image space of a linear transformation (or matrix) is called the **rank** of the linear transformation (or matrix).

The following fundamental theorem gives the relationship between rank and nullity.

THEOREM 5.2

If T is a linear transformation, then

$$\text{Dimension of the domain of } T = \text{rank of } T + \text{nullity of } T$$

If M is a matrix, then

$$\text{Number of columns of } M = \text{rank of } M + \text{nullity of } M$$

(*Note:* Since the number of columns of a matrix is the dimension of its domain, the two parts of this theorem say the same thing.)

We will postpone proving this theorem until Section 5.4. To conclude this section we cite several examples.

EXAMPLE 3 Let $T: \mathbb{R}^2 \to \mathbb{R}^2$ be projection onto the line $y = 2x$. The domain of T is $\mathbb{R}^2$

and has dimension 2. As we saw in Example 1, a basis of the null space of T is

$$\left\{\begin{bmatrix} 2 \\ -1 \end{bmatrix}\right\}$$

and so the nullity of T is 1.

A basis of the image space of T is

$$\left\{\begin{bmatrix} 1 \\ 2 \end{bmatrix}\right\}$$

so the rank of T is 1.

These dimensions agree with the formula in Theorem 5.2 since

$$\begin{aligned} \text{Dimension of domain} &= 2 \\ &= 1 + 1 \\ &= \text{rank of } T + \text{nullity of } T \quad \blacksquare \end{aligned}$$

EXAMPLE 4 In Example 3 of Section 5.1 we studied the matrix

$$M = \begin{bmatrix} 1 & 2 & 3 & 1 \\ -1 & 1 & 0 & 1 \\ 2 & 1 & 1 & -1 \end{bmatrix}$$

We now determine the rank and nullity of M.

A basis of the image space of M is

$$\left\{\begin{bmatrix} 1 \\ -1 \\ 2 \end{bmatrix}, \begin{bmatrix} 2 \\ 1 \\ 1 \end{bmatrix}, \begin{bmatrix} 3 \\ 0 \\ 1 \end{bmatrix}\right\}$$

and so the rank of M is 3. The nullity of M is 1 since the null space has basis

$$\left\{\begin{bmatrix} 5 \\ -1 \\ -3 \\ 6 \end{bmatrix}\right\}$$

We see that Theorem 5.2 holds for this case:

$$\begin{aligned} \text{Rank of } M + \text{nullity of } M &= 3 + 1 \\ &= 4 \\ &= \text{number of columns of } M \quad \blacksquare \end{aligned}$$

EXAMPLE 5 Find the null space and image space of the matrix

$$M = \begin{bmatrix} 1 & -1 & 3 & 2 & 1 \\ 2 & -2 & -4 & 1 & 2 \\ 4 & -4 & 2 & 5 & 4 \end{bmatrix}$$

The first step is matrix reduction.

$$\left[\begin{array}{ccccc|c} 1 & -1 & 3 & 2 & 1 & 0 \\ 2 & -2 & -4 & 1 & 2 & 0 \\ 4 & -4 & 2 & 5 & 4 & 0 \end{array}\right] \xrightarrow[\text{to}]{\text{reduces}} \left[\begin{array}{ccccc|c} 1 & -1 & 3 & 2 & 1 & 0 \\ 0 & 0 & 10 & 3 & 0 & 0 \\ 0 & 0 & 0 & 0 & 0 & 0 \end{array}\right]$$

To find the null space of M, we must solve the equations

$$10x_3 + 3x_4 = 0$$

and
$$x_1 - x_2 + 3x_3 + 2x_4 + x_5 = 0$$

The solutions are
$$x_3 = -(3/10)x_4$$

and
$$x_1 = x_2 - (11/10)x_4 - x_5$$

Therefore the vectors of the null space are all of the form

$$\begin{bmatrix} x_2 - (11/10)x_4 - x_5 \\ x_2 \\ -(3/10)x_4 \\ x_4 \\ x_5 \end{bmatrix} = x_2 \begin{bmatrix} 1 \\ 1 \\ 0 \\ 0 \\ 0 \end{bmatrix} + x_4 \begin{bmatrix} -11/10 \\ 0 \\ -3/10 \\ 1 \\ 0 \end{bmatrix} + x_5 \begin{bmatrix} -1 \\ 0 \\ 0 \\ 0 \\ 1 \end{bmatrix}$$

So a basis for the null space is

$$\left\{ \begin{bmatrix} 1 \\ 1 \\ 0 \\ 0 \\ 0 \end{bmatrix}, \begin{bmatrix} -11 \\ 0 \\ -3 \\ 10 \\ 0 \end{bmatrix}, \begin{bmatrix} -1 \\ 0 \\ 0 \\ 0 \\ 1 \end{bmatrix} \right\}$$

Thus the nullity of M is 3. Since M has 5 columns, using Theorem 5.2, we see that the rank of M is $5 - 3 = 2$. Thus, a basis for the image space of M contains 2 vectors. Since the image space is spanned by the columns of M (not the reduced matrix!), we need only choose two independent columns of M to get a basis for the image space. So a basis for the image space is

$$\left\{ \begin{bmatrix} 1 \\ 2 \\ 4 \end{bmatrix}, \begin{bmatrix} 3 \\ -4 \\ 2 \end{bmatrix} \right\} \blacksquare$$

Figure 5.8 may help you remember Theorem 5.2.

Figure 5.8

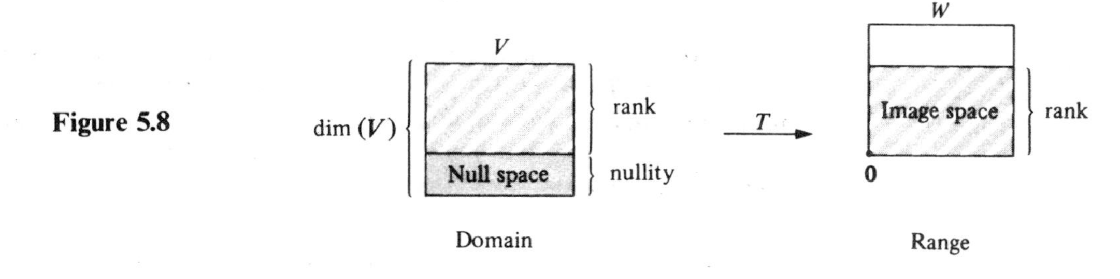

Theorem 5.2 is not all that can be said about the relationship between the domain and the image space of a linear transformation. In fact, in more advanced courses in linear algebra it is shown that the cosets of the null space of a linear transformation themselves form a vector space, and this vector space of cosets is, in many ways, the same as the image space of the transformation. Theorem 5.2 is just the statement that the image space and the vector space of cosets of the null space have the same dimension.

Exercise Set 5.2

For each of the linear transformations in Exercises 1–5, find a basis for the image space and a basis for the null space.

1. $T\colon \mathbb{R}^2 \to \mathbb{R}^2$ is a counterclockwise rotation of 45°.

2. $T\colon \mathbb{R}^2 \to \mathbb{R}^2$ is projection onto the line $y = -3x$.

3. $T\colon \mathbb{R}^3 \to \mathbb{R}^3$

$$\mathbf{v} \to -2\mathbf{v}$$

4. $T\colon \mathbb{R}^3 \to \mathbb{R}^3$ is projection in the direction of the vector $\begin{bmatrix} -1 \\ 0 \\ 0 \end{bmatrix}$.

5. $T\colon \mathbb{R}^3 \to \mathbb{R}^3$ is reflection about the line containing the vector $\begin{bmatrix} -1 \\ 0 \\ 0 \end{bmatrix}$.

6. Verify Theorem 5.2 for each of the linear transformations in Exercises 1–5.

7. Verify Theorem 5.2 for each of the matrices of Exercises 5–9 in Section 5.1.

8. Find the null space and the image space of the matrix

$$M = \begin{bmatrix} 1 & -1 & 2 & 3 \\ 3 & 2 & 1 & 1 \\ 1 & 4 & -3 & -5 \end{bmatrix}$$

See Example 5.

★ **9.** Let $\mathbb{R}[x]$ be the set of all polynomials with real coefficients and D be the differential operator, that is,

$$D\colon \mathbb{R}[x] \to \mathbb{R}[x]$$
$$p(x) \to p'(x)$$

Find the null space of D.

★ **10.** Let $T\colon \mathbb{R}[x] \to \mathbb{R}[x]$ be the function defined by

$$T\colon p(x) \to p(0)$$

a. Show that T is a linear transformation.

b. Find the null space of T.

5.3 Complete Reduction of Matrices: Rank and Nullity

Our goal in this section is to use matrix reduction to get a very useful refinement of Theorem 5.2. We will see how to determine the rank and nullity of a matrix without constructing a basis for either the null space or the image space. We begin our discussion by looking at systems of equations and matrix reduction.

DEFINITION 5.5 A system of linear equations in which the constant term of each equation is 0 is called a **homogeneous** system of linear equations.

As we saw in Section 5.1, to find the null space and image space of a matrix M, it is necessary to solve the homogeneous system of linear equations having M as coefficient matrix. For example, in Example 3 in Section 5.1, to find the null space and image space of the matrix

$$M = \begin{bmatrix} 1 & 2 & 3 & 1 \\ -1 & 1 & 0 & 1 \\ 2 & 1 & 1 & -1 \end{bmatrix}$$

we had to reduce the augmented matrix

$$\left[\begin{array}{rrrr|r} 1 & 2 & 3 & 1 & 0 \\ -1 & 1 & 0 & 1 & 0 \\ 2 & 1 & 1 & -1 & 0 \end{array}\right]$$

During the reduction procedure the column of 0's to the right of the vertical line is unchanged. From now on, for convenience, we will omit the column of 0's when solving systems of homogeneous equations. We will write

$$\begin{bmatrix} 1 & 2 & 3 & 1 \\ -1 & 1 & 0 & 1 \\ 2 & 1 & 1 & -1 \end{bmatrix} \xrightarrow{\text{reduces to}} \begin{bmatrix} 1 & 2 & 3 & 1 \\ 0 & 3 & 3 & 2 \\ 0 & 0 & -2 & -1 \end{bmatrix}$$

It is very important to remember that this is shorthand notation for reduction of augmented matrices where the column to the right of the vertical line consists entirely of 0's.

DEFINITION 5.6 If P is a matrix that can be obtained from the matrix M by a series of applications of the Matrix Reduction Rule (see Section 2.2), then P is said to be **equivalent** to M.

(*Cautionary note:* The word *equivalent* is used in a very technical sense—it does not mean *same*. For this reason we will not use this term very often in this text. Definition 5.6 is given so that if you encounter this term in some other book, you will know what it means).

The following theorem shows what is the same about equivalent matrices. It is just a restatement, in our new terminology, of the fact that the set of solutions of a system of equations is the same as the set of solutions of any system of equations obtained from it using row reduction (see Sections 2.1 and 2.2).

THEOREM 5.3

Let M be a matrix and P be a matrix obtained from M by matrix reduction (P is equivalent to M). Then the null space of P and the null space of M are the same. (*Note:* The image spaces of M and P are generally different.)

Now we consider matrix reduction. Until now, in our applications of matrix reduction it has only been necessary to proceed until one of the unknowns could be found. There are occasions when a more complete reduction is necessary.

DEFINITION 5.7

A matrix P is said to be **completely reduced** if the following conditions on the rows and columns of P hold:

1. *Rows:* P can have only two types of rows—rows in which the first nonzero entry is 1 and rows in which every entry is 0.

2. *Columns:* For row i, the column containing the first nonzero entry of that row (which is 1) is $\mathbf{e}_i$, where

$$\mathbf{e}_1 = \begin{bmatrix} 1 \\ 0 \\ 0 \\ \vdots \\ 0 \\ \vdots \\ 0 \end{bmatrix}, \mathbf{e}_2 = \begin{bmatrix} 0 \\ 1 \\ 0 \\ \vdots \\ 0 \\ \vdots \\ 0 \end{bmatrix}, \ldots, \mathbf{e}_k = \begin{bmatrix} 0 \\ \vdots \\ 0 \\ 1 \\ 0 \\ \vdots \\ 0 \end{bmatrix} \text{ ......} k\text{th coordinate}$$

Moreover, the columns $\mathbf{e}_i$ appear in the order listed.

Note: There may be other columns between the $\mathbf{e}_i$'s.

EXAMPLE 1 The following matrices are completely reduced:

$$\begin{bmatrix} 1 & 0 & 0 \\ 0 & 1 & 0 \\ 0 & 0 & 1 \end{bmatrix}, \quad \begin{bmatrix} 1 & 0 & 1 \\ 0 & 1 & 0 \\ 0 & 0 & 0 \end{bmatrix}, \quad \begin{bmatrix} 1 & 1 \\ 0 & 0 \end{bmatrix}, \quad \begin{bmatrix} 1 & 3 & 0 \\ 0 & 0 & 1 \\ 0 & 0 & 0 \end{bmatrix} \quad \blacksquare$$

EXAMPLE 2 The following matrices are not completely reduced:

$$\begin{bmatrix} 1 & 0 & 0 \\ 0 & 1 & 0 \\ 0 & 1 & 0 \end{bmatrix}, \quad \begin{bmatrix} 1 & 1 & 1 \\ 0 & 0 & 0 \\ 0 & 0 & 1 \end{bmatrix}, \quad \begin{bmatrix} 1 & 0 & 0 \\ 0 & 0 & 0 \\ 0 & 1 & 0 \end{bmatrix} \quad \blacksquare$$

THEOREM 5.4 Every matrix can be completely reduced. The completely reduced matrix obtained from a given matrix is unique and does not depend on the order in which the reduction rules are applied—it is always the same (provided, of course, that the procedure is carried out correctly).

EXAMPLE 3 We illustrate the complete reduction of a matrix.

$$M = \begin{bmatrix} 1 & -1 & 3 & 2 & 1 \\ 2 & -2 & -4 & 1 & 2 \\ 4 & -4 & 2 & 5 & 4 \end{bmatrix} \begin{matrix} (-2) \quad (-4) \\ (1) \quad\quad\;\; \\ \quad\quad (1) \end{matrix} \longrightarrow \begin{bmatrix} 1 & -1 & 3 & 2 & 1 \\ 0 & 0 & -10 & -3 & 0 \\ 0 & 0 & -10 & -3 & 0 \end{bmatrix} \begin{matrix} \\ (-1) \\ (1) \end{matrix} \longrightarrow \begin{bmatrix} 1 & -1 & 3 & 2 & 1 \\ 0 & 0 & -10 & -3 & 0 \\ 0 & 0 & 0 & 0 & 0 \end{bmatrix}$$

This is where we used to stop, but now we continue the reduction procedure to completely reduce M. There are two nonzero rows in the matrix and our object is to get the columns

$$\begin{bmatrix} 1 \\ 0 \\ 0 \end{bmatrix} \quad \text{and} \quad \begin{bmatrix} 0 \\ 1 \\ 0 \end{bmatrix}$$

$$\begin{bmatrix} 1 & -1 & 3 & 2 & 1 \\ 0 & 0 & -10 & -3 & 0 \\ 0 & 0 & 0 & 0 & 0 \end{bmatrix} \begin{matrix} (10) \\ (3) \\ \end{matrix} \longrightarrow \begin{bmatrix} 10 & -10 & 0 & 11 & 10 \\ 0 & 0 & -10 & -3 & 0 \\ 0 & 0 & 0 & 0 & 0 \end{bmatrix} \begin{matrix} (1/10)\longrightarrow \\ (-1/10)\longrightarrow \\ \end{matrix} \begin{bmatrix} 1 & -1 & 0 & 11/10 & 1 \\ 0 & 0 & 1 & 3/10 & 0 \\ 0 & 0 & 0 & 0 & 0 \end{bmatrix}$$

The matrix is completely reduced. $\blacksquare$

EXAMPLE 4 We completely reduce another matrix.

$$M = \begin{bmatrix} 1 & -1 & 2 \\ -1 & 0 & 1 \\ 2 & -5 & 1 \end{bmatrix} \begin{matrix} (1)\urcorner \quad (-2)\urcorner \\ (1)\longrightarrow \\ (1)\longrightarrow \end{matrix} \begin{bmatrix} 1 & -1 & 2 \\ 0 & -1 & 3 \\ 0 & -3 & -3 \end{bmatrix} \begin{matrix} \\ (-3)\urcorner \\ (1)\longrightarrow \end{matrix}$$

$$\begin{bmatrix} 1 & -1 & 2 \\ 0 & -1 & 3 \\ 0 & 0 & -12 \end{bmatrix} \begin{matrix} (6)\longrightarrow \\ (4)\longrightarrow \\ (1)\lrcorner \quad (1)\lrcorner \end{matrix} \begin{bmatrix} 6 & -6 & 0 \\ 0 & -4 & 0 \\ 0 & 0 & -12 \end{bmatrix} \begin{matrix} (-2)\longrightarrow \\ (3)\lrcorner \\ \end{matrix}$$

$$\begin{bmatrix} -12 & 0 & 0 \\ 0 & -4 & 0 \\ 0 & 0 & -12 \end{bmatrix} \begin{matrix} (-1/12)\rightarrow \\ (-1/4)\rightarrow \\ (-1/12)\rightarrow \end{matrix} \begin{bmatrix} 1 & 0 & 0 \\ 0 & 1 & 0 \\ 0 & 0 & 1 \end{bmatrix}$$

The matrix is completely reduced. ■

We now show how completely reduced matrices can be used to strengthen Theorem 5.2. Suppose M is a matrix and P is the completely reduced matrix obtained by reducing M. Then the following are true:

1. The null space of M equals the null space of P, and, therefore, the nullity of M equals the nullity of P.

2. The number of columns of M equals the number of columns of P.

3. By Theorem 5.2,

$$\begin{aligned} \text{Rank of } M &= \text{number of columns of } M - \text{nullity of } M \\ &= \text{number of columns of } P - \text{nullity of } P \\ &= \text{rank of } P \end{aligned}$$

(Again it is emphasized that the image space of M is generally not the same as the image space of P).

Since P is completely reduced, it contains among its columns the k vectors

$$\mathbf{e}_1 = \begin{bmatrix} 1 \\ 0 \\ 0 \\ \vdots \\ 0 \\ \vdots \\ 0 \end{bmatrix}, \quad \mathbf{e}_2 = \begin{bmatrix} 0 \\ 1 \\ 0 \\ \vdots \\ 0 \\ \vdots \\ 0 \end{bmatrix}, \quad \ldots, \quad \mathbf{e}_k = \begin{bmatrix} 0 \\ \vdots \\ 0 \\ 1 \\ 0 \\ \vdots \\ 0 \end{bmatrix} \cdots\cdots k\text{th coordinate}$$

Moreover, no column of P has a nonzero entry below the kth entry. A typical column $\mathbf{c}$ of P is of the form

$$\mathbf{c} = \begin{bmatrix} a_1 \\ a_2 \\ \vdots \\ a_k \\ 0 \\ \vdots \\ 0 \end{bmatrix} = a_1 \begin{bmatrix} 1 \\ 0 \\ 0 \\ \vdots \\ 0 \\ \vdots \\ 0 \end{bmatrix} + a_2 \begin{bmatrix} 0 \\ 1 \\ 0 \\ \vdots \\ 0 \\ \vdots \\ 0 \end{bmatrix} + \cdots + a_k \begin{bmatrix} 0 \\ \vdots \\ 0 \\ 1 \\ 0 \\ \vdots \\ 0 \end{bmatrix} = a_1\mathbf{e}_1 + a_2\mathbf{e}_2 + \cdots + a_k\mathbf{e}_k$$

Thus every column of P is a linear combination of the columns $\mathbf{e}_1, \mathbf{e}_2, \ldots, \mathbf{e}_k$. Since the vectors $\mathbf{e}_1, \mathbf{e}_2, \ldots, \mathbf{e}_k$ are linearly independent, they form a basis of the image space of P (not of the image space of M). Thus

4. The rank of $P = k$

$= $ the number of nonzero rows of P

So we have proved:

THEOREM 5.5 Let M be a matrix and P be the completely reduced matrix obtained by reducing M. Then

1. The rank of $M =$ the number of nonzero rows of P
2. The nullity of $M =$ (number of columns of P) $-$ (number of nonzero rows of P)

EXAMPLE 5 Let M be a matrix. Suppose M reduces to the completely reduced matrix

$$P = \begin{bmatrix} 1 & 1 & 3 \\ 0 & 0 & 0 \\ 0 & 0 & 0 \end{bmatrix}$$

The rank of M is 1, and the nullity of M is 2, by Theorem 5.5. ■

In the discussions above, we have shown that if P is the completely reduced matrix equivalent to M, then the rank of M equals the rank of P. However, the image space of M is not (necessarily) equal to the image space of P. We illustrate this with the following example.

EXAMPLE 6 Let

$$M = \begin{bmatrix} 1 & -1 & 3 & 2 & 1 \\ 2 & -2 & -4 & 1 & 2 \\ 4 & -4 & 2 & 5 & 4 \end{bmatrix}$$

As we saw in Example 3, the completely reduced matrix equivalent to M is

$$P = \begin{bmatrix} 1 & -1 & 0 & 11/10 & 1 \\ 0 & 0 & 1 & 3/10 & 0 \\ 0 & 0 & 0 & 0 & 0 \end{bmatrix}$$

Clearly, the vectors

$$\begin{bmatrix} 1 \\ 0 \\ 0 \end{bmatrix} \quad \text{and} \quad \begin{bmatrix} 0 \\ 1 \\ 0 \end{bmatrix}$$

(which are the first and third columns of P) form a basis for the image space of P. For the reasons discussed in Section 4.3, we see that the vectors

$$\begin{bmatrix} 1 \\ 2 \\ 4 \end{bmatrix} \quad \text{and} \quad \begin{bmatrix} 3 \\ -4 \\ 2 \end{bmatrix}$$

(the first and third columns of M) form a basis for the image space of M. Thus, as is generally true, $\text{rank}(P) = \text{rank}(M)$ (in this case the number is 2). However, the image space of P (which consists of all vectors with third coordinate zero) is different from the image space of M (which is spanned by two vectors with nonzero third coordinates). ■

An application of Theorem 5.5 is:

THEOREM 5.6 If a homogeneous system of linear equations has fewer equations than unknowns, then it has a nontrivial (not all unknowns equal to zero) solution.

To see why this theorem is true, consider the following system of homogeneous equations:

$$a_{11}x_1 + a_{12}x_2 + a_{13}x_3 = 0$$

$$a_{21}x_1 + a_{22}x_2 + a_{23}x_3 = 0$$

The coefficient matrix for this system is

$$M = \begin{bmatrix} a_{11} & a_{12} & a_{13} \\ a_{21} & a_{22} & a_{23} \end{bmatrix}$$

Now the rank of M equals the number of nonzero rows of the completely reduced matrix obtained from M. So the rank of M is less than or equal to the number of rows of M. Thus the rank of M is less than or equal to 2. Using Theorem 5.5, we see that

$$\text{Nullity of } M = 3 - \text{rank of } M$$

So the nullity of $M \geq 1$, which means that there is a nontrivial (different from **0**) vector in the null space of M—that is, the homogeneous system has a nontrivial solution.

We conclude this section by applying some of the ideas we have discussed to square matrices (matrices having the same number of columns as rows).

Let M be an $n \times n$ matrix (n rows and n columns) of rank k. Then the completely reduced matrix P obtained from M has k nonzero rows and $(n - k)$ rows of zeros. Furthermore, among the columns of P are the vectors $\mathbf{e}_1, \mathbf{e}_2, \ldots, \mathbf{e}_k$.

From this we get the following theorem:

THEOREM 5.7 Let M be an $n \times n$ matrix. Then the rank of M is n if and only if the completely reduced matrix obtained from M is

$$I = \begin{bmatrix} \vdots & \vdots & & \vdots \\ \mathbf{e}_1 & \mathbf{e}_2 & \cdots & \mathbf{e}_n \\ \vdots & \vdots & & \vdots \end{bmatrix} = \begin{bmatrix} 1 & 0 & \cdots & 0 \\ 0 & 1 & \cdots & 0 \\ 0 & 0 & \cdots & \vdots \\ \vdots & \vdots & & 0 \\ 0 & 0 & \cdots & 1 \end{bmatrix}$$

Moreover, if the rank of M is less than n, then the completely reduced matrix obtained from M contains at least one row of zeros.

Exercise Set 5.3

In Exercises 1–11, completely reduce each of the matrices given and find its rank and nullity.

1. $\begin{bmatrix} 1 & 1 \\ 1 & 1 \end{bmatrix}$

2. $\begin{bmatrix} -1 & 0 \\ 1 & -1 \end{bmatrix}$

3. $\begin{bmatrix} 1 & 3 & 1 \\ -2 & 4 & 8 \\ 1 & 1 & -1 \end{bmatrix}$

4. $\begin{bmatrix} 1 & 3 & 1 \\ -2 & 4 & 1 \\ 1 & 1 & 2 \end{bmatrix}$

5. $\begin{bmatrix} 1 & -1 & 2 & 3 & 1 \\ 4 & 5 & -2 & 2 & 1 \\ 3 & -1 & 2 & 1 & 1 \end{bmatrix}$

6. $\begin{bmatrix} 0 & 3 \\ 0 & 4 \end{bmatrix}$

7. $\begin{bmatrix} 7 & 3 \\ 4 & -6 \\ 1 & 1 \end{bmatrix}$

8. $\begin{bmatrix} 4 & -6 & 10 \\ 2 & -3 & 5 \\ -2 & 3 & -5 \end{bmatrix}$

9. $\begin{bmatrix} 1 & 3 & 2 & 0 & -2 \\ 0 & 0 & 0 & 1 & 5 \end{bmatrix}$

10. $\begin{bmatrix} 0 & 0 & 0 & 0 \\ 0 & 1 & 2 & 0 \\ 1 & 0 & 3 & -1 \end{bmatrix}$

11. $\begin{bmatrix} 1 & 0 \\ 0 & 1 \\ 0 & 0 \\ 0 & 0 \end{bmatrix}$

12. Determine whether each of the statements below is *true* or *false*. If true, give reasons; and if false, give a counterexample.

a. A nonhomogeneous system of equations having more unknowns than equations always has a solution.

b. If a system of equations has the same number of unknowns as equations and the nullity of the coefficient matrix is zero, then the system has a solution and this solution is unique.

c. Equivalent matrices always have the same image space.

★13. This exercise will simplify applications of Theorem 5.5. Let P be a matrix with k nonzero rows. We will call P in echelon form if among its columns P contains k columns $\mathbf{h}_1, \mathbf{h}_2, \ldots, \mathbf{h}_k$ of the following form:

$$\mathbf{h}_i = \begin{bmatrix} a_1 \\ \vdots \\ a_{i-1} \\ a_i \\ 0 \\ \vdots \\ 0 \end{bmatrix} \qquad \text{where } a_i \neq 0 \text{ and } a_1, a_2, \ldots, a_{i-1} \text{ can be any real numbers.}$$

For example, the matrix from Example 3

$$\begin{bmatrix} 1 & -1 & 3 & 2 & 1 \\ 0 & 0 & 10 & -3 & 0 \\ 0 & 0 & 0 & 0 & 0 \end{bmatrix}$$

is in echelon form. For $\mathbf{h}_1$ and $\mathbf{h}_2$ we can choose the columns

$$\begin{bmatrix} 1 \\ 0 \\ 0 \end{bmatrix} \quad \text{and} \quad \begin{bmatrix} 3 \\ -10 \\ 0 \end{bmatrix}$$

a. Show that if P is in echelon form, then the columns $\mathbf{h}_1, \mathbf{h}_2, \ldots, \mathbf{h}_k$ are a basis for the image space of P.

b. Show that $k =$ rank of P.

c. Show that if M is a matrix that can be reduced to the echelon matrix P, then

Rank of $M =$ the number of nonzero rows of P

★14. **a.** Suppose that $b \neq 0$ and the matrix M is reduced to the matrix P by the operation

$$M = \begin{bmatrix} - m_i - \\ - m_j - \end{bmatrix} \begin{matrix} (a) \\ (b) \end{matrix} \begin{bmatrix} - p_i - \\ - p_j - \end{bmatrix} = P$$

where m_i, m_j, and p_i, p_j are the ith and jth rows of M and P respectively. Express p_i and p_j as linear combinations of m_i and m_j (see Exercise 14 in Section 3.2 for definition of row vectors), and use this expression to find expressions for m_i and m_j as linear combinations of p_i and p_j.

b. Show that if a matrix M can be reduced (in a number of steps) to a matrix P, then each row of M is a linear combination of the rows of P.

c. Let P be the completely reduced matrix equivalent to a matrix M. Show that the nonzero rows of P form a basis for the (row) vector space spanned by the rows of M.

d. Show that the dimension of the vector space spanned by the rows of M (called the **row rank of** M) is equal to the rank of M (sometimes called the **column rank of** M).

*5.4 The Proof of the Rank and Nullity Theorem (Theorem 5.2)

The purpose of this section is to explain why Theorem 5.2 is true. This section is more abstract than most of the other sections of this book. It is optional since no other section of this book will be based on the material in it. However, it contains some fundamental ideas of linear algebra which you may find of interest.

For the convenience of the reader, we restate Theorem 5.2.

THEOREM 5.2

Let $T: V \to W$ be a linear transformation. Then

$$\text{Dimension of } V = \text{rank of } T + \text{nullity of } T$$

Although the theorem is stated for linear transformations from one arbitrary vector space to another, we will only prove it for $V = \mathbb{R}^n$ and $W = \mathbb{R}^m$.

Before beginning the proof, we will point out some of its important features. The whole proof rests on the construction of a special basis for V. We call this basis the **preimage basis**. The elements of the preimage basis are of two types—a set of vectors that is a basis for the null space of T, and a set of vectors of V related to a basis for the image space of T. We construct this basis in such a way that the number of basis vectors belonging to the null space is equal to the nullity of T, and the number of other vectors is equal to the rank of T. From this we easily derive the conclusion of the theorem.

The main work in the proof involves showing that the set of vectors we choose as the preimage basis is indeed a basis for V. To do this, we show that these vectors are linearly independent and that they span V. Since the vectors we deal with are not specific vectors of $\mathbb{R}^n$, we cannot just reduce a matrix to determine if the vectors are linearly independent or if they span V. We must be a little more subtle. In showing both linear independence and spanning, we begin by looking at what happens to the images of these preimage basis vectors under the action of T. From this we get enough information to show that these vectors are indeed a basis for V. Now we start the proof itself.

To construct a preimage basis we begin by letting M be the matrix associated with T. By reducing M we can find a basis for the null space of T, and a basis for the image space consisting of *some* of the columns of M (see Figure 5.9). As before, we let $\mathbf{e}_i$ be the vector of V whose ith entry is 1 and all other entries are 0. We know that if $\mathbf{c}_i$ is the ith column of M, then

$$\mathbf{c}_i = M\mathbf{e}_i$$

Now choose all the vectors $\mathbf{e}_i$ for which $M\mathbf{e}_i = \mathbf{c}_i$ is one of the basis vectors chosen for the image space of T. The preimage basis consists of these $\mathbf{e}_i$'s along with a complete set of basis vectors for the null space of T. We

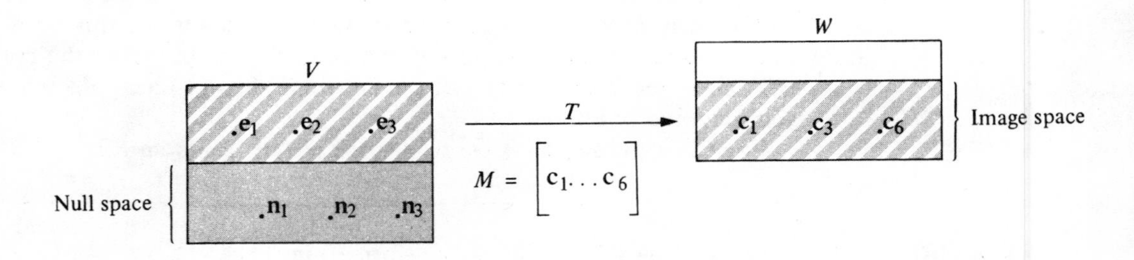

Figure 5.9

will show that this set of vectors is indeed a basis for V. However, before we do this, we will see how this basis is used to prove Theorem 5.2.

The proof is as follows:

$$\begin{aligned}
\text{Dimension of } V &= \text{the number of vectors in any basis for } V \\
&= \text{the number of vectors in a preimage basis for } V \\
&= \text{number of vectors in a basis for null space of } V \\
&\quad + \text{number of } \mathbf{e}_i\text{'s in preimage basis of } T \\
&= \text{nullity of } T \\
&\quad + \text{number of } \mathbf{c}_i\text{'s in a basis for the image space of } T \\
&= \text{nullity of } T + \text{rank of } T
\end{aligned}$$

Except for showing that the set of vectors called a preimage basis is indeed a basis, the theorem is proved.

Now we show that the preimage basis is a basis. To keep the notation from getting out of hand, we will specify certain variables. Let us assume that the matrix M representing T is of the form

$$M = \begin{bmatrix} \cdot & \cdot & \cdot & \cdot & \cdot & \cdot \\ \vdots & \vdots & \vdots & \vdots & \vdots & \vdots \\ \mathbf{c}_1 & \mathbf{c}_2 & \mathbf{c}_3 & \mathbf{c}_4 & \mathbf{c}_5 & \mathbf{c}_6 \\ \vdots & \vdots & \vdots & \vdots & \vdots & \vdots \\ \cdot & \cdot & \cdot & \cdot & \cdot & \cdot \end{bmatrix}$$

Further, let us assume that the columns $\mathbf{c}_1$, $\mathbf{c}_3$, and $\mathbf{c}_6$ are a basis for the image space of T, and a basis for the null space of T is $\{\mathbf{n}_1, \mathbf{n}_2, \mathbf{n}_3\}$. The preimage basis, in this case is

$$\{\mathbf{n}_1, \mathbf{n}_2, \mathbf{n}_3, \mathbf{e}_1, \mathbf{e}_3, \mathbf{e}_6\}$$

To show that this set is a basis, we must show that it is both linearly independent and spans V.

We begin with linear independence. Suppose that

$$\mathbf{0} = a_1\mathbf{n}_1 + a_2\mathbf{n}_2 + a_3\mathbf{n}_3 + b_1\mathbf{e}_1 + b_3\mathbf{e}_3 + b_6\mathbf{e}_6$$

We apply T to both sides of this equation to get

$$T(\mathbf{0}) = T(a_1\mathbf{n}_1 + a_2\mathbf{n}_2 + a_3\mathbf{n}_3 + b_1\mathbf{e}_1 + b_3\mathbf{e}_3 + b_6\mathbf{e}_6)$$

and, since T is linear, this gives

$$\mathbf{0} = a_1T(\mathbf{n}_1) + a_2T(\mathbf{n}_2) + a_3T(\mathbf{n}_3) + b_1T(\mathbf{e}_1) + b_3T(\mathbf{e}_3) + b_6T(\mathbf{e}_6)$$

Now the $\mathbf{n}_i$'s are in the null space of T, so that $T(\mathbf{n}_i) = \mathbf{0}$. Also $T(\mathbf{e}_i) = M(\mathbf{e}_i) = \mathbf{c}_i$. So the equation becomes

$$\mathbf{0} = \mathbf{0} + \mathbf{0} + \mathbf{0} + b_1\mathbf{c}_1 + b_3\mathbf{c}_3 + b_6\mathbf{c}_6$$

Since the $\mathbf{c}_i$'s are a basis for the image space of T, they are linearly independent, and this means that $b_1 = b_3 = b_6 = 0$. Thus the original equation becomes

$$\mathbf{0} = a_1\mathbf{n}_1 + a_2\mathbf{n}_2 + a_3\mathbf{n}_3$$

Now we use the fact that the $\mathbf{n}_i$'s are a basis for the null space of V, and as such are linearly independent. Thus $a_1 = a_2 = a_3 = 0$. Hence we have shown that the only way a linear combination of the elements $\mathbf{n}_1$, $\mathbf{n}_2$, $\mathbf{n}_3$, $\mathbf{e}_1$, $\mathbf{e}_3$, $\mathbf{e}_6$ can be $\mathbf{0}$ is if all the coefficients are zero—that is, these vectors are linearly independent.

Now we show that these vectors span V. To do this we must show that any arbitrary vector $\mathbf{v}$ of V can be expressed as a linear combination of the vectors $\mathbf{n}_1, \mathbf{n}_2, \mathbf{n}_3, \mathbf{e}_1, \mathbf{e}_3, \mathbf{e}_6$. The trick to this proof is not to start with $\mathbf{v}$ directly, but to first express $T(\mathbf{v})$ as a linear combination of $\mathbf{c}_1$, $\mathbf{c}_3$, and $\mathbf{c}_6$. Since $\{\mathbf{c}_1, \mathbf{c}_3, \mathbf{c}_6\}$ is a basis for the image space of T, this is possible. We write

$$T(\mathbf{v}) = b_1\mathbf{c}_1 + b_3\mathbf{c}_3 + b_6\mathbf{c}_6$$

Now we define the vector $\mathbf{v}_0$ of V

$$\mathbf{v}_0 = b_1\mathbf{e}_1 + b_3\mathbf{e}_3 + b_6\mathbf{e}_6$$

The interesting thing about $\mathbf{v}_0$ is that

$$\begin{aligned} T(\mathbf{v}_0) &= T(b_1\mathbf{e}_1 + b_3\mathbf{e}_3 + b_6\mathbf{e}_6) \\ &= b_1T(\mathbf{e}_1) + b_3T(\mathbf{e}_3) + b_6T(\mathbf{e}_6) \qquad \text{since } T \text{ is linear} \\ &= b_1\mathbf{c}_1 + b_3\mathbf{c}_3 + b_6\mathbf{c}_6 \\ &= T(\mathbf{v}) \end{aligned}$$

Therefore, $$T(\mathbf{v} - \mathbf{v}_0) = T(\mathbf{v}) - T(\mathbf{v}_0) = \mathbf{0}$$

which means that $(\mathbf{v} - \mathbf{v}_0)$ is in the null space of T. So we can find scalars a_1, a_2, a_3 such that

$$\mathbf{v} - \mathbf{v}_0 = a_1\mathbf{n}_1 + a_2\mathbf{n}_2 + a_3\mathbf{n}_3$$

Thus $$\begin{aligned} \mathbf{v} &= (a_1\mathbf{n}_1 + a_2\mathbf{n}_2 + a_3\mathbf{n}_3) + \mathbf{v}_0 \\ &= a_1\mathbf{n}_1 + a_2\mathbf{n}_2 + a_3\mathbf{n}_3 + b_1\mathbf{e}_1 + b_3\mathbf{e}_3 + b_6\mathbf{e}_6 \end{aligned}$$

In other words, the preimage basis spans V, and we have shown that the preimage basis is indeed a basis.

EXAMPLE 1 Let $V = \mathbb{R}^3$, $W = \mathbb{R}^2$ and T be the linear transformation represented by the matrix

$$M = \begin{bmatrix} 1 & -2 & 3 \\ -3 & 6 & -9 \end{bmatrix}$$

We will find a preimage basis for V. First we reduce M.

$$\begin{bmatrix} 1 & -2 & 3 \\ -3 & 6 & -9 \end{bmatrix} \xrightarrow{\text{reduces to}} \begin{bmatrix} 1 & -2 & 3 \\ 0 & 0 & 0 \end{bmatrix}$$

which gives the relation $x_1 = 2x_2 - 3x_3$. Thus the null space of V has basis

$$\left\{ \mathbf{n}_1 = \begin{bmatrix} 2 \\ 1 \\ 0 \end{bmatrix}, \quad \mathbf{n}_2 = \begin{bmatrix} -3 \\ 0 \\ 1 \end{bmatrix} \right\}$$

and a basis for the image space is

$$\left\{ \mathbf{c}_1 = \begin{bmatrix} 1 \\ -3 \end{bmatrix} \right\}$$

So the preimage basis is

$$\left\{ \mathbf{n}_1 = \begin{bmatrix} 2 \\ 1 \\ 0 \end{bmatrix}, \quad \mathbf{n}_2 = \begin{bmatrix} -3 \\ 0 \\ 1 \end{bmatrix}, \quad \mathbf{e}_1 = \begin{bmatrix} 1 \\ 0 \\ 0 \end{bmatrix} \right\}$$

We will now show how to express the vector

$$\mathbf{v} = \begin{bmatrix} 1 \\ -1 \\ 3 \end{bmatrix}$$

as a linear combination of the preimage basis vectors to illustrate the technique used in showing that the preimage basis spanned V.

We begin by finding $T(\mathbf{v})$.

$$T(\mathbf{v}) = M(\mathbf{v}) = \begin{bmatrix} 1 & -2 & 3 \\ -3 & 6 & -9 \end{bmatrix} \begin{bmatrix} 1 \\ -1 \\ 3 \end{bmatrix} = \begin{bmatrix} 12 \\ -36 \end{bmatrix} = 12\mathbf{c}_1$$

So the vector $\mathbf{v}_0$ is

$$\mathbf{v}_0 = 12\mathbf{e}_1 = 12 \begin{bmatrix} 1 \\ 0 \\ 0 \end{bmatrix}$$

The vector $\mathbf{v} - \mathbf{v}_0$ is in the null space. We express it as a linear combination of the basis vectors $\mathbf{n}_1, \mathbf{n}_2$.

$$\mathbf{v} - \mathbf{v}_0 = \begin{bmatrix} 1 \\ -1 \\ 3 \end{bmatrix} - 12 \begin{bmatrix} 1 \\ 0 \\ 0 \end{bmatrix} = \begin{bmatrix} -11 \\ -1 \\ 3 \end{bmatrix} = -1 \begin{bmatrix} 2 \\ 1 \\ 0 \end{bmatrix} + 3 \begin{bmatrix} -3 \\ 0 \\ 1 \end{bmatrix}$$

From this we have

$$\mathbf{v} = -1\begin{bmatrix}2\\1\\0\end{bmatrix} + 3\begin{bmatrix}-3\\0\\1\end{bmatrix} + 12\begin{bmatrix}1\\0\\0\end{bmatrix}$$

So $\mathbf{v}$ is a linear combination of the vectors of the preimage basis. ■

EXAMPLE 2 Let T: $\mathbb{R}^5 \to \mathbb{R}^3$ be the linear transformation represented by the matrix

$$M = \begin{bmatrix}1 & -1 & 3 & 2 & 1\\2 & -2 & -4 & 1 & 2\\4 & -4 & 2 & 5 & 4\end{bmatrix}$$

As we saw in Example 5 of Section 5.2, a basis for the null space of T is

$$\left\{\mathbf{n}_1 = \begin{bmatrix}1\\1\\0\\0\\0\end{bmatrix},\ \mathbf{n}_2 = \begin{bmatrix}-11\\0\\-3\\10\\0\end{bmatrix},\ \mathbf{n}_3 = \begin{bmatrix}-1\\0\\0\\0\\1\end{bmatrix}\right\}$$

and a basis for the image space is

$$\left\{\mathbf{c}_1 = \begin{bmatrix}1\\2\\4\end{bmatrix},\ \mathbf{c}_3 = \begin{bmatrix}3\\-4\\2\end{bmatrix}\right\}$$

So the preimage basis is

$$\left\{\mathbf{n}_1 = \begin{bmatrix}1\\1\\0\\0\\0\end{bmatrix},\ \mathbf{n}_2 = \begin{bmatrix}-11\\0\\-3\\10\\0\end{bmatrix},\ \mathbf{n}_3 = \begin{bmatrix}-1\\0\\0\\0\\1\end{bmatrix},\ \mathbf{e}_1 = \begin{bmatrix}1\\0\\0\\0\\0\end{bmatrix},\ \mathbf{e}_3 = \begin{bmatrix}0\\0\\1\\0\\0\end{bmatrix}\right\}$$

This is shown in Figure 5.10.

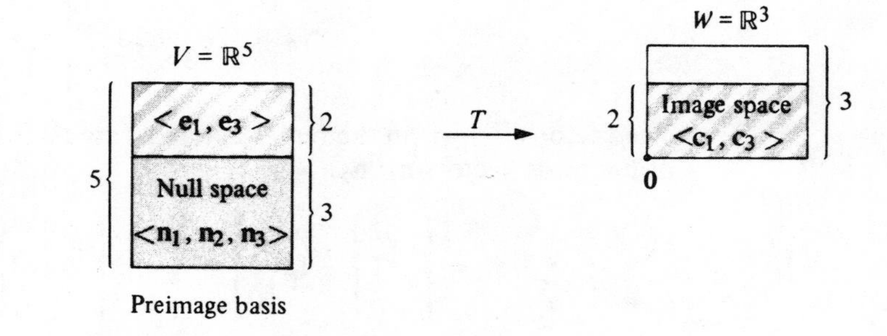

Figure 5.10

Exercise Set 5.4

1–5. For each of the matrices listed in Exercises 5–9 in Exercise Set 5.1, find a preimage basis for the domain of the matrix.

6. Let $M = \begin{bmatrix} 1 & 2 & -3 & 2 \\ 1 & 1 & 1 & -1 \\ 2 & 3 & -2 & 1 \end{bmatrix}$

Find a preimage basis for the domain of M and express the vector

$$\mathbf{v} = \begin{bmatrix} 1 \\ 1 \\ 2 \\ 1 \end{bmatrix}$$

as a linear combination of the vectors of this basis. (Use the method of Example 1).

Applications 5 Coding Theory

Coding theory is the study of methods of adding extra information to messages so that errors that occur during transmission can be corrected. Coding theory does not deal with secret codes, which are more correctly called ciphers. Ciphers were discussed in the Applications Section at the end of Chapter 3.

Coding is not foreign to our everyday experiences. We use a type of code when we spell words over the telephone. The letters *s* and *f* cannot be distinguished when heard over the telephone. So, when spelling words containing these letters, people generally add redundant information such as "*f* as in Frank" or "*s* as in Sam." In this way they make their message unambiguously clear.

Coding theory has become extremely important in recent years. With the advent of space exploration, it has become necessary to send radio messages over vast distances. Error-correcting codes have been used extensively in the space program to correct errors caused by static and other types of interference.

The pictures of man's first landing on the moon were initially televised just as they had been received and were not very clear, since they contained many errors induced during transmission. Later, much clearer computer-corrected pictures were obtained using error-correcting codes.

Coding theory is not only used on a macroscopic level, it is also used on a microscopic scale. Computers store data on microscopic cells in their electronic components. Although these components are quite reliable, there is a chance that the data in one microscopic cell may be changed during storage or retrieval. To prevent the failure of one of these tiny cells from ruining an entire calculation, all data stored in a computer are stored in coded form.

Almost all codes used today are used to transmit data from one computer to another. The basic internal component in an electronic computer is called a flip-flop. It is like a switch in that it can be in one of two possible states—on or off, charged or not charged, magnetized one way or the other, and so on. All information in computers is stored in sequences of flip-flops, some on and some off. Mathematically these sequences correspond to sequences of 0's (off) and 1's (on). The basic internal operation a computer can perform is to either change or not change the state of any of its flip-flops. In terms of 0's and 1's, change corresponds to adding a 1 and not change corresponds to adding a 0. These basic internal computer operations are summarized in Table 5.1.

Table 5.1

flip-flop	off	on
not change	off	on
change	on	off

Writing the information in Table 5.1 in terms of 0's and 1's, gives Table 5.2.

Table 5.2

+	0	1
0	0	1
1	1	0

The type of addition described by Table 5.2 is called **addition modulo 2.** In this section, all calculations will be done using only the numbers 0 and 1 with arithmetic modulo 2. We will use the letter $\mathbb{F}$ to denote the set $\{0, 1\}$ with arithmetic modulo 2.

From the above discussion it is clear that the types of messages that we should consider are sequences of 0's and 1's. By encoding such a message we will mean adding extra 0's and 1's which are used to determine and correct any errors that may have occurred during transmission. By error we will mean that an entry transmitted as a 0 (or a 1) is received as a 1 (or a 0).

In a sense, error-correcting codes are patterned after language. When we find a typographical error in a newspaper article such as

The boy was bitten by a dkg.

we immediately know that the word *dkg* is incorrect because there is no such word in the English language. In coding theory we can recognize an error because only certain sequences of 0's and 1's are code words, and if a word that is not a code word is received, this indicates that an error was introduced during transmission.

Not only can we detect the word *dkg* as incorrect, but we can also correct the error. It is quite obvious that the sentence should read

The boy was bitten by a dog.

However, the sentence

The boy was bitten by a rat.

would make just as much sense, but it is much more probable that only one error was made (that k was written instead of o) than that three errors were made (that d, k, and g were written instead of r, a, and t). In the same way, when we correct codes, we will make our choices as to which entries to change based on probabilistic considerations.

We will discuss one particular family of error-correcting codes called Hamming codes (after R. W. Hamming). The messages we will encode will be sequences of 0's and 1's. In the first example we will consider, these sequences will be broken into groups of four symbols. In encoding we will add additional symbols. This Hamming code will be able to detect and correct up to one error per block of seven symbols. This means that if there are two or more errors, they will either not be detected or they will be incorrectly decoded. However, unless the transmission channel is extremely bad, it is unlikely that more than one of the seven entries in each block will be changed during transmission.

The code works like this: Each word of four entries will be represented by a vector in $\mathbb{F}^4$ (the vector space of four component vectors with entries in $\mathbb{F}$). We will use an encoding matrix M to map this vector into a vector in $\mathbb{F}^7$. The three extra entries will carry the additional information necessary to correct errors in transmission. The encoded message is then transmitted. The message received is a vector in $\mathbb{F}^7$, which is either identical to the vector transmitted or differs from the vector transmitted in at most one entry. We apply a matrix H (after Hamming) called the **syndrome** to the vector received. If the resulting vector is $\mathbf{0}$, then the vector received was (most probably) the vector transmitted; if the resulting vector is not $\mathbf{0}$, we will be able to determine the position in which the error most probably occurred. By changing the entry in this position in the vector received, we obtain the vector transmitted. The final step in the decoding procedure is to remove the redundant entries to obtain the original vector in $\mathbb{F}^4$. The process is illustrated in Figure 5.11.

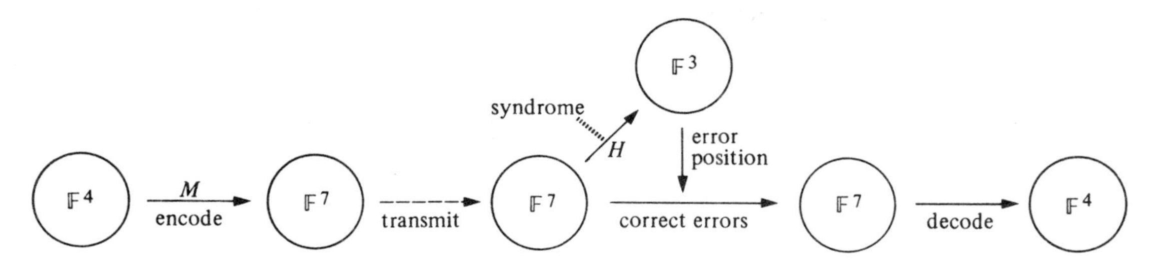

Figure 5.11

At this point we introduce an encoding matrix M and a syndrome matrix H. Later in this section we will show how such matrices are chosen. The encoding matrix we will use is

$$M = \left[\begin{array}{cccc} 1 & 0 & 0 & 0 \\ 0 & 1 & 0 & 0 \\ 0 & 0 & 1 & 0 \\ 0 & 0 & 0 & 1 \\ \hline 1 & 0 & 1 & 1 \\ 1 & 1 & 0 & 1 \\ 1 & 1 & 1 & 0 \end{array}\right] \begin{array}{l} \left.\begin{array}{c} \\ \\ \\ \\ \end{array}\right\} \text{message} \\ \left.\begin{array}{c} \\ \\ \\ \end{array}\right\} \text{code} \end{array}$$

Note that the top four rows of M are the identity matrix and the bottom three rows contain the error correcting information.

The syndrome matrix H is

$$H = \left[\begin{array}{cccc|ccc} 1 & 0 & 1 & 1 & 1 & 0 & 0 \\ 1 & 1 & 0 & 1 & 0 & 1 & 0 \\ 1 & 1 & 1 & 0 & 0 & 0 & 1 \end{array}\right]$$

In the following example we show how the Hamming code works.

EXAMPLE 1 We begin with the four-symbol word 0 1 0 1 and trace its progress through encoding, transmission, error correction, and decoding. We start by writing the message as a vector $\mathbf{w}$ in $\mathbb{F}^4$ and encoding it by applying the matrix M.

$$M\mathbf{w} = \left[\begin{array}{cccc} 1 & 0 & 0 & 0 \\ 0 & 1 & 0 & 0 \\ 0 & 0 & 1 & 0 \\ 0 & 0 & 0 & 1 \\ \hline 1 & 0 & 1 & 1 \\ 1 & 1 & 0 & 1 \\ 1 & 1 & 1 & 0 \end{array}\right] \begin{bmatrix} 0 \\ 1 \\ 0 \\ 1 \end{bmatrix} = \left[\begin{array}{c} 0 \\ 1 \\ 0 \\ 1 \\ \hline 1 \\ 0 \\ 1 \end{array}\right] = \mathbf{b}$$

This vector is then transmitted and during transmission the entry in the fourth position is changed.

$$\mathbf{b} = \left[\begin{array}{c} 0 \\ 1 \\ 0 \\ 1 \\ \hline 1 \\ 0 \\ 1 \end{array}\right] \xrightarrow{\text{transmission}} \left[\begin{array}{c} 0 \\ 1 \\ 0 \\ 0 \\ \hline 1 \\ 0 \\ 1 \end{array}\right] = \mathbf{v}$$

We apply the syndrome matrix H to the vector $\mathbf{v}$ received.

$$H\mathbf{v} = \left[\begin{array}{cccc|ccc} 1 & 0 & 1 & 1 & 1 & 0 & 0 \\ 1 & 1 & 0 & 1 & 0 & 1 & 0 \\ 1 & 1 & 1 & 0 & 0 & 0 & 1 \end{array}\right] \left[\begin{array}{c} 0 \\ 1 \\ 0 \\ 0 \\ \hline 1 \\ 0 \\ 1 \end{array}\right] = \begin{bmatrix} 1 \\ 1 \\ 0 \end{bmatrix} = \mathbf{s}$$

Since $H\mathbf{v} \neq 0$, we know that an error was made. In fact, since $H\mathbf{v}$ is the fourth column of H we have

$$H\mathbf{v} = H\mathbf{e}_4$$

where $\mathbf{e}_4$ is the vector in $\mathbb{F}^7$ that has a 1 in the fourth position and 0's in all the other positions. Since H represents a linear transformation,

$$H(\mathbf{v} - \mathbf{e}_4) = \mathbf{0}$$

Therefore, $(\mathbf{v} - \mathbf{e}_4)$ is a code word and differs only in the fourth position from $\mathbf{v}$. So, to correct the error that was most probably made, we change the entry in the fourth position in the vector received. Thus

$$\text{message received} = \left[\begin{array}{c} 0 \\ 1 \\ 0 \\ 0 \\ \hline 1 \\ 0 \\ 0 \end{array}\right] \xrightarrow[\text{error correction}]{\text{change fourth entry}} \left[\begin{array}{c} 0 \\ 1 \\ 0 \\ 1 \\ \hline 1 \\ 0 \\ 0 \end{array}\right]$$

The final step in the decoding procedure is to remove the redundant entries (last three entries) to obtain the original vector

$$\mathbf{w} = \begin{bmatrix} 0 \\ 1 \\ 0 \\ 1 \end{bmatrix}$$

and write the message as

0101 ■

We summarize the encoding and decoding procedures illustrated in Example 1 by the following steps.

Encode

Step 1 Break the message up into "words" of four symbols each, and convert to vectors in $\mathbb{F}^4$.

Step 2 Apply M, the encoding matrix, to each of these vectors to obtain the code vectors (in $\mathbb{F}^7$).

Transmit the Message

Decode

Step 1 Apply the syndrome H to each vector received.

Step 2 **a.** If the result of applying H to the vector received is $\mathbf{0}$, then no error occurred during transmission.
b. If the result of applying H to the vector received is different from $\mathbf{0}$, it will be one of the columns of H. An error was made in the position corresponding to that column in the vector received.

Step 3 Correct errors.

Step 4 Complete decoding by removing redundant (last 3) entries in corrected vector and writing the message.

In these procedures we have made a number of claims about the Hamming code that still must be verified:

1. That if $\mathbf{b}$ is any code vector, then $H\mathbf{b} = \mathbf{0}$.
2. That if $\mathbf{v}$ is not a code vector, then $H\mathbf{v} \neq \mathbf{0}$.
3. That if $H\mathbf{v} \neq 0$, then $H\mathbf{v}$ is a column of H.

To see why these statements are all true, we use the concepts introduced in this chapter. We begin by describing the code words. The encoding matrix M is a mapping from $\mathbb{F}^4$ to $\mathbb{F}^7$. A vector in $\mathbb{F}^7$ is a code word if it is the image under M of some vector in $\mathbb{F}^4$. In other words, the set of code words is the image space of M. So the set of code words forms a subspace of $\mathbb{F}^7$. Now we find the dimension of this subspace. The image space of M is spanned by the columns of M. It is easy to see that these columns are linearly independent (just look at the first four rows). Hence, the columns of M form a basis for the image space of M (see Figure 5.12). Therefore,

$$4 = \text{rank}(M) = \dim(\text{Im}(M)) = \text{dimension of the vector space of code words}$$

Figure 5.12

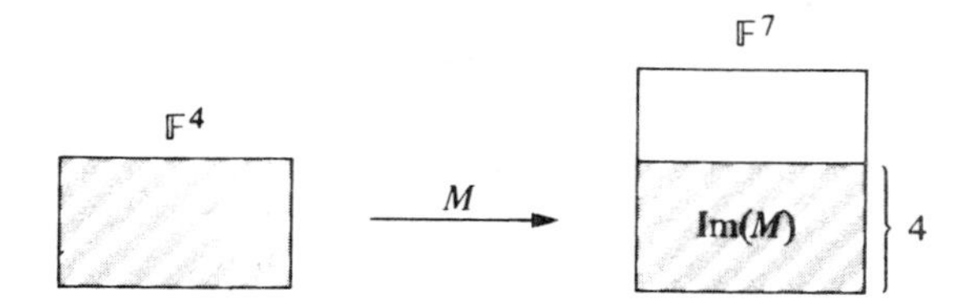

Now we look at the syndrome matrix H. H is a mapping from $\mathbb{F}^7$ to $\mathbb{F}^3$. Since H contains the columns

$$\begin{bmatrix}1\\0\\0\end{bmatrix} \text{ (fifth column)} \quad \begin{bmatrix}0\\1\\0\end{bmatrix} \text{ (sixth column)} \quad \begin{bmatrix}0\\0\\1\end{bmatrix} \text{ (seventh column)}$$

the image space of H spans $\mathbb{F}^3$. Using Theorem 5.2, we have

$$7 = \text{rank of } H + \text{nullity of } H = 3 + \text{nullity of } H.$$

Therefore, the null space of H has dimension 4 (see Figure 5.13).

Figure 5.13

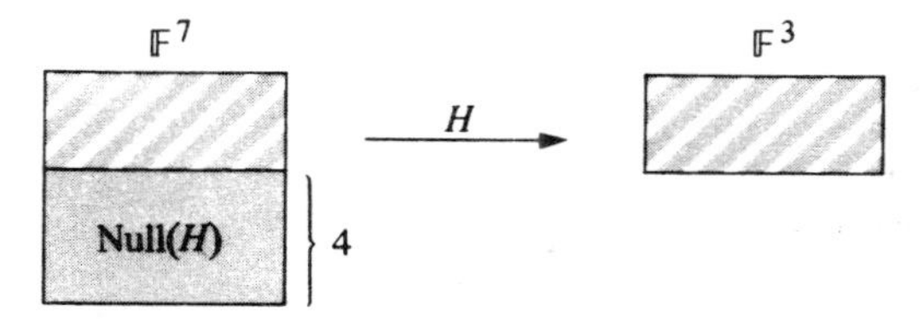

Let us now investigate the relationship between two subspaces of $\mathbb{F}^7$—the image space of M and the null space of H. We will show that each vector in the image space of M is also in the null space of H, that is, we will show that Im(M) is a subset of Null(H). Let $\mathbf{b}$ be an arbitrary vector in Im(M). Then $\mathbf{b} = M\mathbf{w}$ for some vector $\mathbf{w}$ in $\mathbb{F}^4$. That is,

$$M\mathbf{w} = \begin{bmatrix}1&0&0&0\\0&1&0&0\\0&0&1&0\\0&0&0&1\\\hline 1&0&1&1\\1&1&0&1\\1&1&1&0\end{bmatrix}\begin{bmatrix}w_1\\w_2\\w_3\\w_4\end{bmatrix} = \begin{bmatrix}w_1\\w_2\\w_3\\w_4\\\hline w_1+w_3+w_4\\w_1+w_2+w_4\\w_1+w_2+w_3\end{bmatrix} = \mathbf{b}$$

To show that **b** is in the null space of H we apply H to **b**.

$$H\mathbf{b} = \left[\begin{array}{cccc|ccc} 1 & 0 & 1 & 1 & 1 & 0 & 0 \\ 1 & 1 & 0 & 1 & 0 & 1 & 0 \\ 1 & 1 & 1 & 0 & 0 & 0 & 1 \end{array}\right] \left[\begin{array}{c} w_1 \\ w_2 \\ w_3 \\ w_4 \\ \hline w_1 + w_3 + w_4 \\ w_1 + w_2 + w_4 \\ w_1 + w_2 + w_3 \end{array}\right] = \begin{bmatrix} w_1 + w_3 + w_4 + (w_1 + w_3 + w_4) \\ w_1 + w_2 + w_4 + (w_1 + w_2 + w_4) \\ w_1 + w_2 + w_3 + (w_1 + w_2 + w_3) \end{bmatrix} = \begin{bmatrix} 0 \\ 0 \\ 0 \end{bmatrix}$$

The last equality follows because

$$w_i + w_i = \begin{cases} 0 + 0 \text{ if } w_i = 0 \\ 1 + 1 \text{ if } w_i = 1 \end{cases} = \begin{cases} 0 \\ 0 \end{cases} = 0$$

So we have shown that $\mathrm{Im}(M) \subseteq \mathrm{Null}(H)$. But we have also found that both the image space of M and the null space of H have dimension 4 (see Figure 5.14). Therefore

$$\mathrm{Im}(M) = \mathrm{Null}(H)$$

Thus, we have verified our first two claims, that is, **b** is a code word if and only if $H\mathbf{b} = \mathbf{0}$.

Figure 5.14

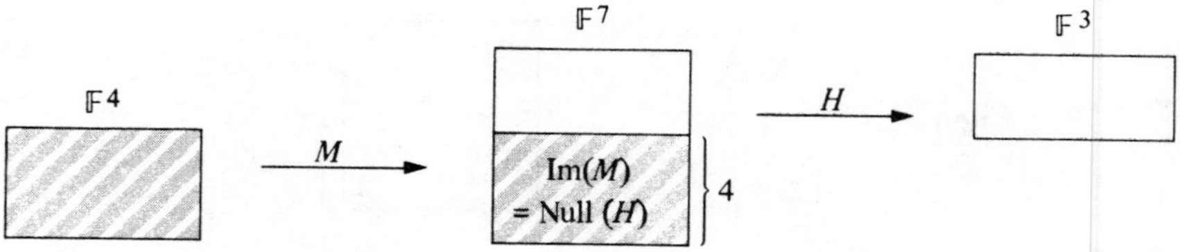

To verify the remaining claim, that if $H\mathbf{v} \neq 0$, then $H\mathbf{v}$ is a column of H, we need only count the number of vectors in the image space of H. The image space of H is $\mathbb{F}^3$, which consists of all three-component vectors with entries of either 0 or 1. Since there are $2^3 = 8$ such vectors and one of these is the zero vector, any vector in $\mathbb{F}^3$ different from $\mathbf{0}$ is one of the seven columns of H.

We now see what happens when errors are introduced during transmission. If an error occurs in the ith position, this means that the entry in the ith position in the received vector **v** is different from the entry in the ith position of the transmitted vector **b**; that is,

$$\mathbf{v} = \mathbf{b} + \mathbf{e}_i$$

Applying H to $\mathbf{v}$ yields

$$\begin{aligned} H\mathbf{v} &= H(\mathbf{b} + \mathbf{e}_i) \\ &= H\mathbf{b} + H\mathbf{e}_i \\ &= \mathbf{0} + (i\text{th column of } H) \end{aligned}$$

This is exactly what was described in Example 1.

Now let us see what happens if two errors are made, one in the ith position and one in the jth position. Then

$$\mathbf{v} = \mathbf{b} + \mathbf{e}_i + \mathbf{e}_j \qquad (i \neq j)$$

and

$$\begin{aligned} H\mathbf{v} &= H(\mathbf{b} + \mathbf{e}_i + \mathbf{e}_j) \\ &= H\mathbf{e}_i + H\mathbf{e}_j \\ &= (i\text{th column of } H) + (j\text{th column of } H) \end{aligned}$$

As mentioned above, there are only eight vectors in the image space of H—the seven columns of H and $\mathbf{0}$. So, $H\mathbf{v}$ will either be $\mathbf{0}$ or a column of H different from either the ith or the jth column. Therefore, the decoding procedure will either detect no errors or it will detect an error in a position different from either of the positions where the errors actually occurred. So, in certain highly improbable instances, the Hamming code can be worse than no code at all.

The Hamming code we have been discussing is one of the smallest Hamming codes. There are infinitely many such codes. This code requires seven letters to transmit four letters of information. Its efficiency rating is 4/7. Larger Hamming codes are more efficient.

To see how to construct a larger Hamming code we reexamine the code we have been discussing. The key to this code is the syndrome matrix H. From H we could determine the code words—the elements of Null(H). By applying H to the vector received, we could determine if and where an error occurred. Even the matrix M could have been constructed from H—it was chosen so that $\text{Im}(M) = \text{Null}(H)$. Moreover, the matrix H was quite easy to construct—its columns were just the seven nonzero vectors of $\mathbb{F}^3$.

Using the ideas discussed in the preceding paragraph, we will sketch the method for constructing larger Hamming codes. Specifically, we will construct a Hamming code having a four-rowed syndrome matrix H. The columns of this matrix will be the $2^4 - 1 = 15$ nonzero vectors of $\mathbb{F}^4$. Although there are a large number of 4×15 matrices with these columns, the matrix which will best suit our needs is

$$H = \left[\begin{array}{ccccccccccc|cccc} 1 & 0 & 1 & 1 & 1 & 0 & 0 & 0 & 1 & 1 & 1 & 1 & 0 & 0 & 0 \\ 1 & 1 & 0 & 1 & 1 & 0 & 1 & 1 & 0 & 0 & 1 & 0 & 1 & 0 & 0 \\ 1 & 1 & 1 & 0 & 1 & 1 & 0 & 1 & 0 & 1 & 0 & 0 & 0 & 1 & 0 \\ 1 & 1 & 1 & 1 & 0 & 1 & 1 & 0 & 1 & 0 & 0 & 0 & 0 & 0 & 1 \end{array}\right]$$

It is clear that $H: \mathbb{F}^{15} \to \mathbb{F}^4$, and since $\mathbf{e}_1, \mathbf{e}_2, \mathbf{e}_3, \mathbf{e}_4$ are columns of H, the image space of H is all of $\mathbb{F}^4$. Hence, the null space of H has dimension $15 - 4 = 11$.

The encoding matrix M is to be chosen so that $\text{Im}(M) = \text{Null}(H)$. So $M: \mathbb{F}^{11} \to \mathbb{F}^{15}$. Since the columns of M form a basis for the image space of M, to find the columns of M we need only construct a basis for the null space of H. Normally this would involve reducing H. However, by our choice of columns for H, we can read the conditions for choosing a basis for the null space of H directly from the rows of H. That is, the vector with coordinates $x_1, \ldots, x_{15}$ is in the null space of H if and only if

$$
\text{(1)} \qquad
\begin{aligned}
(x_1 + x_3 + x_4 + x_5 + x_9 + x_{10} + x_{11}) + x_{12} &= 0 \\
(x_1 + x_2 + x_4 + x_5 + x_7 + x_8 + x_{11}) + x_{13} &= 0 \\
(x_1 + x_2 + x_3 + x_5 + x_6 + x_8 + x_{10}) + x_{14} &= 0 \\
(x_1 + x_2 + x_3 + x_4 + x_6 + x_7 + x_9) + x_{15} &= 0
\end{aligned}
$$

Since the null space of H is equal to the image space of M, these equations represent conditions for a vector to be a column of M. Since $M: \mathbb{F}^{11} \to \mathbb{F}^{15}$, it maps an eleven-entry message vector to a fifteen-entry code vector. The first eleven entries of the code vector are the message itself and the last four entries are the coding information. Equations 1 tell us how to determine the last four entries in the code vector from the eleven digits in the message vector.

EXAMPLE 2 We encode the message 00011100011.

To find the twelfth entry in the code vector we use the first equation in Equations 1:

$$0 = (x_1 + x_3 + x_4 + x_5 + x_9 + x_{10} + x_{11}) + x_{12}$$

which is equivalent to the equation

$$x_{12} = x_1 + x_3 + x_4 + x_5 + x_9 + x_{10} + x_{11} \qquad \text{(since } a + b = 0 \text{ implies } a = b \text{ for } a, b \text{ in } \mathbb{F}\text{)}$$

which becomes

$$x_{12} = 0 + 0 + 1 + 1 + 0 + 1 + 1 = 0$$

In the same way we can use the remaining three equations to obtain x_{13}, x_{14}, and x_{15}. We have

$$x_{13} = 1 \qquad x_{14} = 1 \qquad x_{15} = 0$$

So the code vector corresponding to the message 00011100011 is

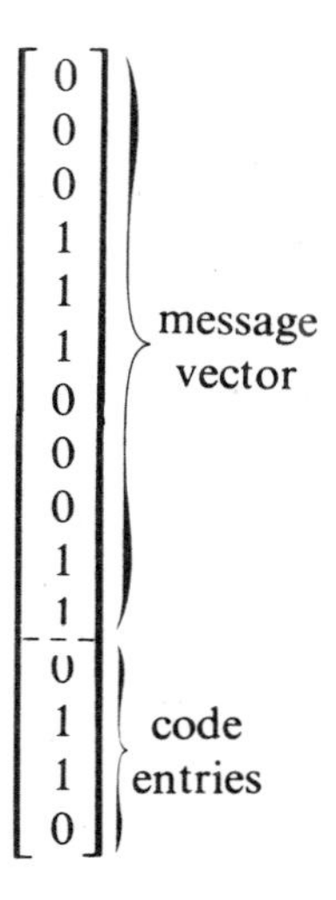

■

To make this more formal, we construct the encoding matrix M. To begin, we examine the syndrome matrix H more closely. The first eleven columns of H consist of all nonzero vectors in $\mathbb{F}^4$ that contain at least two 1's. The last four columns of H are just the 4×4 identity matrix I_4. We will denote the matrix made up of the first eleven columns of H by $\hat{H}$. Hence, we write H as

$$H = [\hat{H} \mid I_4]$$

With the notation introduced above we can put Equations 1 in matrix form as

$$\hat{H}\begin{bmatrix} x_1 \\ \vdots \\ x_{15} \end{bmatrix} + \begin{bmatrix} x_{12} \\ \vdots \\ x_{15} \end{bmatrix} = \begin{bmatrix} 0 \\ 0 \\ 0 \\ 0 \end{bmatrix}$$

or, since we are adding in $\mathbb{F}^4$, Equations 1 can be written as

$$\hat{H}\begin{bmatrix} x_1 \\ \vdots \\ x_{15} \end{bmatrix} = \begin{bmatrix} x_{12} \\ \vdots \\ x_{15} \end{bmatrix}$$

Now we are ready to construct M. M operates as follows:

$$M\colon \mathbb{F}^{11} \to \mathbb{F}^{15}$$

$$\begin{bmatrix} x_1 \\ \vdots \\ x_{11} \end{bmatrix} \mapsto \begin{bmatrix} x_1 \\ \vdots \\ x_{11} \\ \hline x_{12} \\ \vdots \\ x_{15} \end{bmatrix} \begin{matrix} \Big\}\text{message} \\ \\ \Big\}\text{code} \end{matrix}$$

Since

$$\text{message vector} = \begin{bmatrix} x_1 \\ \vdots \\ x_{11} \end{bmatrix} = I \begin{bmatrix} x_1 \\ \vdots \\ x_{11} \end{bmatrix} \quad \text{and} \quad \text{code entries} = \begin{bmatrix} x_{12} \\ \vdots \\ x_{15} \end{bmatrix} = \hat{H} \begin{bmatrix} x_1 \\ \vdots \\ x_{11} \end{bmatrix}$$

(where I is the 11×11 identity matrix), we see that M has the form

$$M = \left[\begin{array}{ccccccc} 1 & 0 & 0 & 0 & \cdots & 0 & 0 & 0 \\ 0 & 1 & 0 & 0 & \cdots & \vdots & \vdots & \vdots \\ 0 & 0 & 1 & 0 & \cdots & & & \\ \vdots & \vdots & \vdots & \vdots & \cdots & & & \\ 0 & 0 & 0 & 0 & \cdots & 0 & 0 & 1 \\ \hline 1 & 0 & 1 & 1 & \cdots & 1 & 1 & 1 \\ 1 & 1 & 0 & 1 & \cdots & 0 & 0 & 1 \\ 1 & 1 & 1 & 0 & \cdots & 0 & 1 & 0 \\ 1 & 1 & 1 & 1 & \cdots & 1 & 0 & 0 \end{array}\right] = \left[\begin{array}{c} I \\ \hline \hat{H} \end{array}\right]$$

The following example illustrates how to decode a message using this Hamming code.

EXAMPLE 3 We decode the message received as

$$1 \quad 0 \quad 0 \quad 1 \quad 1 \quad 1 \quad 0 \quad 0 \quad 0 \quad 1 \quad 1 \quad 0 \quad 1 \quad 1 \quad 0$$

We apply the syndrome matrix H to the vector with these entries:

$$\left[\begin{array}{ccccccccccc|cccc} 1 & 0 & 1 & 1 & 1 & 0 & 0 & 0 & 1 & 1 & 1 & 1 & 0 & 0 & 0 \\ 1 & 1 & 0 & 1 & 1 & 0 & 1 & 1 & 0 & 0 & 1 & 0 & 1 & 0 & 0 \\ 1 & 1 & 1 & 0 & 1 & 1 & 0 & 1 & 0 & 1 & 0 & 0 & 0 & 1 & 0 \\ 1 & 1 & 1 & 1 & 0 & 1 & 1 & 0 & 1 & 0 & 0 & 0 & 0 & 0 & 1 \end{array}\right] \left[\begin{array}{c} 1 \\ 0 \\ 0 \\ 1 \\ 1 \\ 1 \\ 0 \\ 0 \\ 0 \\ 1 \\ 1 \\ \hline 0 \\ 1 \\ 1 \\ 0 \end{array}\right] = \begin{bmatrix} 1 \\ 1 \\ 1 \\ 1 \end{bmatrix}$$

Since the image of the code vector is not the zero vector, there was an error. Moreover, since the image vector is the first column of H, the error occurred in the first position. Hence we decode the message as

$$0\,0\,0\,1\,1\,1\,0\,0\,0\,1\,1$$

(We have changed the first entry and dropped the last four entries that contain the encoding information.) ■

This Hamming code using the four-row syndrome H is considerably more efficient than the Hamming code in which H had only three rows. The efficiency for this code is $11/15 = 0.73$ compared to $4/7 = 0.57$ for the smaller code. Of course this is counterbalanced by the fact that the probability of having two or more errors is considerably higher in a fifteen-letter code than it is in a seven-letter code.

The Hamming codes we have discussed are just one family of error-correcting codes; there are numerous other families of error-correcting codes. Many of these other codes are capable of correcting multiple errors. However, these codes involve mathematical concepts beyond the scope of this text. The article "Coding Theory: A Counterexample To G. H. Hardy's Conception Of Applied Mathematics," by Norman Levinson in the March, 1970 issue of the *American Mathematical Monthly* describes some of these more difficult codes.

Exercises for Applications 5

1. Using the Hamming code with a three-row syndrome matrix, encode the following messages:

a. 1 1 0 1

b. 1 0 1 1 0 1 0 1

2. Decode the following messages using the Hamming code with the three-row syndrome matrix.

a. $\begin{bmatrix} 1 \\ 1 \\ 0 \\ 1 \\ \hline 0 \\ 0 \\ 1 \end{bmatrix}$ **b.** $\begin{bmatrix} 1 \\ 1 \\ 0 \\ 0 \\ \hline 0 \\ 0 \\ 1 \end{bmatrix}$ **c.** $\begin{bmatrix} 0 \\ 0 \\ 0 \\ 0 \\ \hline 1 \\ 1 \\ 0 \end{bmatrix}$

3. Suppose the message 0 1 1 0 were encoded and then transmitted. During transmission, errors occurred in the third and sixth positions. How would this message be decoded (using the three-row syndrome code)?

4. Using the four-row Hamming code, encode the message.

0 0 1 1 0 0 1 1 0 0 1

5. Using the four-row Hamming code decode the message that has entries

0 1 1 0 1 0 0 0 1 0 0 0 0 0 0

6. Find the tenth column of the 15×11 matrix M.

7. Explain why a code that repeated each number would be a poor code. (For example, such a code would encode the message 0 1 0 0 as 00 11 00 00).

8. Write the syndrome for a 5-row Hamming code. Also write the equations that describe the code words and fill in the missing dimensions in the expression

$$\mathbb{F}^{?} \xrightarrow{M} \mathbb{F}^{?} \xrightarrow{H} \mathbb{F}^{5}$$

★ **9.** The 3-row Hamming code can be studied purely as a subspace of $\mathbb{F}^7$ with some special properties. The subspace we consider is the subspace C of code words. Verify that C has the following properties:

a. C is a 4-dimensional *subspace* of $\mathbb{F}^7$.

b. None of the vectors $\mathbf{e}_1, \mathbf{e}_2, \ldots, \mathbf{e}_7$ is a code word.

c. No code word contains exactly two 1's.

d. If $\mathbf{c}$ is a code word, then $\mathbf{c} + \mathbf{e}_i$ is not a code word for $i = 1, 2, \ldots, 7$. (*Hint:* Use parts **a** and **b**.)

e. If $\mathbf{c}_1$ and $\mathbf{c}_2$ are code words and

$$\mathbf{c}_1 + \mathbf{e}_i = \mathbf{c}_2 + \mathbf{e}_j$$

then $\mathbf{c}_1 = \mathbf{c}_2$. (*Hint:* Use parts **a** and **c**.)

f. List all the elements of C.

g. List all the elements of $C + \mathbf{e}_2$.

h. The cosets of C are

$$C, C + \mathbf{e}_1, C + \mathbf{e}_2, \ldots, C + \mathbf{e}_7$$

(*Hint:* Show this by showing that no two cosets of C have any elements in common, and then by finding the total number of elements contained in the cosets listed. Compare this to the total number of elements in $\mathbb{F}^7$.)

Using these results, decoding a message can be implemented as follows: The message received is a vector $\mathbf{v}$ in $\mathbb{F}^7$. Find the coset containing $\mathbf{v}$. If $\mathbf{v}$ is in C, then no error (probably) occurred during transmission. If $\mathbf{v}$ is in the coset $C + \mathbf{e}_i$, then an error (most probably) occurred in the ith entry of $\mathbf{v}$. By changing the ith entry in $\mathbf{v}$ the error is corrected. This type of decoding is called **coset leader decoding**. Explain why coset leader decoding works.

Review Exercises

1. Define the following terms:

a. Null space of a matrix

b. Image space of a matrix

c. Rank of a matrix

d. Nullity of a matrix

e. Completely reduced matrix

2. Let T be a linear transformation from $\mathbb{R}^n$ to $\mathbb{R}^m$. What is the relationship between the rank and the nullity of T?

3. Completely reduce each of the following matrices:

a. $\begin{bmatrix} 1 & -2 & 3 \\ 2 & 3 & -4 \\ 4 & -1 & 2 \end{bmatrix}$

b. $\begin{bmatrix} 1 & 1 & 4 & -3 & 2 \\ 2 & 1 & 1 & 2 & -1 \end{bmatrix}$

c. $\begin{bmatrix} 0 & 1 & 2 \\ 1 & 3 & 1 \end{bmatrix}$

d. $\begin{bmatrix} 1 & 0 & 2 & 0 & 0 \\ 0 & 1 & 3 & -4 & 0 \\ 0 & 0 & 0 & 0 & 1 \end{bmatrix}$

e. $\begin{bmatrix} 1 & 1 & 2 & 1 & 3 \\ 1 & 1 & 2 & 3 & 5 \\ 3 & 3 & 6 & 0 & 6 \end{bmatrix}$

f. $\begin{bmatrix} 1 & -1 & 3 & 0 & 2 \\ -1 & 2 & 3 & 1 & 4 \\ 1 & 0 & 9 & 1 & 8 \end{bmatrix}$

4. a. Find a basis for the null space of each of the matrices of Exercise 3.

b. Find a basis for the image space of each of the matrices of Exercise 3.

c. Find the rank and the nullity of each of the matrices of Exercise 3.

5. Let $T: \mathbb{R}^3 \to \mathbb{R}^3$ be projection onto the line containing the vector

$$\mathbf{v} = \begin{bmatrix} -1 \\ 2 \\ 3 \end{bmatrix}$$

Find a basis for the null space of T and a basis for the image space of T.

6. Let $T: \mathbb{R}^3 \to \mathbb{R}^3$ be reflection about the line containing the vector

$$\mathbf{v} = \begin{bmatrix} -1 \\ 2 \\ 3 \end{bmatrix}$$

Find a basis for the null space of T and a basis for the image space of T.

6 Measurements

In this chapter we see how to find the lengths of vectors and measure the angle between two vectors. This leads us to the definition of the dot product. *We also discuss projections and use projections to find* orthogonal *bases* (*bases in which the basis vectors are mutually perpendicular*).

6.1 Length, Angles, Projections—The Dot Product

This section concerns measurement of lengths of vectors and the angles between vectors. We begin by considering lengths.

If $\mathbf{v} = \begin{bmatrix} x_1 \\ x_2 \end{bmatrix}$ is a vector in $\mathbb{R}^2$, then, by using the Pythagorean theorem, it is quite easy to see that the length of $\mathbf{v}$ is $\sqrt{x_1^2 + x_2^2}$, as shown in Figure 6.1.

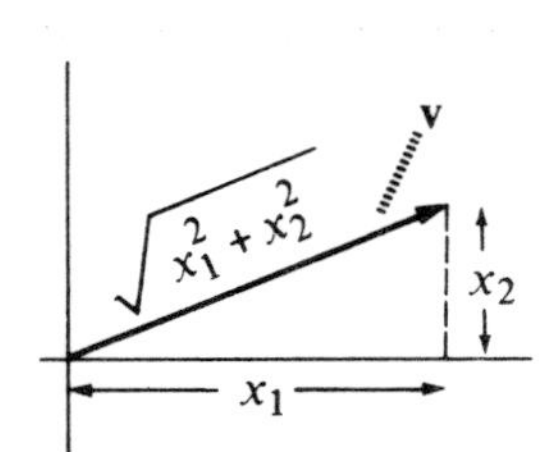

Figure 6.1

For a vector

$$\mathbf{v} = \begin{bmatrix} x_1 \\ x_2 \\ x_3 \end{bmatrix}$$

in $\mathbb{R}^3$, we can apply the Pythagorean theorem twice to find the length of $\mathbf{v}$. So, as we see in Figure 6.2, the length of $\mathbf{v}$ is $\sqrt{x_1^2 + x_2^2 + x_3^2}$.

We can generalize these results to define length for vectors in $\mathbb{R}^n$.

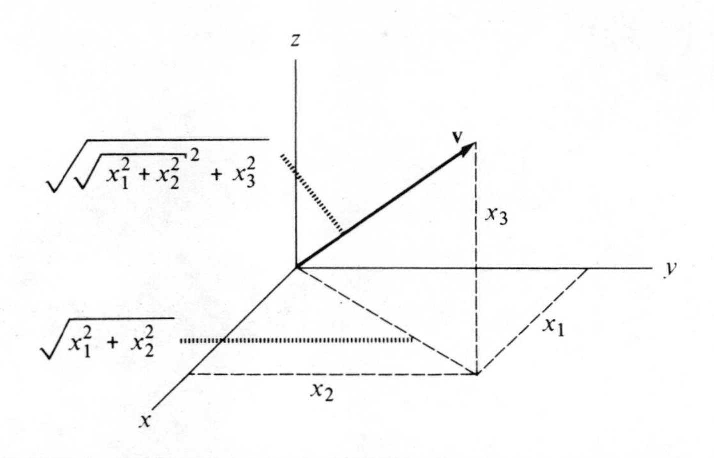

Figure 6.2

DEFINITION 6.1 If

$$\mathbf{v} = \begin{bmatrix} x_1 \\ x_2 \\ \vdots \\ x_n \end{bmatrix}$$

is a vector in $\mathbb{R}^n$, then the length of $\mathbf{v}$ is $\sqrt{x_1^2 + x_2^2 + \cdots + x_n^2}$. The length is denoted by $|\mathbf{v}|$.

EXAMPLE 1 The length of the vector

$$\mathbf{v} = \begin{bmatrix} 1 \\ -3 \\ 2 \\ 1 \\ 3 \end{bmatrix}$$

is $\sqrt{1^2 + (-3)^2 + 2^2 + 1^2 + 3^2} = \sqrt{1 + 9 + 4 + 1 + 9} = \sqrt{24}$ ■

Now we consider the problem of finding the angle between two vectors. We begin by looking at the special case in which

$$\mathbf{v} = \begin{bmatrix} x_1 \\ x_2 \end{bmatrix} \quad \text{and} \quad \mathbf{w} = \begin{bmatrix} y_1 \\ y_2 \end{bmatrix}$$

are vectors of $\mathbb{R}^2$ having length 1.

DEFINITION 6.2 A vector of length one is called a **unit** vector.

Let α be the angle between $\mathbf{v}$ and the x axis, and β be the angle between $\mathbf{w}$ and the x axis. Then, since $\mathbf{v}$ and $\mathbf{w}$ are unit vectors,

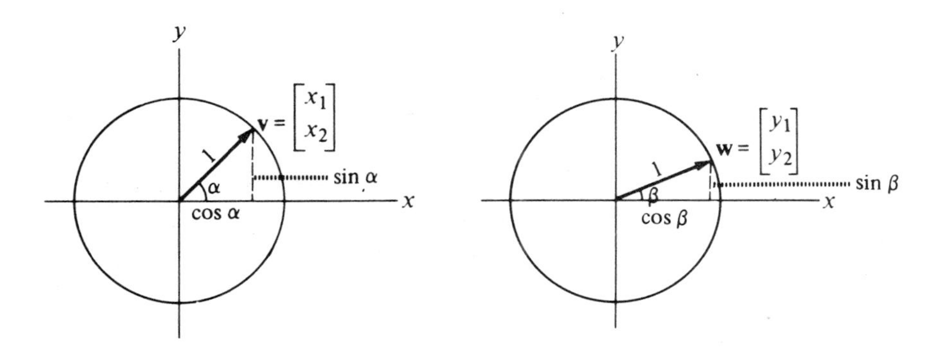

Figure 6.3

$$\mathbf{v} = \begin{bmatrix} x_1 \\ x_2 \end{bmatrix} = \begin{bmatrix} \cos\alpha \\ \sin\alpha \end{bmatrix} \quad \text{and} \quad \mathbf{w} = \begin{bmatrix} y_1 \\ y_2 \end{bmatrix} = \begin{bmatrix} \cos\beta \\ \sin\beta \end{bmatrix}$$

as is shown in Figure 6.3.

Now we are interested in finding the angle $\alpha - \beta$ between $\mathbf{v}$ and $\mathbf{w}$ (see Figure 6.4). We can apply the formula for the cosine of the difference of two angles from trigonometry to obtain:

$$\begin{aligned} \cos(\alpha - \beta) &= \cos\alpha\cos\beta + \sin\alpha\sin\beta \\ &= x_1 y_1 + x_2 y_2 \end{aligned}$$

We conclude that the cosine of the angle between the two unit vectors

$$\mathbf{v} = \begin{bmatrix} x_1 \\ x_2 \end{bmatrix} \quad \text{and} \quad \mathbf{w} = \begin{bmatrix} y_1 \\ y_2 \end{bmatrix}$$

is $(x_1 y_1 + x_2 y_2)$. From this it is quite easy to find the angle between the vectors using trigonometric tables or a calculator.

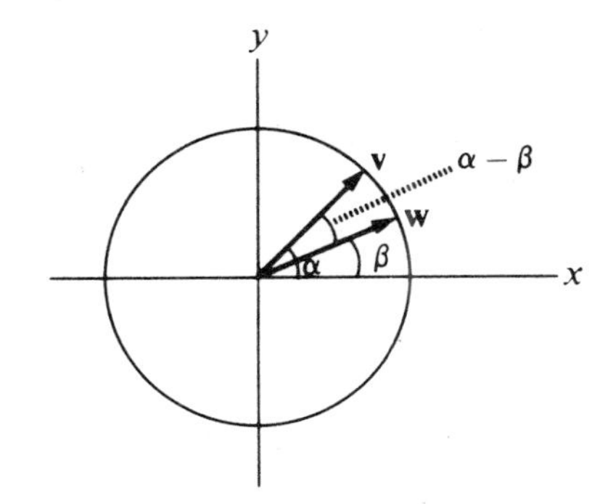

Figure 6.4

EXAMPLE 2 Find the angle between the unit vectors

$$\mathbf{v} = \begin{bmatrix} 3/5 \\ 4/5 \end{bmatrix} \quad \text{and} \quad \mathbf{w} = \begin{bmatrix} 12/13 \\ 5/13 \end{bmatrix}$$

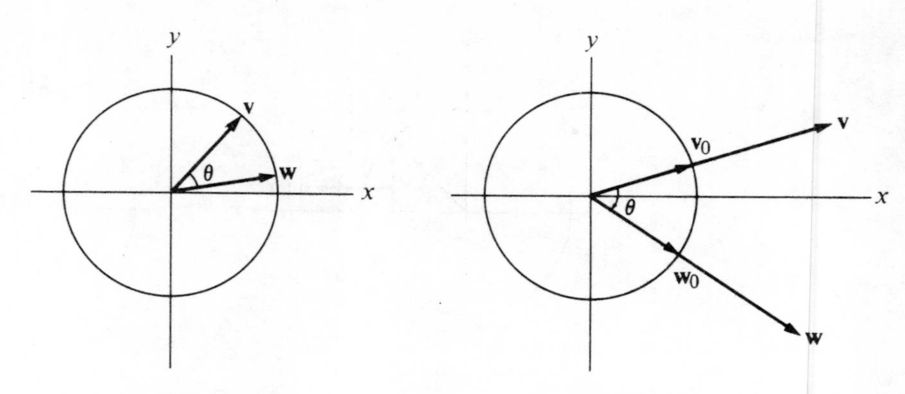

Figure 6.5 **Figure 6.6**

Refer to Figure 6.5. The cosine of the angle θ between these vectors is

$$\cos\theta = (\tfrac{3}{5})(\tfrac{12}{13}) + (\tfrac{4}{5})(\tfrac{5}{13}) = \tfrac{56}{65}$$

From a table or a calculator we find that $\theta = 30.51°$. ■

EXAMPLE 3 Find the cosine of the angle θ between the vectors

$$\mathbf{v} = \begin{bmatrix} 4 \\ 2 \end{bmatrix} \quad \text{and} \quad \mathbf{w} = \begin{bmatrix} 3 \\ -4 \end{bmatrix}$$

Refer to Figure 6.6.

This example is not quite as simple as the previous one, since $\mathbf{v}$ and $\mathbf{w}$ are not unit vectors. In fact

$$|\mathbf{v}| = \sqrt{20} \quad \text{and} \quad |\mathbf{w}| = \sqrt{25}$$

The vectors $\mathbf{v}_0 = \dfrac{1}{\sqrt{20}}\begin{bmatrix} 4 \\ 2 \end{bmatrix}$ and $\mathbf{w}_0 = \dfrac{1}{\sqrt{25}}\begin{bmatrix} 3 \\ -4 \end{bmatrix}$

are unit vectors, and they are parallel to $\mathbf{v}$ and $\mathbf{w}$. Thus the angle between $\mathbf{v}_0$ and $\mathbf{w}_0$ is the same as the angle between $\mathbf{v}$ and $\mathbf{w}$. Therefore

$$\begin{aligned} \cos\theta &= (x \text{ coordinate of } \mathbf{v}_0)(x \text{ coordinate of } \mathbf{w}_0) \\ &\quad + (y \text{ coordinate of } \mathbf{v}_0)(y \text{ coordinate of } \mathbf{w}_0) \\ &= (4/\sqrt{20})(3/\sqrt{25}) + (2/\sqrt{20})(-4/\sqrt{25}) \\ &= (12 - 8)/(\sqrt{20}\sqrt{25}) = 2/(5\sqrt{5}) \end{aligned}$$ ■

The method used in Example 3 to find the cosine of the angle between two nonunit vectors can be applied in general. If

$$\mathbf{v} = \begin{bmatrix} x_1 \\ x_2 \end{bmatrix} \quad \text{and} \quad \mathbf{w} = \begin{bmatrix} y_1 \\ y_2 \end{bmatrix}$$

are vectors of $\mathbb{R}^2$, then the angle between them is the same as the angle between the unit vectors

$$\mathbf{v}_0 = \frac{1}{|\mathbf{v}|}\mathbf{v} = \begin{bmatrix} \dfrac{x_1}{|\mathbf{v}|} \\ \dfrac{x_2}{|\mathbf{v}|} \end{bmatrix} \quad \text{and} \quad \mathbf{w}_0 = \frac{1}{|\mathbf{w}|}\mathbf{w} = \begin{bmatrix} \dfrac{y_1}{|\mathbf{w}|} \\ \dfrac{y_2}{|\mathbf{w}|} \end{bmatrix}$$

The cosine of the angle θ between $\mathbf{v}$ and $\mathbf{w}$ is

$$\cos\theta = \left(\frac{x_1}{|\mathbf{v}|}\right)\left(\frac{y_1}{|\mathbf{w}|}\right) + \left(\frac{x_2}{|\mathbf{v}|}\right)\left(\frac{y_2}{|\mathbf{w}|}\right) = \frac{x_1y_1 + x_2y_2}{|\mathbf{v}||\mathbf{w}|}$$

The numerator $x_1y_1 + x_2y_2$ is an example of a special product that plays a major role in the study of linear algebra.

DEFINITION 6.3

Let $\mathbf{v} = \begin{bmatrix} x_1 \\ x_2 \\ \vdots \\ x_n \end{bmatrix}$ and $\mathbf{w} = \begin{bmatrix} y_1 \\ y_2 \\ \vdots \\ y_n \end{bmatrix}$

be vectors in $\mathbb{R}^n$. The **dot product** $\mathbf{v}\cdot\mathbf{w}$ is defined as

$$\mathbf{v}\cdot\mathbf{w} = x_1y_1 + x_2y_2 + \cdots + x_ny_n$$

EXAMPLE 4 We find the dot product of

$$\mathbf{v} = \begin{bmatrix} 1 \\ 1 \\ 2 \\ 3 \end{bmatrix} \quad \text{and} \quad \mathbf{w} = \begin{bmatrix} -2 \\ 0 \\ 4 \\ 1 \end{bmatrix}$$

Then $\mathbf{v}\cdot\mathbf{w} = (1)(-2) + (1)(0) + (2)(4) + (3)(1) = 9$ ■

Using dot product notation we can rewrite the expression for the cosine of the angle θ between two vectors $\mathbf{v}$ and $\mathbf{w}$ of $\mathbb{R}^2$ as

$$\cos\theta = \frac{\mathbf{v}\cdot\mathbf{w}}{|\mathbf{v}||\mathbf{w}|}$$

and we can use this to give a coordinate-free description of the dot product as

$$\mathbf{v}\cdot\mathbf{w} = |\mathbf{v}||\mathbf{w}|\cos\theta$$

We can also use the dot product to describe the length of a vector.

If

$$\mathbf{v} = \begin{bmatrix} x_1 \\ x_2 \\ \vdots \\ x_n \end{bmatrix}$$

then

$$|\mathbf{v}| = \sqrt{x_1^2 + x_2^2 + \cdots + x_n^2} = \sqrt{\mathbf{v} \cdot \mathbf{v}}$$

Thus

$$|\mathbf{v}|^2 = \mathbf{v} \cdot \mathbf{v}$$

In the same way that we generalized the concept of length we can give meaning to the concept of angle between two vectors in $\mathbb{R}^n$.

DEFINITION 6.4 Let $\mathbf{v}$ and $\mathbf{w}$ be vectors in $\mathbb{R}^n$. Then the angle between $\mathbf{v}$ and $\mathbf{w}$ is an angle such that

$$\cos\theta = \frac{\mathbf{v} \cdot \mathbf{w}}{|\mathbf{v}||\mathbf{w}|}$$

(*Note*: We are using the fact $|\mathbf{v} \cdot \mathbf{w}| \leq |\mathbf{v}||\mathbf{w}|$ (see Exercise 7).)

EXAMPLE 5 Show that the vectors

$$\mathbf{v} = \begin{bmatrix} 1 \\ -1 \\ 3 \end{bmatrix} \quad \text{and} \quad \mathbf{w} = \begin{bmatrix} -9 \\ -3 \\ 2 \end{bmatrix}$$

are perpendicular.

We must show that the angle θ between these two vectors is a right angle. To do this we need only show that $\cos\theta = 0$. Thus, since

$$\cos\theta = \frac{\mathbf{v} \cdot \mathbf{w}}{|\mathbf{v}||\mathbf{w}|} = \frac{(1)(-9) + (-1)(-3) + (3)(2)}{[1^2 + (-1)^2 + 3^2]^{1/2}[(-9)^2 + (-3)^2 + 2^2]^{1/2}} = 0$$

we see that $\mathbf{v}$ and $\mathbf{w}$ are perpendicular. ■

The dot product is not only useful for finding lengths and angles, but can also be used to find projections.

The projection of a vector $\mathbf{v}$ onto a line l can be thought of as the vector $\mathbf{p}$ that lies along the line l and is the shadow cast by $\mathbf{v}$ when a light is placed above (or below) $\mathbf{v}$ so that its rays are perpendicular to the line l (see Figure 6.7).

Although this description gives us an intuitive idea of what we mean by projection, it is not very helpful from a mathematical point of view. To give a more precise definition, let us begin by letting $\mathbf{u}$ be a unit vector on the line l.

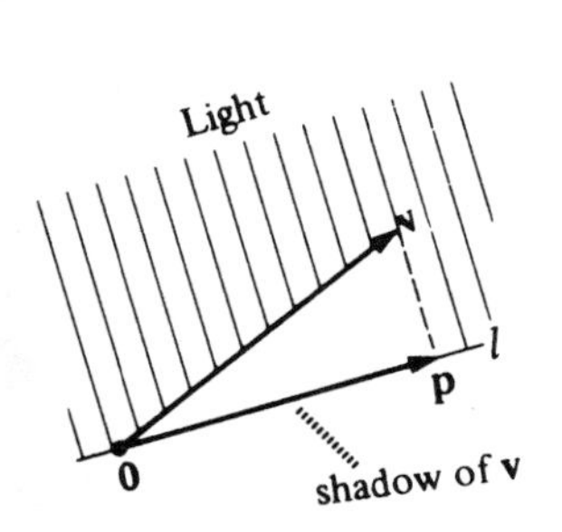

Figure 6.7

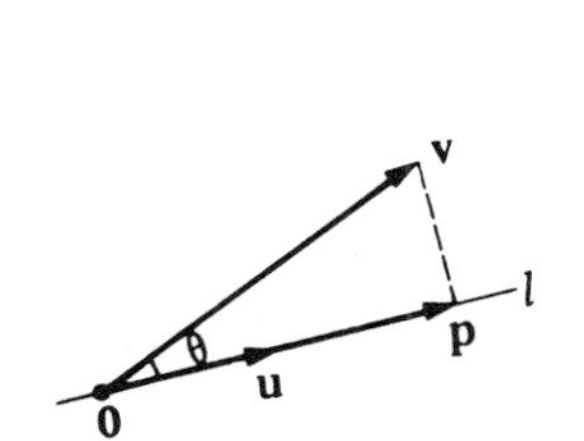

Figure 6.8

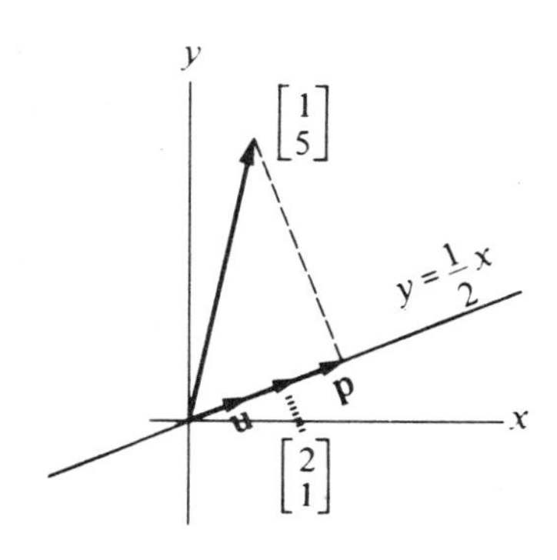

Figure 6.9

Then the projection of **v** along l is a vector parallel to **u**. If θ is the angle between **v** and **u**, then it is easily seen (refer to Figure 6.8) that the length of the projection vector **p** is

$$|\mathbf{p}| = |\mathbf{v}|\cos\theta = |\mathbf{v}||\mathbf{u}|\cos\theta \qquad \text{since } \mathbf{u} \text{ is a unit vector}$$

$$= \mathbf{v}\cdot\mathbf{u}$$

Hence $\mathbf{p} = (\mathbf{v}\cdot\mathbf{u})\mathbf{u}$. (*Note*: There is a minor technical problem when **u** and **p** point in opposite directions, but in this case θ is obtuse and $\cos\theta$ is negative, and things still work out as we want.)

EXAMPLE 6 Find the projection **p** of the vector $\mathbf{v} = \begin{bmatrix}1\\5\end{bmatrix}$ onto the line $y = (1/2)x$. Refer to Figure 6.9.

A vector along the line $y = (1/2)x$ is $\begin{bmatrix}2\\1\end{bmatrix}$. From our discussion, however, we see that we need a unit vector. So we choose

$$\mathbf{u} = \begin{bmatrix}2/\sqrt{5}\\1/\sqrt{5}\end{bmatrix}$$

Then

$$\mathbf{p} = (\mathbf{v}\cdot\mathbf{u})\mathbf{u}$$

$$= \left(\begin{bmatrix}1\\5\end{bmatrix}\cdot\begin{bmatrix}2/\sqrt{5}\\1/\sqrt{5}\end{bmatrix}\right)\begin{bmatrix}2/\sqrt{5}\\1/\sqrt{5}\end{bmatrix} = \frac{7}{\sqrt{5}}\begin{bmatrix}2/\sqrt{5}\\1/\sqrt{5}\end{bmatrix} = \begin{bmatrix}14/5\\7/5\end{bmatrix} \qquad \blacksquare$$

When speaking of projection, we generally speak of the projection of **v** in the direction of **w** instead of the projection of **v** onto the line l. Of course these two expressions mean the same thing when **w** is a vector lying along l. Moreover, when **w** is not a unit vector, we can use the unit vector

$$\mathbf{u} = \frac{1}{|\mathbf{w}|}\mathbf{w}$$

Hence the projection $\mathbf{p}$ in the direction of $\mathbf{w}$ is

$$\begin{aligned}
\mathbf{p} &= (\mathbf{v}\cdot\mathbf{u})\mathbf{u} \\
&= \left(\mathbf{v}\cdot\frac{\mathbf{w}}{|\mathbf{w}|}\right)\frac{\mathbf{w}}{|\mathbf{w}|} \\
&= \frac{\mathbf{w}\cdot\mathbf{v}}{\mathbf{w}\cdot\mathbf{w}}\mathbf{w} \qquad \text{since } |\mathbf{w}|^2 = \mathbf{w}\cdot\mathbf{w}
\end{aligned}$$

Now we generalize the definition of projection to $\mathbb{R}^n$.

DEFINITION 6.5 If $\mathbf{v}$ and $\mathbf{w}$ are vectors of $\mathbb{R}^n$, then the projection $\mathbf{p}$ in the direction of $\mathbf{w}$ is

$$\mathbf{p} = \frac{\mathbf{v}\cdot\mathbf{w}}{\mathbf{w}\cdot\mathbf{w}}\mathbf{w}$$

EXAMPLE 7 Find $\mathbf{p}$, the projection of $\mathbf{v}$ in the direction of $\mathbf{w}$, where

$$\mathbf{v} = \begin{bmatrix} 1 \\ 3 \\ -2 \end{bmatrix} \quad \text{and} \quad \mathbf{w} = \begin{bmatrix} -1 \\ 1 \\ 4 \end{bmatrix}$$

Using Definition 6.5, we have

$$\mathbf{p} = \frac{\mathbf{v}\cdot\mathbf{w}}{\mathbf{w}\cdot\mathbf{w}}\mathbf{w} = \frac{(1)(-1) + (3)(1) + (-2)(4)}{(-1)^2 + 1^2 + 4^2}\begin{bmatrix} -1 \\ 1 \\ 4 \end{bmatrix}$$

$$= \frac{-6}{18}\begin{bmatrix} -1 \\ 1 \\ 4 \end{bmatrix} = \begin{bmatrix} \frac{1}{3} \\ -\frac{1}{3} \\ -\frac{4}{3} \end{bmatrix} \quad \blacksquare$$

The following theorem summarizes the important properties of the dot product.

THEOREM 6.1 Let $\mathbf{v}$, $\mathbf{w}$, $\mathbf{w}_1$, $\mathbf{w}_2$ be vectors in $\mathbb{R}^n$ and c be a real number. Then the dot product has the following properties:

Formal Properties of the Dot Product:

1. $\mathbf{v}\cdot\mathbf{w}$ is a real number
2. $\mathbf{v}\cdot\mathbf{w} = \mathbf{w}\cdot\mathbf{v}$
3. $\mathbf{v}\cdot(\mathbf{w}_1 + \mathbf{w}_2) = \mathbf{v}\cdot\mathbf{w}_1 + \mathbf{v}\cdot\mathbf{w}_2$
4. $\mathbf{v}\cdot(c\mathbf{w}) = c(\mathbf{v}\cdot\mathbf{w})$
5. $\mathbf{v}\cdot\mathbf{v} \geq \mathbf{0}$, and equality holds only when $\mathbf{v} = \mathbf{0}$

Geometric Properties of the Dot Product:

1. $|\mathbf{v}| = \sqrt{\mathbf{v} \cdot \mathbf{v}}$
2. If θ is the angle between $\mathbf{v}$ and $\mathbf{w}$, then $\cos\theta = \dfrac{\mathbf{v} \cdot \mathbf{w}}{|\mathbf{v}||\mathbf{w}|}$
3. The projection $\mathbf{p}$ of $\mathbf{v}$ in the direction of $\mathbf{w}$ is $\mathbf{p} = \dfrac{\mathbf{v} \cdot \mathbf{w}}{\mathbf{w} \cdot \mathbf{w}}\mathbf{w}$

Exercise Set 6.1

1. Find the lengths of the following vectors:

a. $\begin{bmatrix} 1 \\ -1 \end{bmatrix}$ **b.** $(1/2)\begin{bmatrix} 3 \\ 1 \\ 2 \end{bmatrix}$ **c.** $\begin{bmatrix} 1 \\ 1 \\ -1 \\ 2 \end{bmatrix}$

2. Find the cosine of the angle between each of the following pairs of vectors:

a. $\begin{bmatrix} 1 \\ 2 \end{bmatrix}, \begin{bmatrix} 4 \\ 1 \end{bmatrix}$ **b.** $\begin{bmatrix} 1 \\ 2 \end{bmatrix}, \begin{bmatrix} -3 \\ -1 \end{bmatrix}$

c. $\begin{bmatrix} 1 \\ 1 \\ -3 \end{bmatrix}, \begin{bmatrix} 2 \\ 3 \\ 4 \end{bmatrix}$

3. Let $\mathbf{v} = \begin{bmatrix} 1 \\ -2 \\ 2 \end{bmatrix}$

a. Find two vectors parallel to $\mathbf{v}$ with length 1.

b. Find two vectors parallel to $\mathbf{v}$ with length 5.

c. Find two linearly independent vectors perpendicular to $\mathbf{v}$.

4. Do Exercise 3 with $\mathbf{v} = \begin{bmatrix} -1 \\ 1 \\ 2 \end{bmatrix}$

5. Find the projection of $\mathbf{v}$ in the direction of $\mathbf{w}$ where:

a. $\mathbf{v} = \begin{bmatrix} 1 \\ -1 \end{bmatrix}$ and $\mathbf{w} = \begin{bmatrix} 0 \\ 1 \end{bmatrix}$

b. $\mathbf{v} = \begin{bmatrix} 2 \\ -3 \end{bmatrix}$ and $\mathbf{w} = \begin{bmatrix} 4 \\ 2 \end{bmatrix}$

c. $\mathbf{v} = \begin{bmatrix} 1 \\ -1 \\ 2 \end{bmatrix}$ and $\mathbf{w} = \begin{bmatrix} 2 \\ 0 \\ -4 \end{bmatrix}$

6. Let $\mathbf{v} = \begin{bmatrix} a \\ b \end{bmatrix}$ and $\mathbf{v}_p = \begin{bmatrix} -b \\ a \end{bmatrix}$.

a. Show that $|\mathbf{v}| = |\mathbf{v}_p|$ and that $\mathbf{v}$ is perpendicular to $\mathbf{v}_p$.

b. Let $\mathbf{w}$ be a vector in $\mathbb{R}^2$, and θ the angle between $\mathbf{v}$ and $\mathbf{w}$. Show that $|\mathbf{v}_p \cdot \mathbf{w}| = |\mathbf{v}||\mathbf{w}|\sin\theta$.

c. Show that the area of the parallelogram having the vectors $\mathbf{v}$ and $\mathbf{w}$ as two of its sides is $|\mathbf{v}_p \cdot \mathbf{w}|$. Refer to Figure 6.10.

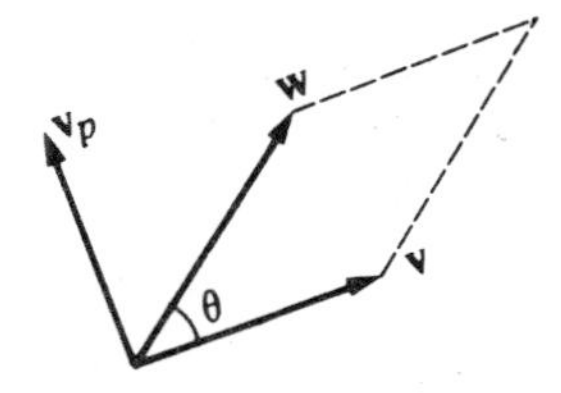

Figure 6.10

★**7.** In this exercise we outline the steps in the proof of the fact that

$$|\mathbf{v} \cdot \mathbf{w}| \le |\mathbf{v}||\mathbf{w}|$$

This is called the **Schwarz inequality** and is an essential fact when we define the cosine of the angle between two vectors in $\mathbb{R}^n$ ($n \ge 4$) to be

$$\cos\theta = \frac{\mathbf{u} \cdot \mathbf{v}}{|\mathbf{u}||\mathbf{v}|}$$

At first glance it may seem that this inequality could be easily proved by just expanding both sides and doing some obvious cancellations. If you try this, you will soon find that things get quite messy and this method does not lead to an obvious proof. The following approach yields a simple but ingenious proof of this result.

a. Let $p(x)$ be the polynomial

$$p(x) = ax^2 + 2bx + c$$

Show that if $p(x) \geq 0$ for all real numbers x, then

$$b^2 - ac \leq 0$$

b. Let $\mathbf{u}, \mathbf{v}$ be vectors in $\mathbb{R}^n$. Show that for any real number x,

$$\begin{aligned} 0 &\leq (x\mathbf{u} + \mathbf{v}) \cdot (x\mathbf{u} + \mathbf{v}) \\ &= (\mathbf{u} \cdot \mathbf{u})x^2 + 2(\mathbf{u} \cdot \mathbf{v})x + (\mathbf{v} \cdot \mathbf{v}) \end{aligned}$$

c. Use parts **a** and **b** to prove that

$$(\mathbf{u} \cdot \mathbf{v})^2 - (\mathbf{u} \cdot \mathbf{u})(\mathbf{v} \cdot \mathbf{v}) \leq 0$$

and then complete the proof of the Schwartz inequality.

8. Let $\mathbf{u}_1, \mathbf{u}_2, \mathbf{u}_3$ be three nonzero vectors of $\mathbb{R}^3$ that are mutually perpendicular. Show that $\{\mathbf{u}_1, \mathbf{u}_2, \mathbf{u}_3\}$ is a basis for $\mathbb{R}^3$.

★ **9.** Let $\mathbf{w}$ be a vector in $\mathbb{R}^n$. Show that the set P of all vectors of $\mathbb{R}^n$ perpendicular to $\mathbf{w}$ is a subspace of $\mathbb{R}^n$.

10. Determine why each of the following statements does not make sense:

a. $\mathbf{u} \cdot (\mathbf{v} \cdot \mathbf{w})$ **b.** $\mathbf{u} + (\mathbf{v} \cdot \mathbf{w})$

c. $M(\mathbf{u} \cdot \mathbf{v})$ where M is a matrix

11. Let $\mathbf{v} = \begin{bmatrix} x \\ y \\ z \end{bmatrix}$ be a vector in $\mathbb{R}^3$, and let α, β, γ be the angles between $\mathbf{v}$ and the x, y, and z axes, respectively. Show that

$$\cos\alpha = x/\sqrt{x^2 + y^2 + z^2}$$
$$\cos\beta = y/\sqrt{x^2 + y^2 + z^2}$$
$$\cos\gamma = z/\sqrt{x^2 + y^2 + z^2}$$

The numbers $\cos\alpha$, $\cos\beta$, and $\cos\gamma$ are called the **direction cosines** of $\mathbf{v}$.

12. a. Let $\mathbf{w}(t) = t\mathbf{w}$ be the equation of a line l through the origin. Show that the projection $\mathbf{p}$ of a vector $\mathbf{v}$ along l has the following properties:

i. $\mathbf{p}$ is on the line l.

ii. The vector $\mathbf{v} - \mathbf{p}$ is perpendicular to every vector on l.

b. Draw a picture showing the vectors $\mathbf{v}$, $\mathbf{w}$, $\mathbf{p}$, $\mathbf{v} - \mathbf{p}$, and l.

c. Let $\mathbf{w}(t_1, t_2) = t_1\mathbf{w}_1 + t_2\mathbf{w}_2$ be the equation of a plane P that contains the origin. The projection $\mathbf{p}$ of a vector $\mathbf{v}$ on the plane P is defined as

$$\mathbf{p} = \left(\frac{\mathbf{v} \cdot \mathbf{w}_1}{\mathbf{w}_1 \cdot \mathbf{w}_1}\right)\mathbf{w}_1 + \left(\frac{\mathbf{v} \cdot \mathbf{w}_2}{\mathbf{w}_2 \cdot \mathbf{w}_2}\right)\mathbf{w}_2$$

Show that $\mathbf{p}$ has the following properties:

i. $\mathbf{p}$ lies in the plane P.

ii. The vector $\mathbf{v} - \mathbf{p}$ is perpendicular to every vector of P.

d. Draw a picture showing the vectors $\mathbf{v}$, $\mathbf{w}$, $\mathbf{p}$, $\mathbf{v} - \mathbf{p}$, and the plane P.

13. a. Show that if $\mathbf{v}$ is a vector in $\mathbb{R}^n$ such that $\mathbf{v}$ is perpendicular to itself, then $\mathbf{v} = \mathbf{0}$.

b. Show that the only vector in $\mathbb{R}^n$ that is perpendicular to every other vector in $\mathbb{R}^n$ is the zero vector.

6.2 Orthonormal Bases

Although there are numerous possible bases for a given vector space, some bases are more useful than others. We illustrate this by looking at two different bases of a subspace of $\mathbb{R}^3$.

The subspace W has two different bases

$$W = \left\langle \mathbf{w}_1 = \begin{bmatrix} 3 \\ -1 \\ -1 \end{bmatrix}, \ \mathbf{w}_2 = \begin{bmatrix} 1 \\ 1 \\ -1 \end{bmatrix} \right\rangle$$

and
$$W = \left\langle \mathbf{u}_1 = \begin{bmatrix} 1/\sqrt{2} \\ -1/\sqrt{2} \\ 0 \end{bmatrix}, \ \mathbf{u}_2 = \begin{bmatrix} 1/\sqrt{3} \\ 1/\sqrt{3} \\ -1/\sqrt{3} \end{bmatrix} \right\rangle$$

It would appear that the basis $\{\mathbf{w}_1, \mathbf{w}_2\}$ is the better basis because its vectors involve no square roots or fractions. This is not the case. The utility of a basis is determined by the ease with which one can find the expression for a vector as a linear combination of the basis vectors. For example, if we wish to express the vector

$$\mathbf{v} = \begin{bmatrix} 1 \\ -5 \\ 2 \end{bmatrix}$$

as a linear combination of $\mathbf{w}_1$ and $\mathbf{w}_2$, we must solve the equation

$$\begin{bmatrix} 1 \\ -5 \\ 2 \end{bmatrix} = x_1 \begin{bmatrix} 3 \\ -1 \\ -1 \end{bmatrix} + x_2 \begin{bmatrix} 1 \\ 1 \\ -1 \end{bmatrix}$$

which involves reducing the matrix

$$\left[\begin{array}{rr|r} 3 & 1 & 1 \\ -1 & 1 & -5 \\ -1 & -1 & 2 \end{array}\right]$$

and is quite tedious.

On the other hand, the vectors $\mathbf{u}_1$ and $\mathbf{u}_2$ that form the other basis for W have two special properties that allow us to easily express $\mathbf{v}$ as a linear combination of $\mathbf{u}_1$ and $\mathbf{u}_2$. These properties are

1. The basis vectors $\mathbf{u}_1$ and $\mathbf{u}_2$ are perpendicular.

2. $\mathbf{u}_1$ and $\mathbf{u}_2$ are unit vectors.

From the first property, we can show that if

$$\mathbf{v} = x_1\mathbf{u}_1 + x_2\mathbf{u}_2$$

then
$$x_1\mathbf{u}_1 = \text{the projection of } \mathbf{v} \text{ in the direction of } \mathbf{u}_1$$
$$x_2\mathbf{u}_2 = \text{the projection of } \mathbf{v} \text{ in the direction of } \mathbf{u}_2$$

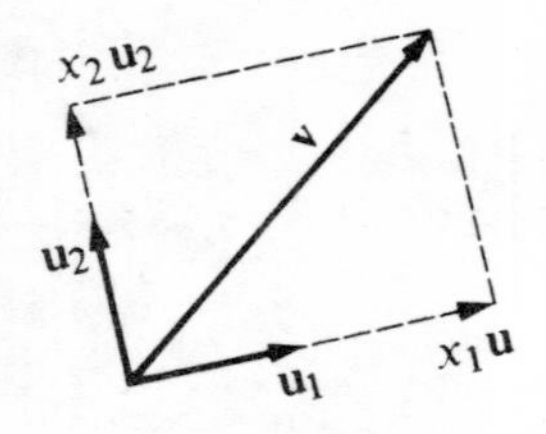

Figure 6.11

To see why this is true, consider Figure 6.11. Since $\mathbf{u}_1$ is perpendicular to $\mathbf{u}_2$, the vector $x_1\mathbf{u}_1$ is the projection of $\mathbf{v}$ in the direction of $\mathbf{u}_1$; similarly, $x_2\mathbf{u}_2$ is the projection of $\mathbf{v}$ in the direction of $\mathbf{u}_2$.

From the second property, we see that $\mathbf{u}_1 \cdot \mathbf{u}_1 = 1$ and $\mathbf{u}_2 \cdot \mathbf{u}_2 = 1$. So

$$x_1\mathbf{u}_1 = (\mathbf{u}_1 \cdot \mathbf{v})\mathbf{u}_1 \quad \text{and} \quad x_2\mathbf{u}_2 = (\mathbf{u}_2 \cdot \mathbf{v})\mathbf{u}_2$$

Therefore
$$\mathbf{v} = (\mathbf{u}_1 \cdot \mathbf{v})\mathbf{u}_1 + (\mathbf{u}_2 \cdot \mathbf{v})\mathbf{u}_2$$

In the following example we show how these ideas can be used in a specific case.

EXAMPLE 1 We express the vector
$$\mathbf{v} = \begin{bmatrix} 1 \\ -5 \\ 2 \end{bmatrix}$$

as a linear combination of the mutually perpendicular unit vectors

$$\mathbf{u}_1 = \begin{bmatrix} 1/\sqrt{2} \\ -1/\sqrt{2} \\ 0 \end{bmatrix} \quad \text{and} \quad \mathbf{u}_2 = \begin{bmatrix} 1/\sqrt{3} \\ 1/\sqrt{3} \\ -1/\sqrt{3} \end{bmatrix}$$

Since
$$\mathbf{u}_1 \cdot \mathbf{v} = \begin{bmatrix} 1/\sqrt{2} \\ -1/\sqrt{2} \\ 0 \end{bmatrix} \cdot \begin{bmatrix} 1 \\ -5 \\ 2 \end{bmatrix} = 6/\sqrt{2}$$

and
$$\mathbf{u}_2 \cdot \mathbf{v} = \begin{bmatrix} 1/\sqrt{3} \\ 1/\sqrt{3} \\ -1/\sqrt{3} \end{bmatrix} \cdot \begin{bmatrix} 1 \\ -5 \\ 2 \end{bmatrix} = -6/\sqrt{3}$$

We have
$$\begin{bmatrix} 1 \\ -5 \\ 2 \end{bmatrix} = (6/\sqrt{2})\begin{bmatrix} 1/\sqrt{2} \\ -1/\sqrt{2} \\ 0 \end{bmatrix} + (-6/\sqrt{3})\begin{bmatrix} 1/\sqrt{3} \\ 1/\sqrt{3} \\ -1/\sqrt{3} \end{bmatrix}$$ ■

DEFINITION 6.6 A basis $\{\mathbf{v}_1, \mathbf{v}_2, \ldots, \mathbf{v}_n\}$ is called **orthogonal** if each basis vector $\mathbf{v}_i$ is perpendicular (orthogonal) to all the other basis vectors.

DEFINITION 6.7 A basis $\{\mathbf{u}_1, \mathbf{u}_2, \ldots, \mathbf{u}_n\}$ is called **orthonormal** if it is orthogonal and each of the $\mathbf{u}_i$'s is a unit vector.

The following theorem generalizes the ideas we have just discussed.

THEOREM 6.2

> Let $\{\mathbf{u}_1, \mathbf{u}_2, \ldots, \mathbf{u}_n\}$ be an orthonormal basis for the vector space W. Then, for any vector $\mathbf{w}$ of W, the unique expression for $\mathbf{w}$ as a linear combination of these basis vectors is
>
> $$\mathbf{w} = (\mathbf{w} \cdot \mathbf{u}_1)\mathbf{u}_1 + (\mathbf{w} \cdot \mathbf{u}_2)\mathbf{u}_2 + \cdots + (\mathbf{w} \cdot \mathbf{u}_n)\mathbf{u}_n$$

This theorem is not only used to express a vector as a linear combination of basis vectors, but also to determine whether or not a vector is in a particular subspace. This is done as follows: Let V be a vector space and W the subspace of V that has the orthonormal basis $\{\mathbf{u}_1, \mathbf{u}_2, \ldots, \mathbf{u}_k\}$. Suppose $\mathbf{v}$ is a vector of V, and we wish to determine if $\mathbf{v}$ is in the subspace W. We form the vector

$$\mathbf{v}^* = (\mathbf{v} \cdot \mathbf{u}_1)\mathbf{u}_1 + (\mathbf{v} \cdot \mathbf{u}_2)\mathbf{u}_2 + \cdots + (\mathbf{v} \cdot \mathbf{u}_k)\mathbf{u}_k$$

The vector $\mathbf{v}^*$ is in W since it is a linear combination of the basis vectors for W (in fact, $\mathbf{v}^*$ is the projection of $\mathbf{v}$ on the subspace W). If $\mathbf{v}$ is in W, then, by Theorem 6.2, $\mathbf{v} = \mathbf{v}^*$. If $\mathbf{v}$ is not in W, then it cannot equal a vector in W, so $\mathbf{v} \neq \mathbf{v}^*$. Thus $\mathbf{v}$ is in W if and only if $\mathbf{v} = \mathbf{v}^*$, as the following example illustrates.

EXAMPLE 2 Let W be the subspace of $\mathbb{R}^3$ with orthonormal basis

$$\left\{ \mathbf{u}_1 = \begin{bmatrix} 1/\sqrt{3} \\ 1/\sqrt{3} \\ 1/\sqrt{3} \end{bmatrix}, \quad \mathbf{u}_2 = \begin{bmatrix} -1/\sqrt{6} \\ 2/\sqrt{6} \\ -1/\sqrt{6} \end{bmatrix} \right\}$$

We will show that the vector $\mathbf{v} = \begin{bmatrix} -3 \\ 5 \\ 4 \end{bmatrix}$ is not in W.

We begin by finding $\mathbf{v}^*$

$$\mathbf{v}^* = (\mathbf{v} \cdot \mathbf{u}_1)\mathbf{u}_1 + (\mathbf{v} \cdot \mathbf{u}_2)\mathbf{u}_2$$

We compute the coefficients

$$\mathbf{v} \cdot \mathbf{u}_1 = \begin{bmatrix} -3 \\ 5 \\ 4 \end{bmatrix} \cdot \begin{bmatrix} 1/\sqrt{3} \\ 1/\sqrt{3} \\ 1/\sqrt{3} \end{bmatrix} = \frac{-3 + 5 + 4}{\sqrt{3}} = \frac{6}{\sqrt{3}}$$

$$\mathbf{v} \cdot \mathbf{u}_2 = \begin{bmatrix} -3 \\ 5 \\ 4 \end{bmatrix} \cdot \begin{bmatrix} -1/\sqrt{6} \\ 2/\sqrt{6} \\ -1/\sqrt{6} \end{bmatrix} = \frac{3 + 10 - 4}{\sqrt{6}} = \frac{9}{\sqrt{6}}$$

So
$$\mathbf{v}^* = (6/\sqrt{3})\begin{bmatrix} 1/\sqrt{3} \\ 1/\sqrt{3} \\ 1/\sqrt{3} \end{bmatrix} + (9/\sqrt{6})\begin{bmatrix} -1/\sqrt{6} \\ 2/\sqrt{6} \\ -1/\sqrt{6} \end{bmatrix} = \begin{bmatrix} 1/2 \\ 5 \\ 1/2 \end{bmatrix}$$

Thus $\mathbf{v} \neq \mathbf{v}^*$ and so $\mathbf{v}$ is not in W. ■

Now that we have seen that orthonormal bases are useful, the question that still remains is how to transform an arbitrary basis of a vector space into an orthonormal basis for that vector space. The main problem is finding a way to get an orthogonal basis for the vector space. Once you have an orthogonal basis, you need only divide each basis vector by its length to obtain an orthonormal basis. So our goal will be to transform a nonorthogonal basis into an orthogonal basis.

We begin with a vector space W of dimension 2 with nonorthogonal basis $\{\mathbf{w}_1, \mathbf{w}_2\}$. To make an orthogonal basis for W, we will replace $\mathbf{w}_2$ by a vector $\mathbf{v}_2$ such that $\{\mathbf{w}_1, \mathbf{v}_2\}$ is an orthogonal basis for W. Therefore, we are looking for a vector $\mathbf{v}_2$ which is perpendicular to $\mathbf{w}_1$ and is a linear combination of $\mathbf{w}_1$ and $\mathbf{w}_2$. Refer to Figure 6.12.

We make use of projections to find $\mathbf{v}_2$. To begin, we form the rectangle with $\mathbf{w}_2$ as diagonal and one side along the line containing $\mathbf{w}_1$, as in Figure 6.13. The vector that forms the side of the rectangle lying along $\mathbf{w}_1$ is $\mathbf{p}$, the projection of $\mathbf{w}_2$ in the direction of $\mathbf{w}_1$. The vector $\mathbf{n}$ along the other side of the rectangle has the property that

$$\mathbf{w}_2 = \mathbf{p} + \mathbf{n} \quad \text{(parallelogram rule for adding vectors)}$$

Hence
$$\mathbf{n} = \mathbf{w}_2 - \mathbf{p}$$

Explicitly
$$\mathbf{n} = \mathbf{w}_2 - \left(\frac{\mathbf{w}_2 \cdot \mathbf{w}_1}{\mathbf{w}_1 \cdot \mathbf{w}_1}\right)\mathbf{w}_1$$

Thus $\mathbf{n}$ is a linear combination of $\mathbf{w}_1$ and $\mathbf{w}_2$, and $\mathbf{n}$ is perpendicular to $\mathbf{w}_1$. Therefore, letting $\mathbf{v}_2 = \mathbf{n}$ gives an orthogonal basis $\{\mathbf{w}_1, \mathbf{v}_2\}$ for W. Dividing each basis vector by its length gives the orthonormal basis

$$\left\{\mathbf{u}_1 = \frac{\mathbf{w}_1}{|\mathbf{w}_1|}, \quad \mathbf{u}_2 = \frac{\mathbf{v}_2}{|\mathbf{v}_2|}\right\}$$

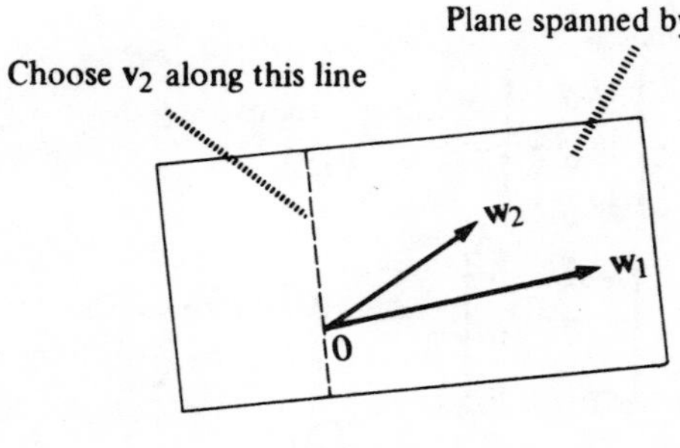

Plane spanned by $\mathbf{w}_1$ and $\mathbf{w}_2$

Choose $\mathbf{v}_2$ along this line

$\mathbf{w}_2$

$\mathbf{w}_1$

0

Figure 6.12

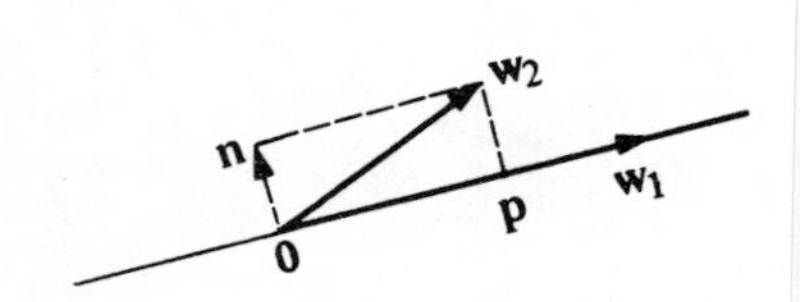

Figure 6.13

DEFINITION 6.8 Let $\mathbf{w}_1$ and $\mathbf{w}_2$ be vectors of $\mathbb{R}^n$ and $\mathbf{p}$ be the projection of $\mathbf{w}_2$ in the direction of $\mathbf{w}_1$. The vector $\mathbf{n} = \mathbf{w}_2 - \mathbf{p}$ is called the **projection of $\mathbf{w}_2$ orthogonal to $\mathbf{w}_1$** (see Figure 6.14).

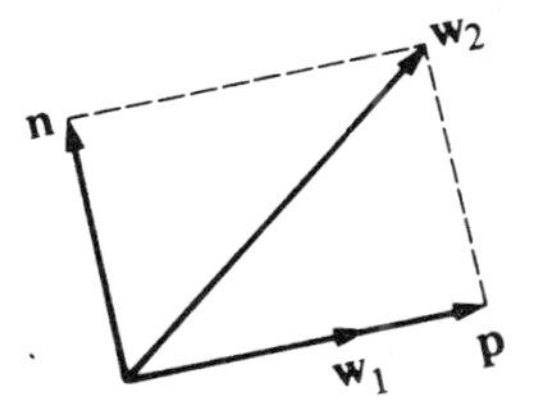

Figure 6.14

EXAMPLE 3 Construct an orthonormal basis for the subspace W of $\mathbb{R}^3$ where

$$W = \left\langle \mathbf{w}_1 = \begin{bmatrix} 1 \\ 1 \\ 2 \end{bmatrix}, \quad \mathbf{w}_2 = \begin{bmatrix} 3 \\ -2 \\ 1 \end{bmatrix} \right\rangle$$

First we find the projection $\mathbf{p}$ of $\mathbf{w}_2$ in the direction of $\mathbf{w}_1$.

$$\mathbf{p} = \left(\frac{\mathbf{w}_2 \cdot \mathbf{w}_1}{\mathbf{w}_1 \cdot \mathbf{w}_1}\right)\mathbf{w}_1 = \frac{(1)(3) + (1)(-2) + (2)(1)}{1^2 + 1^2 + 2^2} \begin{bmatrix} 1 \\ 1 \\ 2 \end{bmatrix} = (1/2)\begin{bmatrix} 1 \\ 1 \\ 2 \end{bmatrix} = \begin{bmatrix} 1/2 \\ 1/2 \\ 1 \end{bmatrix}$$

Next we find $\mathbf{n}$, the projection of $\mathbf{w}_1$ orthogonal to $\mathbf{w}_2$

$$\mathbf{n} = \mathbf{w}_2 - \mathbf{p} = \begin{bmatrix} 3 \\ -2 \\ 1 \end{bmatrix} - \begin{bmatrix} 1/2 \\ 1/2 \\ 1 \end{bmatrix} = \begin{bmatrix} 5/2 \\ -5/2 \\ 0 \end{bmatrix} = \mathbf{v}_2$$

So an orthogonal basis for W is

$$\left\langle \begin{bmatrix} 1 \\ 1 \\ 2 \end{bmatrix}, \begin{bmatrix} 5/2 \\ -5/2 \\ 0 \end{bmatrix} \right\rangle$$

To get an orthonormal basis, divide each of these vectors by its length.

$$\left\{ \mathbf{u}_1 = \begin{bmatrix} 1/\sqrt{6} \\ 1/\sqrt{6} \\ 2/\sqrt{6} \end{bmatrix}, \quad \mathbf{u}_2 = (\sqrt{2}/5)\begin{bmatrix} 5/2 \\ -5/2 \\ 0 \end{bmatrix} = \begin{bmatrix} \sqrt{2}/2 \\ -\sqrt{2}/2 \\ 0 \end{bmatrix} \right\} \quad \blacksquare$$

For subspaces W of dimension 3 the procedure is much the same. Suppose $W = \langle \mathbf{w}_1, \mathbf{w}_2, \mathbf{w}_3 \rangle$. Using the procedure for finding orthogonal bases for subspaces of dimension 2, we can find an orthogonal basis $\{\mathbf{w}_1, \mathbf{v}_2\}$ for the subspace spanned by $\mathbf{w}_1$ and $\mathbf{w}_2$, so that $W = \{\mathbf{w}_1, \mathbf{v}_2, \mathbf{w}_3\}$.

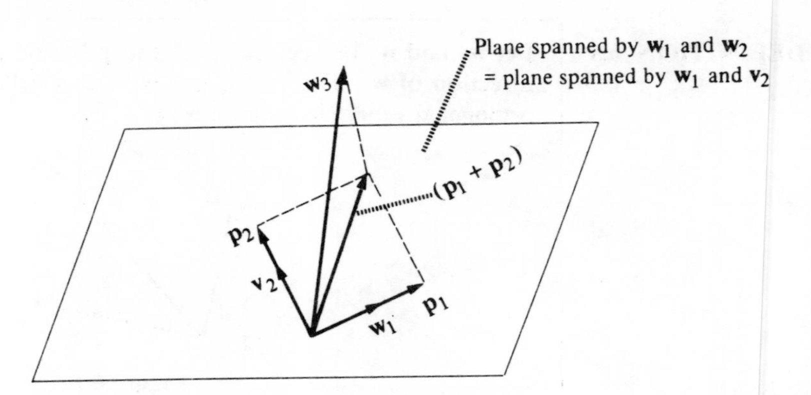

Figure 6.15

We let $\mathbf{p}_1$ be the projection of $\mathbf{w}_3$ (see Figure 6.15) in the direction of $\mathbf{w}_1$, and $\mathbf{p}_2$ be the projection of $\mathbf{w}_3$ in the direction of $\mathbf{v}_2$. The vector $\mathbf{p}_1 + \mathbf{p}_2$ is the projection of $\mathbf{w}_3$ on the plane spanned by $\mathbf{w}_1$ and $\mathbf{v}_2$. Moreover, the vector

$$\mathbf{v}_3 = \mathbf{w}_3 - (\mathbf{p}_1 + \mathbf{p}_2)$$

is perpendicular to the plane spanned by $\mathbf{w}_1$ and $\mathbf{v}_2$ and is also a linear combination of $\mathbf{w}_1$, $\mathbf{w}_2$, and $\mathbf{w}_3$ (see Exercise 12 in Section 6.1). So $\{\mathbf{w}_1, \mathbf{v}_2, \mathbf{v}_3\}$ is an orthogonal basis for W. An orthonormal basis for W is

$$\left\{\mathbf{u}_1 = \frac{\mathbf{w}_1}{|\mathbf{w}_1|},\quad \mathbf{u}_2 = \frac{\mathbf{v}_2}{|\mathbf{v}_2|},\quad \mathbf{u}_3 = \frac{\mathbf{v}_3}{|\mathbf{v}_3|}\right\}$$

This procedure is called the **Gram-Schmidt orthogonalization process.** It can be continued in the same manner to find orthonormal bases for vector spaces of higher dimension.

EXAMPLE 4 Find an orthonormal basis for the space W where

$$W = \left\langle \mathbf{w}_1 = \begin{bmatrix} 1 \\ 1 \\ -1 \\ 0 \end{bmatrix},\quad \mathbf{w}_2 = \begin{bmatrix} 0 \\ 0 \\ 1 \\ 1 \end{bmatrix},\quad \mathbf{w}_3 = \begin{bmatrix} 1 \\ 2 \\ 0 \\ 1 \end{bmatrix} \right\rangle$$

We first find $\mathbf{v}_2$

$$\mathbf{v}_2 = \mathbf{w}_2 - \left(\frac{\mathbf{w}_1 \cdot \mathbf{w}_2}{\mathbf{w}_1 \cdot \mathbf{w}_1}\right)\mathbf{w}_1 = \begin{bmatrix} 0 \\ 0 \\ 1 \\ 1 \end{bmatrix} - (-1/3)\begin{bmatrix} 1 \\ 1 \\ -1 \\ 0 \end{bmatrix} = (1/3)\begin{bmatrix} 1 \\ 1 \\ 2 \\ 3 \end{bmatrix}.$$

Now we find $\mathbf{p}_1$ and $\mathbf{p}_2$

$$\mathbf{p}_1 = \left(\frac{\mathbf{w}_1 \cdot \mathbf{w}_3}{\mathbf{w}_1 \cdot \mathbf{w}_1}\right)\mathbf{w}_1 = \frac{3}{3}\,\mathbf{w}_1 = \begin{bmatrix} 1 \\ 1 \\ -1 \\ 0 \end{bmatrix}$$

$$\mathbf{p}_2 = \left(\frac{\mathbf{v}_2 \cdot \mathbf{w}_3}{\mathbf{v}_2 \cdot \mathbf{v}_2}\right)\mathbf{v}_2 = \left(\frac{6/3}{15/9}\right)\mathbf{v}_2 = \frac{6}{5}\,\mathbf{v}_2 = \frac{2}{5}\begin{bmatrix} 1 \\ 1 \\ 2 \\ 3 \end{bmatrix}$$

Finally, we find $\mathbf{v}_3$

$$\mathbf{v}_3 = \mathbf{w}_3 - (\mathbf{p}_1 + \mathbf{p}_2) = \begin{bmatrix} 1 \\ 2 \\ 0 \\ 1 \end{bmatrix} - \begin{bmatrix} 1 \\ 1 \\ -1 \\ 0 \end{bmatrix} - (2/5)\begin{bmatrix} 1 \\ 1 \\ 2 \\ 3 \end{bmatrix} = (1/5)\begin{bmatrix} -2 \\ 3 \\ 1 \\ -1 \end{bmatrix}$$

So we have an orthogonal basis for W.

$$W = \left\langle \begin{bmatrix} 1 \\ 1 \\ -1 \\ 0 \end{bmatrix},\ (1/3)\begin{bmatrix} 1 \\ 1 \\ 2 \\ 3 \end{bmatrix},\ (1/5)\begin{bmatrix} -2 \\ 3 \\ 1 \\ -1 \end{bmatrix} \right\rangle$$

Dividing each of these vectors by its length gives the orthonormal basis

$$\left\{ (1/\sqrt{3})\begin{bmatrix} 1 \\ 1 \\ -1 \\ 0 \end{bmatrix},\ (1/\sqrt{15})\begin{bmatrix} 1 \\ 1 \\ 2 \\ 3 \end{bmatrix},\ (1/\sqrt{15})\begin{bmatrix} -2 \\ 3 \\ 1 \\ -1 \end{bmatrix} \right\} \quad \blacksquare$$

Exercise Set 6.2

Let W be the subspace of $\mathbb{R}^4$ having the following orthonormal basis:

$$\left\{ \mathbf{u}_1 = (1/2)\begin{bmatrix} 1 \\ -1 \\ 1 \\ 1 \end{bmatrix},\ \mathbf{u}_2 = (1/(2\sqrt{3}))\begin{bmatrix} 1 \\ -1 \\ -3 \\ 1 \end{bmatrix},\ \mathbf{u}_3 = (1/\sqrt{2})\begin{bmatrix} -1 \\ 0 \\ 0 \\ 1 \end{bmatrix} \right\}$$

In Exercises 1–4, determine if the vectors listed belong to W and express those vectors in W as linear combinations of the basis vectors $\mathbf{u}_1$, $\mathbf{u}_2$, $\mathbf{u}_3$.

1. $\begin{bmatrix} 2 \\ -2 \\ -2 \\ 3 \end{bmatrix}$ **2.** $\begin{bmatrix} 1 \\ 5 \\ -2 \\ 0 \end{bmatrix}$

3. $\begin{bmatrix} 2 \\ -4 \\ 0 \\ 6 \end{bmatrix}$ **4.** $\begin{bmatrix} 1 \\ -3 \\ 3 \\ 5 \end{bmatrix}$

In Exercises 5–7 find an orthonormal basis for each of the vector spaces listed.

5. $W = \left\langle \begin{bmatrix} 1 \\ -1 \\ 2 \end{bmatrix}, \begin{bmatrix} 2 \\ 1 \\ 3 \end{bmatrix} \right\rangle$

6. $W = \left\langle \begin{bmatrix} 1 \\ 0 \\ 2 \end{bmatrix}, \begin{bmatrix} 2 \\ -1 \\ -2 \end{bmatrix} \right\rangle$

7. $W = \left\langle \begin{bmatrix} 1 \\ 1 \\ -1 \\ 2 \end{bmatrix}, \begin{bmatrix} 2 \\ 1 \\ 1 \\ 0 \end{bmatrix}, \begin{bmatrix} 13 \\ -2 \\ -1 \\ 1 \end{bmatrix} \right\rangle$

8. Let W be a subspace of a vector space V. Define $W^\perp$ (read "W perp") to be the set of all vectors of V perpendicular to every vector in W.

a. Let $V = \mathbb{R}^3$ and W be the line $\langle \mathbf{w} \rangle$. Give a geometric description of $W^\perp$.

b. Let $V = \mathbb{R}^3$ and W be the plane spanned by the vectors $\mathbf{w}_1$ and $\mathbf{w}_2$. Describe $W^\perp$.

★**c.** Show that for any vector space V and any subspace W, the set $W^\perp$ is always a subspace of V.

★**d.** Show that every vector $\mathbf{v}$ of V can be expressed in the following form

$$\mathbf{v} = \mathbf{w} + \mathbf{p}$$

where $\mathbf{w}$ is in W, and $\mathbf{p}$ is in $W^\perp$.

(*Hint:* Read the discussion following Theorem 6.2).

★**e.** Show that the only vector in both W and $W^\perp$ is $\mathbf{0}$.

f. Use part **e** to show that if

$$\mathbf{w} = \mathbf{w}_1 + \mathbf{p}_1 \quad \text{and} \quad \mathbf{w} = \mathbf{w}_2 + \mathbf{p}_2$$

where $\mathbf{w}_i$ is in W, and $\mathbf{p}_i$ is in $W^\perp$, then

$$\mathbf{w}_1 = \mathbf{w}_2 \quad \text{and} \quad \mathbf{p}_1 = \mathbf{p}_2$$

9. Give an outline of the method for finding an orthonormal basis for a vector space of dimension 4.

10. Let

$$M = [\mathbf{c}_1 \quad \mathbf{c}_2]$$

be a 2×2 matrix such that the columns $\mathbf{c}_1$ and $\mathbf{c}_2$ of M are orthonormal. Show that the following statements are true:

a. For any vectors $\mathbf{v}$, $\mathbf{w}$ of $\mathbb{R}^2$,
$$\mathbf{v} \cdot \mathbf{w} = M\mathbf{v} \cdot M\mathbf{w}.$$

b. For any vector $\mathbf{v}$ in $\mathbb{R}^2$, $|\mathbf{v}| = |M\mathbf{v}|$.

c. For any vectors $\mathbf{v}$, $\mathbf{w}$ of $\mathbb{R}^2$, the cosine of the angle between $\mathbf{v}$ and $\mathbf{w}$ is the same as the cosine of the angle between $M\mathbf{v}$ and $M\mathbf{w}$.

6.3 Projections and Reflections, Scalar Products and Cross Products

We conclude this chapter by gathering together several short topics about products of vectors.

Projections and Reflections

We begin with a discussion of projections and reflections viewed as linear transformations. Although our discussion of projections only dealt with the projection of one specific vector in the direction of another specific vector, projections can also be used to define linear transformations. If $\mathbf{w}_0$ is a specific vector in $\mathbb{R}^n$, then the mapping

$$T: \mathbb{R}^n \to \mathbb{R}^n$$

$$\mathbf{v} \to \text{the projection of } \mathbf{v} \text{ in the direction of } \mathbf{w}_0 = \frac{\mathbf{v} \cdot \mathbf{w}_0}{\mathbf{w}_0 \cdot \mathbf{w}_0} \mathbf{w}_0$$

is a linear transformation (see Exercise 1).

EXAMPLE 1 Let $\mathbf{w}_0 = \begin{bmatrix} 1 \\ 2 \end{bmatrix}$ and $T: \mathbb{R}^2 \to \mathbb{R}^2$ be a projection in the direction of $\mathbf{w}_0$. Find an expression for $T\begin{bmatrix} x \\ y \end{bmatrix}$ and the matrix for T.

$$T\begin{bmatrix} x \\ y \end{bmatrix} = \frac{\begin{bmatrix} x \\ y \end{bmatrix} \cdot \begin{bmatrix} 1 \\ 2 \end{bmatrix}}{\begin{bmatrix} 1 \\ 2 \end{bmatrix} \cdot \begin{bmatrix} 1 \\ 2 \end{bmatrix}} \begin{bmatrix} 1 \\ 2 \end{bmatrix} = \frac{x + 2y}{5} \begin{bmatrix} 1 \\ 2 \end{bmatrix}$$

So $$T\begin{bmatrix} 1 \\ 0 \end{bmatrix} = \frac{1 + 2 \cdot 0}{5} \begin{bmatrix} 1 \\ 2 \end{bmatrix} = \begin{bmatrix} 1/5 \\ 2/5 \end{bmatrix}$$

and $$T\begin{bmatrix} 0 \\ 1 \end{bmatrix} = \frac{0 + 2 \cdot 1}{5} \begin{bmatrix} 1 \\ 2 \end{bmatrix} = \begin{bmatrix} 2/5 \\ 4/5 \end{bmatrix}$$

Therefore, the matrix for T is

$$\begin{bmatrix} 1/5 & 2/5 \\ 2/5 & 4/5 \end{bmatrix} \quad ■$$

Now we discuss reflections. Again we let $\mathbf{w}_0$ be a fixed vector. We wish to find the formula for the *reflection* $\mathbf{v}‡$ of $\mathbf{v}$ about the line containing $\mathbf{w}_0$. taining $\mathbf{w}_0$.

From Figure 6.16 we see that

$$\mathbf{v} = \mathbf{p} + (\mathbf{v} - \mathbf{p}) \qquad \text{where } \mathbf{p} \text{ is the projection of } \mathbf{v} \text{ in the direction of } \mathbf{w}_0$$

Thus, the reflection $\mathbf{v}‡$ is

$$\mathbf{v}‡ = \mathbf{p} - (\mathbf{v} - \mathbf{p}) = 2\mathbf{p} - \mathbf{v}$$

Substituting for the projection $\mathbf{p}$ gives

$$\mathbf{v}‡ = 2\left[\frac{\mathbf{v} \cdot \mathbf{w}_0}{\mathbf{w}_0 \cdot \mathbf{w}_0}\right]\mathbf{w}_0 - \mathbf{v}$$

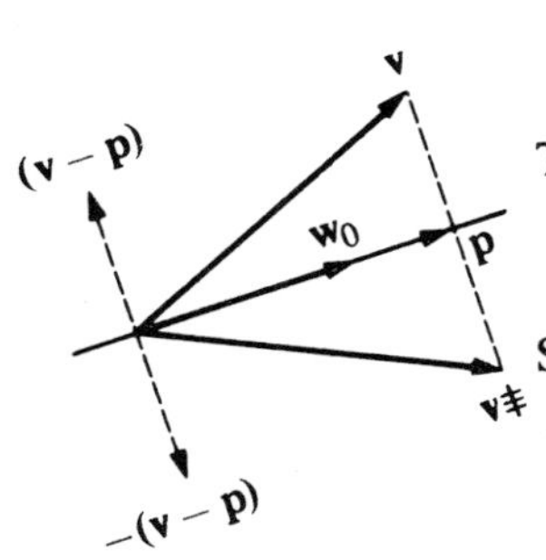

Figure 6.16

We can use this formula to define reflection as a linear transformation.

DEFINITION 6.9

The linear transformation

$$T\colon \mathbb{R}^n \to \mathbb{R}^n$$

$$\mathbf{v} \to 2\left[\frac{\mathbf{v} \cdot \mathbf{w}_0}{\mathbf{w}_0 \cdot \mathbf{w}_0}\right]\mathbf{w}_0 - \mathbf{v}$$

is called **reflection about the line containing $\mathbf{w}_0$**.

EXAMPLE 2 Let $\mathbf{w}_0 = \begin{bmatrix}1\\2\end{bmatrix}$ and $T\colon \mathbb{R}^2 \to \mathbb{R}^2$ be reflection about the line containing $\mathbf{w}_0$. Find an expression for $T\begin{bmatrix}x\\y\end{bmatrix}$ and the matrix for T.

$$T\begin{bmatrix}x\\y\end{bmatrix} = 2\left(\frac{\begin{bmatrix}x\\y\end{bmatrix} \cdot \begin{bmatrix}1\\2\end{bmatrix}}{\begin{bmatrix}1\\2\end{bmatrix} \cdot \begin{bmatrix}1\\2\end{bmatrix}}\right)\begin{bmatrix}1\\2\end{bmatrix} - \begin{bmatrix}x\\y\end{bmatrix} = \frac{2(x+2y)}{5}\begin{bmatrix}1\\2\end{bmatrix} - \begin{bmatrix}x\\y\end{bmatrix} = \begin{bmatrix}-(3/5)x + (4/5)y\\(4/5)x + (3/5)y\end{bmatrix}$$

So

$$T\begin{bmatrix}1\\0\end{bmatrix} = \begin{bmatrix}-(3/5)\\(4/5)\end{bmatrix} \quad \text{and} \quad T\begin{bmatrix}0\\1\end{bmatrix} = \begin{bmatrix}(4/5)\\(3/5)\end{bmatrix}$$

Thus the matrix that represents T is

$$\begin{bmatrix}-(3/5) & (4/5)\\(4/5) & (3/5)\end{bmatrix} \quad \blacksquare$$

Scalar Products

Our second topic in this section is scalar products. We begin by recalling the formal properties of the dot product (see Theorem 6.1).

PROPERTIES OF SCALAR PRODUCTS

1. $\mathbf{v} \cdot \mathbf{w}$ is a real number
2. $\mathbf{v} \cdot \mathbf{w} = \mathbf{w} \cdot \mathbf{v}$
3. $\mathbf{v} \cdot (\mathbf{w}_1 + \mathbf{w}_2) = \mathbf{v} \cdot \mathbf{w}_1 + \mathbf{v} \cdot \mathbf{w}_2$
4. $\mathbf{v} \cdot (c\mathbf{w}) = c(\mathbf{v} \cdot \mathbf{w})$
5. $\mathbf{v} \cdot \mathbf{v} \geq 0$, and equality holds only when $\mathbf{v} = 0$

The vector spaces $\mathbb{R}^n$ are not the only vector spaces on which products satisfying the properties listed above can be defined.

DEFINITION 6.10 If V is a vector space and a product satisfying the properties listed above can be defined on V, then this product is called a **scalar product**. (*Note*: The dot product is a scalar product.)

We now look at an example of a scalar product. Let V be the set of all polynomials with real coefficients. In Example 4 in Section 3.2, we showed that V is a vector space. We define the product $*$ on V as follows: For any polynomials p and q of V

$$p * q = \int_0^1 p(x)q(x)\,dx$$

It is not difficult to see that this product has all the formal properties and is therefore a scalar product.

The importance of this product is that we can use it to define both what we mean by the length of a polynomial and what we mean by the angle between two polynomials. We do this in the same way that we used the dot product to define the meanings of length and angles in $\mathbb{R}^n$ ($n \geq 4$).

We define the length of the polynomial p to be

$$|p| = \sqrt{p * p} = \left\{\int_0^1 [p(x)p(x)]\,dx\right\}^{1/2} = \left[\int_0^1 p^2(x)\,dx\right]^{1/2}$$

and if θ is the angle between p and q, then

$$\cos\theta = \frac{p * q}{|p|\,|q|} = \frac{\int_0^1 [p(x)q(x)]\,dx}{\{\int_0^1 [p(x)]^2\,dx\}^{1/2}\{\int_0^1 [q(x)]^2\,dx\}^{1/2}}$$

EXAMPLE 3 We compute the length of the polynomial $p(x) = x^2$.

$$|p| = \sqrt{p * p} = \left[\int_0^1 (x^2x^2)\,dx\right]^{1/2} = \left(\frac{x^5}{5}\Big|_0^1\right)^{1/2} = (1/\sqrt{5}) \quad \blacksquare$$

EXAMPLE 4 Show that the polynomials $p(x) = x + (1/\sqrt{3})$ and $q(x) = x - (1/\sqrt{3})$ are perpendicular.

It suffices to show that $p * q = 0$.

$$p * q = \int_0^1 [(x + (1/\sqrt{3}))(x - (1/\sqrt{3}))]\,dx$$

$$= \int_0^1 \left(x^2 - \frac{1}{3}\right)dx = \left(\frac{x^3}{3} - \frac{x}{3}\right)\Big|_0^1 = 0$$

So p is perpendicular to q. $\blacksquare$

This scalar product has many important applications. One of the most interesting is the digital analysis of sound used to produce extremely faithful sound reproduction for modern recordings.

Cross Products

The final topic of this section is the cross product. This special product of vectors is defined only in $\mathbb{R}^3$. However, since we live in $\mathbb{R}^3$, it has many important applications to physical situations.

DEFINITION 6.11

The **cross product** of two vectors $\mathbf{v}$ and $\mathbf{w}$ of $\mathbb{R}^3$ is a vector perpendicular to the plane spanned by $\mathbf{v}$ and $\mathbf{w}$. It is denoted by $\mathbf{v} \times \mathbf{w}$. It has length $|\mathbf{v}||\mathbf{w}|\sin\theta$ where θ is the angle between $\mathbf{v}$ and $\mathbf{w}$ (measured from $\mathbf{v}$ to $\mathbf{w}$) and it points in the direction your right thumb would point if you were to curl the fingers of your right hand from $\mathbf{v}$ to $\mathbf{w}$. (See Figure 6.17.)

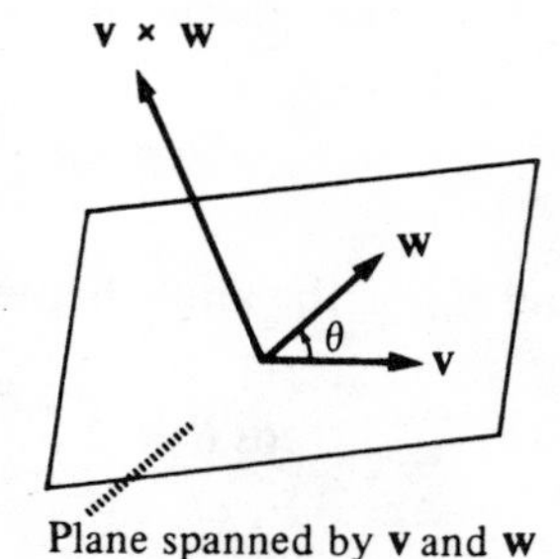

Figure 6.17

The formula for computing the cross product of two vectors is

$$\begin{bmatrix} x_1 \\ x_2 \\ x_3 \end{bmatrix} \times \begin{bmatrix} y_1 \\ y_2 \\ y_3 \end{bmatrix} = \begin{bmatrix} x_2 y_3 - x_3 y_2 \\ -(x_1 y_3 - x_3 y_1) \\ x_1 y_2 - x_2 y_1 \end{bmatrix}$$

EXAMPLE 5

$$\begin{bmatrix} 1 \\ -1 \\ 2 \end{bmatrix} \times \begin{bmatrix} 2 \\ 3 \\ -1 \end{bmatrix} = \begin{bmatrix} (-1)(-1) - (2)(3) \\ -[(1)(-1) - (2)(2)] \\ (1)(3) - (-1)(2) \end{bmatrix} = \begin{bmatrix} -5 \\ 5 \\ 5 \end{bmatrix}$$

To check this calculation you should note that

$$\begin{bmatrix} 1 \\ -1 \\ 2 \end{bmatrix} \cdot \begin{bmatrix} -5 \\ 5 \\ 5 \end{bmatrix} = 0 \quad \text{and} \quad \begin{bmatrix} 2 \\ 3 \\ -1 \end{bmatrix} \cdot \begin{bmatrix} -5 \\ 5 \\ 5 \end{bmatrix} = 0$$

So the cross product $\begin{bmatrix} -5 \\ 5 \\ 5 \end{bmatrix}$

is perpendicular to every vector in the plane spanned by the vectors

$$\begin{bmatrix} 1 \\ -1 \\ 2 \end{bmatrix} \quad \text{and} \quad \begin{bmatrix} 2 \\ 3 \\ -1 \end{bmatrix} \quad \blacksquare$$

This has only been the briefest introduction to the cross product. The cross product is an extremely valuable tool and has many important physical applications. However we will not cover these topics in this book.

Exercise Set 6.3

1. Let $T: \mathbb{R}^2 \to \mathbb{R}^2$ be projection onto the line containing the vector $\mathbf{w}_0 = \begin{bmatrix} a \\ b \end{bmatrix}$. Show that T is a linear transformation.

2. Find the matrix that represents each of the following linear transformations:

a. $T: \mathbb{R}^2 \to \mathbb{R}^2$ is projection onto the line containing the vector $\begin{bmatrix} -1 \\ 1 \end{bmatrix}$.

b. $T: \mathbb{R}^2 \to \mathbb{R}^2$ is reflection about the line containing the vector $\begin{bmatrix} -1 \\ 1 \end{bmatrix}$.

c. $T: \mathbb{R}^2 \to \mathbb{R}^2$ is projection onto the line $y = 3x$.

d. $T: \mathbb{R}^2 \to \mathbb{R}^2$ is reflection about the line $y = 3x$.

e. $T: \mathbb{R}^3 \to \mathbb{R}^3$ is projection onto the line containing the vector

$$\begin{bmatrix} 1 \\ -1 \\ 3 \end{bmatrix}$$

f. $T: \mathbb{R}^3 \to \mathbb{R}^3$ is reflection about the line containing the vector

$$\begin{bmatrix} 1 \\ -1 \\ 3 \end{bmatrix}$$

★ **3.** Show that the product $*$ defined in this section is a scalar product.

★ **4.** Using the product $*$, find:

a. $|p|$ where $p(x) = x^2 + 1$.

b. $\cos\theta$ where θ is the angle between $p(x) = x^2 + 1$ and $q(x) = x^2 - 1$.

5. Find the following cross products:

a. $\begin{bmatrix} 1 \\ 0 \\ 0 \end{bmatrix} \times \begin{bmatrix} 1 \\ 0 \\ 0 \end{bmatrix}$ **b.** $\begin{bmatrix} 1 \\ 0 \\ 0 \end{bmatrix} \times \begin{bmatrix} 0 \\ 1 \\ 0 \end{bmatrix}$

c. $\begin{bmatrix} 0 \\ 1 \\ 0 \end{bmatrix} \times \begin{bmatrix} 0 \\ 0 \\ 1 \end{bmatrix}$ **d.** $\begin{bmatrix} 0 \\ 0 \\ 1 \end{bmatrix} \times \begin{bmatrix} 0 \\ 1 \\ 0 \end{bmatrix}$

e. $\begin{bmatrix} 1 \\ 2 \\ -1 \end{bmatrix} \times \begin{bmatrix} 3 \\ 2 \\ 4 \end{bmatrix}$ **f.** $\begin{bmatrix} -1 \\ 2 \\ 3 \end{bmatrix} \times \begin{bmatrix} 3 \\ -2 \\ 2 \end{bmatrix}$

6. Is $\mathbf{v} \times \mathbf{w}$ equal to $\mathbf{w} \times \mathbf{v}$ for every pair of vectors in $\mathbb{R}^3$? What is the relationship between $\mathbf{v} \times \mathbf{w}$ and $\mathbf{w} \times \mathbf{v}$?

7. Show that the cross product is not associative by showing that

$$(\mathbf{e}_1 \times \mathbf{e}_2) \times \mathbf{e}_2 \neq \mathbf{e}_1 \times (\mathbf{e}_2 \times \mathbf{e}_2)$$

where $\mathbf{e}_1 = \begin{bmatrix} 1 \\ 0 \\ 0 \end{bmatrix}$ and $\mathbf{e}_2 = \begin{bmatrix} 0 \\ 1 \\ 0 \end{bmatrix}$

8. Let $T: \mathbb{R}^n \to \mathbb{R}^n$ be a projection in the direction of the vector $\mathbf{w}$. Find the rank and the nullity of T.

9. Let $T: \mathbb{R}^n \to \mathbb{R}^n$ be a reflection about the line containing the vector $\mathbf{w}$. Find the rank and the nullity of T.

10. Let $T: \mathbb{R}^2 \to \mathbb{R}^2$ be a reflection about the line containing the vector $\mathbf{w}$. Show that there is an orthonormal basis $\{\mathbf{u}_1, \mathbf{u}_2\}$ of $\mathbb{R}^2$ such that

$$T\mathbf{u}_1 = \mathbf{u}_1 \quad \text{and} \quad T\mathbf{u}_2 = -\mathbf{u}_2$$

11. Let $\mathbf{u}$, $\mathbf{v}$, $\mathbf{w}$ be three vectors in $\mathbb{R}^3$ such that $\mathbf{v}$ and $\mathbf{w}$ are not parallel. Show that the vector $\mathbf{u} \times (\mathbf{v} \times \mathbf{w})$ lies in the plane spanned by $\mathbf{v}$ and $\mathbf{w}$.

★ **12.** Let $\mathbf{u} = \begin{bmatrix} \cos\theta \\ \sin\theta \end{bmatrix}$ be a unit vector in $\mathbb{R}^2$ and let T be reflection about the line containing $\mathbf{u}$. Show that the matrix which represents T is

$$M = \begin{bmatrix} \cos 2\theta & \sin 2\theta \\ \sin 2\theta & -\cos 2\theta \end{bmatrix}$$

(*Hint*: Use the trigonometric formula for the cosine of the sum of two angles.)

13. Verify the formula given in Exercise 12 for the following reflections:

a. Reflection about the line $y = x$.

b. Reflection about the x axis.

c. Reflection about the line $y = -x$.

d. Reflection about the line containing the vector $\begin{bmatrix} 1 \\ 2 \end{bmatrix}$.

Applications 6 Analysis of Experimental Data—Least Squares

In this section we see how linear algebra can be used to help analyze experimental data. In many experiments the goal is to determine the relationship between two quantities. For example, you may wish to find the relationship between the amount of corn produced per acre and the amount of fertilizer applied; find the relationship between the amount of alcohol a person consumes and the amount of alcohol in his blood one hour later; or find the relationship between the distance a spring is stretched and the amount of force necessary to stretch it that far. In each of these problems the goal is to establish an empirical relationship between two quantities and then to find a function that accurately expresses this relationship. The problem that occurs in all experimental work is that there is always a certain amount of error due to the intrinsic inaccuracy of all measuring devices. Thus the experimental data gathered may not (and usually does not) agree with the function that represents the exact relationship between the quantities measured.

EXAMPLE 1 In a certain experiment the goal is to find the relationship between the amount of weight attached to the end of a spring and the distance that the spring is stretched. The results of this experiment are summarized in Table 6.1. We plot these points on the graph in Figure 6.18. From this graph it appears that these points all lie close to the line $y = (1/2)x$.

Table 6.1

Weight (in pounds)	1	2	3	4
Distance Stretched (in inches)	0.48	1.01	1.47	2.01

Figure 6.18

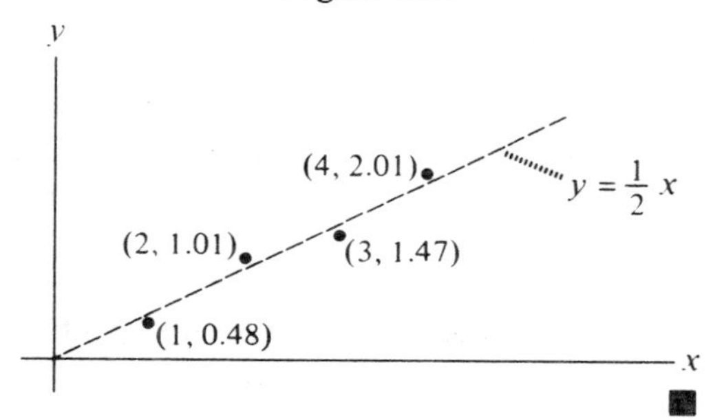

■

The question we discuss in this section is how to choose the straight line that best approximates a set of points obtained from an experiment.

Let us assume that the points (x_1, y_1), (x_2, y_2), ..., (x_n, y_n) were obtained in an experiment. The first question we must answer is what do we mean by "best approximates"? It should mean that we want the line $y = mx + b$, which is, on the average, the closest line to all these points. We will measure the distance between a point (x_i, y_i) and a line $y = mx + b$ by measuring the vertical distance between the point and the line. This is the difference between the y coordinate of the point y_i and the y coordinate $mx_i + b$ of the point on the line having the same x coordinate. That is,

$$|\Delta y_i| = |y_i - (mx_i + b)|$$

We use absolute value because the point (x_i, y_i) may be below the line. See Figure 6.19.

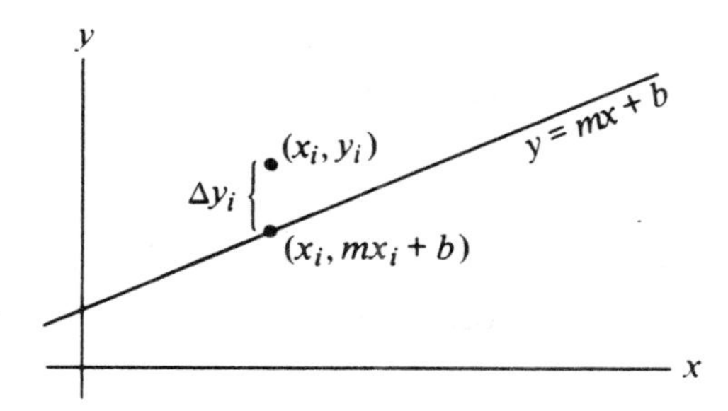

Figure 6.19

We say that the line which is closest to all the experimental points is the line $y = mx + b$ with m and b chosen so that the total of the $|\Delta y_i|$'s is a minimum, that is, we wish to minimize the quantity

$$|\Delta y_1| + |\Delta y_2| + \cdots + |\Delta y_n| \qquad \text{where } \Delta y_i = [y_i - (mx_i + b)]$$

The presence of absolute values in this expression makes it somewhat difficult to deal with. However, the expression attains its minimum at the same point as the expression

$$(\Delta y_1)^2 + (\Delta y_2)^2 + \cdots + (\Delta y_n)^2 \qquad \text{where } \Delta y_i = [y_i - (mx_i + b)]$$

Thus it will suffice to find the m and b that minimize the expression above.
Now we can formally state the problem:

Problem

Given n points $(x_1, y_1), (x_2, y_2), \ldots, (x_n, y_n)$, find numbers m and b so that the quantity

$$(\Delta y_1)^2 + (\Delta y_2)^2 + \cdots + (\Delta y_n)^2 \qquad \text{where } \Delta y_i = [y_i - (mx_i + b)]$$

is minimum.

We can translate this problem into a problem in linear algebra.

Let $$\mathbf{Y} = \begin{bmatrix} y_1 \\ y_2 \\ \vdots \\ y_n \end{bmatrix} \quad \text{and} \quad \mathbf{\Delta}_{m,b} = \begin{bmatrix} \Delta y_1 \\ \Delta y_2 \\ \vdots \\ \Delta y_n \end{bmatrix}$$

Since the square of the length of $\mathbf{\Delta}_{m,b}$ is

$$(\mathbf{\Delta}_{m,b}) \cdot (\mathbf{\Delta}_{m,b}) = (\Delta y_1)^2 + (\Delta y_2)^2 + \cdots + (\Delta y_n)^2$$

and since $$\mathbf{\Delta}_{m,b} = \begin{bmatrix} y_1 - (mx_1 + b) \\ y_2 - (mx_2 + b) \\ \vdots \\ y_n - (mx_n + b) \end{bmatrix} = \mathbf{Y} - \begin{bmatrix} x_1 & 1 \\ x_2 & 1 \\ \vdots & \vdots \\ x_n & 1 \end{bmatrix} \begin{bmatrix} m \\ b \end{bmatrix}$$

we can restate the problem in vector form.

Problem (Vector Form)

Let $$M = \begin{bmatrix} x_1 & 1 \\ x_2 & 1 \\ \vdots & \vdots \\ x_n & 1 \end{bmatrix}, \quad \mathbf{Y} = \begin{bmatrix} y_1 \\ y_2 \\ \vdots \\ y_n \end{bmatrix} \quad \text{and} \quad \mathbf{\Delta}_{m,b} = \mathbf{Y} - M \begin{bmatrix} m \\ b \end{bmatrix}$$

Find numbers m and b so that the vector $\mathbf{\Delta}_{m,b}$ has minimum length.

EXAMPLE 2 We put the data from Example 1 in vector form.

$$\mathbf{Y} = \begin{bmatrix} 0.48 \\ 1.01 \\ 1.47 \\ 2.01 \end{bmatrix} \quad \text{and} \quad M = \begin{bmatrix} 1 & 1 \\ 2 & 1 \\ 3 & 1 \\ 4 & 1 \end{bmatrix}$$

So
$$\Delta_{m,b} = \mathbf{Y} - M\begin{bmatrix} m \\ b \end{bmatrix} = \begin{bmatrix} 0.48 \\ 1.01 \\ 1.47 \\ 2.01 \end{bmatrix} - \begin{bmatrix} 1 & 1 \\ 2 & 1 \\ 3 & 1 \\ 4 & 1 \end{bmatrix}\begin{bmatrix} m \\ b \end{bmatrix} = \begin{bmatrix} 0.48 - (m+b) \\ 1.01 - (2m+b) \\ 1.47 - (3m+b) \\ 2.01 - (4m+b) \end{bmatrix}$$

If we let $m = 0.5$ and $b = 0$ (which corresponds to the line $y = \frac{1}{2}x$), then

$$\Delta_{0.5,0} = \begin{bmatrix} 0.48 - (0.5) \\ 1.01 - 2(0.5) \\ 1.47 - 3(0.5) \\ 2.01 - 4(0.5) \end{bmatrix} = \begin{bmatrix} -0.02 \\ 0.01 \\ -0.03 \\ 0.01 \end{bmatrix}$$

The length of $\Delta_{0.5,0}$ is

$$\sqrt{(-0.02)^2 + (0.01)^2 + (-0.03)^2 + (0.01)^2} = \sqrt{0.0015} = 0.0387$$

We will see in Examples 3 and 4 whether this is the line that best approximates the points given. ■

We apply linear algebra to analyze the vector form of this problem. The matrix M is a matrix with n rows and 2 columns. So $M\colon \mathbb{R}^2 \to \mathbb{R}^n$. Moreover, since the experiment involves choosing different values of x_i and determining the value of the corresponding y_i, we may assume that the x_i's are not all the same. Therefore the column vectors of M

$$\mathbf{c}_1 = \begin{bmatrix} x_1 \\ x_2 \\ \vdots \\ x_n \end{bmatrix} \quad \text{and} \quad \mathbf{c}_2 = \begin{bmatrix} 1 \\ 1 \\ \vdots \\ 1 \end{bmatrix}$$

are linearly independent. Since $M\begin{bmatrix} m \\ b \end{bmatrix} = m\mathbf{c}_1 + b\mathbf{c}_2$ and $\mathbf{c}_1$ and $\mathbf{c}_2$ are linearly independent, M maps every vector in $\mathbb{R}^2$ onto the plane $P = \langle \mathbf{c}_1, \mathbf{c}_2 \rangle$, which is contained in $\mathbb{R}^n$.

The vector we are trying to find is $\mathbf{v}_0 = \begin{bmatrix} m_0 \\ b_0 \end{bmatrix}$ such that the distance from the tip of $M\mathbf{v}_0$ to the tip of $\mathbf{Y}$ is the minimum distance from the tip of $\mathbf{Y}$ to any point in the plane P. See Figure 6.20. From the geometry of this situation, it is easy to see that this occurs when the vector $\Delta_{m_0,b_0} = \mathbf{Y} - M\mathbf{v}_0$ is perpendicular to the plane P.

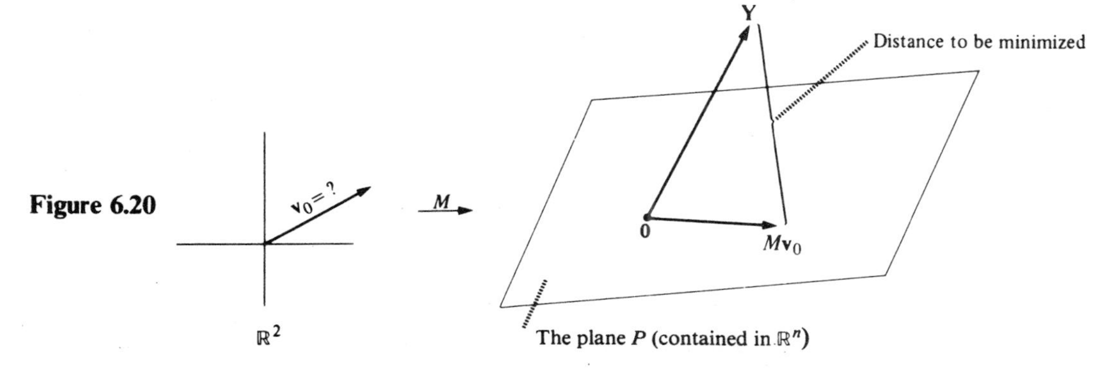

Figure 6.20

If Δ_{m_0, b_0} is perpendicular to the plane P, then

$$\Delta_{m_0, b_0} \cdot \mathbf{w} = 0 \qquad \text{for every vector } \mathbf{w} \text{ of } P$$

Since P is the set of all vectors of the form $M\mathbf{v}$ (for $\mathbf{v}$ in $\mathbb{R}^2$), this condition is equivalent to

$$\textbf{(1)} \qquad \Delta_{m_0, b_0} \cdot M\mathbf{v} = 0 \qquad \text{for all } \mathbf{v} \text{ in } \mathbb{R}^2$$

in other words, $(\mathbf{Y} - M\mathbf{v}_0) \cdot M\mathbf{v} = 0$ for all $\mathbf{v}$ in $\mathbb{R}^2$

EXAMPLE 3 Does the line $y = 0.5x + 0$ best approximate the data from Example 1? As we saw in Example 2, the values of $\mathbf{Y}$, M, and $\Delta_{0.5, 0}$ are:

$$\mathbf{Y} = \begin{bmatrix} 0.48 \\ 1.01 \\ 1.47 \\ 2.01 \end{bmatrix} \qquad M = \begin{bmatrix} 1 & 1 \\ 2 & 1 \\ 3 & 1 \\ 4 & 1 \end{bmatrix} \qquad \Delta_{0.5, 0} = \begin{bmatrix} -0.02 \\ 0.01 \\ -0.03 \\ 0.01 \end{bmatrix}$$

To see if the line $y = 0.5x + 0$ is the line of best approximation to these points we must see if

$$\Delta_{0.5, 0} \cdot M\mathbf{v} = 0 \qquad \text{for all } \mathbf{v} \text{ in } \mathbb{R}^2$$

It is not difficult to see that if

$$\Delta_{0.5, 0} \cdot M \begin{bmatrix} 1 \\ 0 \end{bmatrix} = 0 \qquad \text{and} \qquad \Delta_{0.5, 0} \cdot M \begin{bmatrix} 0 \\ 1 \end{bmatrix} = 0$$

then $\Delta_{0.5, 0} \cdot M\mathbf{v} = 0$ for all $\mathbf{v}$ in $\mathbb{R}^2$. However,

$$\Delta_{0.5, 0} \cdot M \begin{bmatrix} 1 \\ 0 \end{bmatrix} = \begin{bmatrix} -0.02 \\ 0.01 \\ -0.03 \\ 0.01 \end{bmatrix} \cdot \begin{bmatrix} 1 & 1 \\ 2 & 1 \\ 3 & 1 \\ 4 & 1 \end{bmatrix} \begin{bmatrix} 1 \\ 0 \end{bmatrix} = \begin{bmatrix} -0.02 \\ 0.01 \\ -0.03 \\ 0.01 \end{bmatrix} \cdot \begin{bmatrix} 1 \\ 2 \\ 3 \\ 4 \end{bmatrix}$$

$$= -0.02 + 0.02 - 0.09 + 0.04 = -0.05 \neq 0$$

So $y = 0.5x + 0$ is *not* the line that best approximates this data. ■

Now we return to our discussion of equation (1).

$$\textbf{(1)} \qquad \Delta_{m_0, b_0} \cdot M\mathbf{v} = 0 \qquad \text{for all } \mathbf{v} \text{ in } \mathbb{R}^2$$

We use the following technique to solve this equation. The matrix

$$\begin{bmatrix} x_1 & x_2 & \cdots & x_n \\ 1 & 1 & \cdots & 1 \end{bmatrix} = M^t$$

obtained from M by interchanging rows and columns is called the **transpose** of M and is denoted M^t. (A more thorough discussion of transposes is found in Section 8.1.)

An easy computation shows that if

$$\mathbf{w} = \begin{bmatrix} w_1 \\ \vdots \\ w_n \end{bmatrix} \qquad \text{and} \qquad \mathbf{v} = \begin{bmatrix} v_1 \\ v_2 \end{bmatrix}$$

then

$$\mathbf{w} \cdot M\mathbf{v} = \begin{bmatrix} w_1 \\ w_2 \\ \vdots \\ w_n \end{bmatrix} \cdot \begin{bmatrix} x_1 & 1 \\ x_2 & 1 \\ \vdots & \vdots \\ x_n & 1 \end{bmatrix} \begin{bmatrix} v_1 \\ v_2 \end{bmatrix} = \begin{bmatrix} x_1 & x_2 & \cdots & x_n \\ 1 & 1 & \cdots & 1 \end{bmatrix} \begin{bmatrix} w_1 \\ w_2 \\ \vdots \\ w_n \end{bmatrix} \cdot \begin{bmatrix} v_1 \\ v_2 \end{bmatrix} = M^t\mathbf{w} \cdot \mathbf{v}$$

Therefore, using the transpose, equation (1) becomes

$$M^t(\boldsymbol{\Delta}_{m_0, b_0}) \cdot \mathbf{v} = 0 \qquad \text{for all } \mathbf{v} \text{ in } \mathbb{R}^2 \tag{2}$$

Substituting $\mathbf{Y} - M\mathbf{v}_0 = \boldsymbol{\Delta}_{m_0, b_0}$, we have

$$M^t(\mathbf{Y} - M\mathbf{v}_0) \cdot \mathbf{v} = 0 \qquad \text{for all } \mathbf{v} \text{ in } \mathbb{R}^2 \tag{3}$$

Now $M^t(\mathbf{Y} - M\mathbf{v}_0)$ is a vector in $\mathbb{R}^2$, and this equation says that this vector is perpendicular to every vector in $\mathbb{R}^2$. Thus it is perpendicular to itself. Since the zero vector is the only vector perpendicular to itself,

$$M^t(\mathbf{Y} - M\mathbf{v}_0) = \mathbf{0}$$

or equivalently,

$$M^t\mathbf{Y} - M^t(M\mathbf{v}_0) = \mathbf{0}$$

which means that

$$M^t\mathbf{Y} = M^t(M\mathbf{v}_0)$$

At this point we digress to discuss ideas that will be covered in more detail in Chapter 7. As we have just seen, we can find the line that best approximates our data if we can solve the equation

$$M^t\mathbf{Y} = M^t(M\mathbf{v}_0) \tag{4}$$

We study the right hand side of equation **(4)**.

$$M^t(M\mathbf{v}_0) = M^t \begin{bmatrix} x_1 & 1 \\ x_2 & 1 \\ \vdots & \vdots \\ x_n & 1 \end{bmatrix} \begin{bmatrix} m_0 \\ b_0 \end{bmatrix} = M^t \begin{bmatrix} m_0 x_1 + b_0 \\ m_0 x_2 + b_0 \\ \vdots \\ m_0 x_n + b_0 \end{bmatrix}$$

$$= \begin{bmatrix} x_1 & x_2 & \cdots & x_n \\ 1 & 1 & \cdots & 1 \end{bmatrix} m_0 \begin{bmatrix} x_1 \\ x_2 \\ \vdots \\ x_n \end{bmatrix} + b_0 \begin{bmatrix} 1 \\ 1 \\ \vdots \\ 1 \end{bmatrix}$$

To simplify things, we let

$$\mathbf{X} = \begin{bmatrix} x_1 \\ x_2 \\ \vdots \\ x_n \end{bmatrix} \quad \text{and} \quad \mathbf{J} = \begin{bmatrix} 1 \\ 1 \\ \vdots \\ 1 \end{bmatrix}$$

Using this notation, we see that

$$\textbf{(5)} \qquad M^t(M\mathbf{v}_0) = m_0 \begin{bmatrix} \mathbf{X} \cdot \mathbf{X} \\ \mathbf{J} \cdot \mathbf{X} \end{bmatrix} + b_0 \begin{bmatrix} \mathbf{X} \cdot \mathbf{J} \\ \mathbf{J} \cdot \mathbf{J} \end{bmatrix} = \begin{bmatrix} \mathbf{X} \cdot \mathbf{X} & \mathbf{X} \cdot \mathbf{J} \\ \mathbf{J} \cdot \mathbf{X} & \mathbf{J} \cdot \mathbf{J} \end{bmatrix} \begin{bmatrix} m_0 \\ b_0 \end{bmatrix}$$

Now we consider the left hand side of equation **(4)**.

$$\textbf{(6)} \qquad M^t\mathbf{Y} = \begin{bmatrix} x_1 & x_2 & \cdots & x_n \\ 1 & 1 & \cdots & 1 \end{bmatrix} \begin{bmatrix} y_1 \\ y_2 \\ \vdots \\ y_n \end{bmatrix} = \begin{bmatrix} \mathbf{X} \cdot \mathbf{Y} \\ \mathbf{J} \cdot \mathbf{Y} \end{bmatrix}$$

Combining equations (4), (5), and (6) we obtain the following theorem.

THEOREM 6.3 Suppose we are given points $(x_1, y_1), \ldots, (x_n, y_n)$ with distinct x_i's and $n \geq 3$. Then the equation of the line that best approximates these points is

$$y = m_0 x + b_0$$

where m_0 and b_0 satisfy the equation

$$\begin{bmatrix} \mathbf{X} \cdot \mathbf{Y} \\ \mathbf{J} \cdot \mathbf{Y} \end{bmatrix} = \begin{bmatrix} \mathbf{X} \cdot \mathbf{X} & \mathbf{X} \cdot \mathbf{J} \\ \mathbf{J} \cdot \mathbf{X} & \mathbf{J} \cdot \mathbf{J} \end{bmatrix} \begin{bmatrix} m_0 \\ b_0 \end{bmatrix}$$

and

$$\mathbf{X} = \begin{bmatrix} x_1 \\ x_2 \\ \vdots \\ x_n \end{bmatrix} \quad \mathbf{Y} = \begin{bmatrix} y_1 \\ y_2 \\ \vdots \\ y_n \end{bmatrix} \quad \mathbf{J} = \begin{bmatrix} 1 \\ 1 \\ \vdots \\ 1 \end{bmatrix}$$

Using the concepts of the product of two matrices and the inverse of a matrix (which will be discussed in Chapter 7), as well as the relationship between the transpose of a matrix and the dot product (see Section 8.1), we can state Theorem 6.3 more succinctly.

THEOREM 6.3′

Suppose we are given points $(x_1, y_1), \ldots, (x_n, y_n)$ with distinct x_i's and $n \geq 3$. Then the equation of the line that best approximates these points is

$$y = m_0 x + b_0$$

where
$$\begin{bmatrix} m_0 \\ b_0 \end{bmatrix} = (M^t M)^{-1} M^t \mathbf{Y}$$

and
$$M = \begin{bmatrix} x_1 & 1 \\ x_2 & 1 \\ \vdots & \vdots \\ x_n & 1 \end{bmatrix} \qquad \mathbf{Y} = \begin{bmatrix} y_1 \\ y_2 \\ \vdots \\ y_n \end{bmatrix}$$

The following example illustrates Theorem 6.3.

EXAMPLE 4 We use Theorem 6.3 to find the equation of the line which best approximates the data given in Example 1. In this case the vectors **X**, **Y**, and **J** are

$$\mathbf{X} = \begin{bmatrix} 1 \\ 2 \\ 3 \\ 4 \end{bmatrix} \qquad \mathbf{Y} = \begin{bmatrix} 0.48 \\ 1.01 \\ 1.47 \\ 2.01 \end{bmatrix} \qquad \mathbf{J} = \begin{bmatrix} 1 \\ 1 \\ 1 \\ 1 \end{bmatrix}$$

We find that

$$\mathbf{X} \cdot \mathbf{J} = \mathbf{J} \cdot \mathbf{X} = 10 \qquad \mathbf{X} \cdot \mathbf{X} = 30 \qquad \mathbf{J} \cdot \mathbf{Y} = 4.97 \qquad \mathbf{X} \cdot \mathbf{Y} = 14.95 \qquad \mathbf{J} \cdot \mathbf{J} = 4$$

Substituting these values in the equation

$$\begin{bmatrix} \mathbf{X} \cdot \mathbf{Y} \\ \mathbf{J} \cdot \mathbf{Y} \end{bmatrix} = \begin{bmatrix} \mathbf{X} \cdot \mathbf{X} & \mathbf{X} \cdot \mathbf{J} \\ \mathbf{J} \cdot \mathbf{X} & \mathbf{J} \cdot \mathbf{J} \end{bmatrix} \begin{bmatrix} m_0 \\ b_0 \end{bmatrix}$$

yields
$$\begin{bmatrix} 14.95 \\ 4.97 \end{bmatrix} = \begin{bmatrix} 30 & 10 \\ 10 & 4 \end{bmatrix} \begin{bmatrix} m_0 \\ b_0 \end{bmatrix}$$

To solve this equation for $\begin{bmatrix} m_0 \\ b_0 \end{bmatrix}$, we use the elimination technique discussed in Chapter 2.

$$\left[\begin{array}{cc|c} 30 & 10 & 14.95 \\ 10 & 4 & 4.97 \end{array}\right] \begin{array}{l} (1) \\ (-3) \end{array} \rightarrow \left[\begin{array}{cc|c} 30 & 10 & 14.97 \\ 0 & -2 & -0.04 \end{array}\right] \begin{array}{l} (1) \\ (5) \end{array} \rightarrow$$

$$\left[\begin{array}{cc|c} 30 & 0 & 15.15 \\ 0 & -2 & 0.04 \end{array}\right] \begin{array}{l} (1/30) \rightarrow \\ (-1/2) \rightarrow \end{array} \left[\begin{array}{cc|c} 1 & 0 & 0.505 \\ 0 & 1 & -0.02 \end{array}\right]$$

Hence
$$\begin{bmatrix} m_0 \\ b_0 \end{bmatrix} = \begin{bmatrix} 0.505 \\ -0.02 \end{bmatrix}$$

and the line that best approximates our experimental data is

$$y = 0.505x - 0.02$$

(*Note:* The length of the vector $\Delta_{0.505,\,-0.02}$ is 0.0339, which is slightly shorter than 0.0387, which is the length of the vector $\Delta_{0.5,\,0}$—see Example 2.) ■

The method we have discussed for finding the line which best approximates a given set of points is called the **least squares fitting of a straight line.** This technique plays an important role in the statistical analysis of experimental data.

Exercises for Applications 6

For each set of points listed below, sketch the points on a graph, and use your sketch to guess the equation of a line that best approximates the data. Find the line that best approximates the given points using the method of least squares and compare this answer with your guess.

1. $(-1, 3.1), (0, 0.8), (1, -0.8), (2, -3.1)$
2. $(0, 3.1), (2, 6.9), (3, 9.2), (5, 13.0)$
3. $(0, 2.8), (2, 3.9), (6, 6), (10, 7.9), (16, 11.1)$

Review Exercises

1. Define the following terms:

a. Orthogonal basis

b. Orthonormal basis

2. Write down the following formulas:

a. If

$$\mathbf{v} = \begin{bmatrix} x_1 \\ \vdots \\ x_n \end{bmatrix}$$

write the formula for the length of $\mathbf{v}$.

b. If

$$\mathbf{v} = \begin{bmatrix} x_1 \\ \vdots \\ x_n \end{bmatrix} \quad \text{and} \quad \mathbf{w} = \begin{bmatrix} y_1 \\ \vdots \\ y_n \end{bmatrix}$$

write the formula for the cosine of the angle between $\mathbf{v}$ and $\mathbf{w}$.

c. Using the dot product, write the formula for the projection of $\mathbf{v}$ in the direction of $\mathbf{w}$.

d. Using the dot product, write the formula for the reflection of the vector $\mathbf{v}$ about the line containing the vector $\mathbf{w}$.

3. Let $\mathbf{u}_1, \mathbf{u}_2, \mathbf{u}_3, \mathbf{u}_4$ be four nonzero vectors of $\mathbb{R}^4$ that are mutually perpendicular. Show that $\{\mathbf{u}_1, \mathbf{u}_2, \mathbf{u}_3, \mathbf{u}_4\}$ is a basis for $\mathbb{R}^4$.

4. Express $\begin{bmatrix} 1 \\ 0 \end{bmatrix}$ and $\begin{bmatrix} 0 \\ 1 \end{bmatrix}$ as linear combinations of the vectors

$$\begin{bmatrix} 3/5 \\ 4/5 \end{bmatrix} \quad \text{and} \quad \begin{bmatrix} -4/5 \\ 3/5 \end{bmatrix}.$$

5. Find an orthonormal basis of the subspace of $\mathbb{R}^4$ spanned by

$$\begin{bmatrix} 1 \\ 4 \\ -2 \\ 2 \end{bmatrix}, \quad \begin{bmatrix} 3 \\ 4 \\ -3 \\ 0 \end{bmatrix}$$

Show that

$$\begin{bmatrix} 1 \\ -4 \\ 1 \\ -4 \end{bmatrix}$$

is in this subspace.

6. Let W be the subspace of $\mathbb{R}^4$ spanned by the vectors

$$\left\langle \begin{bmatrix} 1 \\ 0 \\ 2 \\ 1 \end{bmatrix}, \begin{bmatrix} 0 \\ 3 \\ 5 \\ 2 \end{bmatrix}, \begin{bmatrix} 1 \\ -1 \\ 2 \\ 0 \end{bmatrix} \right\rangle$$

Find an orthonormal basis for W.

7. Let $T: \mathbb{R}^3 \to \mathbb{R}^3$ be projection onto the line containing the vector

$$\begin{bmatrix} -1 \\ 2 \\ 3 \end{bmatrix}$$

Find the matrix for T.

8. Let $T: \mathbb{R}^3 \to \mathbb{R}^3$ be reflection about the line containing the vector

$$\begin{bmatrix} -1 \\ 2 \\ 3 \end{bmatrix}$$

Find the matrix for T.

9. Find a vector in the plane spanned by

$$\begin{bmatrix} 1 \\ -2 \\ 3 \end{bmatrix} \quad \text{and} \quad \begin{bmatrix} 2 \\ -3 \\ 2 \end{bmatrix}$$

which is perpendicular to the vector

$$\begin{bmatrix} 1 \\ -2 \\ 3 \end{bmatrix}$$

7 Algebra of Linear Transformations and Matrices

To this point we have dealt with only one linear transformation at a time. In this chapter we see how to define the sum of two linear transformations, the product of a scalar and a linear transformation, and the product of two linear transformations. Using these definitions, we define sums, scalar products, and products for matrices. We then see how to determine whether a linear transformation has an inverse, and how to find the inverse of a matrix.

7.1 The Algebra of Linear Transformations

In this section we begin the study of the algebra of linear transformations. We will find that linear transformations can be added together, multiplied by scalars, and multiplied together. In addition, we will find that the algebraic properties of addition and multiplication of linear transformations are quite similar to addition and multiplication of real numbers. We will also find some striking differences.

DEFINITION 7.1

Sum of Two Linear Transformations
Let $T_1: \mathbb{R}^n \to \mathbb{R}^m$ and $T_2: \mathbb{R}^n \to \mathbb{R}^m$ be two linear transformations. We define the sum $T_1 + T_2$ as the function

$$\begin{aligned} T_1 + T_2: \mathbb{R}^n &\to \mathbb{R}^m \\ \mathbf{v} &\to T_1(\mathbf{v}) + T_2(\mathbf{v}) \end{aligned}$$

$T_1 + T_2$ is a linear transformation, see Exercise 11.

We illustrate this definition with the following example.

EXAMPLE 1 Let $T_1: \mathbb{R}^2 \to \mathbb{R}^2$ be projection onto the x-axis and $T_2: \mathbb{R}^2 \to \mathbb{R}^2$ be counterclockwise rotation by 90°. Find $(T_1 + T_2)\begin{bmatrix}1\\2\end{bmatrix}$.

$$(T_1 + T_2)\begin{bmatrix}1\\2\end{bmatrix} = T_1\begin{bmatrix}1\\2\end{bmatrix} + T_2\begin{bmatrix}1\\2\end{bmatrix} = \begin{bmatrix}1\\0\end{bmatrix} + \begin{bmatrix}-2\\1\end{bmatrix} = \begin{bmatrix}-1\\1\end{bmatrix}$$

Figure 7.1 shows the sum $T_1 + T_2$ geometrically.

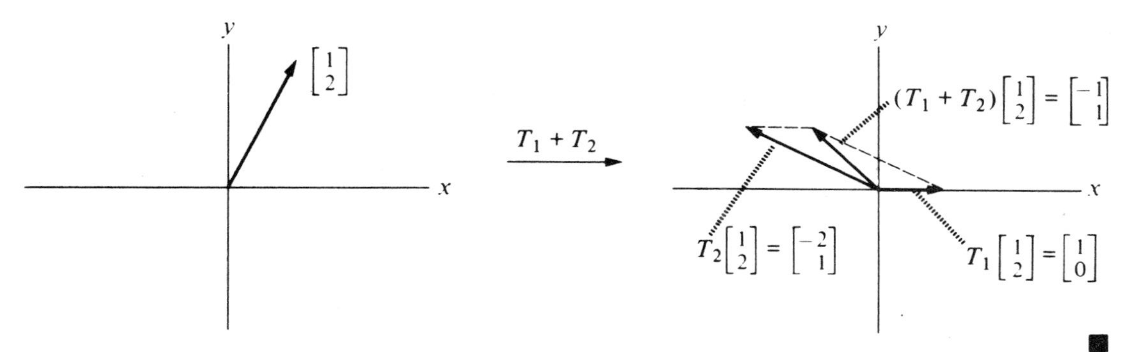

■

Figure 7.1

The definition of multiplying a scalar times a linear transformation is just as straightforward.

DEFINITION 7.2

Product of a Scalar and a Linear Transformation
Let c be a scalar and $T: \mathbb{R}^n \to \mathbb{R}^m$ be a linear transformation. Then cT is the function defined by

$$cT: \mathbb{R}^n \to \mathbb{R}^m$$
$$\mathbf{v} \to c(T\mathbf{v})$$

cT is a linear transformation, see Exercise 11.

Note: With addition and scalar multiplication defined in this way, the set of all linear transformations from $\mathbb{R}^n$ to $\mathbb{R}^m$ is a vector space. (See Exercise 19 in Section 7.2).

EXAMPLE 2 Let $T: \mathbb{R}^2 \to \mathbb{R}^2$ be counterclockwise rotation by 45° and $c = 2$.
The graph of what cT does to a vector $\mathbf{v}$ is shown in Figure 7.2.

Now we consider the operation of multiplication of two linear trans-

Figure 7.2

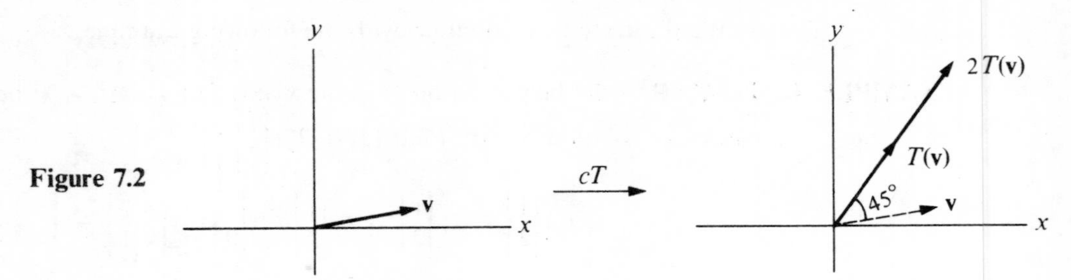

formations. This definition is somewhat more complicated than the definitions of addition and scalar multiplication of linear transformations.

DEFINITION 7.3

Product of Two Linear Transformations

Let $T_1\colon \mathbb{R}^n \to \mathbb{R}^m$ and $T_2\colon \mathbb{R}^p \to \mathbb{R}^n$. We define the product T_1T_2 to be the function defined by

$$T_1T_2\colon \mathbb{R}^p \to \mathbb{R}^m$$
$$\mathbf{v} \to T_1(T_2\,\mathbf{v})$$

T_1T_2 is a linear transformation, see Exercise 11.

In other words, to find T_1T_2 of a vector $\mathbf{v}$, first find $T_2\,\mathbf{v}$ and then apply T_1 to the vector $T_2\,\mathbf{v}$, as seen in Figure 7.3. Be sure to remember that even though we write T_1T_2, we first operate with T_2 and then with T_1.

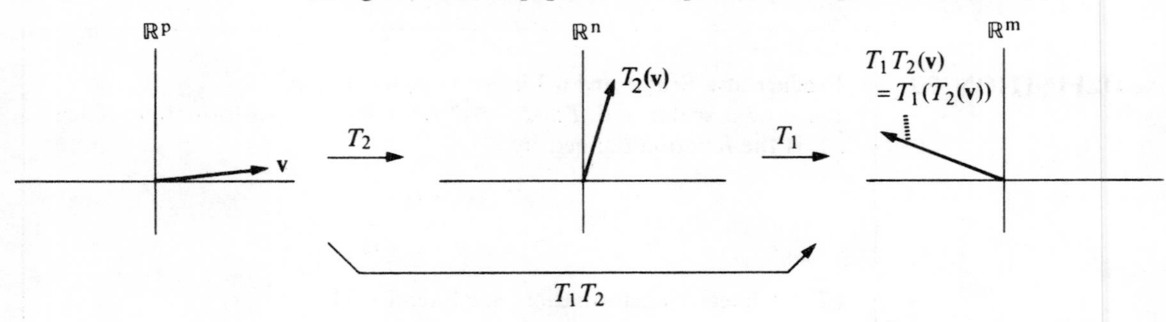

Figure 7.3

EXAMPLE 3 Let $T_1\colon \mathbb{R}^2 \to \mathbb{R}^2$ be counterclockwise rotation by 45° and $T_2\colon \mathbb{R}^2 \to \mathbb{R}^2$ be counterclockwise rotation by 30°. Find T_1T_2.

From Figure 7.4 we see that T_1T_2 is counterclockwise rotation by $30° + 45° = 75°$. ■

EXAMPLE 4 Let $T_1\colon \mathbb{R}^2 \to \mathbb{R}^2$ be counterclockwise rotation by 90° and $T_2\colon \mathbb{R}^2 \to \mathbb{R}^2$ be projection onto the y axis. The action of T_1T_2 on a typical vector is as shown in Figure 7.5.

Since T_1T_2 is a linear transformation, we can find the matrix that represents it. As we saw in Section 3.3, the columns of this matrix are $T_1T_2\begin{bmatrix}1\\0\end{bmatrix}$ and $T_1T_2\begin{bmatrix}0\\1\end{bmatrix}$

From Figure 7.6 we see that the matrix representing T_1T_2 is $\begin{bmatrix}0 & -1\\0 & 0\end{bmatrix}$ ■

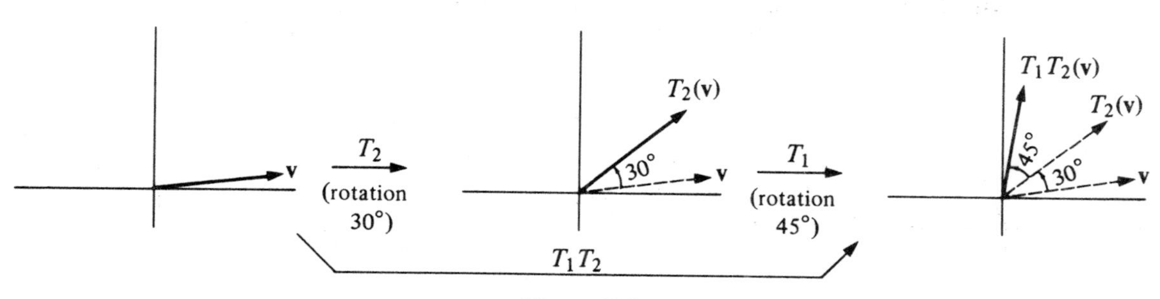

Figure 7.4

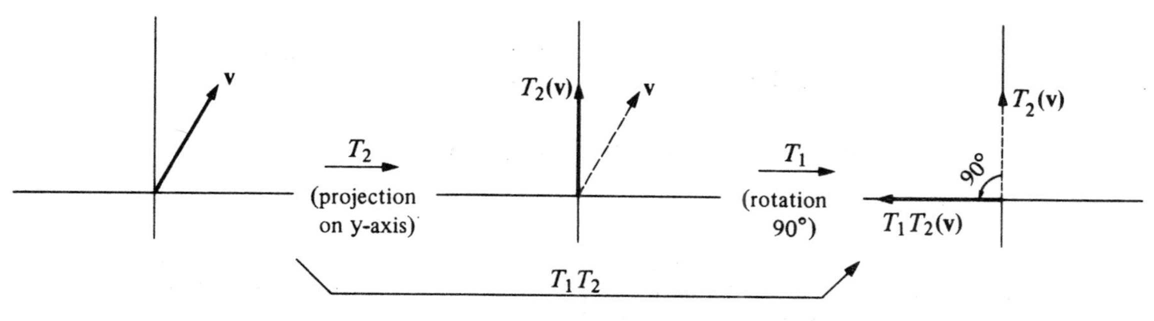

Figure 7.5

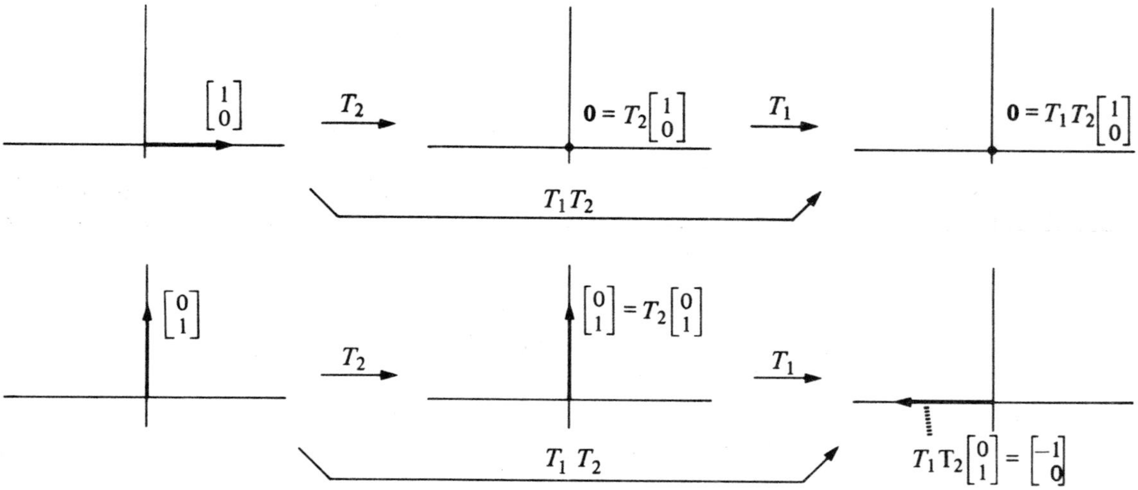

Figure 7.6

Note: The order in which you multiply linear transformations is very important—in general T_1T_2 is different from T_2T_1. In the next example we see what happens when we multiply the linear transformations of Example 4 in the opposite order.

EXAMPLE 5 Let T_1 and T_2 be as in Example 4. We find $T_2 T_1$.

For a typical vector **v** we find $T_2 T_1$ as in Figure 7.7.

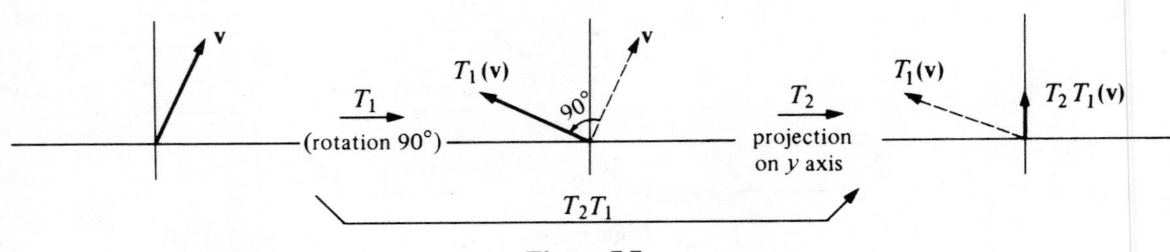

Figure 7.7

By finding $T_2 T_1 \begin{bmatrix} 1 \\ 0 \end{bmatrix}$ and $T_2 T_1 \begin{bmatrix} 0 \\ 1 \end{bmatrix}$

we find the matrix representing $T_2 T_1$ as in Figure 7.8.

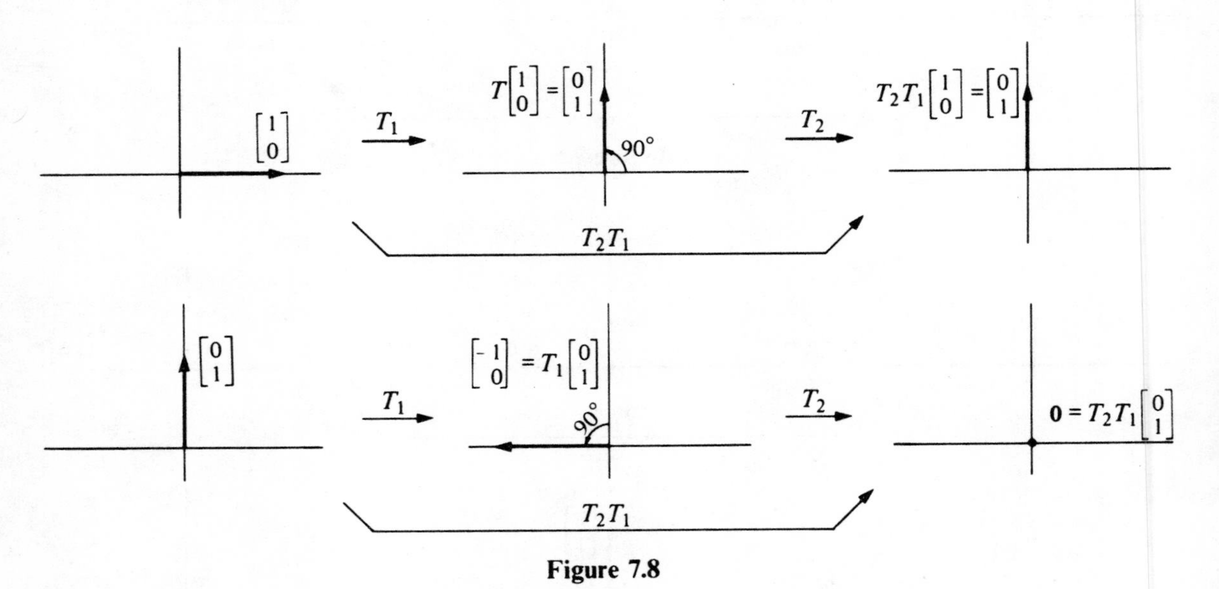

Figure 7.8

Thus the matrix representing $T_2 T_1$ is $\begin{bmatrix} 0 & 0 \\ 1 & 0 \end{bmatrix}$

Therefore, we see that $T_1 T_2 \neq T_2 T_1$. ■

In the theorem that follows we will summarize some of the similarities between the algebra of linear transformations and the algebra of real numbers.

THEOREM 7.1

Algebraic Properties of Linear Transformations

Let T_1, T_2, T_3 be linear transformations (chosen so that the spaces on which they operate are of the proper dimension so that the following statements make sense). Then the following statements hold:

1. Associative law of addition:

$$T_1 + (T_2 + T_3) = (T_1 + T_2) + T_3$$

2. Commutative law of addition:

$$T_1 + T_2 = T_2 + T_1$$

3. Scalar multiplication properties:

$$c(T_1 T_2) = (cT_1)T_2 = T_1(cT_2)$$

4. Associative law of multiplication:

$$T_1(T_2 T_3) = (T_1 T_2)T_3$$

5. There is an additive identity (a zero element). Let $T_0\colon \mathbb{R}^n \to \mathbb{R}^m$ be the transformation mapping each vector $\mathbf{v}$ to $\mathbf{0}$. Then

$$T_0 + T = T$$

Moreover, $T_0 T = T T_0 = T_0$.

6. There is a multiplicative identity for transformations from $\mathbb{R}^n$ to $\mathbb{R}^n$. Let $I\colon \mathbb{R}^n \to \mathbb{R}^n$ be the linear transformation mapping each vector $\mathbf{v}$ to itself ($I\colon \mathbf{v} \to \mathbf{v}$). Then I is called the identity transformation and

$$IT = TI = T$$

for every linear transformation $T\colon \mathbb{R}^n \to \mathbb{R}^n$.

7. Distributive laws:

$$T_1(T_2 + T_3) = T_1 T_2 + T_1 T_3$$

$$(T_1 + T_2)T_3 = T_1 T_3 + T_2 T_3$$

Although the algebra of linear transformations is similar to the algebra of real numbers in many ways, there are some important differences that should be noted.

Pathological Properties of the Algebra of Linear Transformations

1. Multiplication of linear transformations is generally not commutative: As we saw in Examples 4 and 5, $T_1T_2 \neq T_2T_1$. (For a further consequence of this, see Exercise 9.)

2. The cancellation laws do not usually hold. (In Example 6 we will show that $T_1T = T_2T$ does not imply that $T_1 = T_2$. Similarly, in Exercise 8 you will show that $TT_1 = TT_2$ does not imply that $T_1 = T_2$.)

EXAMPLE 6 Let

$T: \mathbb{R}^2 \to \mathbb{R}^2$ be projection onto the x axis

$T_1: \mathbb{R}^2 \to \mathbb{R}^2$ be counterclockwise rotation of 90°

$T_2: \mathbb{R}^2 \to \mathbb{R}^2$ be reflection about the line $y = x$

We will show that $T_1T = T_2T$ even though $T_1 \neq T_2$.

To do this we see what T_1T (Figure 7.9) and T_2T (Figure 7.10) do to an arbitrary vector $\mathbf{v} = \begin{bmatrix} x \\ y \end{bmatrix}$.

$$T_1T\begin{bmatrix} x \\ y \end{bmatrix} = T_1\left(T\begin{bmatrix} x \\ y \end{bmatrix}\right) = T_1\begin{bmatrix} x \\ 0 \end{bmatrix} = \begin{bmatrix} 0 \\ x \end{bmatrix} \qquad T_2T\begin{bmatrix} x \\ y \end{bmatrix} = T_2\left(T\begin{bmatrix} x \\ y \end{bmatrix}\right) = T_2\begin{bmatrix} x \\ 0 \end{bmatrix} = \begin{bmatrix} 0 \\ x \end{bmatrix}$$

Thus the cancellation law does not hold (in general) for linear transformations.

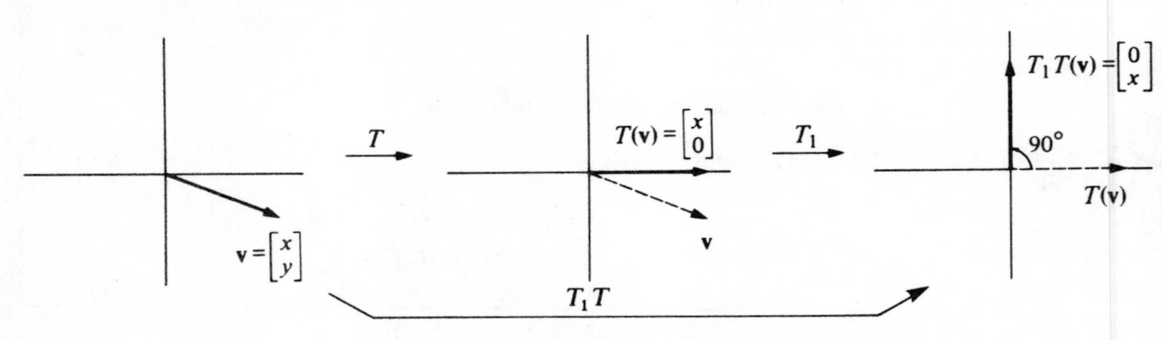

Figure 7.9

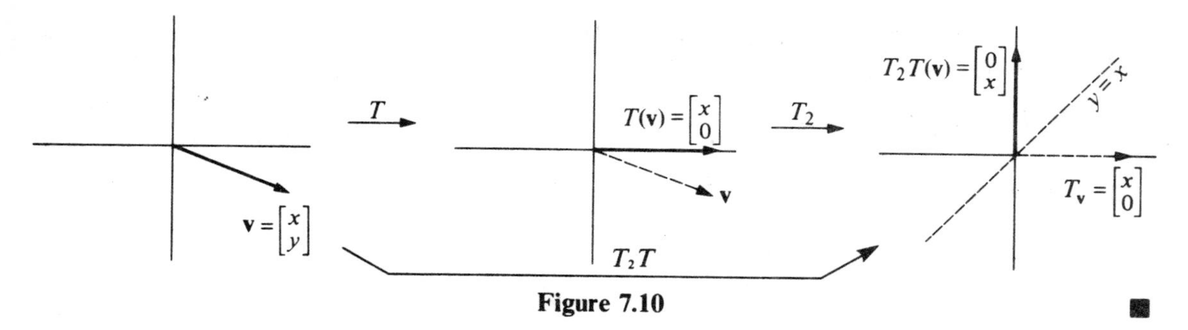

Figure 7.10

■

Exercise Set 7.1

Let

$T_1\colon \mathbb{R}^2 \to \mathbb{R}^2$ be counterclockwise rotation of 90°.

$T_2\colon \mathbb{R}^2 \to \mathbb{R}^2$ be reflection about the line $y = x$.

$T_3\colon \mathbb{R}^2 \to \mathbb{R}^2$ be projection onto the line $y = -3x$.

In Exercises 1–6 find matrices that represent the given linear transformations

1. $T_1 + T_2$ **2.** T_1T_2

3. T_2T_1 (Is this the same as your answer to Exercise 2?)

4. T_3T_1 **5.** $T_1(T_2 + T_3)$

6. $T_1T_2 + T_1T_3$ (Is this the same as your answer to Exercise 5?)

7. a. Write the matrix that represents the linear transformation $T_0\colon \mathbb{R}^2 \to \mathbb{R}^2$.

b. Write the matrix that represents the linear transformation $I\colon \mathbb{R}^2 \to \mathbb{R}^2$.

c. Show that $IT = TI = T$ and $T_0T = T_0$ where T is the linear transformation represented by the matrix

$$\begin{bmatrix} a_{11} & a_{12} \\ a_{21} & a_{22} \end{bmatrix}$$

8. Let

$T\colon \mathbb{R}^2 \to \mathbb{R}^2$ be projection onto the line $y = x$

$T_1\colon \mathbb{R}^2 \to \mathbb{R}^2$ be reflection about the line $y = x$

$T_2\colon \mathbb{R}^2 \to \mathbb{R}^2$ be the identity transformation I

Show that $TT_1 = TT_2$ but that $T_1 \neq T_2$.

9. For any pair of real numbers the following identity holds

$$(a + b)^2 = (a + b)(a + b) = a^2 + 2ab + b^2$$

a. Let

$T_1\colon \mathbb{R}^2 \to \mathbb{R}^2$ be counterclockwise rotation of 90°.

$T_2\colon \mathbb{R}^2 \to \mathbb{R}^2$ be projection onto the y axis

Show that $(T_1 + T_2)^2 \neq T_1^2 + 2T_1T_2 + T_2^2$

(*Note*: $(T_1 + T_2)^2$ is defined to be $(T_1 + T_2)(T_1 + T_2)$.)

b. Find a general expression for $(T_1 + T_2)^2$ and explain why this expression is different from the expression in part **a**.

10. Let $T_1\colon \mathbb{R}^2 \to \mathbb{R}^3$ and $T_2\colon \mathbb{R}^2 \to \mathbb{R}^2$ be linear transformations. Explain why we can define T_1T_2 but cannot define T_2T_1. What are the dimensions n and m such that $T_1T_2\colon \mathbb{R}^n \to \mathbb{R}^m$?

11. a. Let $T_1, T_2\colon \mathbb{R}^n \to \mathbb{R}^m$ be linear transformations. Show that $T_1 + T_2$ is a linear transformation.

b. Let $T\colon \mathbb{R}^n \to \mathbb{R}^m$ be a linear transformation and c a scalar. Show that cT is a linear transformation.

c. Let $T_1: \mathbb{R}^n \to \mathbb{R}^m$ and $T_2: \mathbb{R}^m \to \mathbb{R}^k$. Show that $T_2 T_1$ is a linear transformation from $\mathbb{R}^n$ to $\mathbb{R}^k$.

12. Let T_1, T_2 be linear transformations. Show that

$$\text{Null space of } T_2 \subseteq \text{null space of } T_1 T_2$$

★13. Let T_1, T_2 be linear transformations. Show that

$$\text{Null space of } T_1 T_2 = \{\mathbf{v} \mid T_2(\mathbf{v}) \text{ is in the null space of } T_1\}$$

See Figure 7.11.

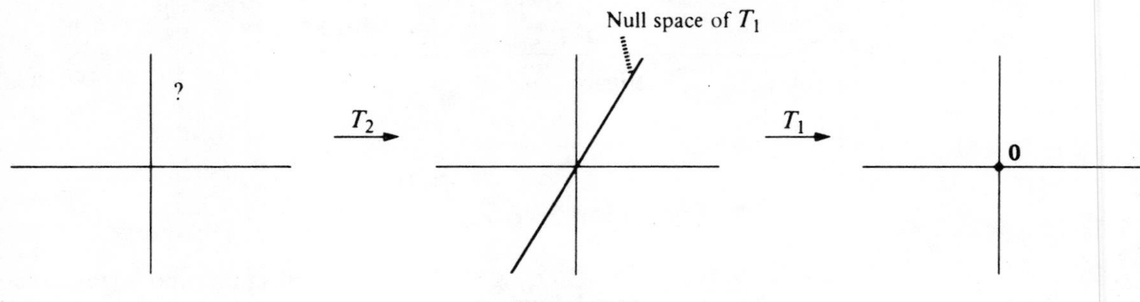

Figure 7.11

7.2 The Algebra of Matrices

In this section we will see how the definitions of addition, scalar multiplication, and multiplication of linear transformations can be used to define addition, scalar multiplication, and multiplication of matrices. The resulting sums, scalar products, and products of the matrices are the matrices corresponding to the sums, scalar products, and products of the linear transformations represented by the matrices.

We begin with addition. Let $T_1: \mathbb{R}^n \to \mathbb{R}^m$ and $T_2: \mathbb{R}^n \to \mathbb{R}^m$ be linear transformations and M_1 and M_2 be the matrices representing T_1 and T_2. Let T equal $T_1 + T_2$ and M be the matrix representing T. We want to find an expression for the coordinates of M in terms of the coordinates of M_1 and M_2. We know that the columns of M_1 and M_2 are

$$M_1 = \begin{bmatrix} T_1\mathbf{e}_1 & T_1\mathbf{e}_2 & \cdots & T_1\mathbf{e}_n \end{bmatrix} \qquad M_2 = \begin{bmatrix} T_2\mathbf{e}_1 & T_2\mathbf{e}_2 & \cdots & T_2\mathbf{e}_n \end{bmatrix}$$

where $\mathbf{e}_i$ represents the vector in $\mathbb{R}^n$ with a 1 in the ith position and a 0 in all other positions.

Moreover, since M represents $T_1 + T_2$, the columns of M are

$$M = \begin{bmatrix} (T_1 + T_2)\mathbf{e}_1 & (T_1 + T_2)\mathbf{e}_2 & \cdots & (T_1 + T_2)\mathbf{e}_n \end{bmatrix}$$

However, since T is linear, $(T_1 + T_2)\mathbf{e}_i = T_1\mathbf{e}_i + T_2\mathbf{e}_i$, so we see that

$$M = \begin{bmatrix} (T_1\mathbf{e}_1 + T_2\mathbf{e}_1) & (T_1\mathbf{e}_2 + T_2\mathbf{e}_2) & \cdots & (T_1\mathbf{e}_n + T_2\mathbf{e}_n) \end{bmatrix}$$

In other words, each column of M is obtained by adding the corresponding columns of M_1 and M_2.

DEFINITION 7.4

Sum of Two Matrices
Let M_1 and M_2 be two $m \times n$ matrices (m rows and n columns). The **sum** $M_1 + M_2$ is the matrix whose entries are the sums of the corresponding entries of M_1 and M_2.

EXAMPLE 1

$$\begin{bmatrix} 1 & 2 & -3 \\ 2 & 1 & 1 \\ 2 & 2 & 1 \end{bmatrix} + \begin{bmatrix} 3 & -2 & 2 \\ 0 & 1 & -5 \\ 1 & 1 & -1 \end{bmatrix} = \begin{bmatrix} (1+3) & (2-2) & (-3+2) \\ (2+0) & (1+1) & (1-5) \\ (2+1) & (2+1) & (1-1) \end{bmatrix}$$

$$= \begin{bmatrix} 4 & 0 & -1 \\ 2 & 2 & -4 \\ 3 & 3 & 0 \end{bmatrix} \quad \blacksquare$$

The definition for scalar multiplication of matrices is just as simple.

DEFINITION 7.5

Product of a Scalar and a Matrix
If M is a matrix and c is a scalar, the matrix cM is the matrix in which each entry is c times as large as the corresponding entry of M.

In Exercise 18 you will show that if M is the matrix representing T, then cM is the matrix that represents cT.

EXAMPLE 2

$$3\begin{bmatrix} 1 & -1 & 2 \\ 2 & 2 & 3 \\ 3 & 1 & 1 \end{bmatrix} = \begin{bmatrix} 3 & -3 & 6 \\ 6 & 6 & 9 \\ 9 & 3 & 3 \end{bmatrix} \quad \blacksquare$$

From the work just completed, you might assume that to find the product of two matrices you merely find the product of the corresponding entries. However, you must remember that we want to define the product of two matrices so that it will represent the linear transformation that is the product of the linear transformations represented by the two matrices. It turns out that simply multiplying corresponding entries does not give the desired result.

To see what the entries of the product matrix should be, we begin by letting T_1 and T_2 be the linear transformations represented by the matrices M_1 and M_2 respectively. If M is the matrix representing the product T_1T_2, then the jth column of M is $(T_1T_2)\mathbf{e}_j$. By definition

$$(T_1T_2)\mathbf{e}_j = T_1(T_2\,\mathbf{e}_j)$$

and since M_1 and M_2 represent T_1 and T_2,

$$T_1(T_2\,\mathbf{e}_j) = M_1(M_2\,\mathbf{e}_j)$$

Since $M_2(\mathbf{e}_j)$ is the jth column of M_2, we see that

$$\begin{bmatrix} j\text{th} \\ \text{column} \\ \text{of} \\ M \end{bmatrix} = M_1 \begin{bmatrix} j\text{th} \\ \text{column} \\ \text{of} \\ M_2 \end{bmatrix}$$

To help clarify this, we consider the following example.

EXAMPLE 3 Let $\quad M_1 = \begin{bmatrix} 1 & 2 \\ -1 & 1 \end{bmatrix} \quad$ and $\quad M_2 = \begin{bmatrix} -3 & 1 \\ 2 & 4 \end{bmatrix}$

We will find $M = M_1M_2$.

$$\text{First column of } M = T_1(T_2\,\mathbf{e}_1) = M_1(M_2\,\mathbf{e}_1) = M_1\left(\begin{bmatrix} -3 & 1 \\ 2 & 4 \end{bmatrix}\begin{bmatrix} 1 \\ 0 \end{bmatrix}\right)$$

$$= M_1\begin{bmatrix} -3 \\ 2 \end{bmatrix} = \begin{bmatrix} 1 & 2 \\ -1 & 1 \end{bmatrix}\begin{bmatrix} -3 \\ 2 \end{bmatrix} = \begin{bmatrix} 1 \\ 5 \end{bmatrix}$$

$$\text{Second column of } M = T_1(T_2\,\mathbf{e}_2) = M_1(M_2\,\mathbf{e}_2) = M_1\left(\begin{bmatrix} -3 & 1 \\ 2 & 4 \end{bmatrix}\begin{bmatrix} 1 \\ 0 \end{bmatrix}\right)$$

$$= M_1\begin{bmatrix} 1 \\ 4 \end{bmatrix} = \begin{bmatrix} 1 & 2 \\ -1 & 1 \end{bmatrix}\begin{bmatrix} 1 \\ 4 \end{bmatrix} = \begin{bmatrix} 9 \\ 3 \end{bmatrix}$$

Thus

$$M = M_1M_2 = \begin{bmatrix} 1 & 9 \\ 5 & 3 \end{bmatrix} \qquad ■$$

We examine this procedure a little more closely to see how to determine a particular entry in the product M of the two matrices M_1 and M_2. We will see how to compute the entry in the ith row and jth column of M.

$$M = \begin{bmatrix} & \vdots & \\ \cdots & \times & \cdots \\ & \vdots & \end{bmatrix} \begin{matrix} \\ i \\ \\ \end{matrix} \qquad (\text{column } j)$$

The jth column of M is found by multiplying M_1 times the jth column of M_2.

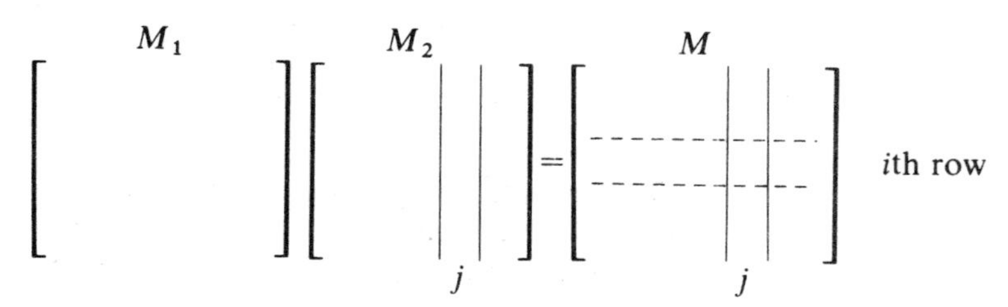

To find the ith entry in the product of a matrix times a vector, you multiply the ith row of the matrix times the vector.

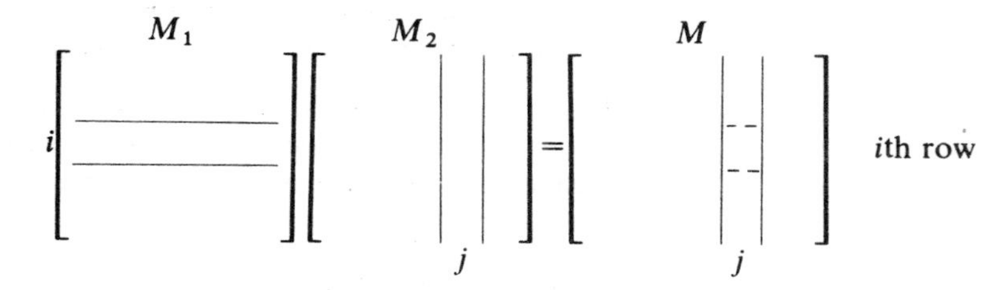

Therefore, the entry in the ith row and jth column of M is obtained by multiplying the ith row of M_1 times the jth column of M_2. We restate this in the following definition.

DEFINITION 7.6

Product of Two Matrices
Let M_1 be an $m \times n$ matrix and M_2 be an $n \times p$ matrix. The **product** matrix $M = M_1M_2$ is the $m \times p$ matrix in which the entry in the ith row and jth column is obtained by multiplying the ith row of M_1 times the jth column of M_2 (for $1 \leq i \leq m$ and $1 \leq j \leq p$).

Remark: The process of matrix multiplication is not extremely difficult. However, it will take some practice for most students to learn to multiply matrices quickly and accurately. When computing a particular entry of the product of two matrices, start at the left end of a row of the matrix on the left and the top of a column of the matrix on the right and compute the product of the first entries. Then move to the second entries and compute their product and add this to the product of the first two entries. Move to the third entries, compute their product and add this to the sum of the other two. Continue in this way until you come to the end of the row and column and write the total in the proper position in the product matrix (note that the matrix product can only be defined when the length of the rows of the matrix on the left is the same as the length of the columns of the matrix on the right).

EXAMPLE 4 We find the product of two matrices.

$$\begin{bmatrix} 1 & 2 & -1 & 3 \\ 2 & 1 & 1 & 2 \\ -1 & 1 & 0 & 3 \end{bmatrix}\begin{bmatrix} 1 & 3 \\ 1 & -1 \\ 0 & -1 \\ 3 & 2 \end{bmatrix} = \begin{bmatrix} 12 & 8 \\ 9 & 8 \\ 9 & 2 \end{bmatrix} \quad \blacksquare$$

To conclude this section, we state the following theorem to reemphasize the relationship between the definitions for sums, scalar products, and products for matrices and linear transformations.

THEOREM 7.2

1. Let $T_1: \mathbb{R}^n \to \mathbb{R}^m$ and $T_2: \mathbb{R}^n \to \mathbb{R}^m$ be linear transformations. If M_1 and M_2 are the matrices representing T_1 and T_2, then the sum $M_1 + M_2$ (defined above) is the matrix representing $T_1 + T_2$.

2. Let $T: \mathbb{R}^n \to \mathbb{R}^m$ be a linear transformation and let c be a scalar. If M is the matrix representing T, then the product cM (defined above) is the matrix representing cT.

3. Let $T_1: \mathbb{R}^m \to \mathbb{R}^n$ and $T_2: \mathbb{R}^p \to \mathbb{R}^m$. If M_1 and M_2 are respectively the matrices representing T_1 and T_2, then the product M_1M_2 (as defined above) is the matrix representing T_1T_2.

Note: Since each matrix represents a linear transformation, all the conclusions of Theorem 7.1 are valid for matrices as well as linear transformations. Moreover, the pathological properties of linear transformations discussed in Section 7.1 are also pathological properties of matrices.

Exercise Set 7.2

In Exercises 1–13 find the matrix products indicated.

1. $\begin{bmatrix} 1 & 2 \\ 0 & 1 \end{bmatrix}\begin{bmatrix} 1 & 2 & -3 \\ 2 & 1 & 1 \end{bmatrix}$

2. $\begin{bmatrix} 1 & 2 & -3 \\ 2 & 1 & 1 \end{bmatrix}\begin{bmatrix} 1 & 2 \\ 0 & 1 \end{bmatrix}$

3. $\begin{bmatrix} -1 & 3 & 4 \\ 2 & 1 & 1 \\ -3 & 1 & 2 \end{bmatrix}\begin{bmatrix} 1 & 0 & -2 \\ 3 & 2 & -4 \\ 1 & -1 & 2 \end{bmatrix}$

4. $\begin{bmatrix} 1 & 2 & -3 & 4 \\ 2 & 1 & -1 & 1 \\ 1 & -1 & -1 & 2 \\ 2 & 2 & -3 & 5 \end{bmatrix}\begin{bmatrix} 1 & 2 & 3 \\ -2 & 2 & 4 \\ 3 & 1 & 2 \\ 4 & -2 & 2 \end{bmatrix}$

5. $\left(\begin{bmatrix} 1 & 2 \\ -1 & 3 \end{bmatrix}\begin{bmatrix} 2 & 1 \\ 1 & 3 \end{bmatrix}\right)\begin{bmatrix} 5 & -2 \\ 2 & 1 \end{bmatrix}$

6. $\begin{bmatrix} 1 & 2 \\ -1 & 3 \end{bmatrix}\left(\begin{bmatrix} 2 & 1 \\ 1 & 3 \end{bmatrix}\begin{bmatrix} 5 & -2 \\ 2 & 1 \end{bmatrix}\right)$

(Is this answer the same as the answer to Exercise 5? Should it be?)

7. $\begin{bmatrix} 1 & -1 \\ 2 & 3 \end{bmatrix}\begin{bmatrix} 2 & 2 \\ -1 & 3 \end{bmatrix}$

8. $\begin{bmatrix} 2 & 2 \\ -1 & 3 \end{bmatrix}\begin{bmatrix} 1 & -1 \\ 2 & 3 \end{bmatrix}$ (Is this answer the same as the answer to Exercise 7? Should it be?)

9. $[1 \quad 3 \quad 4]\begin{bmatrix} 2 \\ 3 \\ 5 \end{bmatrix}$ **10.** $\begin{bmatrix} 1 & -1 \\ 2 & 4 \\ -1 & 2 \end{bmatrix}\begin{bmatrix} 2 & 7 \\ 0 & 1 \\ 1 & -7 \end{bmatrix}$

11. $[1 \quad 3 \quad -2 \quad 7]\begin{bmatrix} 1 & 0 & 2 & 3 \\ 6 & 1 & -3 & -2 \\ 3 & -1 & 1 & -2 \\ 1 & 1 & 0 & -3 \end{bmatrix}$

12. $\begin{bmatrix} 1 & 0 & 0 & 0 \\ 0 & 0 & 0 & 1 \\ 0 & 0 & 1 & 0 \\ 0 & 1 & 0 & 0 \end{bmatrix}\begin{bmatrix} a_{11} & a_{12} & a_{13} & a_{14} \\ a_{21} & a_{22} & a_{23} & a_{24} \\ a_{31} & a_{32} & a_{33} & a_{34} \\ a_{41} & a_{42} & a_{43} & a_{44} \end{bmatrix}$

13. $\begin{bmatrix} a_{11} & a_{12} & a_{13} & a_{14} \\ a_{21} & a_{22} & a_{23} & a_{24} \\ a_{31} & a_{32} & a_{33} & a_{34} \\ a_{41} & a_{42} & a_{43} & a_{44} \end{bmatrix}\begin{bmatrix} 1 & 0 & 0 & 0 \\ 0 & 0 & 0 & 1 \\ 0 & 0 & 1 & 0 \\ 0 & 1 & 0 & 0 \end{bmatrix}$

14. Let P_{ij} be the matrix obtained from the $n \times n$ identity matrix by interchanging the ith and jth columns.

a. If M is an $m \times n$ matrix, show that the matrix MP_{ij} is the matrix obtained from M by interchanging the ith and jth columns of M (see Exercise 13).

b. If M is an $n \times m$ matrix, show that the matrix $P_{ij}M$ is the matrix obtained from M by interchanging the ith and jth rows of M (see Exercise 12).

15. Let B be the matrix obtained by multiplying the ith column of the $n \times n$ identity matrix by the real number b.

a. Let M be an $m \times n$ matrix. Show that the matrix MB is the matrix obtained by multiplying the ith column of M by b.

b. Let M be an $n \times m$ matrix. Describe the matrix BM.

16. Express each of the linear transformations T_1, T_2, T_3 of Exercises 1–5 in Section 7.1 as a matrix and use these matrices to compute the sums and products listed. Compare the matrices you get in this way to the ones you got when you originally did each exercise.

17. a. Let $M = \begin{bmatrix} 0 & 1 \\ 0 & 0 \end{bmatrix}$

Show that even though M is not the zero matrix,

$$M^2 = \begin{bmatrix} 0 & 0 \\ 0 & 0 \end{bmatrix} \qquad M^2 \text{ is defined as } MM.$$

b. Let $M = \begin{bmatrix} 0 & 1 \\ -1 & 0 \end{bmatrix}$

Show that $M^4 = I$.

18. Show that if $T: \mathbb{R}^n \to \mathbb{R}^m$ is a linear transformation and M is the matrix representing T, then cM is the matrix representing cT for any scalar c.

19. a. Let $\mathcal{M}_{2\times 2}$ be the set of all 2×2 matrices (with entries in $\mathbb{R}$). Show that $\mathcal{M}_{2\times 2}$ is a vector space under the operations of matrix addition and scalar multiplication (see Section 3.2).

b. Show that a basis of $\mathcal{M}_{2\times 2}$ is

$$\left\{\begin{bmatrix} 1 & 0 \\ 0 & 0 \end{bmatrix}, \begin{bmatrix} 0 & 1 \\ 0 & 0 \end{bmatrix}, \begin{bmatrix} 0 & 0 \\ 1 & 0 \end{bmatrix}, \begin{bmatrix} 0 & 0 \\ 0 & 1 \end{bmatrix}\right\}$$

c. Show that if $\mathcal{M}_{n\times m}$ is the set of all $n \times m$ matrices, then $\mathcal{M}_{n\times m}$ is a vector space of dimension nm.

★ **20.** Let the vectors

$$\mathbf{a} = \begin{bmatrix} a_1 \\ a_2 \\ \vdots \\ a_n \end{bmatrix} \quad \text{and} \quad \mathbf{b} = \begin{bmatrix} b_1 \\ b_2 \\ \vdots \\ b_n \end{bmatrix}$$

be linearly independent. Let

$$M = \begin{bmatrix} a_1 & a_2 & \cdots & a_n \\ b_1 & b_2 & \cdots & b_n \end{bmatrix} \quad \text{and} \quad N = \begin{bmatrix} a_1 & b_1 \\ a_2 & b_2 \\ \vdots & \vdots \\ a_n & b_n \end{bmatrix}$$

Show that Null space of $MN = \{\mathbf{0}\}$

(*Hint*: Express MN in terms of dot products, row reduce MN and then convert dot products to lengths of vectors.)

21. Let $T: \mathbb{R}^2 \to \mathbb{R}^2$ be counterclockwise rotation by 120°.

a. Find the matrix M that represents T.

b. Show that $M^3 = I$.

c. Find the complex number $c = a_0 + b_0 i$ such

that multiplication by c induces the linear transformation T on the vector space of complex numbers (see Example 5 in Section 3.2 or Applications 1).

d. Show that $c \neq 1$, but $c^3 = 1$. That is, c is a cube root of unity.

7.3 Inverses of Linear Transformations and Matrices

Since linear transformations are functions, we can look at them from that point of view. One important question we ask when dealing with a function is whether it has an inverse. Some important examples of functions with inverses that you have probably already studied are: The inverse of the function $f(x) = x^3$ is the function $g(x) = \sqrt[3]{x}$, and the inverse of the function $f(x) = 10^x$ is the function $g(x) = \log_{10} x$. In this section we will use the special properties of linear transformations to determine which linear transformations and matrices have inverses and how to find their inverses.

Suppose $T: V \to W$ is a linear transformation. We have studied problems in which a fixed vector $\mathbf{w}_0$ in W is given and we wish to find all vectors $\mathbf{v}$ in V for which

$$T\mathbf{v} = \mathbf{w}_0$$

As we saw in Theorem 5.1, if this equation has a solution, then the set of all vectors satisfying the equation is a coset $\mathbf{v}_0 + N$ where N is the null space of T.

Using this theorem, we can form a correspondence, which we will call the **inverse correspondence**, between the vectors $\mathbf{w}$ of the image space of T and those vectors $\mathbf{v}$ in the domain of T such that $T(\mathbf{v}) = \mathbf{w}$.

$$\mathbf{w} \xrightarrow{\text{corresponds to}} (\text{all vectors } \mathbf{v} \text{ such that } T(\mathbf{v}) = \mathbf{w})$$

EXAMPLE 1 Let $T: \mathbb{R}^2 \to \mathbb{R}^2$ be projection onto the x axis. Then the image space of T is the x axis. A typical vector of the image space of T is

$$\mathbf{w}_0 = \begin{bmatrix} x_0 \\ 0 \end{bmatrix}$$

We see that the set of all vectors $\mathbf{v}$ for which

$$T\mathbf{v} = \begin{bmatrix} x_0 \\ 0 \end{bmatrix}$$

is the set of all vectors of the form

$$\begin{bmatrix} x_0 \\ y \end{bmatrix} \qquad \text{for any real number } y$$

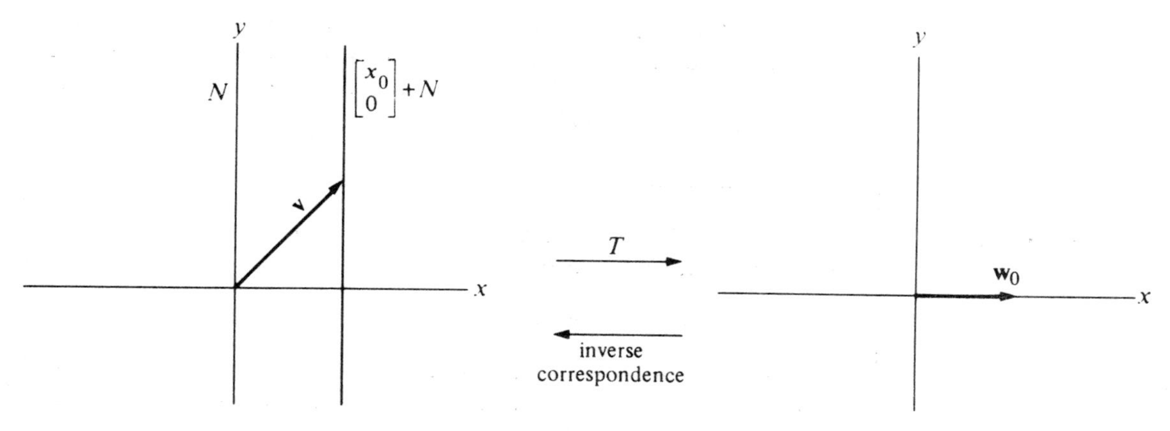

Figure 7.12

The null space of T is $\left\langle \begin{bmatrix} 0 \\ 1 \end{bmatrix} \right\rangle$. So the set of vectors corresponding to $\mathbf{w}_0$ is the coset $\begin{bmatrix} x_0 \\ 0 \end{bmatrix} + \left\langle \begin{bmatrix} 0 \\ 1 \end{bmatrix} \right\rangle$. So the inverse correspondence is

$$\mathbf{w} = \begin{bmatrix} x_0 \\ 0 \end{bmatrix} \xrightarrow{\text{corresponds to}} \text{all vectors in the coset } \begin{bmatrix} x_0 \\ 0 \end{bmatrix} + \left\langle \begin{bmatrix} 0 \\ 1 \end{bmatrix} \right\rangle$$

Note: The inverse correspondence defined here is not a function from $\mathbb{R}^2$ to $\mathbb{R}^2$. It fails to satisfy either of the conditions necessary for a correspondence to be a function (Section 1.1), that is

1. This correspondence is not defined for all vectors of W—it is only defined for those vectors in the image space of T.

2. For a given vector $\mathbf{w}$ of the image space of T, there are many (not just one) vectors to which it corresponds (all vectors with the same x coordinate as $\mathbf{w}$).

■

Now we will see what conditions a linear transformation must satisfy so that the inverse correspondence is a function. The two properties that any correspondence from W to V must satisfy to be a function are:

1. For each $\mathbf{w}$ in W, there is a vector $\mathbf{v}$ in V that corresponds to $\mathbf{w}$.

2. There is only one vector $\mathbf{v}$ in V corresponding to each $\mathbf{w}$ in W.

Since T is a linear transformation, we can translate these as follows:

1. The image space of T equals W.

2. The null space of T is $\langle \mathbf{0} \rangle$. (This follows because if $\mathbf{w}$ corresponds

to two vectors $\mathbf{v}_1$ and $\mathbf{v}_2$, this means that $T\mathbf{v}_1 = \mathbf{w}$ and $T\mathbf{v}_2 = \mathbf{w}$. As we saw in Theorem 5.1, this means that $\mathbf{v}_1 - \mathbf{v}_2$ is in the null space of T. So to ensure that only one vector can correspond to $\mathbf{w}$, the null space of T must be $\langle \mathbf{0} \rangle$.)

DEFINITION 7.7

Let $T: V \to W$ be a linear transformation. If the inverse correspondence between W and V given by

$$\mathbf{w} \xrightarrow{\text{corresponds to}} (\text{all vectors } \mathbf{v} \text{ such that } T\mathbf{v} = \mathbf{w})$$

is a function, then this function is called the **inverse** of T and is denoted T^{-1}. A linear transformation that has an inverse is said to be **invertible**.

Note: If $T: V \to W$ is an invertible linear transformation, then by looking at Figure 7.13 and reading clockwise from the upper left, we see that $T^{-1}(T\mathbf{v}) = T^{-1}\mathbf{w} = \mathbf{v}$. Moreover, from the figure, starting at the bottom right and reading clockwise, we see that $T(T^{-1}\mathbf{w}) = T\mathbf{v} = \mathbf{w}$. Thus, $T^{-1}T = I$ and $TT^{-1} = I$, where I is the identity transformation.

Making a slight refinement on the conditions for a linear transformation to be invertible gives the following theorem.

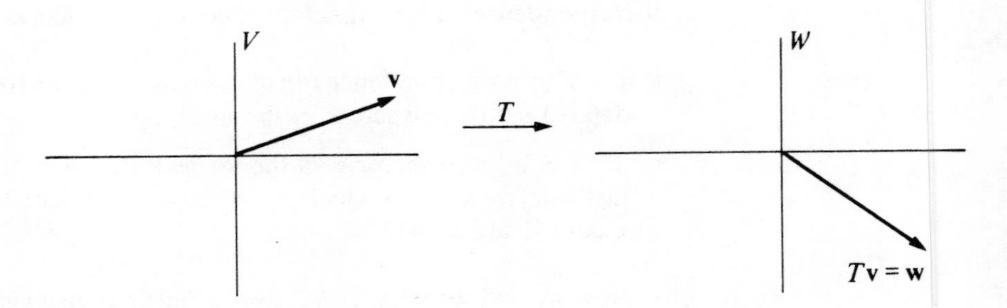

Figure 7.13

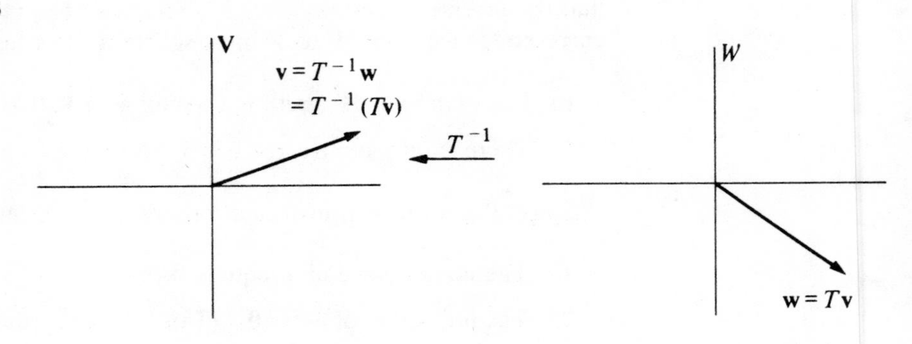

THEOREM 7.3

> Let $T: V \to W$ be a linear transformation. Then T has an inverse if and only if both of the following conditions hold:
>
> **1.** Rank of T = dimension of W
>
> **2.** Nullity of $T = 0$

EXAMPLE 2 Show that if $T: \mathbb{R}^2 \to \mathbb{R}^3$ is a linear transformation, then T is not invertible.

Figure 7.14

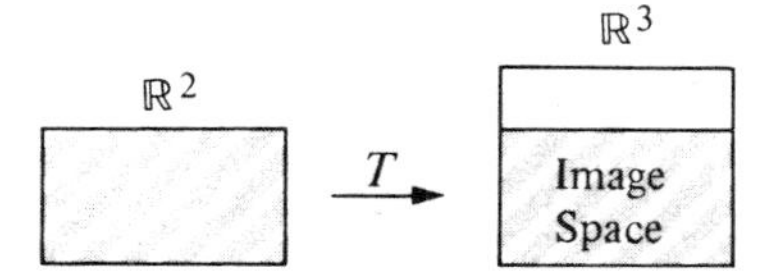

By definition, the rank of T equals the dimension of the image space of T (see Figure 7.14). Since the image space of T is spanned by the two vectors

$$T\begin{bmatrix}1\\0\end{bmatrix} \quad \text{and} \quad T\begin{bmatrix}0\\1\end{bmatrix}$$

it has dimension at most 2. So

$$\text{Rank of } T \leq 2 \neq 3 = \text{dimension of } \mathbb{R}^3$$

Thus T does not satisfy Condition 1 of Theorem 7.3, and therefore is not invertible. ■

EXAMPLE 3 Show that if $T: \mathbb{R}^3 \to \mathbb{R}^2$ is a linear transformation, then T is not invertible.

Figure 7.15

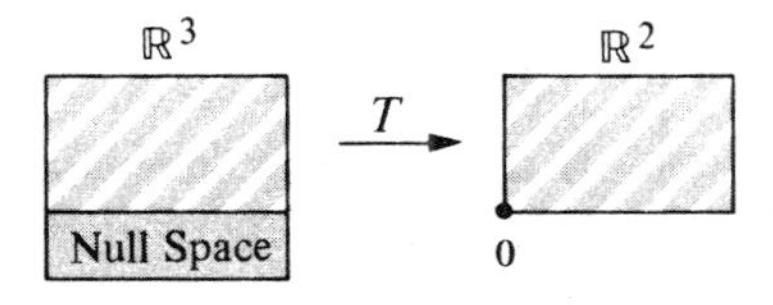

Since the range of T has dimension 2, the image space of T has dimension less than or equal to 2—that is, rank $T \leq 2$ (see Figure 7.15). By Theorem 5.2.

$$\text{Rank of } T + \text{nullity of } T = \text{dimension of } \mathbb{R}^3 = 3$$

Since the rank of T is at most 2, we see that the nullity of T must be at least 1. Hence, T does not satisfy condition 2 of Theorem 7.3. ■

Generalizing the ideas discussed in Examples 2 and 3 gives:

THEOREM 7.4

The linear transformation $T\colon \mathbb{R}^n \to \mathbb{R}^m$ is not invertible if $m \neq n$. (*Note:* This does not say that T is invertible if $n = m$. As we have seen in Example 1, this is not always the case.)

We apply Theorem 5.2 to this result to get a better version of Theorem 7.3. If $T\colon \mathbb{R}^n \to \mathbb{R}^n$ is a linear transformation such that the nullity of T is 0, then since

$$\text{Rank of } T + \text{nullity of } T = n$$

we have

$$\text{Rank of } T + 0 = n$$

From this we have

THEOREM 7.5

A linear transformation $T\colon \mathbb{R}^n \to \mathbb{R}^m$ is invertible if and only if $n = m$ and either the nullity of T is 0 or the rank of $T = n$.

EXAMPLE 4 Let $T\colon \mathbb{R}^2 \to \mathbb{R}^2$ be reflection about the line $y = 2x$. Is this transformation invertible?

Since the nullity of T is 0 (see Example 2 in Section 5.2), T is invertible. ■

Now let us look at matrices. A matrix is invertible if and only if the linear transformation it represents is invertible. Translating Theorem 7.5 to matrix language, we have

THEOREM 7.5′

A matrix M is invertible if and only if M is a square matrix and either the nullity of M is 0 or the rank of M is equal to the number of columns of M.

Although this is an elegant theorem, it only tells us when a matrix has an inverse—it does not tell us what the inverse is. We will now see how to find the inverse of a matrix. Suppose $T\colon \mathbb{R}^2 \to \mathbb{R}^2$ is an invertible linear transformation (see Figure 7.16) and M is its matrix. We will see how to find M^{-1}. The first column of M^{-1} is $T^{-1}\begin{bmatrix}1\\0\end{bmatrix}$, that is, the vector $\mathbf{v}_1$ such that

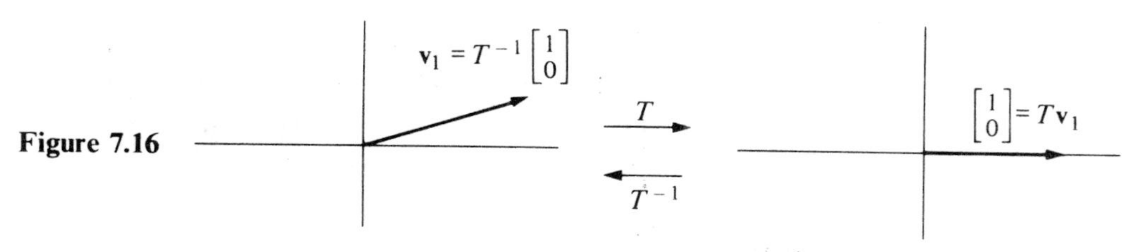

Figure 7.16

$$T\mathbf{v}_1 = \begin{bmatrix}1\\0\end{bmatrix}$$

Thus $\mathbf{v}_1$ is the solution of the equation

$$M\mathbf{v}_1 = \begin{bmatrix}1\\0\end{bmatrix}$$

Similarly, the second column of M^{-1} is the vector $\mathbf{v}_2$ such that

$$M\mathbf{v}_2 = \begin{bmatrix}0\\1\end{bmatrix}$$

Now we see how to find $\mathbf{v}_1$ and $\mathbf{v}_2$ for a specific example.

EXAMPLE 5 We find the inverse of the matrix

$$M = \begin{bmatrix}1 & -2\\-1 & 1\end{bmatrix}$$

The first column $\mathbf{v}_1$ of M^{-1} is the solution of the equation

$$M\mathbf{v}_1 = \begin{bmatrix}1 & -2\\-1 & 1\end{bmatrix}\mathbf{v}_1 = \begin{bmatrix}1\\0\end{bmatrix}$$

To find $\mathbf{v}_1$, we reduce the augmented matrix associated with this equation.

$$\left[\begin{array}{rr|r}1 & -2 & 1\\-1 & 1 & 0\end{array}\right] \begin{array}{l}(1)\\(1)\end{array}\longrightarrow \left[\begin{array}{rr|r}1 & -2 & 1\\0 & -1 & 1\end{array}\right] \begin{array}{l}(1)\\(-2)\end{array}\longrightarrow$$

$$\left[\begin{array}{rr|r}1 & 0 & -1\\0 & -1 & 1\end{array}\right] \quad (-1)\longrightarrow \left[\begin{array}{rr|r}1 & 0 & -1\\0 & 1 & -1\end{array}\right]$$

Thus $\mathbf{v}_1 = \begin{bmatrix}-1\\-1\end{bmatrix}$

Similarly, the second column $\mathbf{v}_2$ of M^{-1} is the solution of the equation

$$M\mathbf{v}_2 = \begin{bmatrix}1 & -2\\-1 & 1\end{bmatrix}\mathbf{v}_2 = \begin{bmatrix}0\\1\end{bmatrix}$$

We reduce the augmented matrix associated with this equation.

$$\left[\begin{array}{rr|r} 1 & -2 & 0 \\ -1 & 1 & 1 \end{array}\right] \begin{array}{l} (1) \\ (1) \end{array} \longrightarrow \left[\begin{array}{rr|r} 1 & -2 & 0 \\ 0 & -1 & 1 \end{array}\right] \begin{array}{l} (1) \\ (-2) \end{array} \longrightarrow$$

$$\left[\begin{array}{rr|r} 1 & 0 & -2 \\ 0 & -1 & 1 \end{array}\right] \quad (-1) \longrightarrow \left[\begin{array}{rr|r} 1 & 0 & -2 \\ 0 & 1 & -1 \end{array}\right]$$

So $\mathbf{v}_2 = \begin{bmatrix} -2 \\ -1 \end{bmatrix}$

Therefore $$M^{-1} = \begin{bmatrix} -1 & -2 \\ -1 & -1 \end{bmatrix}$$

To check our calculations we verify that $M^{-1}M = I$.

$$\begin{bmatrix} -1 & -2 \\ -1 & -1 \end{bmatrix} \begin{bmatrix} 1 & -2 \\ -1 & 1 \end{bmatrix} = \begin{bmatrix} 1 & 0 \\ 0 & 1 \end{bmatrix} \qquad \blacksquare$$

If we examine the computations involved in finding M^{-1} in Example 5, we see that they consisted of reducing the two augmented matrices

$$\left[\begin{array}{rr|r} 1 & -2 & 1 \\ -1 & 1 & 0 \end{array}\right] \quad \text{and} \quad \left[\begin{array}{rr|r} 1 & -2 & 0 \\ -1 & 1 & 1 \end{array}\right]$$

These augmented matrices have the same coefficient matrix (the matrix to the left of the vertical line), and we can use the same steps in the reduction procedure for both of them. Therefore, to be more efficient we can reduce the doubly augmented matrix

$$\left[\begin{array}{rr|rr} 1 & -2 & 1 & 0 \\ -1 & 1 & 0 & 1 \end{array}\right] = \left[\begin{array}{c|c} M & I \end{array}\right]$$

which, using the same reduction steps, gives

$$\left[\begin{array}{rr|rr} 1 & 0 & -1 & -2 \\ 0 & 1 & -1 & -1 \end{array}\right] = \left[\begin{array}{c|c} I & M^{-1} \end{array}\right]$$

Generalizing this idea gives the following theorem.

THEOREM 7.6

Let M be an invertible matrix. To find M^{-1}, reduce the augmented matrix

$$\left[\begin{array}{c|c} M & I \end{array}\right]$$

to get

$$\left[\begin{array}{c|c} I & M^{-1} \end{array}\right]$$

Note: To use Theorem 7.6, it is not necessary to first determine if the $n \times n$ matrix M is invertible. If M is not invertible, then the rank of M is less than n and from Theorem 5.7 we see that the completely reduced matrix equivalent to M is not the identity matrix. Therefore you will not be able to reduce the matrix

$$\left[\begin{array}{c|c} M & I \end{array}\right]$$

to obtain an identity matrix to the left of the vertical line—instead you will end up with at least one row of zeros in the completely reduced matrix to the left of the line.

EXAMPLE 6 We find the inverse of $M = \begin{bmatrix} 1 & 1 & 1 \\ -1 & 2 & 1 \\ 3 & 1 & 2 \end{bmatrix}$

$$\left[\begin{array}{rrr|rrr} 1 & 1 & 1 & 1 & 0 & 0 \\ -1 & 2 & 1 & 0 & 1 & 0 \\ 3 & 1 & 2 & 0 & 0 & 1 \end{array}\right] \begin{matrix} (1) \quad (-3) \\ (1) \\ \quad (1) \end{matrix} \longrightarrow \left[\begin{array}{rrr|rrr} 1 & 1 & 1 & 1 & 0 & 0 \\ 0 & 3 & 2 & 1 & 1 & 0 \\ 0 & -2 & -1 & -3 & 0 & 1 \end{array}\right] \begin{matrix} \\ (2) \\ (3) \end{matrix} \longrightarrow$$

$$\left[\begin{array}{rrr|rrr} 1 & 1 & 1 & 1 & 0 & 0 \\ 0 & 3 & 2 & 1 & 1 & 0 \\ 0 & 0 & 1 & -7 & 2 & 3 \end{array}\right] \begin{matrix} \quad (1) \\ (1) \\ (-2) \quad (-1) \end{matrix} \longrightarrow \left[\begin{array}{rrr|rrr} 1 & 1 & 0 & 8 & -2 & -3 \\ 0 & 3 & 0 & 15 & -3 & -6 \\ 0 & 0 & 1 & -7 & 2 & 3 \end{array}\right] \begin{matrix} (-3) \\ (1) \\ \\ \end{matrix} \longrightarrow$$

$$\left[\begin{array}{rrr|rrr} -3 & 0 & 0 & -9 & 3 & 3 \\ 0 & 3 & 0 & 15 & -3 & -6 \\ 0 & 0 & 1 & -7 & 2 & 3 \end{array}\right] \begin{matrix} (-\frac{1}{3}) \longrightarrow \\ (\frac{1}{3}) \longrightarrow \\ \\ \end{matrix} \left[\begin{array}{rrr|rrr} 1 & 0 & 0 & 3 & -1 & -1 \\ 0 & 1 & 0 & 5 & -1 & -2 \\ 0 & 0 & 1 & -7 & 2 & 3 \end{array}\right]$$

So

$$M^{-1} = \begin{bmatrix} 3 & -1 & -1 \\ 5 & -1 & -2 \\ -7 & 2 & 3 \end{bmatrix}$$

To check we show that $M^{-1}M = I$.

$$\begin{bmatrix} 3 & -1 & -1 \\ 5 & -1 & -2 \\ -7 & 2 & 3 \end{bmatrix} \begin{bmatrix} 1 & 1 & 1 \\ -1 & 2 & 1 \\ 3 & 1 & 2 \end{bmatrix} = \begin{bmatrix} 1 & 0 & 0 \\ 0 & 1 & 0 \\ 0 & 0 & 1 \end{bmatrix} \quad \blacksquare$$

EXAMPLE 7 We show what happens when we try to use Theorem 7.6 to invert a noninvertible matrix. Let $M = \begin{bmatrix} 1 & -2 \\ -2 & 4 \end{bmatrix}$.

$$\left[\begin{array}{rr|rr} 1 & -2 & 1 & 0 \\ -2 & 4 & 0 & 1 \end{array}\right] \begin{matrix} (2) \\ (1) \end{matrix} \longrightarrow \left[\begin{array}{rr|rr} 1 & -2 & 1 & 0 \\ 0 & 0 & 2 & 1 \end{array}\right]$$

Since the matrix to the left of the line is completely reduced and is not the identity matrix, we conclude that M does not have an inverse. $\blacksquare$

Now we discuss the inverse of the product of two linear transformations or matrices.

THEOREM 7.7

1. If $T_1: \mathbb{R}^n \to \mathbb{R}^n$ and $T_2: \mathbb{R}^n \to \mathbb{R}^n$ are invertible linear transformations, then their product T_1T_2 is invertible and its inverse is

$$(T_1T_2)^{-1} = T_2^{-1}T_1^{-1}$$

2. If M_1 and M_2 are invertible $n \times n$ matrices, then their product M_1M_2 is invertible and its inverse is

$$(M_1M_2)^{-1} = M_2^{-1}M_1^{-1}$$

It is easy to see why this theorem is true. Using the associative law several times we have

$$(T_2^{-1}T_1^{-1})(T_1T_2) = T_2^{-1}(T_1^{-1}T_1)T_2 = T_2^{-1}(I)T_2 = T_2^{-1}T_2 = I$$

Cautionary Note: Since multiplication of linear transformations (and matrices) is not commutative, $T_2^{-1}T_1^{-1}$ is usually different from $T_1^{-1}T_2^{-1}$. Therefore, in general

$$(T_1T_2)^{-1} \neq T_1^{-1}T_2^{-1} \qquad \text{and} \qquad (M_1M_2)^{-1} \neq M_1^{-1}M_2^{-1}$$

Examples of this are given as Exercises 8 and 9.

We digress briefly to discuss some relationships between invertible linear transformations and matrix row operations. If we consider a typical row operation of the form

$$M = \begin{bmatrix} (\text{row } i) \\ (\text{row } j) \end{bmatrix} \begin{matrix} (a) \\ (b) \end{matrix} \longrightarrow \begin{bmatrix} (\text{row } i) \\ a(\text{row } i) + b(\text{row } j) \end{bmatrix} = P$$

we can think of it as operating on each of the columns of M by mapping the ith column of M to the ith column of P. As such, this row operation represents the linear transformation

$$\rho: \quad \mathbb{R}^n \to \mathbb{R}^n$$

$$\begin{bmatrix} x_{1k} \\ \vdots \\ x_{ik} \\ \vdots \\ x_{jk} \\ \vdots \\ x_{nk} \end{bmatrix} \to \begin{bmatrix} x_{1k} \\ \vdots \\ x_{ik} \\ \vdots \\ ax_{ik} + bx_{jk} \\ \vdots \\ x_{nk} \end{bmatrix}$$

Moreover, if $E(\rho)$ is the matrix which represents the linear transformation ρ, then (by Theorem 3.2)

$$E(\rho) = \begin{bmatrix} \rho(\mathbf{e}_1) & \rho(\mathbf{e}_2) & \cdots & \rho(\mathbf{e}_n) \end{bmatrix} = \rho\left(\begin{bmatrix} \mathbf{e}_1 & \mathbf{e}_2 & \cdots & \mathbf{e}_n \end{bmatrix}\right) = \rho(I)$$

where the $\mathbf{e}_i$'s are the vectors with a 1 in the ith position and 0's elsewhere, and I is the $n \times n$ identity matrix.

EXAMPLE 8 Let

$$M = \begin{bmatrix} 2 & -1 & 2 \\ 3 & 1 & 2 \\ 1 & 2 & -2 \end{bmatrix}$$

We will find the matrix $E(\rho)$ which corresponds to the following row operation:

$$M = \begin{bmatrix} 2 & -1 & 2 \\ 3 & 1 & 2 \\ 1 & 2 & -2 \end{bmatrix} \xrightarrow[(2)]{(-3)} \begin{bmatrix} 2 & -1 & 2 \\ 0 & 5 & -2 \\ 1 & 2 & -2 \end{bmatrix} = P$$

To find $E(\rho)$ we apply ρ to I to obtain

$$I = \begin{bmatrix} 1 & 0 & 0 \\ 0 & 1 & 0 \\ 0 & 0 & 1 \end{bmatrix} \xrightarrow[(2)]{(-3)} \begin{bmatrix} 1 & 0 & 0 \\ -3 & 2 & 0 \\ 0 & 0 & 1 \end{bmatrix} = \rho(I) = E(\rho)$$

We verify that $E(\rho)M = P$.

$$E(\rho)M = \begin{bmatrix} 1 & 0 & 0 \\ -3 & 2 & 0 \\ 0 & 0 & 1 \end{bmatrix} \begin{bmatrix} 2 & -1 & 2 \\ 3 & 1 & 2 \\ 1 & 2 & -2 \end{bmatrix} = \begin{bmatrix} 2 & -1 & 2 \\ 0 & 5 & -2 \\ 1 & 2 & -2 \end{bmatrix}$$

which is equal to P, as we wanted. ■

Therefore we have shown that if $\rho(M)$ is the matrix obtained from M by applying the row operation ρ to M, then

$$\rho(M) = E(\rho)M$$

Moreover, the result of applying a series of row operations to M can be obtained by multiplying M on the left by a series of $E(\rho)$ matrices. In Exercise 17 it is shown that each $E(\rho)$ matrix is invertible. Since the product of invertible matrices is invertible (Theorem 7.7), this yields the following theorem.

THEOREM 7.8 If P and M are equivalent matrices (that is P can be obtained from M by a series of row operations), then there is an invertible matrix E such that

$$EM = P$$

Exercise Set 7.3

1. For each of the linear transformations given below describe and sketch a graph of the inverse correspondences as in Example 1. Also indicate which of these linear transformations have inverses and what the inverse transformations are.

a. $T\colon \mathbb{R}^2 \to \mathbb{R}^2$ is counterclockwise rotation by 45°.

b. $T\colon \mathbb{R}^2 \to \mathbb{R}^2$ is projection onto the line $y = 2x$.

c. $T\colon \mathbb{R}^2 \to \mathbb{R}^2$ is reflection about the line $y = -x$.

d. $T\colon \mathbb{R}^2 \to \mathbb{R}^2$
$$\mathbf{v} \to -2\mathbf{v}$$

Find the inverse of each of the matrices in Exercises 2–7, if possible.

2. $\begin{bmatrix} 2 & -2 \\ -1 & 3 \end{bmatrix}$

3. $\begin{bmatrix} 1 & 1 & 2 \\ 2 & -3 & 3 \end{bmatrix}$

4. $\begin{bmatrix} 1 & 1 \\ 2 & 1 \\ 1 & 3 \end{bmatrix}$

5. $\begin{bmatrix} 1 & -1 & 0 \\ 2 & 1 & 3 \\ 3 & -1 & 2 \end{bmatrix}$

6. $\begin{bmatrix} 1 & -1 & 0 \\ 2 & 1 & 3 \\ 3 & -1 & -2 \end{bmatrix}$

7. $\begin{bmatrix} 1 & -1 & 2 & 1 \\ 2 & 2 & -1 & 3 \\ 3 & -2 & 2 & 1 \\ 1 & 1 & -1 & 1 \end{bmatrix}$

8. Let

$T_1\colon \mathbb{R}^2 \to \mathbb{R}^2$ be reflection about the line $y = x$.
$T_2\colon \mathbb{R}^2 \to \mathbb{R}^2$ be counterclockwise rotation of 90°.

a. Find the matrices that represent T_1^{-1}, T_2^{-1}, $(T_1T_2)^{-1}$, $T_1^{-1}T_2^{-1}$ and $T_2^{-1}T_1^{-1}$.

b. Show that $(T_1T_2)^{-1}$ is $T_2^{-1}T_1^{-1}$.

c. Show that $T_1^{-1}T_2^{-1}$ is not $(T_1T_2)^{-1}$.

9. Let

$$M_1 = \begin{bmatrix} 1 & -2 \\ 1 & -1 \end{bmatrix} \quad \text{and} \quad M_2 = \begin{bmatrix} 2 & -2 \\ -1 & 3 \end{bmatrix}$$

a. Find M_1M_2, M_1^{-1}, M_2^{-1}, $M_2^{-1}M_1^{-1}$, $M_1^{-1}M_2^{-1}$.

b. Show that

$$M_2^{-1}M_1^{-1} = (M_1M_2)^{-1} \neq M_1^{-1}M_2^{-1}$$

10. Let M_1, M_2, M_3 be invertible $n \times n$ matrices. Find $(M_1M_2M_3)^{-1}$.

★ **11.** Let $T_1\colon \mathbb{R}^n \to \mathbb{R}^n$ and $T_2\colon \mathbb{R}^n \to \mathbb{R}^n$ be linear transformations.

a. Suppose the null space of T_2 contains a nonzero vector $\mathbf{u}$. Show that $\mathbf{u}$ is in the null space of T_1T_2.

b. Show that if the nullity of T_1 is not zero, then there is a nonzero vector in the null space of T_1T_2.

c. Show that if either of the linear transformations T_1 or T_2 is not invertible, then their product T_1T_2 is not invertible.

★ **12.** Let $\mathbb{R}[x]$ be the vector space of all polynomials with real coefficients, and D the differentiation operator (linear transformation). Does D have an inverse? Explain the inverse correspondence for D.

13. Consider the matrix

$$M = \begin{bmatrix} -7 & -6 \\ 18 & 14 \end{bmatrix}$$

a. Show that M satisfies the equation

$$M^2 - 7M + 10I = 0$$ where 0 represents the zero matrix.

b. Show that

$$M(M - 7I) = -10I$$

c. Use part **b** to express M^{-1} as a sum of scalar multiples of powers of M and I.

d. Check your answer in part **c**.

★ **14.** Generalize the ideas you used to solve Exercise 13. Show that if M is a square matrix satisfying an equation of the form

$$M^n + a_{n-1}M^{n-1} + \cdots + a_1 M + a_0 I = 0$$

then if $a_0 \neq 0$, M^{-1} exists. Also find an expression for M^{-1}.

15. a. Show that a matrix with a column of zeros is not invertible.

b. Show that a matrix with a row of zeros is not invertible.

16. Suppose that M and P are equivalent matrices (that is, P can be obtained from M by using row reduction). Show that M is invertible if and only if P is invertible.

★ **17.** Consider the following three types of matrix row operations:

i. Multiplying a row of a matrix by a nonzero constant c.

ii. Interchanging two rows of a matrix.

iii. Replacing a row of a matrix by the sum of itself and another row.

These operations are called **elementary row operations**.

a. Show that each of these elementary row operations has an inverse operation and find the inverse operation for each of them.

b. Find the matrices associated with each of the elementary row operations (these matrices are called **elementary matrices**) and their inverses.

c. Show that every row operation of the type

$$\begin{bmatrix} \text{row } i \\ \text{row } j \end{bmatrix} \begin{matrix} (a) \\ (b) \end{matrix} \longrightarrow \begin{bmatrix} \text{row } i \\ a(\text{row } i) + b(\text{row } j) \end{bmatrix}$$

is a product of elementary row operations.

d. Show that every row operation is invertible.

★ **18.** Let V be a vector space and $T: V \to V$ be a linear transformation.

a. If W is a subspace of V, show that $T(W)$, (the set of all vectors of the form $T\mathbf{w}$ where $\mathbf{w}$ is an element of W) is also a subspace of V.

b. Give an example of a vector space V, an invertible linear transformation T, and a subspace W such that $T(W) \neq W$.

c. If W is a subspace of V and T is invertible, show that $\dim(W) = \dim(T(W))$.

★ **19.** Let M be an $n \times n$ invertible matrix.

a. Show that if A is an $m \times n$ matrix, then

$$\operatorname{rank}(AM) = \operatorname{rank}(A)$$

b. Show that if B is an $n \times m$ matrix, then

$$\operatorname{rank}(MB) = \operatorname{rank}(B)$$

(*Hint*: Use Exercise 18).

Applications 7 Fun and Games

Not all applications of linear algebra need be serious. Linear algebra can be used to play games, such as the computer game called Merlin® made by Parker Brothers. One of the games Merlin plays is called Magic Square. This game is played with a square array of nine lights. At any stage in the game some of the lights are on and some are off. Touching any one of the nine lights changes the pattern of lights in a predictable way.

In a typical game, for example, lights 1, 2, 3, 8, 9 may be on. If light 1 is touched, then the pattern of lights is changed to the pattern having lights 3, 4, 5, 8, 9 on, as Figure 7.17 shows.

Initial pattern

Touch light 1

New pattern

Figure 7.17

The way the game is played is this: Merlin randomly selects an initial pattern of lights. The object is to change this pattern to the pattern shown in Figure 7.18 by touching some combination of the lights.

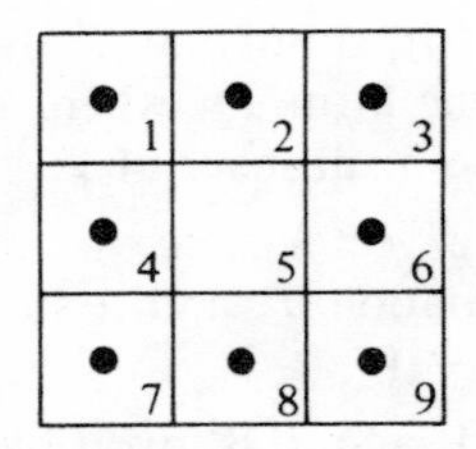

Figure 7.18

Playing this game can be amusing as well as frustrating. Sometimes it is quite easy to find the right combination of buttons, and other times it seems impossible. However, it is possible to completely analyze this game so that no matter which pattern Merlin initially gives, the final pattern can be obtained in at most nine moves. The key to discovering Merlin's secret is finding the inverse of a certain matrix.

We begin our mathematical analysis of Merlin by determining how touching each of the lights changes the pattern of lights. By a little experimenting, it can be seen (if you don't have a Merlin, you'll have to take this on faith) that no matter which pattern of lights you start with, touching a given light always affects the same subset of the nine lights. All the lights in this subset that were on, are turned off; and all those lights in the subset that were off, are turned on; all the lights not in this subset are left alone. For example, the subset of lights affected by touching light 1 is $\{1, 2, 4, 5\}$. As we saw in Figure 7.17, after touching light 1, lights 1 and 2 (which were on) were turned off, and lights 4 and 5 (which were off) were turned on; the remaining lights were unaffected. Table 7.1 lists the subsets affected by each of Merlin's nine lights.

Table 7.1

Light	Subset affected
1	$\{1, 2, 4, 5\}$
2	$\{1, 2, 3\}$
3	$\{2, 3, 5, 6\}$
4	$\{1, 4, 7\}$
5	$\{2, 4, 5, 6, 8\}$
6	$\{3, 6, 9\}$
7	$\{4, 5, 7, 8\}$
8	$\{7, 8, 9\}$
9	$\{5, 6, 8, 9\}$

To mathematically analyze Merlin we convert the ideas about patterns of lights and changes in those patterns into mathematical statements.

To begin, let us use a vector **p** to represent a pattern of lights. If a light in the pattern is on, we will put a 1 in the corresponding position in the vector; and if a light is off, we will put a 0 in the corresponding position in the vector. An example of this is shown in Figure 7.19.

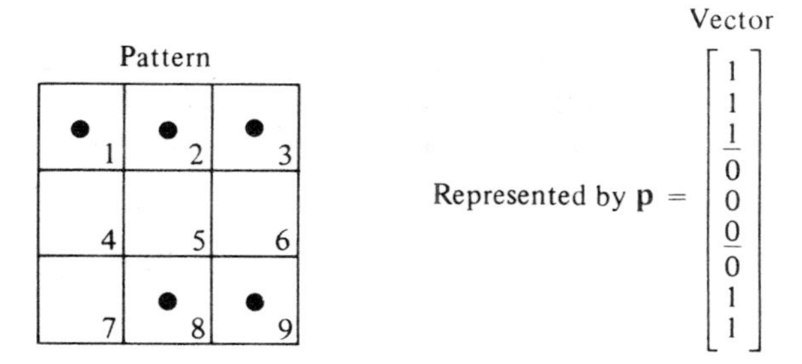

Figure 7.19

We also want to represent the way Merlin operates to change from one pattern to another in terms of 0's and 1's. As Table 7.1 shows, each light affects a certain subset of the nine lights when it is touched. To each light we will associate the vector having 1's in positions corresponding to the elements in the subset affected, and 0's in the positions corresponding to the elements not in this subset. For example, since touching light 1 affects lights 1, 2, 4, 5, we have

$$\begin{bmatrix} 1 \\ 1 \\ 0 \\ 1 \\ 1 \\ 0 \\ 0 \\ 0 \\ 0 \end{bmatrix} \quad \text{is associated with light 1}$$

Suppose we concentrate on a given light L. We investigate the effect on this light of touching some other light. Let S denote the subset of lights affected by touching the other light. There are four possibilities:

Initial State of L	Is L In S?/Not in S?	Final State of L
off	not in S	off
off	in S	on
on	not in S	on
on	in S	off

These can be expressed more succinctly as

	Not in S	In S
off	off	on
on	on	off

where the first column represents the initial state of L and the second two columns represent the final states of L (depending on whether L lies in S or not). Rewriting this in terms of 0's and 1's we have

$\oplus$	0	1
0	0	1
1	1	0

This table represents a special kind of addition called **addition modulo 2** (this is the same addition that was discussed in Applications 5). We will denote it by $\oplus$ to distinguish it from ordinary addition. This is the addition we shall use throughout this section. You should note that computations are extremely simple with this type of arithmetic since the only numbers are 0 and 1. (Since a given light can only be off or on, these are all the numbers we will need.) We are now ready to use addition modulo 2 to describe how Merlin works.

If we start with the pattern that has lights 1, 2, 3, 8, 9 on and touch light 1, this changes lights 1, 2, 4, 5. In terms of vectors of 0's and 1's and addition modulo 2 we can write this as

$$\begin{array}{ccccc} \text{Initial pattern} & & \text{Touch light 1} & & \text{New pattern} \\ \begin{bmatrix} 1 \\ 1 \\ 1 \\ \hline 0 \\ 0 \\ 0 \\ \hline 0 \\ 1 \\ 1 \end{bmatrix} & \oplus & \begin{bmatrix} 1 \\ 1 \\ 0 \\ \hline 1 \\ 1 \\ 0 \\ \hline 0 \\ 0 \\ 0 \end{bmatrix} & = & \begin{bmatrix} 0 \\ 0 \\ 1 \\ \hline 1 \\ 1 \\ 0 \\ \hline 0 \\ 1 \\ 1 \end{bmatrix} \end{array}$$

We now must translate the information in Table 7.1 into vector notation. We will let the vector $\mathbf{c}_i$ be the vector corresponding to the changes produced by touching light i. We list the vectors $\mathbf{c}_i$ below.

$$\mathbf{c}_1 = \begin{bmatrix} 1 \\ 1 \\ 0 \\ \overline{1} \\ 1 \\ 0 \\ \overline{0} \\ 0 \\ 0 \end{bmatrix} \quad \mathbf{c}_2 = \begin{bmatrix} 1 \\ 1 \\ 1 \\ \overline{0} \\ 0 \\ 0 \\ \overline{0} \\ 0 \\ 0 \end{bmatrix} \quad \mathbf{c}_3 = \begin{bmatrix} 0 \\ 1 \\ 1 \\ \overline{0} \\ 1 \\ 1 \\ \overline{0} \\ 0 \\ 0 \end{bmatrix} \quad \mathbf{c}_4 = \begin{bmatrix} 1 \\ 0 \\ 0 \\ \overline{1} \\ 0 \\ 0 \\ \overline{1} \\ 0 \\ 0 \end{bmatrix} \quad \mathbf{c}_5 = \begin{bmatrix} 0 \\ 1 \\ 0 \\ \overline{1} \\ 1 \\ 1 \\ \overline{0} \\ 1 \\ 0 \end{bmatrix} \quad \mathbf{c}_6 = \begin{bmatrix} 0 \\ 0 \\ 1 \\ \overline{0} \\ 0 \\ 1 \\ \overline{0} \\ 0 \\ 1 \end{bmatrix} \quad \mathbf{c}_7 = \begin{bmatrix} 0 \\ 0 \\ 0 \\ \overline{1} \\ 1 \\ 0 \\ \overline{1} \\ 1 \\ 0 \end{bmatrix} \quad \mathbf{c}_8 = \begin{bmatrix} 0 \\ 0 \\ 0 \\ \overline{0} \\ 0 \\ 0 \\ \overline{1} \\ 1 \\ 1 \end{bmatrix} \quad \mathbf{c}_9 = \begin{bmatrix} 0 \\ 0 \\ 0 \\ \overline{0} \\ 1 \\ 1 \\ \overline{0} \\ 1 \\ 1 \end{bmatrix}$$

Using this notation, we can describe what happens to a pattern represented by the vector $\mathbf{p}$ if light i is touched

$$\begin{array}{ccc} \text{Initial pattern} & \xrightarrow{\text{becomes}} & \text{New pattern} \\ \mathbf{p} & & \mathbf{p} \oplus \mathbf{c}_i \end{array}$$

The effect of touching several lights is just to add the vectors corresponding to these lights to the initial pattern. For example, touching lights 2, 4, and 7 would change the pattern $\mathbf{p}$ as follows:

$$\begin{array}{ccc} \text{Initial pattern} & \xrightarrow{\text{becomes}} & \text{New pattern} \\ \mathbf{p} & & \mathbf{p} \oplus \mathbf{c}_2 \oplus \mathbf{c}_4 \oplus \mathbf{c}_7 \end{array}$$

We are now able to predict the effect of touching any series of lights on an arbitrary pattern. This is not the best description, since we must have the values of the vectors $\mathbf{c}_1, \mathbf{c}_2, \ldots, \mathbf{c}_9$ given to do any calculation. It would be much better if this information (the values of the vectors $\mathbf{c}_i$) could be contained in the equation itself. To obtain such an equation, we use a technique we have used many times before. We define the matrix C:

$$C = \begin{bmatrix} \mathbf{c}_1 & \mathbf{c}_2 & \cdots & \mathbf{c}_9 \end{bmatrix}$$

To express the sum of a certain set of the $\mathbf{c}_i$'s, we just apply the matrix C to the vector having 1's in the positions corresponding to the vectors in this set and 0's in the remaining positions. For example

$$\mathbf{c}_2 \oplus \mathbf{c}_4 \oplus \mathbf{c}_7 = C \begin{bmatrix} 0 \\ 1 \\ 0 \\ \overline{1} \\ 0 \\ 0 \\ \overline{1} \\ 0 \\ 0 \end{bmatrix}$$

Now we are finally in a position to describe the game of Magic Square. Given an initial pattern $\mathbf{p}_0$, the object of the game is to find a vector $\mathbf{v}$ so that

$$\text{(1)} \qquad \mathbf{p}_0 \oplus C\mathbf{v} = \mathbf{p}_f$$

where $\mathbf{p}_f$ is the vector having 1's in all but the fifth position (it represents the final pattern).

To solve equation (1) for $\mathbf{v}$, we begin by adding $\mathbf{p}_0$ to both sides

$$\text{(2)} \qquad \mathbf{p}_0 \oplus (\mathbf{p}_0 \oplus C\mathbf{v})$$

$$= \mathbf{p}_0 \oplus \mathbf{p}_f = (\mathbf{p}_0 \oplus \mathbf{p}_0) + C\mathbf{v} = C\mathbf{v} \qquad \text{since } \mathbf{p}_0 \oplus \mathbf{p}_0 = \mathbf{0}.$$

If C is invertible (modulo 2), then we can solve for $\mathbf{v}$ to obtain

$$\text{(3)} \qquad \mathbf{v} = C^{-1}(\mathbf{p}_0 \oplus \mathbf{p}_f)$$

If C is not invertible (modulo 2), there will be some patterns $\mathbf{p}_0$ which cannot be changed to $\mathbf{p}_f$ (see Exercise 3).

The way to invert the matrix C, as well as determine if it has an inverse, is to use the row reduction method described in Section 7.3; that is

$$\left[C \,\middle|\, I\right] \xrightarrow[\text{reduce}]{\text{row}} \left[I \,\middle|\, C^{-1}\right]$$

As you have probably noticed, this requires the reduction of a 9×18 matrix. This is the bad news. The good news is that we are working in arithmetic modulo 2 and only have two numbers to deal with.

The main chore in such a large reduction is not the row operations themselves, but the rewriting of the matrix after each operation. Since all our matrices can have only two numbers as coefficients, we can avoid this problem. To do this draw a 9×18 matrix of square boxes on a large sheet of paper as shown in Figure 7.20. Place a marker (a poker chip, a penny, a button, etc.) on each square that corresponds to a position in the matrix

$$\left[C \,\middle|\, I\right]$$

that contains a 1 and leave the squares corresponding to 0's empty. Now carry out the row operations by adding or removing markers as necessary (you will need about 70 markers for this calculation). Figure 7.21 shows how to use the first row as a pivot row to eliminate the first entry in the second row. The simple rule to follow in replacing a row by the sum of itself and another (pivot) row is that the new row is obtained by changing the entry in each box below (or above) each marker in the pivot row; the other entries are left fixed. This procedure should take less than 30 minutes.

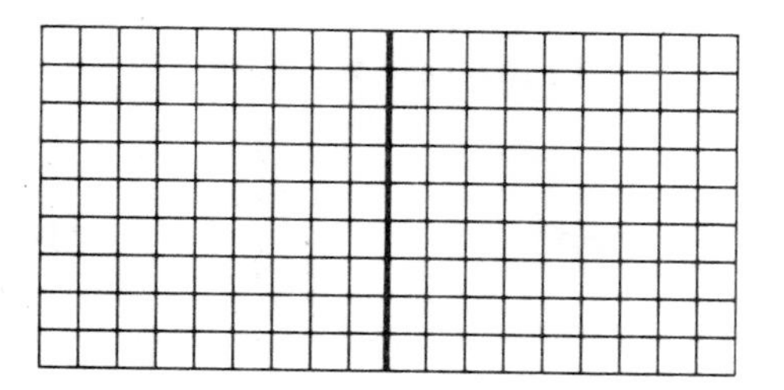

Figure 7.20

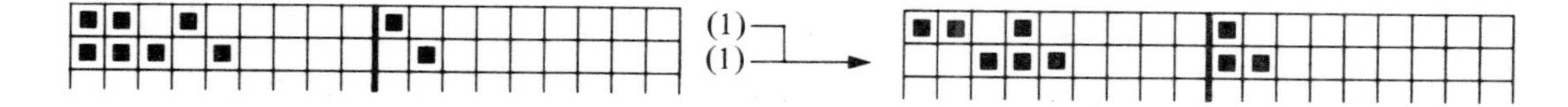

Figure 7.21

You should find that C^{-1} exists and is

$$C^{-1} = \begin{bmatrix} 1 & 0 & 1 & 0 & 0 & 1 & 1 & 1 & 0 \\ 1 & 1 & 1 & 1 & 1 & 1 & 0 & 0 & 0 \\ 1 & 0 & 1 & 1 & 0 & 0 & 0 & 1 & 1 \\ 1 & 1 & 0 & 1 & 1 & 0 & 1 & 1 & 0 \\ 1 & 0 & 1 & 0 & 1 & 0 & 1 & 0 & 1 \\ 0 & 1 & 1 & 0 & 1 & 1 & 0 & 1 & 1 \\ 1 & 1 & 0 & 0 & 0 & 1 & 1 & 0 & 1 \\ 0 & 0 & 0 & 1 & 1 & 1 & 1 & 1 & 1 \\ 0 & 1 & 1 & 1 & 0 & 0 & 1 & 0 & 1 \end{bmatrix}.$$

The following example summarizes the results of this section by showing how to play Magic Square efficiently.

EXAMPLE 1 Suppose the initial pattern in a game of Magic Square is as shown in Figure 7.22. Determine which lights to press to obtain the final pattern shown in Figure 7.23.

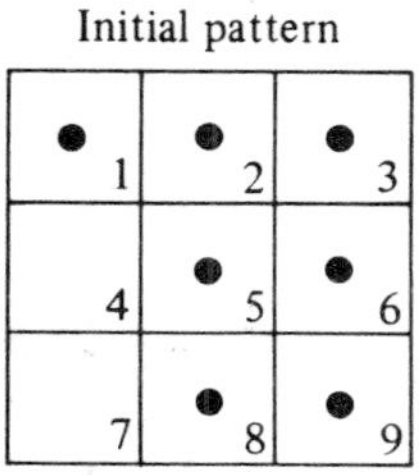

Initial pattern

● 1	● 2	● 3
4	● 5	● 6
7	● 8	● 9

Figure 7.22

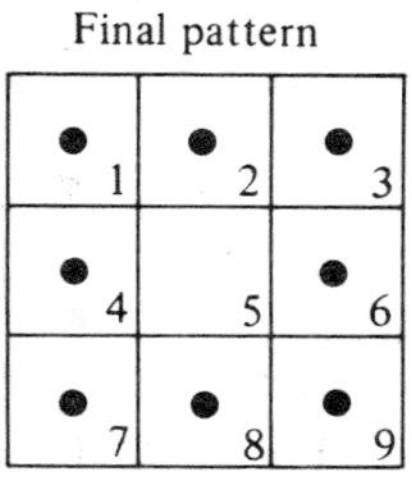

Final pattern

● 1	● 2	● 3
● 4	5	● 6
● 7	● 8	● 9

Figure 7.23

In this case we have

$$\mathbf{p}_0 = \begin{bmatrix} 1 \\ 1 \\ 1 \\ 0 \\ 1 \\ 1 \\ 0 \\ 1 \\ 1 \end{bmatrix}, \quad \mathbf{p}_f = \begin{bmatrix} 1 \\ 1 \\ 1 \\ 1 \\ 0 \\ 1 \\ 1 \\ 1 \\ 1 \end{bmatrix}, \quad \text{and} \quad \mathbf{p}_0 \oplus \mathbf{p}_f = \begin{bmatrix} 0 \\ 0 \\ 0 \\ 1 \\ 1 \\ 0 \\ 1 \\ 0 \\ 0 \end{bmatrix}$$

The vector that represents the solution of this problem is

$$\mathbf{v} = C^{-1}(\mathbf{p}_0 \oplus \mathbf{p}_f) = \begin{bmatrix} 1 \\ 0 \\ 1 \\ 1 \\ 0 \\ 1 \\ 1 \\ 1 \\ 0 \end{bmatrix}$$

Hence, touching lights 1, 3, 4, 6, 7, and 8 will change $\mathbf{p}_0$ to $\mathbf{p}_f$ and win the game. ■

Exercises for Applications 7

1. Suppose the initial pattern in a game of Magic Square is as shown in Figure 7.24. What lights should you touch in order to obtain the final pattern?

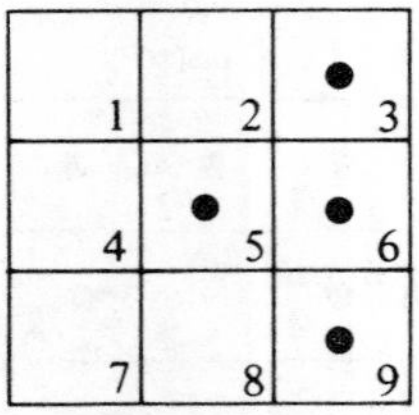

1	2	● 3
4	● 5	● 6
7	8	● 9

Figure 7.24

2. Suppose there is a small version of Magic Square with only four squares instead of nine, and that the effect of touching each of the lights is

Light	Subset Affected
1	{1, 3}
2	{1, 2}
3	{2, 4}
4	{2, 3, 4}

a. If the initial pattern is as shown in Figure 7.25, what is the effect of touching light 4?

b. Write the vectors $\mathbf{c}_1, \mathbf{c}_2, \mathbf{c}_3, \mathbf{c}_4$ that represent the change in pattern produced by touching each of the lights. Also write the matrix C.

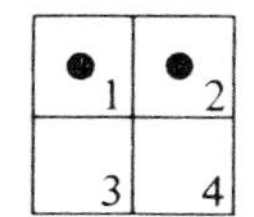

Figure 7.25

c. Find C^{-1}

d. Using C^{-1} determine which lights to touch to go from pattern $\mathbf{p}_0$ to pattern $\mathbf{p}_f$ in Figure 7.26.

$\mathbf{p}_0$

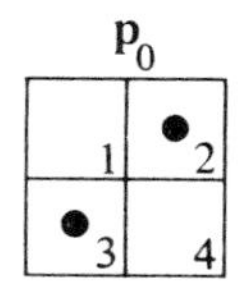

$\mathbf{p}_f$

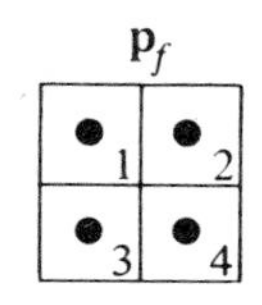

Figure 7.26

3. Suppose that in a four-square version of Magic Square the effect of touching each of the lights is

Light	Subset affected
1	{1, 3}
2	{1, 2}
3	{2, 4}
4	{3, 4}

a. Find the transition matrix C.

b. Show that C^{-1} does not exist.

c. Show that it is impossible to go from pattern $\mathbf{p}_0$ in Figure 7.27 to pattern $\mathbf{p}_f$.

d. Show that there are two different ways of going from pattern $\mathbf{p}_0$ to pattern $\mathbf{p}_f$ in Figure 7.28 in such a way that no light is touched more than once.

e. Find the null space of C and the image space of C. Explain the relationship between these spaces and parts **b**, **c**, and **d** of this exercise.

$\mathbf{p}_0$

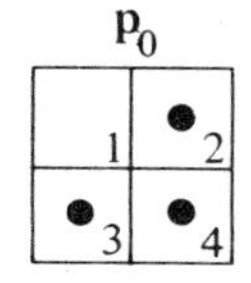

$\mathbf{p}_f$

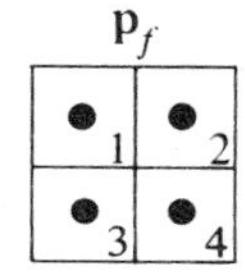

Figure 7.27

$\mathbf{p}_0$

$\mathbf{p}_f$

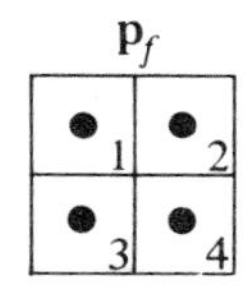

Figure 7.28

Review Exercises

1. In each of the following, either give an example satisfying the condition listed, or explain why no such example can be given.

a. M and N are matrices. The product MN is defined, but the product NM is not defined.

b. M is a 3×3 matrix of rank 3 and M is not invertible.

c. M is a 3×3 matrix of nullity 3 and M is not invertible.

2. Find the following matrix products if possible:

a. $\begin{bmatrix} 1 & 0 \\ 2 & 3 \end{bmatrix} \begin{bmatrix} 2 & -2 & 1 & 0 \\ 0 & 2 & 3 & 0 \end{bmatrix}$

b. $\begin{bmatrix} 3 & 1 & -4 \\ 1 & 1 & 0 \\ -1 & 1 & 0 \end{bmatrix} \begin{bmatrix} 1 & -2 & -1 \\ 6 & -1 & 0 \\ 1 & 0 & 0 \end{bmatrix}$

c. $\begin{bmatrix} 1 & -3 & 4 \\ 5 & 0 & 9 \\ -2 & -1 & 6 \end{bmatrix} \begin{bmatrix} 2 & 0 & 7 \\ 1 & 1 & -4 \end{bmatrix}$

d. $\begin{bmatrix} 1 & 0 & 2 & 7 \\ -2 & 3 & 6 & 0 \end{bmatrix} \begin{bmatrix} 3 & -5 \\ 1 & 1 \\ -2 & 3 \\ 7 & 0 \end{bmatrix}$

3. Find the inverses of the following matrices:

a. $\begin{bmatrix} 0 & 0 & 1 \\ 1 & 1 & -2 \\ 1 & 0 & -2 \end{bmatrix}$

b. $\begin{bmatrix} 1 & 2 & 1 \\ -1 & 0 & 3 \\ 1 & 3 & 0 \end{bmatrix}$

c. $\begin{bmatrix} 3 & 1 & 1 \\ 1 & -1 & 1 \\ 0 & 2 & 1 \end{bmatrix}$

4. Find B^2 where $B = \begin{bmatrix} 1 & 2 & 3 \\ -2 & 4 & -2 \\ -3 & 2 & 0 \end{bmatrix}$

5. Consider the following linear transformations:

$T_1: \mathbb{R}^2 \to \mathbb{R}^2$ is reflection about the line $y = -x$
$T_2: \mathbb{R}^2 \to \mathbb{R}^2$ is projection onto the line $y = x$

Find the matrices that represent the following linear transformations:

a. $T_1 T_2$ b. $T_2 T_1$
c. T_1^2 d. T_2^2
e. T_1^{-1} if it exists. f. T_2^{-1} if it exists.

6. Determine which of the following statements are true and which are false. If the statement is true explain why it is true, and if it is false, give a counterexample.

a. $(M_1 M_2)^{-1} = M_1^{-1} M_2^{-1}$

b. If $T: \mathbb{R}^5 \to \mathbb{R}^4$, then T is not invertible.

c. Suppose M is a square matrix and there are two distinct vectors $\mathbf{v}_1$ and $\mathbf{v}_2$ for which $M\mathbf{v}_1 = M\mathbf{v}_2$, then M is not invertible.

d. If M is an $n \times n$ matrix and $\text{rank}(M) = n$, then M is invertible.

e. If M is invertible and $M\mathbf{v} = \mathbf{w}$, then $\mathbf{v} = M^{-1}\mathbf{w}$.

f. $(M_1 M_2)^2 = M_1^2 M_2^2$ for every pair of $n \times n$ matrices M_1 and M_2.

g. $P^{-1}MP = M$ for every pair of $n \times n$ matrices M and P where P is invertible.

h. If M is a square matrix such that the rank of M is equal to the nullity of M, then M is invertible.

7. Let M_1 and M_2 be matrices such that the product $M_1 M_2$ is defined. Show that

a. $\text{rank}(M_1 M_2) \leq \text{rank}(M_1)$

b. $\text{rank}(M_1 M_2) \leq \text{rank}(M_2)$

8 Measurement and Linear Transformations

This chapter is concerned with two major topics. The first is to determine which linear transformations from $\mathbb{R}^n$ to $\mathbb{R}^n$ preserve shape; the second is to find the amount by which the size (area or volume) of an object is magnified (or shrunk) under the action of a linear transformation. The discussion of the second topic leads us to the definition of the determinant.

8.1 Orthogonal Linear Transformations and Transposes

In this section and the section that follows we will look at how geometric objects are changed under the action of a linear transformation (or matrix). We begin with two examples.

EXAMPLE 1 Let T be the linear transformation represented by the matrix $M = \begin{bmatrix} 1 & -1 \\ 1 & 2 \end{bmatrix}$. We see what T does to the unit square—the square with sides $\begin{bmatrix} 1 \\ 0 \end{bmatrix}$ and $\begin{bmatrix} 0 \\ 1 \end{bmatrix}$. We see in Figure 8.1 that the sides

$$\begin{bmatrix} 1 \\ 0 \end{bmatrix} \quad \text{and} \quad \begin{bmatrix} 0 \\ 1 \end{bmatrix}$$

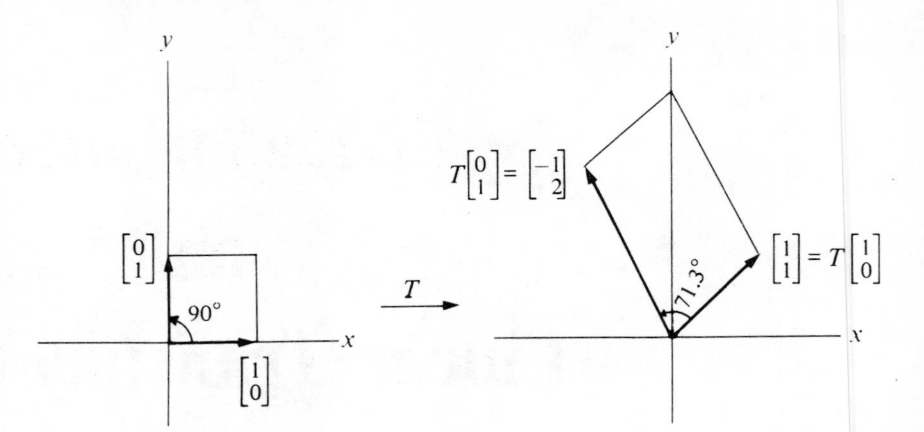

Figure 8.1

of the unit square are mapped onto the sides

$$\begin{bmatrix}1\\1\end{bmatrix} \quad \text{and} \quad \begin{bmatrix}-1\\2\end{bmatrix}$$

of a parallelogram. In fact, each point (vector) in this square is mapped onto a point of the parallelogram (see Exercise 5). We compare some measurements of the square to those of the corresponding parts of the parallelogram:

$$\text{Length of } \begin{bmatrix}1\\0\end{bmatrix} = 1 \qquad \text{Length of } T\begin{bmatrix}1\\0\end{bmatrix} = \sqrt{2} \doteq 1.41$$

$$\text{Length of } \begin{bmatrix}0\\1\end{bmatrix} = 1 \qquad \text{Length of } T\begin{bmatrix}0\\1\end{bmatrix} = \sqrt{5} \doteq 2.24$$

$$\text{Angle between } \begin{bmatrix}1\\0\end{bmatrix} \text{ and } \begin{bmatrix}0\\1\end{bmatrix} = 90° \quad \text{Angle between } T\begin{bmatrix}1\\0\end{bmatrix} \text{ and } T\begin{bmatrix}0\\1\end{bmatrix} \doteq 71.3°$$

(We found this by using the formula in Section 6.1.)

So we see that the unit square is both twisted and stretched by the action of this linear transformation. ■

Now we look at our second example.

EXAMPLE 2 Let T be the linear transformation represented by the matrix

$$M = \begin{bmatrix}3/5 & -4/5\\4/5 & 3/5\end{bmatrix}$$

Figure 8.2 shows what this transformation does to the unit square. Comparing lengths and angles, we have:

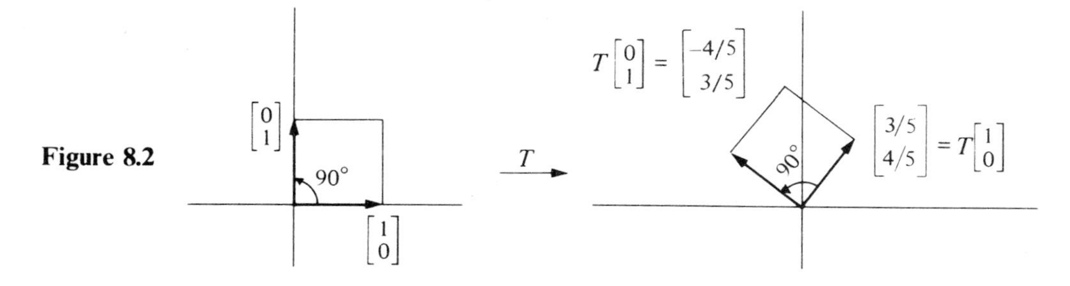

Figure 8.2

$$\text{Length of } \begin{bmatrix}1\\0\end{bmatrix} = 1 \qquad \text{Length of } T\begin{bmatrix}1\\0\end{bmatrix} = \sqrt{(\tfrac{3}{5})^2 + (\tfrac{4}{5})^2} = 1$$

$$\text{Length of } \begin{bmatrix}0\\1\end{bmatrix} = 1 \qquad \text{Length of } T\begin{bmatrix}0\\1\end{bmatrix} = \sqrt{(-\tfrac{4}{5})^2 + (\tfrac{3}{5})^2} = 1$$

$$\text{Angle between } \begin{bmatrix}1\\0\end{bmatrix} \text{ and } \begin{bmatrix}0\\1\end{bmatrix} = 90^\circ \quad \text{Angle between } T\begin{bmatrix}1\\0\end{bmatrix} \text{ and } T\begin{bmatrix}0\\1\end{bmatrix} = 90^\circ$$

(Their dot product is 0.)

Thus, although the action of T changes the position of the unit square, it does not change its shape. ■

Our goal in this section will be to give a criterion for determining when a linear transformation is shape preserving like the one in Example 2.

First we must define what we mean by shape preserving.

DEFINITION 8.1

If $T: \mathbb{R}^n \to \mathbb{R}^n$ is a linear transformation, then T is **shape preserving** if it has these two properties:

1. $|\mathbf{v}| = |T\mathbf{v}|$ for every vector $\mathbf{v}$ of $\mathbb{R}^n$.
2. The angle between $\mathbf{v}_1$ and $\mathbf{v}_2$ is the same as the angle between $T\mathbf{v}_1$ and $T\mathbf{v}_2$ for every pair of vectors $\mathbf{v}_1$ and $\mathbf{v}_2$ of $\mathbb{R}^n$.

Each of these two properties involves quantities that are computed by using the dot product. The length of a vector $\mathbf{v}$ is

$$|\mathbf{v}| = \sqrt{\mathbf{v} \cdot \mathbf{v}}$$

and the angle θ between two vectors is found by using the formula

$$\cos\theta = \frac{\mathbf{v}_1 \cdot \mathbf{v}_2}{|\mathbf{v}_1||\mathbf{v}_2|}$$

Thus a linear transformation T is shape preserving if

$$\mathbf{v}_1 \cdot \mathbf{v}_2 = T\mathbf{v}_1 \cdot T\mathbf{v}_2 \qquad \text{for every pair of vectors } \mathbf{v}_1, \mathbf{v}_2 \text{ of } \mathbb{R}^n.$$

The converse is also true.

The shape-preserving linear transformations are given a special name—orthogonal.

DEFINITION 8.2

> A linear transformation $T\colon \mathbb{R}^n \to \mathbb{R}^n$ is said to be an **orthogonal linear transformation** if for every pair of vectors $\mathbf{v}_1, \mathbf{v}_2$ in $\mathbb{R}^n$
>
> $$\mathbf{v}_1 \cdot \mathbf{v}_2 = T\mathbf{v}_1 \cdot T\mathbf{v}_2$$
>
> A matrix is said to be an **orthogonal** matrix if it represents an orthogonal linear transformation.

We will now see how to recognize orthogonal matrices.

If M is an $n \times n$ orthogonal matrix, then for each vector $\mathbf{e}_i$ (where $\mathbf{e}_i$ is the vector whose ith entry is 1 and all other entries are 0), we have

$$1 = \mathbf{e}_i \cdot \mathbf{e}_i = M\mathbf{e}_i \cdot M\mathbf{e}_i = \mathbf{c}_i \cdot \mathbf{c}_i \qquad \text{where } \mathbf{c}_i \text{ is the } i\text{th column of } M$$

So each column of M is a unit vector. Moreover, if $i \neq j$, then

$$0 = \mathbf{e}_i \cdot \mathbf{e}_j = M\mathbf{e}_i \cdot M\mathbf{e}_j = \mathbf{c}_i \cdot \mathbf{c}_j$$

So the columns of M are mutually orthogonal. Therefore, we have shown that if M is orthogonal, then the columns of M are an orthonormal set of vectors.

We are not done yet because we have not shown the converse—namely, that if the columns of an $n \times n$ matrix M are an orthonormal set, then M is orthogonal.

Before we can attack this part of the problem, we must take a detailed look at the relationship between matrices and the dot product. For simplicity, we consider the special case of 2×2 matrices. However, you will be able to see that the same ideas work for any square matrix. Let

$$M = \begin{bmatrix} m_{11} & m_{12} \\ m_{21} & m_{22} \end{bmatrix} \qquad \mathbf{v}_1 = \begin{bmatrix} x_1 \\ y_1 \end{bmatrix} \qquad \mathbf{v}_2 = \begin{bmatrix} x_2 \\ y_2 \end{bmatrix}$$

then

$$\begin{aligned} M\mathbf{v}_1 \cdot \mathbf{v}_2 &= \begin{bmatrix} m_{11} & m_{12} \\ m_{21} & m_{22} \end{bmatrix} \begin{bmatrix} x_1 \\ y_1 \end{bmatrix} \cdot \begin{bmatrix} x_2 \\ y_2 \end{bmatrix} \\ &= \begin{bmatrix} m_{11}x_1 + m_{12}y_1 \\ m_{21}x_1 + m_{22}y_1 \end{bmatrix} \cdot \begin{bmatrix} x_2 \\ y_2 \end{bmatrix} \\ &= (m_{11}x_1 + m_{12}y_1)x_2 + (m_{21}x_1 + m_{22}y_1)y_2 \end{aligned}$$

We rearrange the last expression by factoring out x_1 and y_1 to get

$$M\mathbf{v}_1 \cdot \mathbf{v}_2 = x_1(m_{11}x_2 + m_{21}y_2) + y_1(m_{12}x_2 + m_{22}y_2)$$

which can be written as

$$M\mathbf{v}_1 \cdot \mathbf{v}_2 = \begin{bmatrix} x_1 \\ y_1 \end{bmatrix} \cdot \begin{bmatrix} m_{11}x_2 + m_{21}y_2 \\ m_{12}x_2 + m_{22}y_2 \end{bmatrix} = \begin{bmatrix} x_1 \\ y_1 \end{bmatrix} \cdot \begin{bmatrix} m_{11} & m_{21} \\ m_{12} & m_{22} \end{bmatrix} \begin{bmatrix} x_2 \\ y_2 \end{bmatrix}$$

Be sure to notice that the matrix in the last equation is not M. It is the matrix whose first *row* has the same entries as the first *column* of M, and whose second *row* has the same entries as the second *column* of M. Such matrices have a special name.

DEFINITION 8.3

Let M be a matrix. The **transpose** M^t of M is the matrix obtained from M by interchanging rows and columns.

$$M = \begin{bmatrix} \mathbf{c}_1 & \mathbf{c}_2 & \cdots & \mathbf{c}_n \end{bmatrix} \qquad M^t = \begin{bmatrix} \mathbf{c}_1^t \\ \mathbf{c}_2^t \\ \vdots \\ \mathbf{c}_n^t \end{bmatrix}$$

where $\mathbf{c}_i^t$ is the row whose entries are the same as the column $\mathbf{c}_i$.

We can use our new notation to summarize our results. If M is a 2×2 matrix, then $M\mathbf{v}_1 \cdot \mathbf{v}_2 = \mathbf{v}_1 \cdot M^t\mathbf{v}_2$. This is a special case of the following theorem.

THEOREM 8.1

Let M be an $n \times n$ matrix and $\mathbf{v}_1$, $\mathbf{v}_2$ be vectors in $\mathbb{R}^n$. Then

$$M\mathbf{v}_1 \cdot \mathbf{v}_2 = \mathbf{v}_1 \cdot M^t\mathbf{v}_2$$

EXAMPLE 3 We illustrate Theorem 8.1 by showing that $M\mathbf{v}_1 \cdot \mathbf{v}_2 = \mathbf{v}_1 \cdot M^t\mathbf{v}_2$ for

$$M = \begin{bmatrix} 2 & -3 & 1 \\ 3 & 1 & -5 \\ -1 & -1 & 3 \end{bmatrix} \qquad \mathbf{v}_1 = \begin{bmatrix} 1 \\ 0 \\ 0 \end{bmatrix} \qquad \mathbf{v}_2 = \begin{bmatrix} 0 \\ 1 \\ 0 \end{bmatrix}$$

We compute $M\mathbf{v}_1 \cdot \mathbf{v}_2$:

$$\begin{bmatrix} 2 & -3 & 1 \\ 3 & 1 & -5 \\ -1 & -1 & 3 \end{bmatrix} \begin{bmatrix} 1 \\ 0 \\ 0 \end{bmatrix} \cdot \begin{bmatrix} 0 \\ 1 \\ 0 \end{bmatrix} = \begin{bmatrix} 2 \\ 3 \\ -1 \end{bmatrix} \cdot \begin{bmatrix} 0 \\ 1 \\ 0 \end{bmatrix} = 3$$

We compute $\mathbf{v}_1 \cdot M^t\mathbf{v}_2$:

$$\begin{bmatrix}1\\0\\0\end{bmatrix} \cdot \begin{bmatrix}2 & 3 & -1\\-3 & 1 & -1\\1 & -5 & 3\end{bmatrix}\begin{bmatrix}0\\1\\0\end{bmatrix} = \begin{bmatrix}1\\0\\0\end{bmatrix} \cdot \begin{bmatrix}3\\1\\-5\end{bmatrix} = 3$$

Hence, in this case, $M\mathbf{v}_1 \cdot \mathbf{v}_2 = \mathbf{v}_1 \cdot M^t\mathbf{v}_2$. ■

Now we return to our original problem. We want to show that if M is an $n \times n$ matrix and the columns of M form an orthonormal set of vectors, then M is orthogonal—that is, we must show that $M\mathbf{v}_1 \cdot M\mathbf{v}_2 = \mathbf{v}_1 \cdot \mathbf{v}_2$ for every pair of vectors $\mathbf{v}_1$, $\mathbf{v}_2$ of $\mathbb{R}^n$.

Now
$$M\mathbf{v}_1 \cdot M\mathbf{v}_2 = M\mathbf{v}_1 \cdot (M\mathbf{v}_2)$$

So using Theorem 8.1, we get

$$M\mathbf{v}_1 \cdot M\mathbf{v}_2 = \mathbf{v}_1 \cdot M^t(M\mathbf{v}_2)$$

and using the definition of matrix product, this becomes

$$M\mathbf{v}_1 \cdot M\mathbf{v}_2 = \mathbf{v}_1 \cdot (M^tM)\mathbf{v}_2$$

Now let us look at M^tM:

$$M^tM = \begin{bmatrix}\mathbf{c}_1^t\\\mathbf{c}_2^t\\\vdots\\\mathbf{c}_n^t\end{bmatrix}\begin{bmatrix}\mathbf{c}_1 & \mathbf{c}_2 & \cdots & \mathbf{c}_n\end{bmatrix} = \begin{bmatrix}\mathbf{c}_1\cdot\mathbf{c}_1 & \mathbf{c}_1\cdot\mathbf{c}_2 & \cdots & \mathbf{c}_1\cdot\mathbf{c}_n\\\mathbf{c}_2\cdot\mathbf{c}_1 & \mathbf{c}_2\cdot\mathbf{c}_2 & \cdots & \mathbf{c}_2\cdot\mathbf{c}_n\\\vdots & \vdots & & \vdots\\\mathbf{c}_n\cdot\mathbf{c}_1 & \mathbf{c}_n\cdot\mathbf{c}_2 & \cdots & \mathbf{c}_n\cdot\mathbf{c}_n\end{bmatrix}$$

Since the vectors $\mathbf{c}_1, \mathbf{c}_2, \ldots, \mathbf{c}_n$ are orthonormal we see that

$$M^tM = \begin{bmatrix}1 & 0 & \cdots & 0\\0 & 1 & \cdots & \vdots\\0 & 0 & \cdots & 0\\\vdots & \vdots & \cdots & 0\\0 & 0 & \cdots & 1\end{bmatrix} = I$$

Therefore,
$$M\mathbf{v}_1 \cdot M\mathbf{v}_2 = \mathbf{v}_1 \cdot I\mathbf{v}_2 = \mathbf{v}_1 \cdot \mathbf{v}_2$$

Summarizing these results gives

THEOREM 8.2

An $n \times n$ matrix M is orthogonal if and only if any one of the following equivalent conditions holds:

1. For every pair of vectors $\mathbf{v}_1$, $\mathbf{v}_2$ of $\mathbb{R}^n$, $\mathbf{v}_1 \cdot \mathbf{v}_2 = M\mathbf{v}_1 \cdot M\mathbf{v}_2$ (This is just the definition.)

2. The columns of M form an orthonormal set of vectors.

3. $M^tM = I$ (This is what we proved in the course of proving Condition 2.)

4. $M^t = M^{-1}$ (This is what Condition 3 means.)

EXAMPLE 4 Use Condition 3 of Theorem 8.2 to show that the following matrix is orthogonal.

$$M = \begin{bmatrix} 1/\sqrt{5} & -2/\sqrt{5} & 0 \\ 2/\sqrt{6} & 1/\sqrt{6} & 1/\sqrt{6} \\ -2/\sqrt{30} & -1/\sqrt{30} & 5/\sqrt{30} \end{bmatrix}$$

$$M^tM = \begin{bmatrix} 1/\sqrt{5} & 2/\sqrt{6} & -2/\sqrt{30} \\ -2/\sqrt{5} & 1/\sqrt{6} & -1/\sqrt{30} \\ 0 & 1/\sqrt{6} & 5/\sqrt{30} \end{bmatrix} \begin{bmatrix} 1/\sqrt{5} & -2/\sqrt{5} & 0 \\ 2/\sqrt{6} & 1/\sqrt{6} & 1/\sqrt{6} \\ -2/\sqrt{30} & -1/\sqrt{30} & 5/\sqrt{30} \end{bmatrix} = \begin{bmatrix} 1 & 0 & 0 \\ 0 & 1 & 0 \\ 0 & 0 & 1 \end{bmatrix}. \blacksquare$$

It should be noted that although the main thrust of this section was the proof of Theorem 8.2, the result of Theorem 8.1 was our main tool. The relationship between the transpose of a matrix and the dot product is a fundamental relationship of linear algebra.

We conclude this section with a theorem about the transpose of a product of matrices. You should note that the transpose of a product behaves the same way the inverse of a product behaves—the order of the factors is reversed.

THEOREM 8.3 Let M and N be $n \times n$ matrices. Then $(MN)^t = N^tM^t$.

A discussion of the proof of this theorem is contained in Exercises 6 and 7.

Exercise Set 8.1

1. Determine which of the following matrices are orthogonal:

a. $\begin{bmatrix} -1 & 0 \\ 0 & 1 \end{bmatrix}$

b. $\begin{bmatrix} 1/\sqrt{2} & -1/\sqrt{2} \\ 1/\sqrt{2} & 1/\sqrt{2} \end{bmatrix}$

c. $\begin{bmatrix} 0 & -1 & 2 \\ -2 & 0 & 1 \\ -1 & -3 & 0 \end{bmatrix}$

d. $\begin{bmatrix} 1/\sqrt{6} & -1/\sqrt{14} & 2/\sqrt{5} \\ -2/\sqrt{6} & 2/\sqrt{14} & 1/\sqrt{5} \\ -1/\sqrt{6} & -3/\sqrt{14} & 0 \end{bmatrix}$

e. $\begin{bmatrix} 1/\sqrt{6} & -1/\sqrt{30} & 2/\sqrt{5} \\ -2/\sqrt{6} & 2/\sqrt{30} & 1/\sqrt{5} \\ -1/\sqrt{6} & -5/\sqrt{30} & 0 \end{bmatrix}$

2. Determine which of the following linear transformations are orthogonal.

a. $T: \mathbb{R}^2 \to \mathbb{R}^2$ is counterclockwise rotation of 45°.

b. $T: \mathbb{R}^2 \to \mathbb{R}^2$ is reflection about the line $y = 2x$.

c. $T: \mathbb{R}^2 \to \mathbb{R}^2$ is projection onto the x axis.

d. $T: \mathbb{R}^2 \to \mathbb{R}^2$
$\mathbf{v} \to 2\mathbf{v}$

3. Let $M = \begin{bmatrix} 1 & 2 \\ 3 & -5 \end{bmatrix}$ $\quad N = \begin{bmatrix} -2 & 5 \\ 1 & -3 \end{bmatrix}$

Verify Theorem 8.3 by showing that $(MN)^t = N^tM^t$ for these particular matrices.

4. Let $\quad M = \begin{bmatrix} 1 & 2 & -3 \\ -1 & 0 & 2 \\ 2 & 3 & 5 \end{bmatrix}$

$$\mathbf{v}_1 = \begin{bmatrix} 1 \\ 2 \\ -4 \end{bmatrix} \quad \mathbf{v}_2 = \begin{bmatrix} 3 \\ -1 \\ 3 \end{bmatrix}$$

Use a direct calculation to verify Theorem 8.1 by showing that

$$M\mathbf{v}_1 \cdot \mathbf{v}_2 = \mathbf{v}_1 \cdot M^t\mathbf{v}_2$$

5. Let $\mathbf{v}_1$ and $\mathbf{v}_2$ be two nonparallel vectors of $\mathbb{R}^2$. Consider the parallelogram having $\mathbf{v}_1$ and $\mathbf{v}_2$ as two of its sides.

a. Show that each point on the side of this parallelogram opposite $\mathbf{v}_1$ is the tip of a vector of the form

$$\mathbf{v}_2 + s\mathbf{v}_1 \qquad \text{where } 0 \le s \le 1$$

b. Show that each point on the side of the parallelogram opposite $\mathbf{v}_2$ is the tip of a vector of the form

$$\mathbf{v}_1 + t\mathbf{v}_2 \qquad \text{where } 0 \le t \le 1$$

c. Show that every point of this parallelogram is the tip of some vector of the form

$$s\mathbf{v}_1 + t\mathbf{v}_2 \qquad \text{where } \begin{cases} 0 \le s \le 1 \\ 0 \le t \le 1 \end{cases}$$

d. If T is a linear transformation from $\mathbb{R}^2$ to $\mathbb{R}^2$, show that each point (vector) inside or on the edges of the parallelogram formed by $\mathbf{v}_1$ and $\mathbf{v}_2$ is mapped onto a point (vector) inside or on the edges of the parallelogram formed by $T\mathbf{v}_1$ and $T\mathbf{v}_2$.

★**6. a.** For any matrix M show that $\mathbf{e}_i \cdot M\mathbf{e}_j$ is the entry m_{ij} in the ith row and jth column of M.

b. Use part **a** to show that if $\mathbf{v}_1 \cdot M\mathbf{v}_2 = \mathbf{v}_1 \cdot N\mathbf{v}_2$ for every pair of vectors $\mathbf{v}_1$ and $\mathbf{v}_2$, then $M = N$.

★**7. a.** Let M and N be two $n \times n$ matrices. By using Theorem 8.1, show that

$$(MN)\mathbf{v}_1 \cdot \mathbf{v}_2 = \mathbf{v}_1 \cdot (MN)^t\mathbf{v}_2$$

and also that

$$(MN)\mathbf{v}_1 \cdot \mathbf{v}_2 = M(N\mathbf{v}_1) \cdot \mathbf{v}_2 = \mathbf{v}_1 \cdot N^tM^t\mathbf{v}_2$$

b. Using part **a** and Exercise 6, show that

$$(MN)^t = N^tM^t$$

8. Suppose $T: \mathbb{R}^2 \to \mathbb{R}^2$ is an orthogonal linear transformation with the property that there is a nonzero vector $\mathbf{u}$ such that $T\mathbf{u} = \mathbf{u}$.

a. If $\mathbf{v}$ is a nonzero vector perpendicular to $\mathbf{u}$, show that $T\mathbf{v} = \mathbf{v}$ or $-\mathbf{v}$.

b. If $T\mathbf{v} = \mathbf{v}$, find T.

c. If $T\mathbf{v} = -\mathbf{v}$, show that T is a reflection about a line and find that line.

9. a. Let M_0 be the matrix that represents reflection about the x axis in $\mathbb{R}^2$, and let $M_{\pi/4}$ be the matrix that represents reflection about the line $y = x$. Show that the product $M_0 M_{\pi/4}$ of these matrices represents a rotation of $-\pi/2$.

b. Let M_0 be the matrix that represents reflection about the x axis in $\mathbb{R}^2$, and let M_θ be reflection about the line through the origin that makes an angle θ with the positive x axis. Show that the matrix $M_0 M_\theta$ represents a rotation by an angle -2θ.

★**10.** Using the formula derived in Exercise 12 in Section 6.3 for the matrix representing a reflection in $\mathbb{R}^2$, show that the product of two reflections in $\mathbb{R}^2$ is a rotation. (*Hint:* You will need the trigonometric formulas for the sine and cosine of the difference of two angles as well as the result of Theorem 3.3.) Compare your result to the results of Example 9 in Section 3.3 and Exercise 9 above.

11. Let M be a square matrix. Then M is said to be **symmetric** if $M = M^t$, and M is said to be **antisymmetric** if $M = -M^t$. Determine which of the following matrices are symmetric, antisymmetric, or neither:

a. $M = \begin{bmatrix} 1 & 2 \\ 2 & 3 \end{bmatrix}$ **b.** $M = \begin{bmatrix} 2 & 1 \\ 3 & 2 \end{bmatrix}$

c. $M = \begin{bmatrix} 1 & -2 \\ 2 & 3 \end{bmatrix}$

d. $M = \begin{bmatrix} 1 & 4 & -3 \\ 4 & -1 & 2 \\ -3 & 2 & 7 \end{bmatrix}$

e. $M = \begin{bmatrix} 7 & -5 & 3 \\ 5 & -2 & -1 \\ -3 & 1 & 21 \end{bmatrix}$

12. Using the definitions of symmetric and antisymmetric matrices given in Exercise 11, prove the following statements:

a. If M is an arbitrary square matrix, then the matrix $S = M + M^t$ is a symmetric matrix.

b. If M is an arbitrary square matrix, then the matrix $A = M - M^t$ is an antisymmetric matrix.

c. Show that every square matrix can be written as the sum of a symmetric matrix and an antisymmetric matrix.

8.2 Determinants on $\mathbb{R}^2$

As we saw in Section 8.1, linear transformations can change the shape of an object. In addition, most linear transformations change the size of objects—either magnifying them or shrinking them. In this section we will see how to determine how much the size of objects is changed by the action of a linear transformation. We will only discuss linear transformations from $\mathbb{R}^2$ to $\mathbb{R}^2$ in this section and generalize these ideas to $\mathbb{R}^n$ in the next section.

Let $T: V \to W$ be a linear transformation from $\mathbb{R}^2$ to $\mathbb{R}^2$, and let

$$M = \begin{bmatrix} m_{11} & m_{12} \\ m_{21} & m_{22} \end{bmatrix}$$

be the matrix which represents T. Consider a region $\mathscr{R}$ in the plane V (see Figure 8.3). Each point (tip of a vector) in this region is mapped by T onto a point of the region $T(\mathscr{R})$ in the plane W. The unit square with sides $\mathbf{e}_1$ and $\mathbf{e}_2$ of V is mapped onto the parallelogram with sides $\mathbf{c}_1 = \begin{bmatrix} m_{11} \\ m_{21} \end{bmatrix}$ and $\mathbf{c}_2 = \begin{bmatrix} m_{12} \\ m_{22} \end{bmatrix}$ of W. Moreover, since T is linear, each of the unit squares of the grid covering V is mapped onto a parallelogram (congruent to the one with sides $\mathbf{c}_1$ and $\mathbf{c}_2$) of the grid of parallelograms covering W. Therefore, the number of squares needed to cover $\mathscr{R}$ is equal to the number of parallelograms needed to cover $T(\mathscr{R})$. Since each of these unit squares has area 1, we see that the area of the region $\mathscr{R}$ is just the number of squares needed to cover $\mathscr{R}$. If s squares cover R, then s parallelograms are needed to cover $T(\mathscr{R})$. So the area of $T(\mathscr{R})$ is sA where A is the

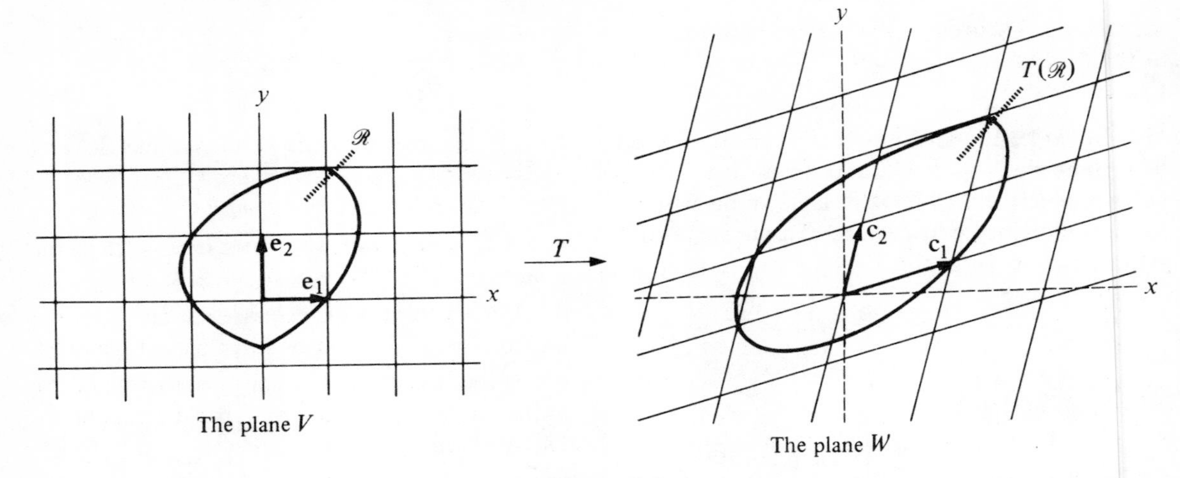

Figure 8.3

area of the parallelogram with sides $\mathbf{c}_1$ and $\mathbf{c}_2$—that is:

$$\text{Area of } T(\mathscr{R}) = (\text{area of } \mathscr{R})(\text{area of parallelogram with sides } \mathbf{c}_1, \mathbf{c}_2)$$

or equivalently, the factor by which the area of $\mathscr{R}$ is changed (magnified or shrunk) by the action of T is equal to the area of the parallelogram with sides

$$T\begin{bmatrix}1\\0\end{bmatrix} \quad \text{and} \quad T\begin{bmatrix}0\\1\end{bmatrix}$$

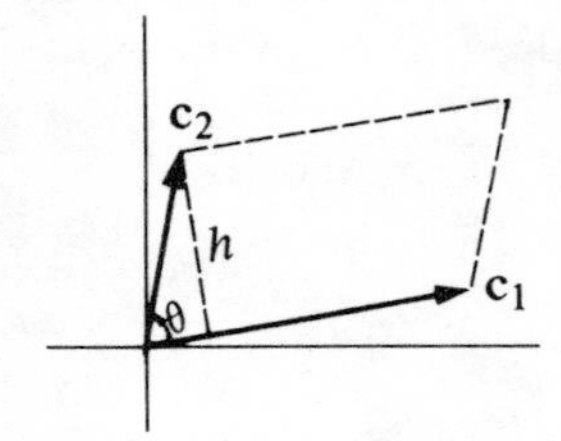

Figure 8.4

Now we see how to compute the area of this parallelogram (see Figure 8.4). The formula for the area of a parallelogram is

$$\text{Area} = (\text{base})(\text{height})$$

The length of the base is $|\mathbf{c}_1|$ and the height h is $|\mathbf{c}_2| \sin \theta$. So

$$\text{Area} = |\mathbf{c}_1||\mathbf{c}_2| \sin \theta$$

There is a technical problem here. By convention, the angle θ between the two vectors $\mathbf{c}_1$ and $\mathbf{c}_2$ is measured in the direction from $\mathbf{c}_1$ to $\mathbf{c}_2$. If $0 < \theta < 180°$, then $\sin \theta$ is positive and the formula for area given above is correct. However, if $-180° < \theta < 0$, then $\sin \theta$ is negative and the quantity $|\mathbf{c}_1||\mathbf{c}_2| \sin \theta$ represents the negative of the area of the parallelogram. Of course we could solve the problem by saying that the absolute value of the quantity $|\mathbf{c}_1||\mathbf{c}_2| \sin \theta$ represents the area of the parallelogram, but by doing this we would lose some information about the linear transformation T. If the quantity $|\mathbf{c}_1||\mathbf{c}_2| \sin \theta$ is positive, then this means that the orientation of the vectors $\mathbf{c}_1 = T\mathbf{e}_1$ and $\mathbf{c}_2 = T\mathbf{e}_2$ is the same as the orientation of the vectors $\mathbf{e}_1$ and $\mathbf{e}_2$; and if this quantity is negative, then the orientation of the vectors $\mathbf{c}_1$ and $\mathbf{c}_2$ is opposite that of $\mathbf{e}_1$ and $\mathbf{e}_2$. This is shown in Figure 8.5.

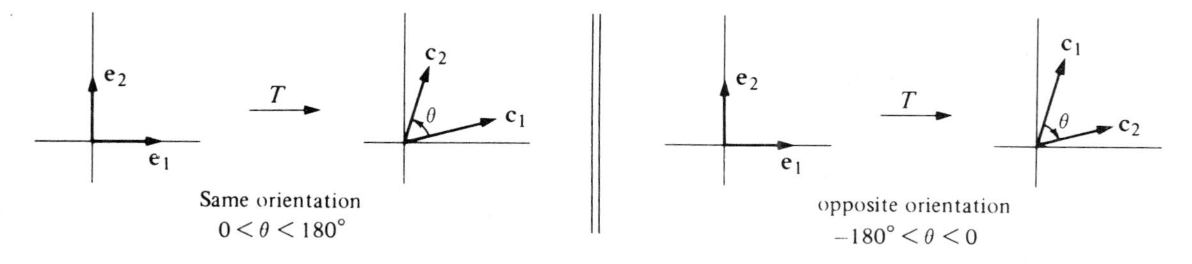

Figure 8.5

We will call the quantity $|\mathbf{c}_1||\mathbf{c}_2| \sin \theta$ the **oriented area** of the parallelogram with sides $\mathbf{c}_1$ and $\mathbf{c}_2$.

For aid in computing the oriented area of this parallelogram, we introduce the vector $\mathbf{c}_1^\perp$ (read "$\mathbf{c}_1$ perp"):

$$\mathbf{c}_1 = \begin{bmatrix} m_{11} \\ m_{21} \end{bmatrix} \qquad \mathbf{c}_1^\perp = \begin{bmatrix} -m_{21} \\ m_{11} \end{bmatrix}$$

The vector $\mathbf{c}_1^\perp$ has two important properties:

1. Length

$$|\mathbf{c}_1^\perp| = \sqrt{(-m_{21})^2 + m_{11}^2} + \sqrt{m_{11}^2 + m_{21}^2} = |\mathbf{c}_1|$$

2. Direction

$$\mathbf{c}_1^\perp \cdot \mathbf{c}_1 = m_{11}m_{21} - m_{21}m_{11} = 0$$

So $\mathbf{c}_1^\perp$ is perpendicular to $\mathbf{c}_1$.

Now we compute the oriented area of the parallelogram using $\mathbf{c}_1^\perp$. If we let α be the angle between $\mathbf{c}_2$ and $\mathbf{c}_1^\perp$, then by looking at Figure 8.6 we see

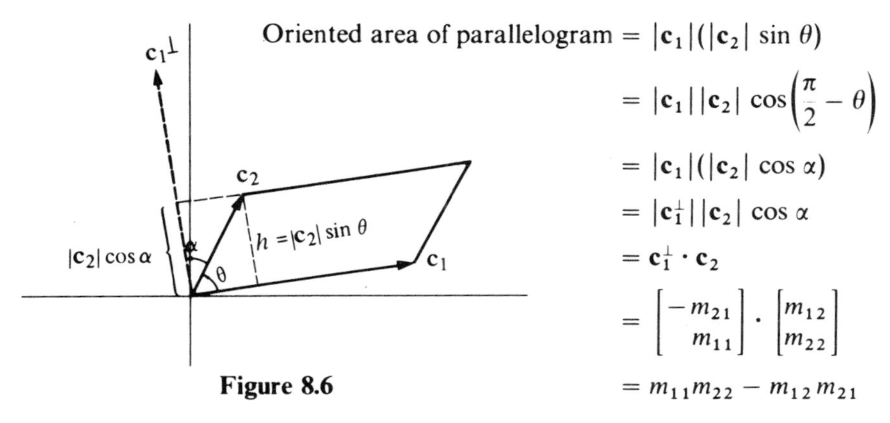

$$\begin{aligned}\text{Oriented area of parallelogram} &= |\mathbf{c}_1|(|\mathbf{c}_2| \sin \theta) \\ &= |\mathbf{c}_1||\mathbf{c}_2| \cos\left(\frac{\pi}{2} - \theta\right) \\ &= |\mathbf{c}_1|(|\mathbf{c}_2| \cos \alpha) \\ &= |\mathbf{c}_1^\perp||\mathbf{c}_2| \cos \alpha \\ &= \mathbf{c}_1^\perp \cdot \mathbf{c}_2 \\ &= \begin{bmatrix} -m_{21} \\ m_{11} \end{bmatrix} \cdot \begin{bmatrix} m_{12} \\ m_{22} \end{bmatrix} \\ &= m_{11}m_{22} - m_{12}m_{21}\end{aligned}$$

Figure 8.6

This quantity is very important and is given a special name.

DEFINITION 8.4 Let

$$M = \begin{bmatrix} m_{11} & m_{12} \\ m_{21} & m_{22} \end{bmatrix}$$

The quantity $(m_{11}m_{22} - m_{12}m_{21})$ is called the **determinant** of M. It is denoted by

$$\det(M) \text{ or } \quad |M| \quad \text{ or } \quad \begin{vmatrix} m_{11} & m_{12} \\ m_{21} & m_{22} \end{vmatrix}$$

If T is a linear transformation, then the determinant of T is defined to be $\det(M)$, where M is the matrix that represents T.

We summarize our results:

THEOREM 8.4 Let $T: \mathbb{R}^2 \to \mathbb{R}^2$ be a linear transformation. Then

1. The absolute value of $\det(M)$ is the factor by which the area of objects is changed under the action of T.

2. The sign of $\det(M)$ indicates the orientation of the vectors $T\mathbf{e}_1$ and $T\mathbf{e}_2$. If $\det(M)$ is positive, the vectors $T\mathbf{e}_1$ and $T\mathbf{e}_2$ have the same orientation as $\mathbf{e}_1$ and $\mathbf{e}_2$; and if $\det(M)$ is negative then the orientation of $T\mathbf{e}_1$ and $T\mathbf{e}_2$ is opposite that of $\mathbf{e}_1$ and $\mathbf{e}_2$.

EXAMPLE 1 Let $M = \begin{bmatrix} 1 & -3 \\ -2 & 1 \end{bmatrix}$. Then $\det(M) = (1)(1) - (-3)(-2) = -5$. So the area of the parallelogram with sides $M\begin{bmatrix} 1 \\ 0 \end{bmatrix} = \begin{bmatrix} 1 \\ -2 \end{bmatrix}$ and $M\begin{bmatrix} 0 \\ 1 \end{bmatrix} = \begin{bmatrix} -3 \\ 1 \end{bmatrix}$ is 5, and the orientation of the vectors $\begin{bmatrix} 1 \\ -2 \end{bmatrix}$ and $\begin{bmatrix} -3 \\ 1 \end{bmatrix}$ is opposite that of the vectors $\begin{bmatrix} 1 \\ 0 \end{bmatrix}$ and $\begin{bmatrix} 0 \\ 1 \end{bmatrix}$, since the sign $\det(M)$ is negative (see Figure 8.7). ■

Now we see what it means for the determinant of a matrix to be zero. If M is a 2×2 matrix and $\det(M) = 0$, this means that the area of the parallelogram formed by the vectors $M\mathbf{e}_1$ and $M\mathbf{e}_2$ is zero. This can happen in either of two ways: Either $M\mathbf{e}_1$ or $M\mathbf{e}_2$ is the zero vector, or $M\mathbf{e}_1$ and $M\mathbf{e}_2$ are parallel and thus the parallelogram formed by these two vectors is just a line segment which has area zero. In either case, we see that the vectors $M\mathbf{e}_1$ and $M\mathbf{e}_2$ do not span $\mathbb{R}^2$, which means that the rank of M is less than 2. Conversely, if the rank of a 2×2 matrix is less than 2, then the parallelogram formed by its column vectors has zero area. Combining these results with Theorems 5.2 and 7.5′ we have:

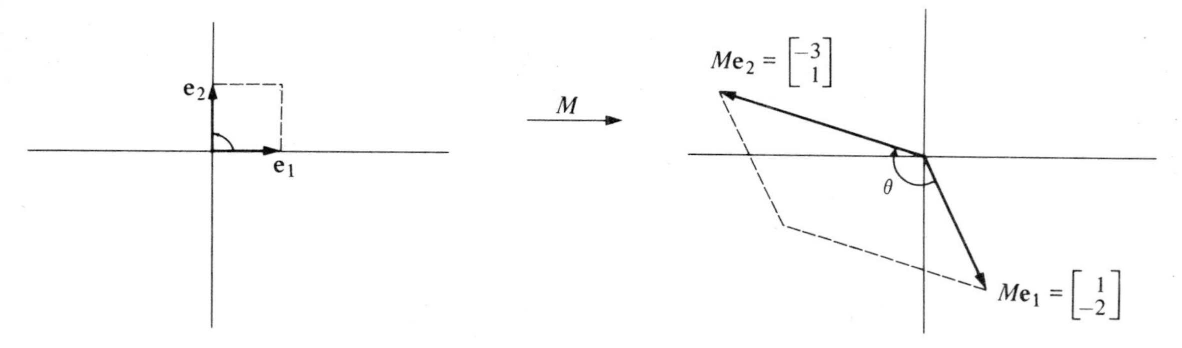

Figure 8.7

THEOREM 8.5 Let M be a 2×2 matrix. The following four conditions are equivalent (that is, if one is true, they are all true).

1. $\det(M) = 0$
2. Rank of $M < 2$
3. Nullity of $M \neq 0$
4. M is not invertible.

We illustrate Theorem 8.5.

EXAMPLE 2 Let

$$M = \begin{bmatrix} 1 & -2 \\ -2 & 4 \end{bmatrix}$$

Then

$$\det(M) = (1)(4) - (-2)(-2) = 0$$

The "parallelogram" formed by the column vectors $\begin{bmatrix} 1 \\ -2 \end{bmatrix}$ and $\begin{bmatrix} -2 \\ 4 \end{bmatrix}$ is a line segment and so it has zero area (see Figure 8.8). The rank of M is the dimension of $\left\langle \begin{bmatrix} 1 \\ -2 \end{bmatrix}, \begin{bmatrix} -2 \\ 4 \end{bmatrix} \right\rangle$ which is 1, and the nullity of M is $(2 - 1) = 1$.

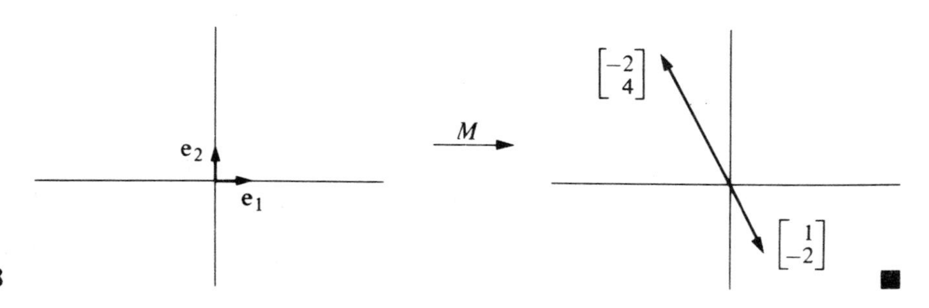

Figure 8.8

■

Now we look at determinants of sums and products of matrices. As we will see in Example 3, the determinant of the sum of two matrices is not equal to the sum of the determinants of the two matrices.

EXAMPLE 3 Let

$$M_1 = \begin{bmatrix} 1 & 0 \\ 0 & 0 \end{bmatrix} \quad \text{and} \quad M_2 = \begin{bmatrix} 0 & 0 \\ 0 & 1 \end{bmatrix}$$

Then

$$\det(M_1) + \det(M_2) = 0 + 0 = 0$$

But

$$\det(M_1 + M_2) = \det\left(\begin{bmatrix} 1 & 0 \\ 0 & 1 \end{bmatrix}\right) = 1$$

So

$$\det(M_1) + \det(M_2) \neq \det(M_1 + M_2) \quad \blacksquare$$

Determinants of products behave much more nicely than sums. Let T_1 and T_2 be linear transformations from $\mathbb{R}^2$ to $\mathbb{R}^2$.

As we see in Figure 8.9 T_2 magnifies the area of the unit square by $|\det(T_2)|$ and then T_1 magnifies the area of the parallelogram (which is the image of the unit square under T_2) by $|\det(T_1)|$. Thus the total magnification is $|\det(T_1)|\,|\det(T_2)|$. Moreover, since the sign of the determinant tells us whether the orientation of the vectors remains the same or is reversed, it is clear that the sign of $\det(T_1)\det(T_2)$ is the same as the sign of $\det(T_1 T_2)$. This gives us the following theorem.

THEOREM 8.6

Let $T_1\colon \mathbb{R}^2 \to \mathbb{R}^2$ and $T_2\colon \mathbb{R}^2 \to \mathbb{R}^2$ be linear transformations. Then

$$\det(T_1 T_2) = \det(T_1)\det(T_2)$$

If M_1 and M_2 are 2×2 matrices, then

$$\det(M_1 M_2) = \det(M_1)\det(M_2).$$

EXAMPLE 4 We give a concrete example of Theorem 8.6. Let

$$M_1 = \begin{bmatrix} 1 & 2 \\ -1 & 3 \end{bmatrix} \quad \text{and} \quad M_2 = \begin{bmatrix} 2 & 5 \\ -3 & 1 \end{bmatrix}$$

Then

$$M_1 M_2 = \begin{bmatrix} 1 & 2 \\ -1 & 3 \end{bmatrix}\begin{bmatrix} 2 & 5 \\ -3 & 1 \end{bmatrix} = \begin{bmatrix} -4 & 7 \\ -11 & -2 \end{bmatrix}$$

Computing determinants gives

$$\det(M_1) = 5 \qquad \det(M_2) = 17 \qquad \det(M_1 M_2) = 85$$

So we have

$$85 = \det(M_1 M_2) = (5)(17) = \det(M_1)\det(M_2) \quad \blacksquare$$

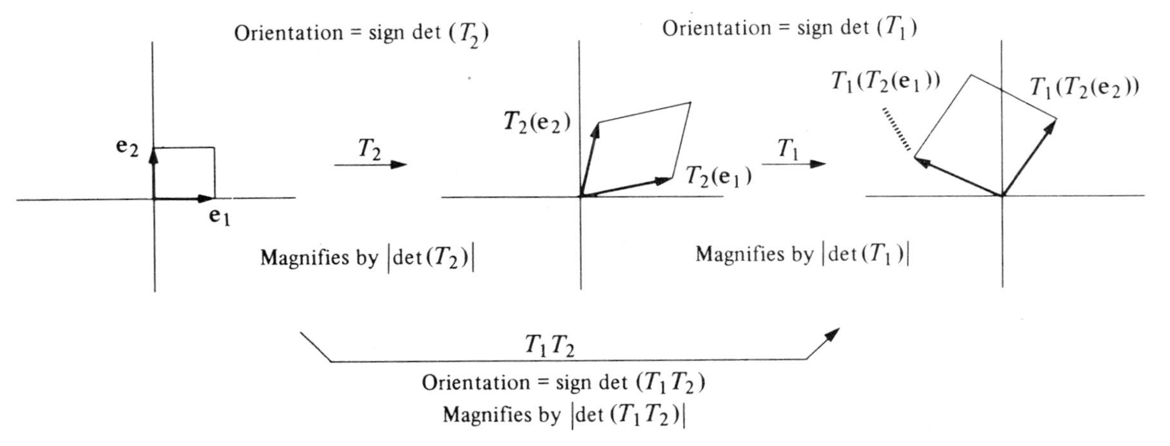

Figure 8.9

Determinants occur in a number of unexpected ways in the study of linear algebra. We will now see our first instance of this when we find a general formula for the inverse of a 2×2 matrix.

Consider the matrix
$$M = \begin{bmatrix} m_{11} & m_{12} \\ m_{21} & m_{22} \end{bmatrix}$$

We wish to find a general formula for M^{-1}. To attack this problem we begin by trying to find a matrix N such that $MN = I$. Instead of trying to find N directly, we solve the problem in two steps. First we look for a matrix N^* such that

$$MN^* = \begin{bmatrix} m_{11} & m_{12} \\ m_{21} & m_{22} \end{bmatrix} \begin{bmatrix} n_{11} & n_{12} \\ n_{21} & n_{22} \end{bmatrix} = \begin{bmatrix} a & 0 \\ 0 & b \end{bmatrix}$$

Once we have found N^* it is easy to get N—just divide the first column of N^* by a and the second column of N^* by b. The conditions on the entries of N^* are

$$0 = [m_{21} \quad m_{22}] \begin{bmatrix} n_{11} \\ n_{21} \end{bmatrix} = \begin{bmatrix} m_{21} \\ m_{22} \end{bmatrix} \cdot \begin{bmatrix} n_{11} \\ n_{21} \end{bmatrix} \qquad \text{lower left}$$

$$0 = [m_{11} \quad m_{12}] \begin{bmatrix} n_{12} \\ n_{22} \end{bmatrix} = \begin{bmatrix} m_{11} \\ m_{12} \end{bmatrix} \cdot \begin{bmatrix} n_{12} \\ n_{22} \end{bmatrix} \qquad \text{upper right}$$

So we see that we need to find a vector perpendicular to the given vector in each case. Using the ideas we used when we found the vector $\mathbf{c}_1^{\perp}$ to compute the area of a parallelogram, we see that

$$0 = \begin{bmatrix} m_{21} \\ m_{22} \end{bmatrix} \cdot \begin{bmatrix} m_{22} \\ -m_{21} \end{bmatrix} \qquad \text{so we choose} \qquad \begin{bmatrix} n_{11} \\ n_{21} \end{bmatrix} = \begin{bmatrix} m_{22} \\ -m_{21} \end{bmatrix}$$

$$0 = \begin{bmatrix} m_{11} \\ m_{12} \end{bmatrix} \cdot \begin{bmatrix} -m_{12} \\ m_{11} \end{bmatrix} \qquad \text{so we choose} \qquad \begin{bmatrix} n_{12} \\ n_{22} \end{bmatrix} = \begin{bmatrix} -m_{12} \\ m_{11} \end{bmatrix}$$

Therefore, we can choose N^* to be

$$N^* = \begin{bmatrix} m_{22} & -m_{12} \\ -m_{21} & m_{11} \end{bmatrix}$$

Then
$$MN^* = \begin{bmatrix} m_{11} & m_{12} \\ m_{21} & m_{22} \end{bmatrix} \begin{bmatrix} m_{22} & -m_{12} \\ -m_{21} & m_{11} \end{bmatrix}$$
$$= \begin{bmatrix} (m_{11}m_{22} - m_{12}m_{22}) & 0 \\ 0 & (m_{11}m_{22} - m_{12}m_{21}) \end{bmatrix}$$

So, the determinant occurs unexpectedly as the diagonal entry of the product of M and N^*. Hence, $MN^* = (\det M)I$. So

$$M\left(\frac{1}{\det(M)} N^*\right) = I.$$

Thus
$$M^{-1} = \frac{1}{\det(M)} N^* = \frac{1}{\det(M)} \begin{bmatrix} m_{22} & -m_{12} \\ -m_{21} & m_{11} \end{bmatrix}$$

For reference, we state this result as a theorem.

THEOREM 8.7

> Let
> $$M = \begin{bmatrix} m_{11} & m_{12} \\ m_{21} & m_{22} \end{bmatrix}$$
> be a matrix with $\det(M) \neq 0$. Then the inverse of M is
> $$M^{-1} = \frac{1}{\det(M)} \begin{bmatrix} m_{22} & -m_{12} \\ -m_{21} & m_{11} \end{bmatrix}$$

EXAMPLE 5 We use Theorem 8.7 to find the inverse of the matrix

$$M = \begin{bmatrix} 1 & -1 \\ 2 & 3 \end{bmatrix}$$

The determinant of M is 5. So

$$M^{-1} = \frac{1}{5} \begin{bmatrix} 3 & 1 \\ -2 & 1 \end{bmatrix} = \begin{bmatrix} 3/5 & 1/5 \\ -2/5 & 1/5 \end{bmatrix}$$ ■

We conclude this section by stating a theorem that tells how row operations on a matrix affect its determinant.

THEOREM 8.8

> Let M be a 2×2 matrix.
>
> **1.** If two rows of M are interchanged, then the determinant of the resulting matrix is $-\det(M)$.
>
> **2.** If $$\text{Row } r_j \xrightarrow{\text{is replaced by}} a(\text{row } r_i) + b(\text{row } r_j)$$
>
> then the determinant of the resulting matrix is $b(\det(M))$.

The proof of this theorem follows easily from Theorem 8.6 and is discussed in Exercises 6 and 7.

The following examples illustrate Theorem 8.8.

EXAMPLE 6 Let

$$M = \begin{bmatrix} 2 & -1 \\ 3 & -4 \end{bmatrix}.$$

Then $\det(M) = -5$. If we interchange the rows of M, we obtain the matrix

$$\begin{bmatrix} 3 & -4 \\ 2 & -1 \end{bmatrix}$$

which has determinant $+5 = -\det(M)$. ■

EXAMPLE 7 Let M be as in Example 6. If we replace the second row of M using the following row operation

$$\begin{bmatrix} 2 & -1 \\ 3 & -4 \end{bmatrix} \begin{matrix} (-3) \\ (2) \end{matrix} \longrightarrow \begin{bmatrix} 2 & -1 \\ 0 & -5 \end{bmatrix}$$

the determinant of the resulting matrix is $-10 = 2(\det(M))$. ■

Exercise Set 8.2

1. Find the determinants of the following matrices:

a. $\begin{bmatrix} 1 & 3 \\ -4 & 6 \end{bmatrix}$ **b.** $\begin{bmatrix} 5 & -1 \\ 2 & 7 \end{bmatrix}$

c. $\begin{bmatrix} -2 & 13 \\ 1 & 7 \end{bmatrix}$ **d.** $\begin{bmatrix} 3/5 & -4/5 \\ 4/5 & 3/5 \end{bmatrix}$

2. Find the area of the parallelogram with vertices $(0, 0)$, $(5, 2)$, $(-1, 7)$, and $(4, 9)$.

3. Use Theorem 8.7 to find the inverse of each of the matrices in Exercise 1.

4. Use Theorems 8.5 and 8.6 to give a different proof of the fact that if M_1 and M_2 are 2×2 matrices, then (M_1M_2) is invertible if and only if both M_1 and M_2 are invertible.

5. Using an explicit computation, show that $\det(M_1M_2) = \det(M_1)\det(M_2)$ when

$$M_1 = \begin{bmatrix} 1 & -1 \\ 2 & 3 \end{bmatrix} \quad \text{and} \quad M_2 = \begin{bmatrix} 3 & 2 \\ 5 & 1 \end{bmatrix}$$

6. a. Show that the row operation

$$\begin{bmatrix} M \end{bmatrix} \begin{matrix} (a) \\ (b) \end{matrix} \longrightarrow \begin{bmatrix} P \end{bmatrix}$$

can be accomplished by multiplying M on the

left by the matrix

$$B = \begin{bmatrix} 1 & 0 \\ a & b \end{bmatrix}$$

b. Show that the row operation

$$\begin{bmatrix} M \end{bmatrix} \xrightarrow{(a),\,(b)} \begin{bmatrix} Q \end{bmatrix}$$

can be accomplished by multiplying M on the left by the matrix

$$A = \begin{bmatrix} a & b \\ 0 & 1 \end{bmatrix}$$

c. Show that $b[\det(M)] = \det(P)$ and that $a[\det(M)] = \det(Q)$. This proves part 2 of Theorem 8.8.

7. Prove part 1 of Theorem 8.8.

8. Let M be a 2×2 matrix. Show that $\det(M) = \det(M^t)$, where M^t is the transpose of M.

★**9. a.** Show that the unit circle (in $\mathbb{R}^2$) is the set of all points that are the tips of vectors of the form

$$\mathbf{v}(t) = \begin{bmatrix} \cos t \\ \sin t \end{bmatrix}$$

b. Let M be the matrix $M = \begin{bmatrix} 2 & 0 \\ 0 & 1 \end{bmatrix}$

Show that if $\mathbf{v}(t)$ is a vector whose tip lies on the unit circle, then $M\mathbf{v}(t)$ is a vector whose tip lies on the ellipse $y^2 + x^2/2^2 = 1$.

c. Use Theorem 8.4 to find the area of the ellipse $y^2 + x^2/2^2 = 1$.

10. Let M be a 2×2 matrix.

a. If M is invertible, show that $\det(M^{-1}) = 1/\det(M)$

b. If M is invertible, show that $\det(M^{-1}NM) = \det(N)$ for any 2×2 matrix N. (Does $M^{-1}NM$ always equal N?)

8.3 Determinants on $\mathbb{R}^n$

As we saw in Section 8.2, the determinant of a 2×2 matrix M represents the oriented area of the parallelogram that is the image of the unit square under M. We can also define the determinant of a square matrix in $\mathbb{R}^3$ and higher dimensional spaces. In $\mathbb{R}^3$ the determinant of a 3×3 matrix represents the oriented volume of the parallelepiped that is the image of the unit cube under the action of M. In this section we show how to compute the determinant of an $n \times n$ matrix for $n \geq 3$ and discuss the analogues of the theorems of Section 8.2 for higher dimensional spaces. The proof of the fact that the determinant defined below actually does represent an oriented volume in $\mathbb{R}^3$ will be postponed until Section 8.4.

The definition of determinant that we give is recursive in the sense that to find the determinant of an $n \times n$ matrix, it is necessary to know the determinants of a number of its $(n-1) \times (n-1)$ submatrices.

DEFINITION 8.5

Let M be an $n \times n$ matrix. The **cofactor** M_{ij} of M is the determinant of the $(n-1) \times (n-1)$ matrix obtained from M by deleting the ith row and jth column. A matrix obtained from M by deleting a row and a column is called a **minor** of M.

EXAMPLE 1 Let

$$M = \begin{bmatrix} 1 & 3 & 5 \\ -1 & -2 & -3 \\ 2 & 4 & -4 \end{bmatrix}$$

Then

$$M_{11} = \begin{vmatrix} \not{1} & \not{3} & \not{5} \\ -\not{1} & -2 & -3 \\ \not{2} & 4 & -4 \end{vmatrix} = \begin{vmatrix} -2 & -3 \\ 4 & -4 \end{vmatrix} = 20$$

$$M_{23} = \begin{vmatrix} 1 & 3 & \not{5} \\ \not{-1} & \not{-2} & \not{-3} \\ 2 & 4 & -\not{4} \end{vmatrix} = \begin{vmatrix} 1 & 3 \\ 2 & 4 \end{vmatrix} = -2$$

$$M_{32} = \begin{vmatrix} 1 & \not{3} & 5 \\ -1 & -\not{2} & -3 \\ \not{2} & \not{4} & \not{-4} \end{vmatrix} = \begin{vmatrix} 1 & 5 \\ -1 & -3 \end{vmatrix} = 2$$

■

Now we define the determinant of a 3×3 matrix.

DEFINITION 8.6

Determinant of a 3 × 3 matrix: Let $M = \begin{bmatrix} m_{11} & m_{12} & m_{13} \\ m_{21} & m_{22} & m_{23} \\ m_{31} & m_{32} & m_{33} \end{bmatrix}$

Then

$$\det(M) = m_{11}M_{11} - m_{12}M_{12} + m_{13}M_{13}$$
$$= m_{11}\begin{vmatrix} m_{22} & m_{23} \\ m_{32} & m_{33} \end{vmatrix} - m_{12}\begin{vmatrix} m_{21} & m_{23} \\ m_{31} & m_{33} \end{vmatrix} + m_{13}\begin{vmatrix} m_{21} & m_{22} \\ m_{31} & m_{32} \end{vmatrix}$$

EXAMPLE 2 Compute $\det(M)$ where $M = \begin{bmatrix} 5 & 3 & 2 \\ -1 & 4 & 1 \\ 2 & -1 & 1 \end{bmatrix}$

$$\det(M) = m_{11}M_{11} - m_{12}M_{12} + m_{13}M_{13}$$
$$= 5\begin{vmatrix} 4 & 1 \\ -1 & 1 \end{vmatrix} - 3\begin{vmatrix} -1 & 1 \\ 2 & 1 \end{vmatrix} + 2\begin{vmatrix} -1 & 4 \\ 2 & -1 \end{vmatrix}$$
$$= 5(5) - 3(-3) + 2(-7) = 20$$

■

DEFINITION 8.7

Determinant of $n \times n$ matrix: Let $M = \begin{bmatrix} m_{11} & m_{12} & \cdots & m_{1n} \\ m_{21} & m_{22} & \cdots & m_{2n} \\ \vdots & \vdots & & \vdots \\ m_{n1} & m_{n2} & \cdots & m_{nn} \end{bmatrix}$

Then the determinant of M is given by the alternating sum:

$$\det(M) = m_{11}M_{11} - m_{12}M_{12} + m_{13}M_{13} - m_{14}M_{14} + \cdots \pm m_{1n}M_{1n}$$

Note: The method of finding determinants given above is generally called **expansion by minors**. In other treatments of this subject the determinant is defined in a more abstract way and then expansion by minors is introduced as a computational technique.

Now we will see how the results that we obtained for 2×2 matrices in Section 8.2 can be generalized to $n \times n$ matrices. We begin by looking at the determinant of the product of two matrices. As we will show in Section 8.4, determinants represent oriented volumes in $\mathbb{R}^3$ and generalized oriented volumes in $\mathbb{R}^n$, for $n \geq 4$. Therefore, the ideas used in $\mathbb{R}^2$ to prove Theorem 8.7 can be applied in higher dimensions to give the following theorem.

THEOREM 8.9

Let M_1 and M_2 be $n \times n$ matrices, then

$$\det(M_1 M_2) = \det(M_1)\det(M_2)$$

Note: As we saw in Example 3 in Section 8.2, in general

$$\det(M_1 + M_2) \neq \det(M_1) + \det(M_2)$$

Theorem 8.9 can be used to prove the following theorem (see Exercises 6 and 8).

THEOREM 8.10

Let M be an $n \times n$ matrix.

1. If P is the matrix obtained from M by the row operation

$$M = \begin{bmatrix} (\text{row } i) \\ (\text{row } j) \end{bmatrix} \begin{matrix} (a) \\ (b) \end{matrix} \longrightarrow \begin{bmatrix} (\text{row } i) \\ a(\text{row } i) + b(\text{row } j) \end{bmatrix} = P$$

then $b \cdot \det(M) = \det(P)$.

2. If two rows of M are interchanged, then the determinant of the resulting matrix is $-\det(M)$.

A practical use of Theorem 8.10 is given in the following example. To compute the determinant of a 4×4 matrix using just Definition 8.7 involves computing four 3×3 determinants, and this in turn involves computing twelve 2×2 determinants. A more efficient method for computing such a determinant is given below.

EXAMPLE 3 We find the determinant of the matrix

$$M = \begin{bmatrix} 2 & 1 & -1 & 3 \\ -1 & 1 & 2 & 4 \\ 2 & 1 & 3 & -2 \\ 3 & 2 & 1 & 1 \end{bmatrix}$$

We row reduce M to obtain a matrix that has only one nonzero entry in the first column.

$$M = \begin{bmatrix} 2 & 1 & -1 & 3 \\ -1 & 1 & 2 & 4 \\ 2 & 1 & 3 & -2 \\ 3 & 2 & 1 & 1 \end{bmatrix} \quad \begin{matrix} (1)\ (-1)\ (-3) \\ \text{②} \\ \text{①} \\ \text{②} \end{matrix}$$

$$\begin{bmatrix} 2 & 1 & -1 & 3 \\ 0 & 3 & 3 & 11 \\ 0 & 0 & 4 & -5 \\ 0 & 1 & 5 & -7 \end{bmatrix} = P$$

By applying Theorem 8.10 three times (once for each row changed) we have

$$\det(M) = \frac{1}{(2)(1)(2)}\det(P) = (1/4)\det(P)$$

Since the cofactors P_{12}, P_{13}, and P_{14} each involve matrices with a column of zeros, each of these cofactors is 0. Therefore,

$$\det(M) = (1/4)\det(P) = (1/4)(p_{11}P_{11} - p_{12}P_{12} + p_{13}P_{13} - p_{14}P_{14})$$

$$= (1/4)(p_{11}P_{11}) = (1/4)\left(2\begin{vmatrix} 3 & 3 & 11 \\ 0 & 4 & -5 \\ 1 & 5 & -7 \end{vmatrix}\right)$$

$$= (1/2)[3(-3) - 3(5) + 11(-4)] = -34 \qquad \blacksquare$$

One of the most useful features of the determinant is that it can be used to tell whether an $n \times n$ matrix is invertible or not. The results we proved as Theorem 8.5 for 2×2 matrices are true for $n \times n$ matrices (for the same reasons).

THEOREM 8.11 Let M be an $n \times n$ matrix. Then the following statements are equivalent (if one is true, they all are):

1. $\det(M) = 0$ **2.** Rank of $M < n$

3. Nullity of $M \neq 0$ **4.** M is not invertible

DEFINITION 8.8

A matrix M that satisfies any one of the conditions (and therefore all the conditions) of Theorem 8.11 is called **singular**.

EXAMPLE 4 Show that the following matrix does not have an inverse.

$$M = \begin{bmatrix} 1 & -1 & 3 \\ 2 & 2 & -4 \\ 3 & 1 & -1 \end{bmatrix}$$

By Theorem 8.11 it suffices to show that $\det(M) = 0$

$$\det(M) = 1(-2+4) - (-1)(-2+12) + 3(2-6) = 2 + 10 - 12 = 0$$

Thus, M is a singular matrix and is not invertible. ■

Determinants are not only useful for determining if a matrix has an inverse, but also in finding the inverse of a matrix. We now state the three-dimensional version of Theorem 8.7.

THEOREM 8.12

Let M be a 3×3 matrix with $\det(M) \neq 0$. The following procedure can be used to find the inverse of M.

1. Form the matrix A whose ij entry is the cofactor M_{ij} (for $1 \leq i,j \leq 3$) with plus or minus sign affixed as indicated:

$$A = \begin{bmatrix} M_{11} & -M_{12} & M_{13} \\ -M_{21} & M_{22} & -M_{23} \\ M_{31} & -M_{32} & M_{33} \end{bmatrix}$$

2. Find $C = A^t$ (where A^t is the transpose of A)—C is called the **classical adjoint** of M.
3. The inverse of M is $(1/\det(M))C$

Note: The procedure for finding the inverse of a 4×4 matrix is just the four-dimensional version of this. However, since constructing the matrix A would involve calculating sixteen 3×3 determinants, this is not a very practical method. In general, this method is efficient only for 2×2 or 3×3 matrices (even when you have a computer). For higher dimensions it is best to use the row reduction technique described in Section 7.3.

EXAMPLE 5 We use Theorem 8.12 to find the inverse of the following matrix:

$$M = \begin{bmatrix} 1 & 2 & -1 \\ 3 & 1 & 1 \\ 2 & -1 & 1 \end{bmatrix}$$

We begin by finding the determinant of M.

$$\det(M) = 1(2) - 2(1) - 1(-5) = 5 \neq 0$$

Thus M is invertible.

1. We form the matrix of cofactors: $A = \begin{bmatrix} 2 & -(1) & -5 \\ -(1) & 3 & -(-5) \\ 3 & -(4) & -5 \end{bmatrix}$

2. We find the classical adjoint: $C = A^t = \begin{bmatrix} 2 & -1 & 3 \\ -1 & 3 & -4 \\ -5 & 5 & -5 \end{bmatrix}$

3. We find M^{-1}:

$$M^{-1} = \frac{1}{\det(M)} C = \begin{bmatrix} 2/5 & -1/5 & 3/5 \\ -1/5 & 3/5 & -4/5 \\ -1 & 1 & -1 \end{bmatrix} \quad \blacksquare$$

We conclude this section by seeing how part 2 of Theorem 8.10 can be applied to find different ways of computing determinants.

We consider the matrix $M = \begin{bmatrix} m_{11} & m_{12} & m_{13} \\ m_{21} & m_{22} & m_{23} \\ m_{31} & m_{32} & m_{33} \end{bmatrix}$

By interchanging the first two rows of M we get

$$M^* = \begin{bmatrix} m_{21} & m_{22} & m_{23} \\ m_{11} & m_{12} & m_{13} \\ m_{31} & m_{32} & m_{33} \end{bmatrix}$$

Using Theorem 8.10 and expansion by minors, we have

$$\begin{aligned} \det(M) = -\det(M^*) &= -[m_{21}(m_{12}m_{33} - m_{13}m_{32}) - m_{22}(m_{11}m_{33} - m_{13}m_{31}) \\ &\quad + m_{23}(m_{11}m_{32} - m_{12}m_{31})] \\ &= -[m_{21}M_{21} - m_{22}M_{22} + m_{23}M_{23}] \end{aligned}$$

So

$$\det(M) = -m_{21}M_{21} + m_{22}M_{22} - m_{23}M_{23}$$

Therefore we have shown that the determinant of M can be computed by expanding M along the second row. (Note that the pattern of signs is reversed.)

A similar expansion can be given for the determinant in terms of the third row—only in this case the signs are not reversed. In general we have:

THEOREM 8.13 Let M be an $n \times n$ matrix. Denote the entry in the ith row and jth column of M by m_{ij}. Then

$$\begin{aligned}\det(M) &= m_{11}M_{11} - m_{12}M_{12} + m_{13}M_{13} - \cdots \pm m_{1n}M_{1n}\\ &= -m_{21}M_{21} + m_{22}M_{22} - m_{23}M_{23} + \cdots \mp m_{2n}M_{2n}\\ &= m_{31}M_{31} - m_{32}M_{32} + m_{33}M_{33} - \cdots \pm m_{3n}M_{3n}\end{aligned}$$

and so on. In general, the sign associated with the term $m_{ij}M_{ij}$ is positive if $i+j$ is even and negative if $i+j$ is odd. A more concise way of writing the above equations is

$$\det(M) = \sum_{j=1}^{n} (-1)^{i+j} m_{ij} M_{ij}$$

We illustrate this theorem with the following example.

EXAMPLE 6 We compute the determinant of the same matrix in Example 2, by expanding along the second row.

$$M = \begin{bmatrix} 5 & 3 & 2\\ -1 & 4 & 1\\ 2 & -1 & 1\end{bmatrix}$$

$$\begin{aligned}\det(M) &= -m_{21}M_{21} + m_{22}M_{22} - m_{23}M_{23}\\ &= -(-1)\begin{vmatrix} 3 & 2\\ -1 & 1\end{vmatrix} + 4\begin{vmatrix} 5 & 2\\ 2 & 1\end{vmatrix} - 1\begin{vmatrix} 5 & 3\\ 2 & -1\end{vmatrix}\\ &= 1(5) + 4(1) - 1(-11) = 20\end{aligned}$$

which is the same answer we got when we expanded M along the first row in Example 2. ■

Determinants may also be computed by expanding along a column of a matrix. This follows from the fact (which we do not prove) that the determinant of a matrix and the determinant of its transpose are the same. We state these ideas formally in the following theorem.

THEOREM 8.14 Let

$$M = \begin{bmatrix} m_{11} & m_{12} & \cdots & m_{1n}\\ \vdots & \vdots & & \vdots\\ m_{n1} & m_{n2} & \cdots & m_{nn}\end{bmatrix}$$

Then

1. $\det(M) = \det(M^t)$

2. (Column expansion) $\det(M) = \sum_{i=1}^{n} (-1)^{i+j} m_{ij} M_{ij}$

Note: For $j = 1$, this is expansion down the first column, that is,

$$\det(M) = m_{11}M_{11} - m_{21}M_{21} + m_{31}M_{31} - \cdots \pm m_{n1}M_{n1}$$

EXAMPLE 7 Compute the determinant of the matrix

$$M = \begin{bmatrix} 5 & 3 & 2 \\ -1 & 4 & 1 \\ 2 & -1 & 1 \end{bmatrix}$$

by expansion along the first column.

$$\begin{aligned} \det(M) &= 5M_{11} - (-1)M_{21} + 2M_{31} \\ &= 5\begin{vmatrix} 4 & 1 \\ -1 & 1 \end{vmatrix} - (-1)\begin{vmatrix} 3 & 2 \\ -1 & 1 \end{vmatrix} + 2\begin{vmatrix} 3 & 2 \\ 4 & 1 \end{vmatrix} \\ &= 5(5) - (-1)(5) + 2(-5) = 20 \end{aligned}$$

This is the same answer we obtained when we expanded the determinant along the first row in Example 2. ■

Exercise Set 8.3

1. Using expansion by minors, find the determinant of each of the matrices listed below:

a. $M = \begin{bmatrix} 1 & 2 & -3 \\ 5 & -1 & 1 \\ 2 & 4 & -6 \end{bmatrix}$

b. $M = \begin{bmatrix} -1 & 3 & 2 \\ 5 & 1 & 1 \\ -3 & -2 & 7 \end{bmatrix}$

c. $M = \begin{bmatrix} 0 & -2 & 3 \\ 3 & 4 & 7 \\ 5 & -1 & 2 \end{bmatrix}$

d. $M = \begin{bmatrix} -3 & 5 & 1 \\ 2 & 1 & 5 \\ 3 & -4 & 2 \end{bmatrix}$

e. $M = \begin{bmatrix} 1 & -1 & 2 & 0 \\ 0 & 1 & 1 & 3 \\ 2 & -4 & 0 & 5 \\ 1 & 1 & -2 & 3 \end{bmatrix}$

f. $M = \begin{bmatrix} 2 & -1 & 3 & 2 \\ -1 & 4 & 0 & 5 \\ 2 & 3 & -2 & 7 \\ -1 & -1 & 4 & -5 \end{bmatrix}$

g. $M = \begin{bmatrix} 2 & 3 & -5 \\ 0 & -1 & 12 \\ 0 & 0 & -3 \end{bmatrix}$

h. $M = \begin{bmatrix} 3 & 0 & 0 \\ 1 & -4 & 0 \\ 15 & -7 & 2 \end{bmatrix}$

i. $M = \begin{bmatrix} 2 & 0 & 0 \\ 0 & 3 & 0 \\ 0 & 0 & -1 \end{bmatrix}$

j. $M = \begin{bmatrix} 0 & 0 & 2 \\ 0 & 3 & 0 \\ -1 & 0 & 0 \end{bmatrix}$ (see Theorem 8.10)

2. Use Theorem 8.12 to find the inverses of each of the nonsingular 3×3 matrices of Exercise 1.

3. For each of the matrices of Exercise 1, compute its determinant by using the method of Example 3. Compare your answers with the answers you got using expansion by minors.

4. a. Find the determinant of each of the following matrices:

$$\begin{bmatrix} d_1 & 0 \\ a & d_2 \end{bmatrix} \quad \begin{bmatrix} d_1 & 0 & 0 \\ a & d_2 & 0 \\ b & c & d_3 \end{bmatrix} \quad \begin{bmatrix} d_1 & 0 & 0 & 0 \\ a & d_2 & 0 & 0 \\ b & c & d_3 & 0 \\ d & e & f & d_4 \end{bmatrix}$$

b. A matrix is called **lower triangular** if all the entries above its main diagonal (the diagonal that goes like this [╲]) are zero. Show that the determinant of a lower triangular matrix is just the product of the entries on its main diagonal.

★**5. a.** Show that if two rows of a matrix are the same, then the determinant of the matrix is zero. (*Hint:* Use Theorem 8.10.)

b. Using the fact that the determinant of a matrix represents an oriented volume, explain why a matrix in which two columns are the same has determinant zero.

★**6.** Let $M = \begin{bmatrix} m_{11} & m_{12} & m_{13} \\ m_{21} & m_{22} & m_{23} \\ m_{31} & m_{32} & m_{33} \end{bmatrix}$

a. Use Theorem 8.9 to show that interchanging the first two rows of M gives a matrix with determinant $-\det(M)$. *Hint:* Consider the determinant of the product BM where

$$B = \begin{bmatrix} 0 & 1 & 0 \\ 1 & 0 & 0 \\ 0 & 0 & 1 \end{bmatrix}$$

b. Show that the matrix

$$M^* = \begin{bmatrix} m_{31} & m_{32} & m_{33} \\ m_{11} & m_{12} & m_{13} \\ m_{21} & m_{22} & m_{23} \end{bmatrix}$$

has the same determinant as M.

7. Expand the determinant of the following matrix along each of its three rows and three columns and show that the result is the same in all cases.

$$M = \begin{bmatrix} -1 & 2 & 3 \\ 2 & 1 & 4 \\ 5 & -1 & -4 \end{bmatrix}$$

★**8.** Let M be a 3×3 matrix.

a. Show that the matrix P obtained from M by the row operation

$$M = \begin{bmatrix} (\text{row } 1) \\ (\text{row } 2) \\ (\text{row } 3) \end{bmatrix} \begin{matrix} (a) \\ (b) \\ \end{matrix} \longrightarrow$$

$$\begin{bmatrix} (\text{row } 1) \\ a(\text{row } 1) + b(\text{row } 2) \\ (\text{row } 3) \end{bmatrix} = P$$

is $P = BM$ where

$$B = \begin{bmatrix} 1 & 0 & 0 \\ a & b & 0 \\ 0 & 0 & 1 \end{bmatrix}$$

b. Find a matrix C so that the matrix Q obtained from M by the row operation

$$M = \begin{bmatrix} (\text{row } 1) \\ (\text{row } 2) \\ (\text{row } 3) \end{bmatrix} \begin{matrix} (b) \\ \\ (a) \end{matrix} \longrightarrow$$

$$\begin{bmatrix} b(\text{row } 1) + a(\text{row } 3) \\ (\text{row } 2) \\ (\text{row } 3) \end{bmatrix} = Q$$

is $Q = CM$.

c. Prove part 1 of Theorem 8.10. (*Hint:* Use Theorem 8.9.)

★**9.** Let f_1, f_2, f_3 be twice differentiable functions defined on the interval (a, b), and 0_f be the function which is identically zero on this interval.

a. Show that if c_1, c_2, c_3 are constants such that

$$c_1 f_1 + c_2 f_2 + c_3 f_3 = 0_f$$

then
$$c_1 f_1' + c_2 f_2' + c_3 f_3' = 0_f$$

and
$$c_1 f_1'' + c_2 f_2'' + c_3 f_3'' = 0_f$$

b. Let W be the matrix (called the **Wronskian** matrix)

$$W = \begin{bmatrix} f_1 & f_2 & f_3 \\ f_1' & f_2' & f_3' \\ f_1'' & f_2'' & f_3'' \end{bmatrix} \quad \text{where } f_i' \text{ is the derivative of } f$$

Prove that if $\det(W(x)) \neq 0$ for some x in (a, b), then the functions f_1, f_2, f_3 are linearly independent.

10. a. Let M be an $n \times n$ matrix such that all entries of M are integers. If $\det(M) = 1$, show that all entries of M^{-1} are integers.

b. Find a matrix M in which all entries are integers, but M^{-1} does not have only integer entries.

★ **11.** Determinants can be used to determine if three points in $\mathbb{R}^2$ are collinear. Prove the following statements:

a. The point (x_1, y_1) is on the line $ax + by + c = 0$ if and only if

$$\begin{bmatrix} x_1 & y_1 & 1 \end{bmatrix} \begin{bmatrix} a \\ b \\ c \end{bmatrix} = [0]$$

b. The line (in $\mathbb{R}^2$) that joins the points (x_1, y_1) and (x_2, y_2) has the equation $ax + by + c = 0$ if and only if the vector

$$\begin{bmatrix} a \\ b \\ c \end{bmatrix}$$

is in the null space of the matrix

$$\begin{bmatrix} x_1 & y_1 & 1 \\ x_2 & y_2 & 1 \end{bmatrix}$$

c. The three points (x_1, y_1), (x_2, y_2), (x_3, y_3) lie on the same line $ax + by + c = 0$ if and only if the vector

$$\begin{bmatrix} a \\ b \\ c \end{bmatrix}$$

is in the null space of the matrix

$$\begin{bmatrix} x_1 & y_1 & 1 \\ x_2 & y_2 & 1 \\ x_3 & y_3 & 1 \end{bmatrix}$$

d. The points (x_1, y_1), (x_2, y_2), (x_3, y_3) are collinear if and only if

$$\begin{vmatrix} x_1 & y_1 & 1 \\ x_2 & y_2 & 1 \\ x_3 & y_3 & 1 \end{vmatrix} = 0$$

e. The equation of the line (in $\mathbb{R}^2$) that contains the points (x_1, y_1) and (x_2, y_2) is

$$\begin{vmatrix} x_1 & y_1 & 1 \\ x_2 & y_2 & 1 \\ x & y & 1 \end{vmatrix} = 0$$

*8.4 Volumes Associated with 3 × 3 Matrices and the Use of Determinants to Invert Matrices—Proofs

In this section we discuss the proofs of two theorems. The first proof shows that the determinant of a 3 × 3 matrix represents (up to the sign) a volume, and the second is the proof of Theorem 8.12, the method for finding the inverse of a 3 × 3 matrix using determinants. The work in this section is somewhat more abstract than that of most of the other sections of the book. It is optional since no other sections of this text are based on the material covered here. However, it does contain some fundamental ideas of linear algebra which you may find of interest.

Volumes

THEOREM 8.15 Let M be a 3×3 matrix. The determinant $\det(M)$ represents the volume (up to a sign) of the parallelepiped that is the image of the unit cube under M.

We begin our discussion of the proof of this theorem by looking at a special case. Let

$$M = \begin{bmatrix} m_{11} & 0 & 0 \\ m_{21} & m_{22} & m_{23} \\ m_{31} & m_{32} & m_{33} \end{bmatrix}$$

The parallelepiped that is the image of the unit cube under M is the parallelepiped that has the three column vectors of M as three of its edges (see Figure 8.10).

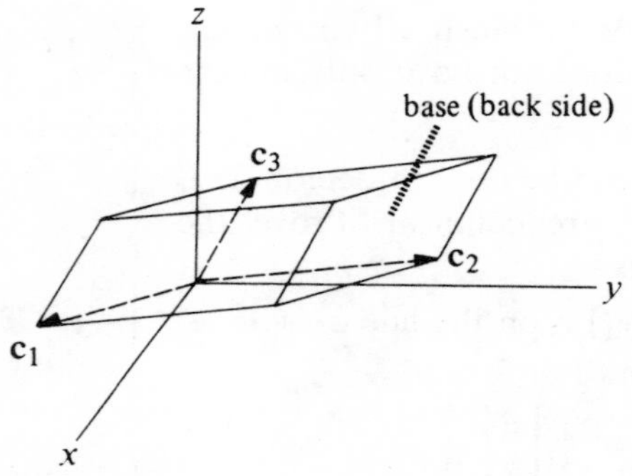

Figure 8.10

In this particular case the vectors

$$\mathbf{c}_2 = \begin{bmatrix} 0 \\ m_{22} \\ m_{32} \end{bmatrix} \quad \text{and} \quad \mathbf{c}_3 = \begin{bmatrix} 0 \\ m_{23} \\ m_{33} \end{bmatrix}$$

lie in the yz plane. We will consider the face containing these two vectors to be the base of the parallelepiped. The height of this solid is just the (perpendicular) distance between the base and the face opposite it. Since both of these faces are parallel to the yz plane, they are perpendicular to the x axis. So the height (up to a sign) of this parallelepiped is just the x coordinate of the vector

$$\mathbf{c}_1 = \begin{bmatrix} m_{11} \\ m_{21} \\ m_{31} \end{bmatrix}$$

Therefore, we can find the volume (up to a sign) of the parallelepiped having edges $\mathbf{c}_1$, $\mathbf{c}_2$, and $\mathbf{c}_3$ as follows:

$$\begin{aligned}
\text{Volume} &= (\text{area of base})(\text{height}) \\
&= \begin{vmatrix} m_{22} & m_{23} \\ m_{32} & m_{33} \end{vmatrix} (m_{11}) \qquad \text{up to a sign} \\
&= m_{11}M_{11} \\
&= m_{11}M_{11} - 0M_{12} + 0M_{13} \\
&= \det(M)
\end{aligned}$$

So we have proved the following special case of Theorem 8.15.

THEOREM 8.16 If

$$M = \begin{bmatrix} m_{11} & 0 & 0 \\ m_{21} & m_{22} & m_{23} \\ m_{31} & m_{32} & m_{33} \end{bmatrix}$$

then the absolute value of $\det(M)$ represents the volume of the parallelepiped that is the image of the unit cube under M.

The way we will attack the general problem of proving Theorem 8.15 is to show how to convert an arbitrary matrix to a matrix having zeros in the second and third positions of the top row in such a way that we can keep track of how the determinant changes. To do this we will use column operations. These are analogues of the row operations we have used so extensively.

We will discuss two types of column operations. The first is multiplying a column of a matrix by a constant a. It is easy to see that if a column of a matrix is multiplied by a scalar a, then the volume of the parallelepiped associated with the resulting matrix is a times as large as the volume of the parallelepiped associated with the original matrix.

The other column operation we consider is adding a multiple of one column to another. This does not change the volume of the associated parallelepiped. To see why this is true, we examine this in a slightly more general context. We consider the parallelepiped with edges $\mathbf{c}_1$, $\mathbf{c}_2$, $\mathbf{c}_3$ as shown in Figure 8.11. For convenience, we will call the face having edges $\mathbf{c}_1$ and $\mathbf{c}_2$ the base. Then the height of the parallelepiped is the (perpendicular) distance between the plane spanned by $\mathbf{c}_1$ and $\mathbf{c}_2$ (this is the plane containing the base) and the plane parallel to this plane and containing the tip of the vector $\mathbf{c}_3$ (this is the plane containing the face of the parallelepiped opposite the base).

Now all parallelepipeds that have the parallelogram formed by $\mathbf{c}_1$ and $\mathbf{c}_2$ as base and fit between these two planes (have a face in each plane) have

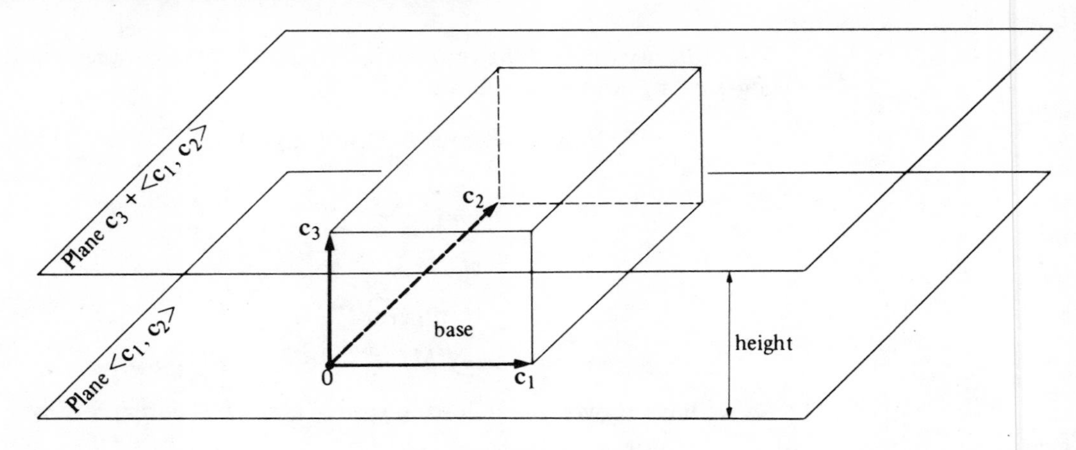

Figure 8.11

the same height and the same base. Therefore, they have the same volume. The vector equation of the plane that contains the face of this parallelepiped opposite the base is

$$\mathbf{v} = \mathbf{c}_3 + (a\mathbf{c}_1 + b\mathbf{c}_2) \qquad \text{where } a \text{ and } b \text{ are scalars}$$

(see Section 4.1). Therefore any parallelepiped that has edges $\mathbf{c}_1$, $\mathbf{c}_2$, and $\mathbf{v}$, where the tip of $\mathbf{v}$ is in this plane, has the same base and the same height as the parallelepiped with edges $\mathbf{c}_1$, $\mathbf{c}_2$, and $\mathbf{c}_3$. So the parallelepiped having edges $\mathbf{c}_1$, $\mathbf{c}_2$, $\mathbf{c}_3$ and the parallelepiped having edges $\mathbf{c}_1$, $\mathbf{c}_2$, $(\mathbf{c}_3 + a\mathbf{c}_1 + b\mathbf{c}_2)$ have the same volume. Restating this in terms of matrices, we have: Adding a multiple of one or more columns of a matrix to another column of that matrix does not change the volume of the associated parallelepiped.

Now we introduce notation for column operations. This notation is like the notation for row operations. We denote the column operation of replacing the jth column $\mathbf{c}_j$ of a matrix by $(a\mathbf{c}_i + b\mathbf{c}_j)$ where a and b are scalars and $\mathbf{c}_i$ is the ith column of the matrix, by

$$\begin{matrix} a & b \end{matrix}$$

$$\begin{bmatrix} \mathbf{c}_i & \mathbf{c}_j \end{bmatrix} \quad \begin{bmatrix} \mathbf{c}_i & (a\mathbf{c}_i + b\mathbf{c}_j) \end{bmatrix}$$

EXAMPLE 1 We illustrate a typical column operation.

$$\begin{matrix} 2 & 3 \end{matrix}$$

$$\begin{bmatrix} 3 & -2 & 2 \\ 2 & 1 & 2 \\ 4 & 3 & 5 \end{bmatrix} \quad \begin{bmatrix} 3 & 0 & 2 \\ 2 & 7 & 2 \\ 4 & 17 & 5 \end{bmatrix}$$

■

We will now see how the volume of the parallelepiped associated with a 3 × 3 matrix is changed by a column operation.

The operation

$$\begin{matrix} a & b \end{matrix}$$

$$\begin{bmatrix} \mathbf{c}_i & \mathbf{c}_j \end{bmatrix} \qquad \begin{bmatrix} \mathbf{c}_i & (a\mathbf{c}_i + b\mathbf{c}_j) \end{bmatrix}$$

can be expressed as a sequence of two simple column operations as

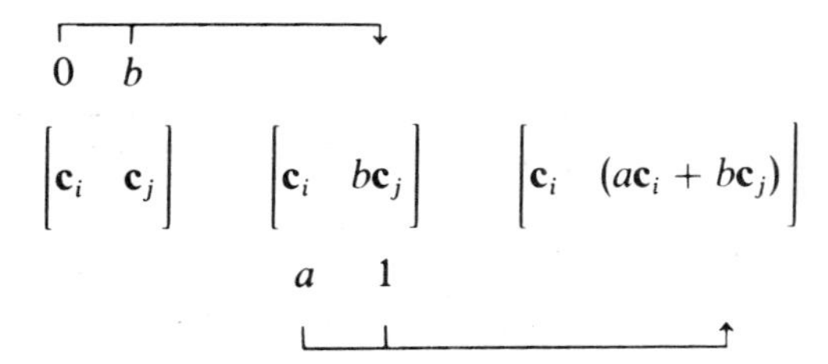

If we let v stand for the volume of the parallelepiped associated with the matrix on the left, then we see that the volume of the parallelepiped associated with the second matrix is bv since we have multiplied one of the columns by b. Furthermore, since the second simple column operation is adding a multiple of one of the columns of the second matrix to another column, the volume of the parallelepipeds associated with the second and third matrices are the same. So we have proved Theorem 8.17.

THEOREM 8.17 If the 3 × 3 matrix M is column reduced to the matrix P by the following operation

$$\begin{matrix} a & \textcircled{b} \end{matrix}$$

$$M = \begin{bmatrix} \mathbf{c}_i & \mathbf{c}_j \end{bmatrix} \qquad \begin{bmatrix} \mathbf{c}_i & (a\mathbf{c}_i + b\mathbf{c}_j) \end{bmatrix} = P$$

then the volume of the parallelepiped associated with P is b times as large as the volume of the parallelepiped associated with M.

Theorem 8.17 is the tool we need to find the volume of the parallelepiped associated with an arbitrary 3 × 3 matrix.

Let

$$M = \begin{bmatrix} m_{11} & m_{12} & m_{13} \\ m_{21} & m_{22} & m_{23} \\ m_{31} & m_{32} & m_{33} \end{bmatrix} \qquad \text{For technical reasons assume } m_{11} \neq 0.$$

To compute the volume of the parallelepiped associated with M we use column operations to reduce M to a matrix P having 0 as its second and third entries of the first row. We then can use Theorems 8.16 and 8.17 to find the volume of the parallelepiped associated with M.

We column reduce M twice:

$$-m_{13} \quad \textcircled{m_{11}}$$

$$-m_{12} \quad \textcircled{m_{11}}$$

$$M = \begin{bmatrix} m_{11} & m_{12} & m_{13} \\ m_{21} & m_{22} & m_{23} \\ m_{31} & m_{32} & m_{33} \end{bmatrix} \quad \begin{bmatrix} m_{11} & 0 & 0 \\ m_{21} & (m_{11}m_{22} - m_{12}m_{21}) & (m_{11}m_{23} - m_{13}m_{21}) \\ m_{31} & (m_{11}m_{32} - m_{12}m_{31}) & (m_{11}m_{33} - m_{13}m_{31}) \end{bmatrix} = P$$

So, applying Theorem 8.17, we have

(1) Volume of parallelepiped associated with P

$$= m_{11}^2 \text{ (volume of parallelepiped associated with } M)$$

Using Theorem 8.16 we find that the volume (up to a sign) of the parallelepiped associated with P is

$$m_{11}[(m_{11}m_{22} - m_{12}m_{21})(m_{11}m_{33} - m_{13}m_{31}) - (m_{11}m_{23} - m_{13}m_{21})(m_{11}m_{32} - m_{12}m_{31})]$$

Expressing this as a polynomial in powers of m_{11} gives:

$$m_{11}[m_{11}^2(m_{22}m_{33} - m_{23}m_{32}) + m_{11}(-m_{22}m_{13}m_{31} - m_{12}m_{21}m_{33} + m_{23}m_{12}m_{31} + m_{32}m_{13}m_{21}) + 1(m_{12}m_{21}m_{13}m_{31} - m_{13}m_{21}m_{12}m_{31})]$$

This becomes

$$m_{11}^2[m_{11}(m_{22}m_{33} - m_{23}m_{32}) + m_{13}(m_{21}m_{32} - m_{22}m_{31}) - m_{12}(m_{21}m_{33} - m_{23}m_{31})]$$

which is equal to $m_{11}^2(m_{11}M_{11} - m_{12}M_{12} + m_{13}M_{13})$

So we see that the volume (up to a sign) associated with the matrix P is

(2) $$m_{11}^2(m_{11}M_{11} - m_{12}M_{12} + m_{13}M_{13}) = m_{11}^2(\det(M))$$

Therefore, from Equations (1) and (2) we have

$$m_{11}^2(\text{volume associated with } M) = \text{volume associated with } P$$
$$= m_{11}^2\,|\det(M)|$$

So, $\quad |\det(M)| = \text{volume of the parallelepiped associated with } M$

This concludes the proof of the fact that if M is a 3×3 matrix, then the absolute value of the determinant of M represents the volume of the parallelepiped formed by the columns of M. Using the same arguments as in Section 8.2, it is easily shown that $|\det(M)|$ is also the factor by which volume is magnified (shrunk) under the action of M.

Inverses

Our second major topic is showing why the method of inverting 3×3 matrices using determinants (Theorem 8.12) works. Let M be a nonsingular 3×3 matrix. To find the inverse of M, we begin as in the 2×2 case (Theorem 8.7), by trying to find a matrix N such that

$$MN = \begin{bmatrix} d_1 & 0 & 0 \\ 0 & d_2 & 0 \\ 0 & 0 & d_3 \end{bmatrix} \tag{3}$$

Once such a matrix N is found, it will be a simple matter to find M^{-1}—just divide each column of N by the d_i in the corresponding column.

To find N, we break M up into its rows. Each row of M is itself a 1×3 matrix. Let the rows of M be R_1, R_2, and R_3.

$$M = \begin{bmatrix} m_{11} & m_{12} & m_{13} \\ m_{21} & m_{22} & m_{23} \\ m_{31} & m_{32} & m_{33} \end{bmatrix} = \begin{bmatrix} R_1 \\ R_2 \\ R_3 \end{bmatrix}$$

We break the matrix N down into its columns

$$N = \begin{bmatrix} \mathbf{c}_1 & \mathbf{c}_2 & \mathbf{c}_3 \end{bmatrix}$$

With N and M in this form, equation (3) becomes

$$MN = \begin{bmatrix} R_1 \\ R_2 \\ R_3 \end{bmatrix}\begin{bmatrix} \mathbf{c}_1 & \mathbf{c}_2 & \mathbf{c}_3 \end{bmatrix} = \begin{bmatrix} d_1 & 0 & 0 \\ 0 & d_2 & 0 \\ 0 & 0 & d_3 \end{bmatrix}$$

This equation is equivalent to the following nine equations:

$$\begin{array}{lll} R_1\mathbf{c}_1 = d_1 & R_1\mathbf{c}_2 = 0 & R_1\mathbf{c}_3 = 0 \\ R_2\mathbf{c}_1 = 0 & R_2\mathbf{c}_2 = d_2 & R_2\mathbf{c}_3 = 0 \\ R_3\mathbf{c}_1 = 0 & R_3\mathbf{c}_2 = 0 & R_3\mathbf{c}_3 = d_3 \end{array}$$

To put this more concisely, we are looking for vectors $\mathbf{c}_1$, $\mathbf{c}_2$, $\mathbf{c}_3$ such that

$\mathbf{c}_2$ and $\mathbf{c}_3$ are in the null space of R_1 (first row)

$\mathbf{c}_1$ and $\mathbf{c}_3$ are in the null space of R_2 (second row)

$\mathbf{c}_1$ and $\mathbf{c}_2$ are in the null space of R_3 (third row)

Although finding three such vectors seems to be just as difficult as our original problem, it is actually much simpler. To see this we use a very neat application of the determinant.

The matrix

$$M(2, 2, 3) = \begin{bmatrix} R_2 \\ R_2 \\ R_3 \end{bmatrix} = \begin{bmatrix} m_{21} & m_{22} & m_{23} \\ m_{21} & m_{22} & m_{23} \\ m_{31} & m_{32} & m_{33} \end{bmatrix}$$

has two rows that are the same and so (see Exercise 5 in Section 8.3)

$$\det(M(2, 2, 3)) = 0$$

Similarly, we see that if

$$M(3, 2, 3) = \begin{bmatrix} R_3 \\ R_2 \\ R_3 \end{bmatrix} = \begin{bmatrix} m_{31} & m_{32} & m_{33} \\ m_{21} & m_{22} & m_{23} \\ m_{31} & m_{32} & m_{33} \end{bmatrix}$$

then

$$\det(M(3, 2, 3)) = 0$$

Expanding $\det(M(2, 2, 3))$ along the top row gives

$$\begin{aligned} 0 &= \det(M(2, 2, 3)) \\ &= m_{21}(m_{22}m_{33} - m_{23}m_{32}) \\ &\quad - m_{22}(m_{21}m_{33} - m_{23}m_{31}) \\ &\quad + m_{23}(m_{21}m_{32} - m_{22}m_{31}) \\ &= m_{21}M_{11} - m_{22}M_{12} + m_{23}M_{13} \end{aligned}$$

where M_{11}, M_{12}, and M_{13} are cofactors of the original matrix M.

Similarly, expanding $\det(M(3, 2, 3))$ gives

$$0 = m_{31}M_{11} - m_{32}M_{12} + m_{33}M_{13}$$

Writing these equations as the product of a 1×3 matrix and a column vector we have

$$[0] = [m_{21} \quad m_{22} \quad m_{23}]\begin{bmatrix} M_{11} \\ -M_{12} \\ M_{13} \end{bmatrix} \quad \text{and} \quad [0] = [m_{31} \quad m_{32} \quad m_{33}]\begin{bmatrix} M_{11} \\ -M_{12} \\ M_{13} \end{bmatrix}$$

So we see that the vector $\begin{bmatrix} M_{11} \\ -M_{12} \\ M_{13} \end{bmatrix}$

is in the null space of each of the row matrices R_2 and R_3. So we choose

$$\mathbf{c}_1 = \begin{bmatrix} M_{11} \\ -M_{12} \\ M_{13} \end{bmatrix}$$

In the same way we can expand the determinants of the singular matrices

$$M(1, 1, 3) = \begin{bmatrix} R_1 \\ R_1 \\ R_3 \end{bmatrix} = \begin{bmatrix} m_{11} & m_{12} & m_{13} \\ m_{11} & m_{12} & m_{13} \\ m_{31} & m_{32} & m_{33} \end{bmatrix}$$

and

$$M(3, 1, 3) = \begin{bmatrix} R_3 \\ R_1 \\ R_3 \end{bmatrix} = \begin{bmatrix} m_{31} & m_{32} & m_{33} \\ m_{11} & m_{12} & m_{13} \\ m_{31} & m_{32} & m_{33} \end{bmatrix}$$

to show that the vector $\mathbf{c}_2 = \begin{bmatrix} M_{21} \\ -M_{22} \\ M_{23} \end{bmatrix}$

is in the null space of each of the 1×3 matrices R_1 and R_3.

Similarly, we can see that

$$\mathbf{c}_3 = \begin{bmatrix} M_{31} \\ -M_{32} \\ M_{33} \end{bmatrix}$$

is in the null spaces of R_1 and R_2 by expanding the determinants of the matrices

$$M(1, 1, 2) = \begin{bmatrix} R_1 \\ R_1 \\ R_2 \end{bmatrix} = \begin{bmatrix} m_{11} & m_{12} & m_{13} \\ m_{11} & m_{12} & m_{13} \\ m_{21} & m_{22} & m_{23} \end{bmatrix}$$

and

$$M(2, 1, 2) = \begin{bmatrix} R_2 \\ R_1 \\ R_2 \end{bmatrix} = \begin{bmatrix} m_{21} & m_{22} & m_{23} \\ m_{11} & m_{12} & m_{13} \\ m_{21} & m_{22} & m_{23} \end{bmatrix}$$

So we see that the matrix N is

$$N = \begin{bmatrix} M_{11} & M_{21} & M_{31} \\ -M_{12} & -M_{22} & -M_{32} \\ M_{13} & M_{23} & M_{33} \end{bmatrix}$$

To find M^{-1} we need only find d_1, d_2, and d_3.

$$[d_1] = R_1 \mathbf{c}_1 = [m_{11} \quad m_{12} \quad m_{13}] \begin{bmatrix} M_{11} \\ -M_{12} \\ M_{13} \end{bmatrix} = [\det(M)]$$

$$[d_2] = R_2 \mathbf{c}_2 = [m_{21} \quad m_{22} \quad m_{23}] \begin{bmatrix} M_{21} \\ -M_{22} \\ M_{23} \end{bmatrix} = [-\det(M)]$$

(This is expansion along the second row of M—see Theorem 8.13.)

$$[d_3] = R_3 \mathbf{c}_3 = [m_{31} \quad m_{32} \quad m_{33}] \begin{bmatrix} M_{31} \\ -M_{32} \\ M_{33} \end{bmatrix} = [\det(M)]$$

(This is expansion along the third row of M.)

So $MN = \begin{bmatrix} \det(M) & 0 & 0 \\ 0 & -\det(M) & 0 \\ 0 & 0 & \det(M) \end{bmatrix} = \det(M) \begin{bmatrix} 1 & 0 & 0 \\ 0 & -1 & 0 \\ 0 & 0 & 1 \end{bmatrix}$

Thus

$$M^{-1} = \frac{1}{\det(M)} \begin{bmatrix} \mathbf{c}_1 & -\mathbf{c}_2 & \mathbf{c}_3 \end{bmatrix} = \frac{1}{\det(M)} \begin{bmatrix} M_{11} & -M_{21} & M_{31} \\ -M_{12} & M_{22} & -M_{32} \\ M_{13} & -M_{23} & M_{33} \end{bmatrix}$$

which is Theorem 8.12.

Exercise Set 8.4

1. Column reduce each of the matrices below to obtain matrices with zeros in the second and third position of the top row. Then find the determinant of the original matrix and the determinant of the reduced matrix by expansion by minors. Finally, compute the determinant of the original matrix by the method of Theorem 8.17 and compare your answer with the answer you got by expansion by minors.

a. $\begin{bmatrix} 1 & 2 & -3 \\ 4 & 5 & -1 \\ 2 & 1 & 3 \end{bmatrix}$ **b.** $\begin{bmatrix} 2 & -5 & 3 \\ 1 & -1 & 1 \\ 4 & 5 & 2 \end{bmatrix}$

★ **2.** For 3×3 matrices there is another method for showing that the determinant represents a (oriented) volume. Since this method involves the cross product, it cannot be generalized to higher dimensions like the method discussed in this section.

a. Let $\mathbf{v}_1$, $\mathbf{v}_2$, $\mathbf{v}_3$ be three vectors of $\mathbb{R}^3$. Show that the product

$$(\mathbf{v}_1 \times \mathbf{v}_2) \cdot \mathbf{v}_3$$

represents the volume (up to a sign) of the parallelepiped formed by $\mathbf{v}_1$, $\mathbf{v}_2$, and $\mathbf{v}_3$. (*Note:* This product is called the **scalar triple product**.)

b. Use expansion by minors to show that if

$$M = \begin{bmatrix} \mathbf{v}_1 & \mathbf{v}_2 & \mathbf{v}_3 \end{bmatrix}$$

then

$$\det(M) = (\mathbf{v}_1 \times \mathbf{v}_2) \cdot \mathbf{v}_3$$

c. Show that $\det(M)$ represents the volume (up to a sign) of the parallelepiped having the column vectors of M as three of its edges.

Applications 8 The Jacobian

In this section we see how the determinant can be applied to a problem of integral calculus. One important aspect of our discussion is that ideas about linear functions will be applied to a question about nonlinear functions.

In what follows we give a detailed discussion of how to convert a double integral given in rectangular coordinates to an equivalent double integral in polar coordinates. This discussion will be fairly general so that the method for converting a double integral in rectangular coordinates to a double integral in any other coordinate system will be an easy generalization of this technique.

We begin by recalling that if P is a point in the plane $\mathbb{R}^2$ with rectangular coordinates (x, y) and polar coordinates (r, θ) (see Figure 8.12), then the relationship between the polar and rectangular coordinates of P is:

$$x = r \cos \theta \qquad \text{and} \qquad y = r \sin \theta$$

Figure 8.12

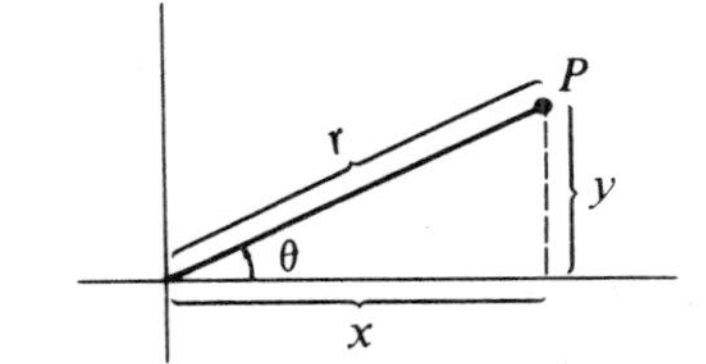

$$P = \begin{cases} (x, y) \text{ in rectangular coordinates} \\ (r, \theta) \text{ in polar coordinates} \end{cases}$$

The discussion that follows will be clearer if we refer to the function $(r \cos \theta)$ that gives the x coordinate of the point P as $\alpha(r, \theta)$, and the function $(r \sin \theta)$ which gives the y coordinate of P as $\beta(r, \theta)$, that is

$$x = \alpha(r, \theta) \qquad \text{and} \qquad y = \beta(r, \theta)$$

Now we can think of the relationship between the polar and rectangular coordinates of points in the plane in a slightly different way. We assume that we have two copies of the plane $\mathbb{R}^2$ (see Figure 8.13). In one, the horizontal axis is called the r axis and the vertical axis is called the θ axis. In the other copy of $\mathbb{R}^2$, the horizontal axis is called the x axis and the vertical axis is called the y axis, as is usually done.

The relationship between the polar coordinates of points and the rectangular coordinates of those points can be thought of as a function C

Figure 8.13

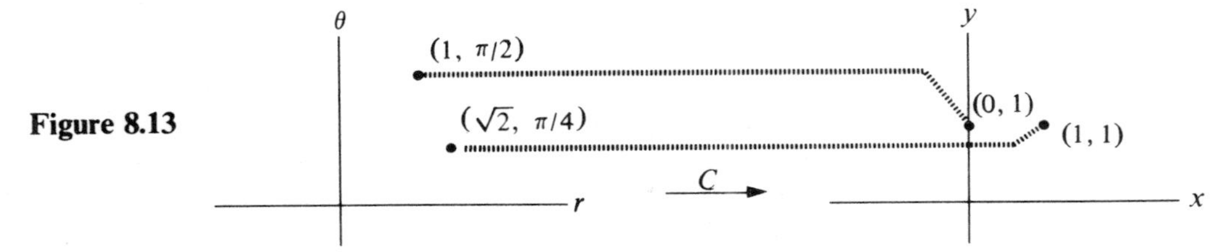

from the points in the $r\theta$ plane to points in the xy plane. So, for example

$$C(1, \pi/2) = (1 \cos(\pi/2), 1 \sin(\pi/2)) = (0, 1)$$

and $$C(\sqrt{2}, \pi/4) = (\sqrt{2} \cos(\pi/4), \sqrt{2} \sin(\pi/4)) = (1, 1)$$

and, in general, $C(r, \theta) = (r \cos \theta, r \sin \theta) = (\alpha(r, \theta), \beta(r, \theta))$

The function C is not linear, but we can ask a question about C that we asked about linear functions; namely, how is the area of a region $\mathscr{S}$ in the $r\theta$ plane related to the area of the image $\mathscr{R}$ of that region under the action of C? See Figure 8.14.

Figure 8.14

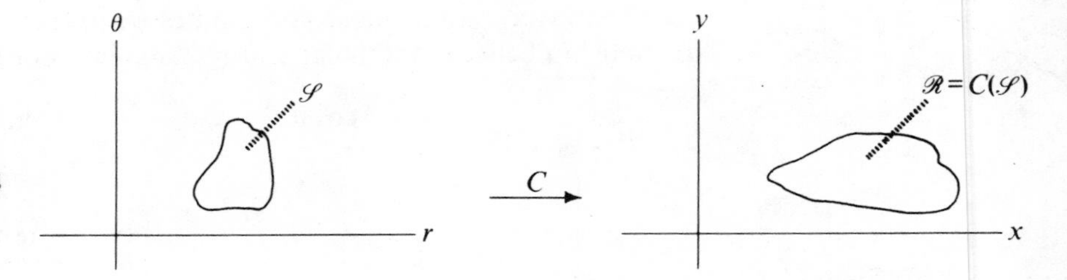

For our purposes it will be better to ask this question in a slightly different way: How is the area of a region $\mathscr{R}$ in the xy plane related to the area of the region $\mathscr{S}$ from which it came (under the action of C)? (Here we need the fact that C^{-1} exists, at least locally). In the following example we compare the areas of several regions $\mathscr{R}$ in the xy plane to the areas of the regions $\mathscr{S}$ in the $r\theta$ plane from which they came (under C).

EXAMPLE 1 We consider several squares in the xy plane and find the regions $\mathscr{S}$ in the $r\theta$ plane that are mapped by C onto these squares. We will compare the areas of these squares to the areas of the regions in the $r\theta$ plane from which they came (under the action of C).

To find the vertices of the regions, we use the formulas for converting rectangular coordinates to polar coordinates (this is really finding C^{-1}). If (x, y) is a point in the first quadrant, then the polar coordinates of this point are

$$r = \sqrt{x^2 + y^2} \qquad \theta = \arctan(y/x)$$

a. Let $\mathscr{R}$ be the square with vertices (1, 0), (2, 0), (2, 1), and (1, 1). The region $\mathscr{S}$ such that $C(\mathscr{S}) = \mathscr{R}$ is shown in Figure 8.15. We note that $\mathscr{S}$ is somewhat rectangular in shape having width approximately 1 and height approximately $\frac{2}{3}$. Thus the area of $\mathscr{S}$ is approximately $\frac{2}{3}$. So the area of the region $\mathscr{R}$ is about $1\frac{1}{2}$ times as large as the area of the region $\mathscr{S}$.

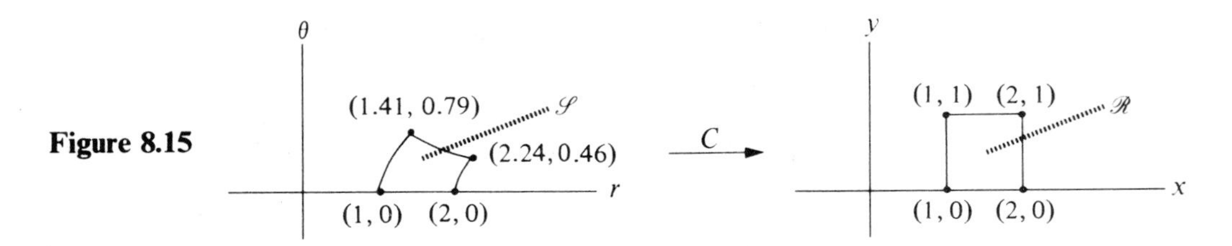

Figure 8.15

b. Let $\mathscr{R}$ be the square with vertices (3, 0), (4, 0), (4, 1), and (3, 1). The region $\mathscr{S}$ such that $C(\mathscr{S}) = \mathscr{R}$ is shown in Figure 8.16. Again we note that $\mathscr{S}$ is somewhat rectangular in shape having width approximately 1 and height approximately $\frac{1}{3}$. So the area of $\mathscr{S}$ is approximately $\frac{1}{3}$. Therefore, the area of the square $C(\mathscr{S}) = \mathscr{R}$ is about three times as large as the area of the region $\mathscr{S}$.

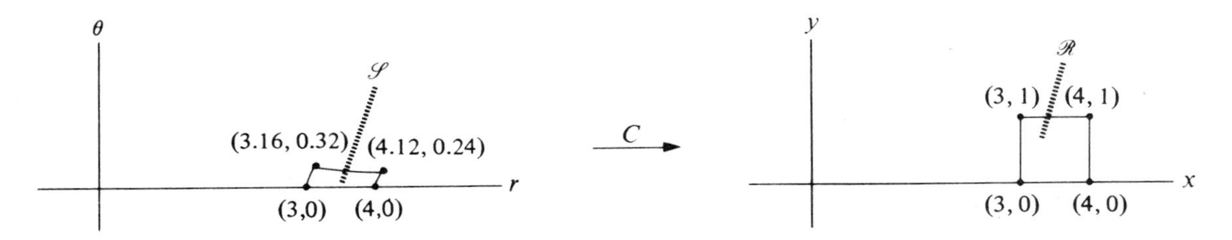

Figure 8.16

c. Let $\mathscr{R}$ be the square with vertices (5, 0), (6, 0), (6, 1), and (5, 1). The region $\mathscr{S}$ such that $C(\mathscr{R}) = \mathscr{S}$ is shown in Figure 8.17. As before we note that $\mathscr{S}$ is somewhat rectangular and has height approximately $\frac{1}{5}$ and width approximately 1. So the area of $\mathscr{S}$ is approximately $\frac{1}{5}$. Therefore, the area of the square $C(\mathscr{S}) = \mathscr{R}$ is about five times as large as the area of the region $\mathscr{S}$.

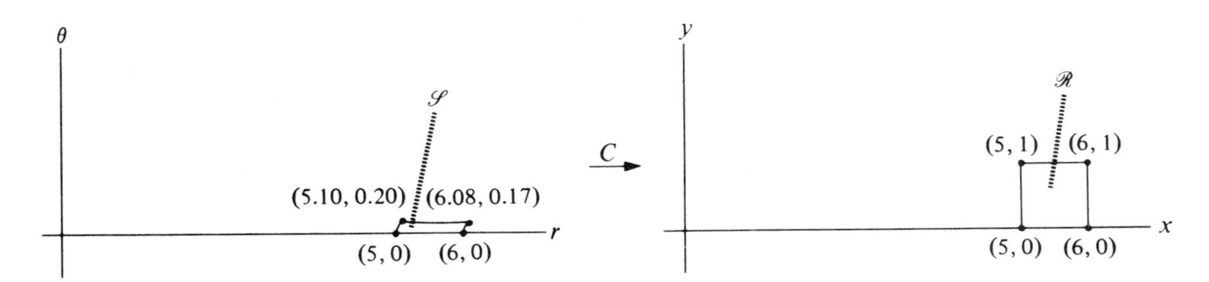

Figure 8.17

From the empirical evidence of this example, we might suspect that the magnification under the action of C increases as we move along the r axis—in fact, it appears as if the relationship between the area of a region $\mathscr{S}$ and the area of the region $C(\mathscr{S})$ is

$$\text{Area of } C(\mathscr{S}) = r(\text{area of } \mathscr{S})$$

We will now analyze this question more rigorously and show that this is indeed the case.

Let $\mathscr{R}$ be the rectangle with vertices (x_1, y_1), (x_2, y_1), (x_2, y_2), and (x_1, y_2). Furthermore, let (r_1, θ_1) and (r_2, θ_2) be the points such that

$$C(r_1, \theta_1) = (\alpha(r_1, \theta_1), \beta(r_1, \theta_1)) = (x_1, y_1)$$
$$C(r_2, \theta_2) = (\alpha(r_2, \theta_2), \beta(r_2, \theta_2)) = (x_2, y_2)$$

Refer to Figure 8.18.

Figure 8.18

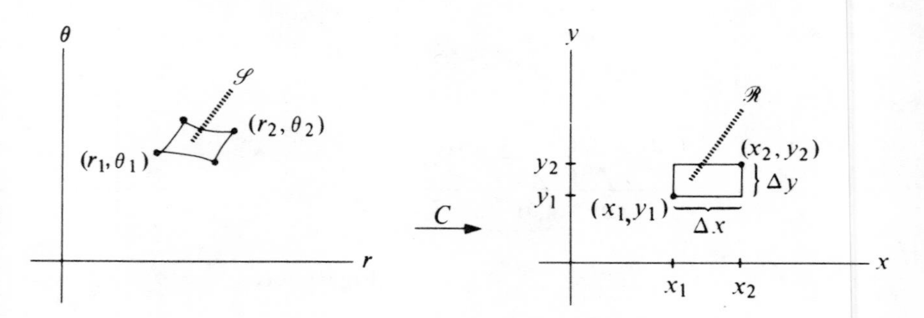

The way we will find the relationship between the areas of the regions $\mathscr{S}$ and $\mathscr{R} = C(\mathscr{S})$ is first to express each of the quantities Δx and Δy in terms of r and θ, and then to use these expressions to write the area $\Delta x \Delta y$ of $\mathscr{R}$ in terms of r and θ.

We begin with Δx:

$$\Delta x = x_2 - x_1 = \alpha(r_2, \theta_2) - \alpha(r_1, \theta_1)$$
$$= \alpha(r_2, \theta_2) + (-\alpha(r_1, \theta_2) + \alpha(r_1, \theta_2)) - \alpha(r_1, \theta_1)$$

(1) $$\Delta x = (\alpha(r_2, \theta_2) - \alpha(r_1, \theta_2)) + (\alpha(r_1, \theta_2) - \alpha(r_1, \theta_1))$$

At this point we need to use a special case of the Mean Value Theorem of calculus. For the convenience of the reader we state this theorem.

THEOREM

Special Case of the Mean Value Theorem

Let $f(u, v)$ be a function of two variables such that the functions $f(u, v_0)$ and $f(u_0, v)$ are continuous for $u_1 \leq u \leq u_2$ and $v_1 \leq v \leq v_2$ and differentiable for $u_1 < u < u_2$ and $v_1 < v < v_2$ (where u_0 and v_0 are fixed). Then

$$f(u_2, v_0) - f(u_1, v_0) = f_u(u^*, v_0)(u_2 - u_1) \qquad \text{where } u_1 < u^* < u_2$$

and $$f(u_0, v_2) - f(u_0, v_1) = f_v(u_0, v^*)(v_2 - v_1) \qquad \text{where } v_1 < v^* < v_2$$

(*Note:* f_u and f_v are the partial derivatives of f with respect to u and v, respectively.)

Applying this theorem with $f = \alpha$, $u = r$, and $v_0 = \theta_2$, we have

$$\alpha(r_2, \theta_2) - \alpha(r_1, \theta_2) = \alpha_r(r^*, \theta_2)(r_2 - r_1) \qquad \text{where } r_1 < r^* < r_2 \tag{2}$$

and $\alpha_r(r^*, \theta_2)$ is the partial derivative of α with respect to r evaluated at the point (r^*, θ_2).

Similarly,

$$\alpha(r_1, \theta_2) - \alpha(r_1, \theta_1) = \alpha_\theta(r_1, \theta^*)(\theta_2 - \theta_1) \qquad \text{where } \theta_1 < \theta^* < \theta_2 \tag{3}$$

From equations **(1)**, **(2)**, and **(3)** we have

$$\Delta x = \alpha_r(r^*, \theta_2)\Delta r + \alpha_\theta(r_1, \theta^*)\Delta\theta \qquad \text{where } \begin{aligned} \Delta r &= r_2 - r_1 \\ \Delta\theta &= \theta_2 - \theta_1 \end{aligned} \tag{4}$$

In exactly the same way we can show that

$$\Delta y = \beta_r(r^{**}, \theta_2)\Delta r + \beta_\theta(r_1, \theta^{**})\Delta\theta \tag{5}$$

We have finally come to the place where we can apply linear algebra. We express the relationships given in equations **(4)** and **(5)** in matrix form as

$$\begin{bmatrix} \Delta x \\ \Delta y \end{bmatrix} = \begin{bmatrix} \alpha_r(r^*, \theta_2) & \alpha_\theta(r_1, \theta^*) \\ \beta_r(r^{**}, \theta_2) & \beta_\theta(r_1, \theta^{**}) \end{bmatrix} \begin{bmatrix} \Delta r \\ \Delta\theta \end{bmatrix} = M \begin{bmatrix} \Delta r \\ \Delta\theta \end{bmatrix} \tag{6}$$

This is a relationship of the form

$$\mathbf{w} = M\mathbf{v}$$

which we have studied extensively (see Figure 8.19).

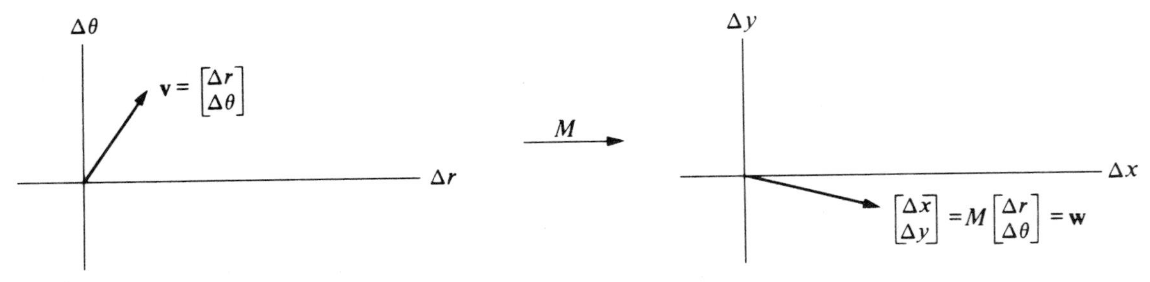

Figure 8.19

However, there is one significant difference. In our previous work, the matrix M was a fixed matrix and did not change. In this case the matrix M can change with each new choice of points (r_1, θ_1) and (r_2, θ_2). However, it can be shown (using advanced calculus) that when Δr and $\Delta\theta$ are sufficiently small, this equation becomes

$$\begin{bmatrix}\Delta x\\ \Delta y\end{bmatrix} \doteq \begin{bmatrix}\alpha_r(r,\theta) & \alpha_\theta(r,\theta)\\ \beta_r(r,\theta) & \beta_\theta(r,\theta)\end{bmatrix}\begin{bmatrix}\Delta r\\ \Delta\theta\end{bmatrix}$$

and this matrix behaves like an ordinary matrix. What this means is that the **absolute value of the determinant of M represents the amount by which the area of regions is magnified (or shrunk) under the action of M.** In particular, we have

$$\Delta x \Delta y \doteq |\det(M)|\Delta r \Delta\theta$$

See Figure 8.20.

Figure 8.20

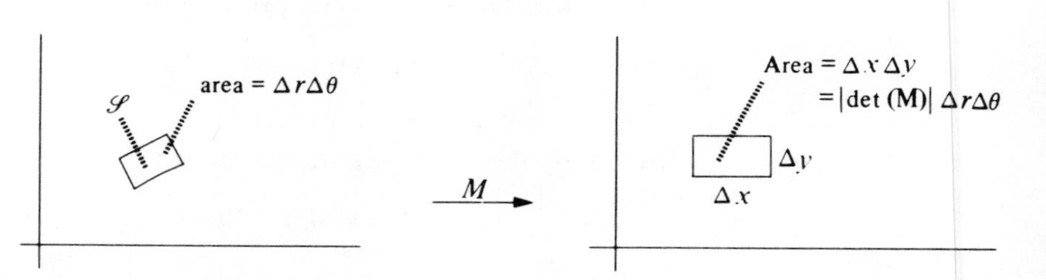

We compute the various partial derivatives involved in M. Since $\alpha(r, \theta) = r\cos\theta$ and $\beta(r, \theta) = r\sin\theta$,

$$\alpha_r(r,\theta) = \frac{\partial}{\partial r}(r\cos\theta) = \cos\theta \quad \text{and} \quad \alpha_\theta(r,\theta) = \frac{\partial}{\partial\theta}(r\cos\theta) = -r\sin\theta$$

$$\beta_r(r,\theta) = \frac{\partial}{\partial r}(r\sin\theta) = \sin\theta \quad \text{and} \quad \beta_\theta(r,\theta) = \frac{\partial}{\partial\theta}(r\sin\theta) = r\cos\theta$$

So

$$|\det(M)| = \begin{vmatrix}\cos\theta & -r\sin\theta\\ \sin\theta & r\cos\theta\end{vmatrix} = |(\cos\theta)(r\cos\theta) - (-r\sin\theta)(\sin\theta)| = |r|$$

Thus

$$|\det(M)| = |r|$$

and

$$\Delta x\Delta y \doteq r\Delta r\Delta\theta \qquad (\text{for } r \geq 0)$$

These ideas can be applied to other changes of coordinates provided that the coordinate change function C satisfies certain conditions. We state these conditions below.

General Hypotheses on C:

We assume that the function $C: \mathbb{R}^2 \to \mathbb{R}^2$

$$(u, v) \to (\alpha(u, v), \beta(u, v))$$

which relates the uv coordinates of a point of $\mathbb{R}^2$ to the xy coordinates of that same point satisfies the following conditions:

1. C is a one-to-one function, that is, C carries pairs of distinct points in the uv plane to pairs of distinct points in the xy plane.
2. The functions α and β are continuous and the partial derivatives α_u, α_v, β_u, β_v exist and are continuous.

For functions C satisfying these general hypotheses, we obtain the following result.

THEOREM 8.18

If the function C that relates the uv coordinates of a point to the xy coordinates of that point satisfies the general hypotheses on C given above, then the quantity

$$J = \left| \det \begin{bmatrix} \alpha_u & \alpha_v \\ \beta_u & \beta_v \end{bmatrix} \right|$$

represents the local magnification of area under the mapping C. In other words, when Δx and Δy are small

$$\Delta x \Delta y \doteq J \Delta u \Delta v$$

DEFINITION 8.9

The quantity J defined in Theorem 8.18 is called the **Jacobian** of the coordinate change function C.

Now we will see how the Jacobian can be applied to double integration. We first see how it can be used to convert a double integral in rectangular coordinates to a double integral in polar coordinates. The integral

$$\iint_{\mathscr{R}} f(x, y)\, dx\, dy$$

is defined to be the limit, as Δx_i and Δy_j approach zero, of sums of the form

$$\sum_{i, j} f(x_i^*, x_j^*) \Delta x_i \Delta y_j$$

where $\Delta x_i = (x_{i+1} - x_i)$, $\Delta y_j = (y_{j+1} - y_j)$, and (x_i^*, y_j^*) is a point in the rectangle $\mathscr{R}_{ij}$ (see Figure 8.21). The sum is taken over all rectangles $\mathscr{R}_{ij}$ that cover the region $\mathscr{R}$.

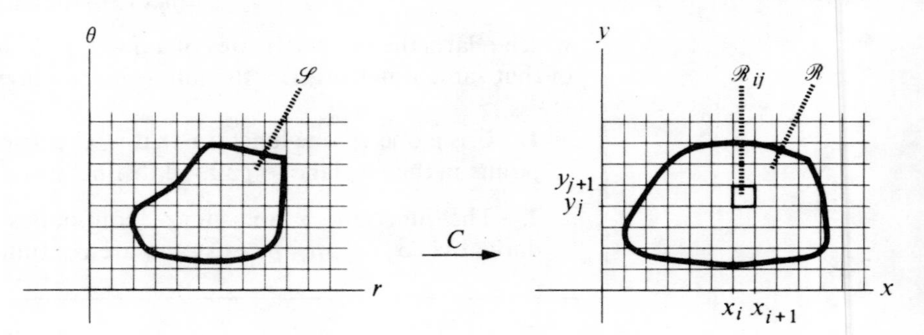

Figure 8.21

We now replace each of the entries in this sum with the equivalent entry in the $r\theta$ coordinates. This sum becomes

$$\sum_{ij} f(x_i^*, y_j^*)\Delta x_i \Delta y_j \doteq \sum_{ij} f(r_i^* \cos \theta_i^*, r_j^* \sin \theta_j^*) r \Delta r_i \Delta \theta_j$$

which is equivalent to the double integral

$$\iint_{\mathscr{S}} f(r \cos \theta, r \sin \theta) r \, dr \, d\theta$$

where $\mathscr{S}$ is the region in the $r\theta$ plane such that $C(\mathscr{S}) = \mathscr{R}$.

We give an example to show how this works.

EXAMPLE 2 The double integral $$\int_0^1 \int_0^{\sqrt{1-x^2}} 1 \, dy \, dx$$

gives the area of the part of the unit circle lying in the first quadrant (see Figure 8.22). Therefore, the value of this double integral should be $\pi/4$.

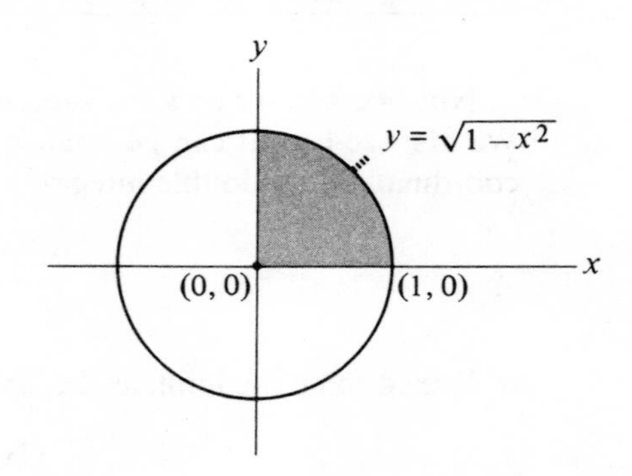

Figure 8.22

We evaluate this integral by changing to polar coordinates.

$$\int_0^1 \int_0^{\sqrt{1-x^2}} 1\, dy\, dx = \int_0^{\pi/2} \int_0^1 1r\, dr\, d\theta$$

$$= \int_0^{\pi/2} (r^2/2)\Big|_0^1 \, d\theta$$

$$= \int_0^{\pi/2} (1/2)\, d\theta$$

$$= \theta/2 \Big|_0^{\pi/2}$$

$$= \pi/4 \quad \blacksquare$$

The method used to find the rule for substitution in polar coordinates can be generalized to find rules for substitution in double integrals for other coordinate systems. The following theorem gives a formal statement of how this is accomplished.

THEOREM 8.19 If the function C that relates the uv coordinates of a point in $\mathbb{R}^2$ to the xy coordinates of that point satisfies the general hypotheses on C given above, then

$$\iint_{\mathscr{R}} f(x, y)\, dy\, dx = \iint_{\mathscr{S}} f(\alpha(u, v), \beta(u, v))J\, du\, dv$$

where $\quad J = \left| \det \begin{bmatrix} \alpha_u & \alpha_v \\ \beta_u & \beta_v \end{bmatrix} \right| \quad$ and $\quad C(\mathscr{S}) = \mathscr{R}$

In particular, the area of the region $\mathscr{R}$ in the xy coordinates is given by the integral

$$\text{Area of } \mathscr{R} = \iint_{\mathscr{R}} 1\, dy\, dx = \iint_{\mathscr{S}} J\, du\, dv$$

In the following example we show how to apply Theorem 2 to evaluate a double integral.

EXAMPLE 3 We wish to evaluate the double integral

$$\iint_{\mathscr{R}} x\, dy\, dx$$

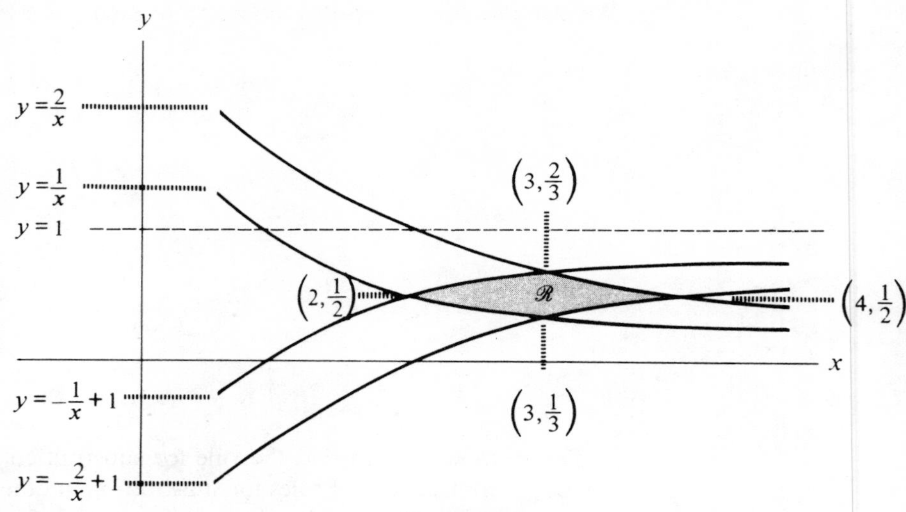

Figure 8.23

where $\mathscr{R}$ is the region bounded by the curves: $y = 1/x$, $y = 2/x$, $y = -1/x + 1$, and $y = -2/x + 1$.

See Figure 8.23.

We choose new coordinates u and v where

$$x = \alpha(u, v) = u \qquad \text{and} \qquad y = \beta(u, v) = v/u$$

Therefore, in this case, the coordinate change mapping C which converts uv coordinates to xy coordinates is

$$C(u, v) = (u, u/v) = (x, y)$$

We now convert this double integral to a double integral in the uv coordinate system:

The first-step is to find the region $\mathscr{S}$ in the uv plane which is mapped by C onto the region $\mathscr{R}$. The boundaries of $\mathscr{S}$ are the curves which are mapped by C onto the boundaries of $\mathscr{R}$. To find the boundaries of $\mathscr{S}$, we write each of the curves which define the boundaries of $\mathscr{R}$ in terms of u and v.

The equation $y = 1/x$ becomes $v/u = 1/u$ or $v = 1$.

The equation $y = 2/x$ becomes $v/u = 2/u$ or $v = 2$.

The equation $y = -1/x + 1$ becomes $v/u = -1/u + 1$ or $u = v + 1$.

The equation $y = -2/x + 1$ becomes $v/u = -2/u + 1$ or $u = v + 2$.

So $\mathscr{S}$ is the parallelogram defined by the inequalities

Figure 8.24

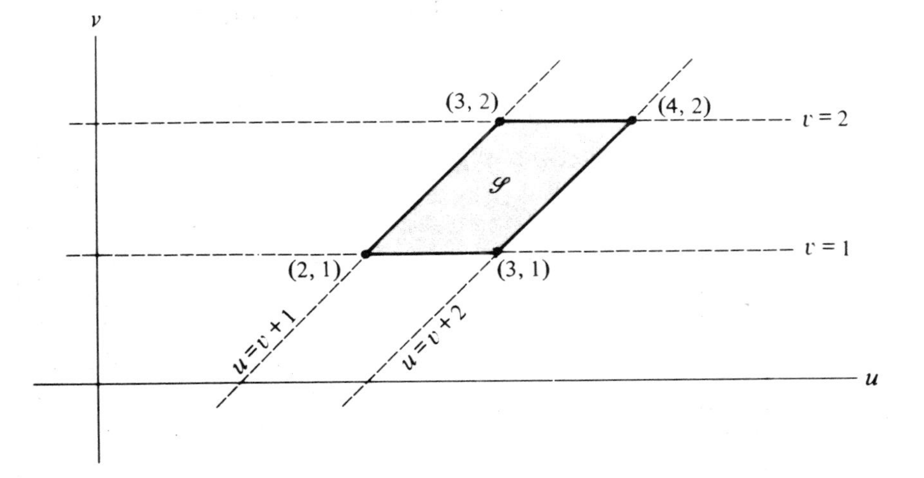

$$(v + 1) \leq u \leq (v + 2) \qquad \text{and} \qquad 1 \leq v \leq 2$$

See Figure 8.24.

The next step is to compute the Jacobian of the coordinate change function C. For this we will need the various partial derivatives of α and β.

$$\text{Since} \quad \alpha = u, \qquad \alpha_u = \frac{\partial u}{\partial u} = 1 \qquad \text{and} \qquad \alpha_v = \frac{\partial u}{\partial v} = 0$$

$$\text{Since} \quad \beta = \frac{v}{u}, \qquad \beta_u = \frac{\partial(v/u)}{\partial u} = -\frac{v}{u^2} \quad \text{and} \quad \beta_v = \frac{\partial(v/u)}{\partial v} = \frac{1}{u}$$

From this we find that the Jacobian J is

$$J = \begin{vmatrix} \alpha_u & \alpha_v \\ \beta_u & \beta_v \end{vmatrix} = \begin{vmatrix} 1 & 0 \\ -v/u^2 & 1/u \end{vmatrix} = |1/u|$$

The final step is to convert the double integral in rectangular coordinates to the equivalent double integral in the uv coordinates.

$$\iint_{\mathscr{R}} x\,dy\,dx = \int_{v=1}^{2} \int_{u=v+1}^{v+2} uJ\,du\,dv$$

$$= \int_{v=1}^{2} \int_{u=v+1}^{v+2} (u)(1/u)\,du\,dv$$

$$= \int_{1}^{2} \int_{v+1}^{v+2} 1\,du\,dv$$

$$= \int_{1}^{2} 1\,dv$$

$$= 1 \quad ■$$

Exercises for Applications 8

1. Let $x = \dfrac{u - v}{2} \qquad y = \dfrac{u + v}{2}$

and let $\mathscr{R}$ be the region shown in Figure 8.25.

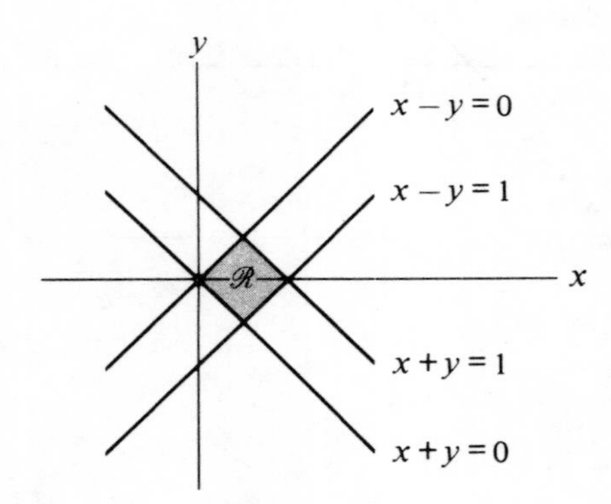

Figure 8.25

a. Find the region $\mathscr{S}$ (see Figure 8.26) in the uv plane that is mapped by C onto $\mathscr{R}$ where

$$C(u, v) = \left(\frac{u - v}{2}, \frac{u + v}{2}\right) = (\alpha(u, v), \beta(u, v))$$

b. Find the matrix M and the Jacobian $J = |\det(M)|$ where

$$M = \begin{bmatrix} \alpha_u(u, v) & \alpha_v(u, v) \\ \beta_u(u, v) & \beta_v(u, v) \end{bmatrix}$$

c. Express the area of $\mathscr{R}$ with a double integral in the xy coordinates and then use the Jacobian to convert this integral to a double integral in the uv coordinates. Evaluate this integral.

d. Compute the area of $\mathscr{R}$ directly by noting that $\mathscr{R}$ is a rectangle and two sides of $\mathscr{R}$ are the vectors

$$\begin{bmatrix} \frac{1}{2} \\ \frac{1}{2} \end{bmatrix} \quad \text{and} \quad \begin{bmatrix} \frac{1}{2} \\ -\frac{1}{2} \end{bmatrix}.$$

e. Compare the answers to parts **c** and **d**.

2. Let $\mathscr{R}$ be the region defined in Exercise 1. Evaluate

$$\iint_{\mathscr{R}} (x + y)^3 (x - y)^2 \, dy \, dx$$

in two ways—directly, and as an integral with respect to u and v.

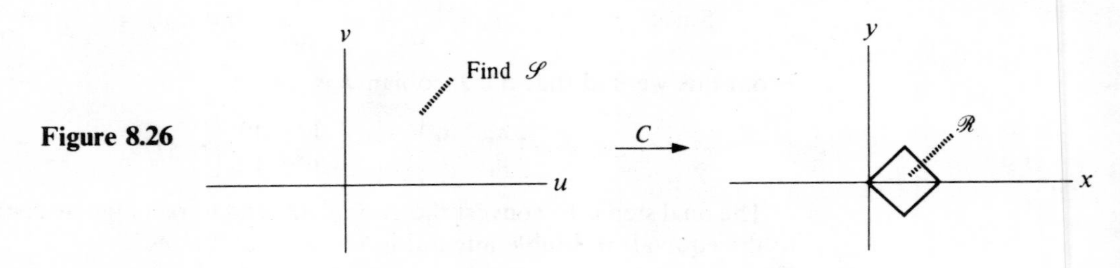

Figure 8.26

Review Exercises

1. Fill in the missing word in the following statements:

a. A matrix that represents a linear transformation which preserves the shape of objects is called a(n) ______________ matrix.

b. A basis for a vector space that consists of mutually perpendicular unit vectors is called a(n) ______________ basis.

c. If the columns of a square matrix are mutually perpendicular unit vectors, then the matrix is called a(n) ______________ matrix.

d. The absolute value of the ______________ of a matrix is the factor by which the size of objects is changed under the action of that matrix.

e. The determinant of the product of two matrices ___is/is not___ the product of the determinants of the two matrices.

f. The determinant of the sum of two matrices ___is/is not___ the sum of the determinants of the two matrices.

g. The formula for "moving a matrix around" the dot product is $M\mathbf{v} \cdot \mathbf{w} = \mathbf{v} \cdot$ ______________.

2. Determine which of the following matrices are orthogonal.

a. $\begin{bmatrix} 1 & 1 \\ -1 & 1 \end{bmatrix}$ **b.** $\begin{bmatrix} 1/\sqrt{2} & 1/\sqrt{2} \\ -1/\sqrt{2} & 1/\sqrt{2} \end{bmatrix}$

c. $\begin{bmatrix} 1/\sqrt{2} & 0 & 2/\sqrt{5} \\ -1/\sqrt{2} & 0 & 1/\sqrt{5} \\ 0 & 1 & -2/\sqrt{5} \end{bmatrix}$

d. $\begin{bmatrix} 1/\sqrt{2} & 2/3 & 1/\sqrt{18} \\ 0 & -1/3 & 4/\sqrt{18} \\ -1/\sqrt{2} & 2/3 & 1/\sqrt{18} \end{bmatrix}$

3. a. Show that if M is an orthogonal matrix, then $\det(M) = \pm 1$.

b. Find an orthogonal matrix M such that $\det(M) = 1$.

c. Find an orthogonal matrix M such that $\det(M) = -1$.

4. Compute the determinants of the following matrices.

a. $\begin{bmatrix} 1 & 1 & 1 \\ 2 & 3 & 4 \\ 3 & 2 & 1 \end{bmatrix}$ **b.** $\begin{bmatrix} 2 & -2 & 0 & 4 \\ -3 & 3 & 1 & 2 \\ 1 & 0 & -2 & 5 \\ 1 & -1 & 3 & 0 \end{bmatrix}$

5. Use determinants to find the inverses (if they exist) for the following matrices:

a. $\begin{bmatrix} 2 & 1 \\ 1 & 1 \end{bmatrix}$ **b.** $\begin{bmatrix} 1 & 2 & 1 \\ -1 & 1 & 2 \\ 3 & 1 & -3 \end{bmatrix}$

c. $\begin{bmatrix} 1 & 2 & 0 \\ -1 & 2 & 2 \\ 4 & 2 & -3 \end{bmatrix}$

6. Suppose that M is an $n \times n$ matrix and $\det(M) = 0$. What does this say about:

a. Rank of M?

b. Nullity of M?

c. Invertibility of M?

d. Existence and uniqueness of solutions of equations of the form $M\mathbf{x} = \mathbf{w}$?

7. Let M be an $n \times n$ matrix with $M^2 = I$.

a. Show that $\det(M) \neq 0$

b. What are the possible values of $\det(M)$?

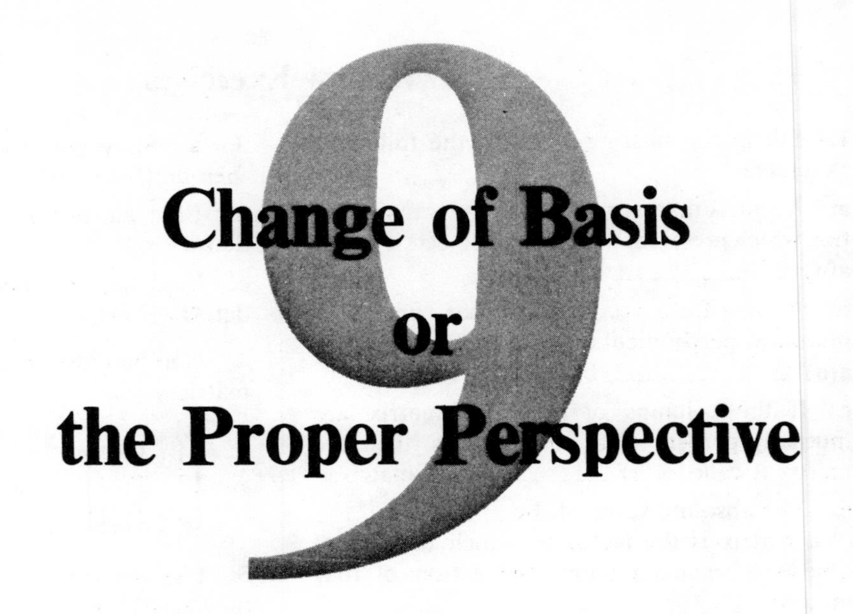

Change of Basis or the Proper Perspective

In this chapter we employ all the concepts and techniques developed in the previous eight chapters and analyze the action of a linear transformation on a vector space. We will find that a great many linear transformations can be described in a very simple way.

Until now, we have always considered the basis for a vector space to be fixed and described the action of a linear transformation in terms of its effects on the vectors of this basis. In many instances things become clearer if we choose a different basis—a basis related to the linear transformation we are studying. To see how to choose such a basis, we will look at the interaction between a linear transformation and the vector space on which it operates (from the point of view of the linear transformation). That is, we will choose a basis for the vector space so that it agrees (in a special way) with the linear transformation.

The following analogy is intended to illustrate a case in which the standard coordinates are not the best ones. Suppose a town is laid out on one side of a river that runs from northeast to southwest. As Figure 9.1 shows, all the roads in this town are laid out either parallel or perpendicular

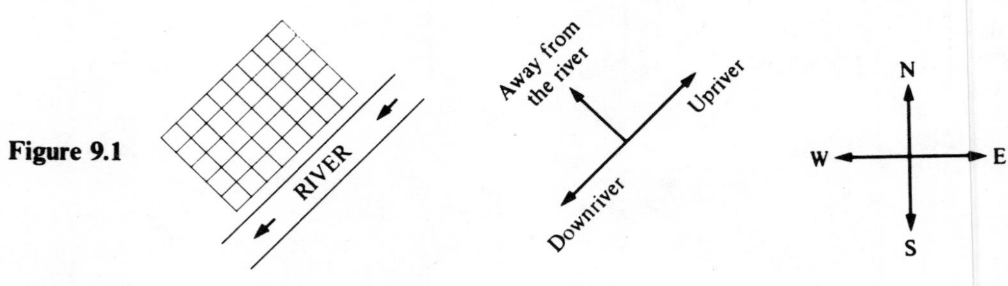

Figure 9.1

to the river. For convenience, the townspeople talk of going upriver instead of going northeast, downriver instead of southwest, and away from the river instead of northwest. The "natural" coordinate system for this town is one based on the direction of the river, not the standard north, south, east, west coordinate system.

Similarly, the standard basis vectors (which lie along the x and y axes) may not be the best basis for describing the action of a linear transformation. For example, consider the linear transformation $T: \mathbb{R}^2 \to \mathbb{R}^2$, *which is reflection about the line* $y = 2x$. *If we describe T in terms of the standard basis we have (see Example 2 in Section 6.3)*

$$T\mathbf{e}_1 = (-3/5)\mathbf{e}_1 + (4/5)\mathbf{e}_2 \qquad T\mathbf{e}_2 = (4/5)\mathbf{e}_1 + (3/5)\mathbf{e}_2$$

However, if we choose the more "natural" basis

$$\mathbf{u}_1 = \begin{bmatrix} 1 \\ 2 \end{bmatrix} \quad \text{and} \quad \mathbf{u}_2 = \begin{bmatrix} -2 \\ 1 \end{bmatrix}$$

T can be described by the very simple equations

$$T\mathbf{u}_1 = \mathbf{u}_1 \qquad T\mathbf{u}_2 = -\mathbf{u}_2$$

In this chapter we will learn how to find such natural bases (which will be called bases of characteristic vectors), *how to use these bases to better understand the action of a linear transformation, and how to apply these very important ideas.*

9.1 Change of Basis

The concept of basis was introduced in Sections 1.5 and 4.3. In this section we will look at this topic again. Until now, when dealing with the vector space $\mathbb{R}^n$, we have usually expressed vectors as linear combinations of the standard basis vectors

$$\mathbf{e}_1 = \begin{bmatrix} 1 \\ 0 \\ 0 \\ \vdots \\ 0 \end{bmatrix}, \quad \mathbf{e}_2 = \begin{bmatrix} 0 \\ 1 \\ 0 \\ \vdots \\ 0 \end{bmatrix}, \quad \ldots, \quad \mathbf{e}_n = \begin{bmatrix} 0 \\ \vdots \\ 0 \\ 0 \\ 1 \end{bmatrix}$$

In many instances a much better understanding of the mathematical situation can be obtained by using bases other than the standard basis. The problem we will discuss is: If a vector $\mathbf{v}$ is expressed as a linear combination of the vectors of one basis, how can we find the expression for $\mathbf{v}$ as a linear combination of the vectors of another basis. To keep things from getting too confusing, we will work with a specific example.

Let $V = \mathbb{R}^2$. We consider two bases of $\mathbb{R}^2$. The standard basis

$$\left\{\mathbf{e}_1 = \begin{bmatrix} 1 \\ 0 \end{bmatrix}, \quad \mathbf{e}_2 = \begin{bmatrix} 0 \\ 1 \end{bmatrix}\right\}$$

and the new basis $$\left\{\mathbf{u}_1 = \begin{bmatrix} 1 \\ 2 \end{bmatrix}, \quad \mathbf{u}_2 = \begin{bmatrix} -3 \\ 3 \end{bmatrix}\right\}$$

For any vector $\mathbf{v}$ of V, we can express $\mathbf{v}$ either as a linear combination of $\mathbf{e}_1$ and $\mathbf{e}_2$

$$\mathbf{v} = s_1\mathbf{e}_1 + s_2\mathbf{e}_2$$

or as a linear combination of $\mathbf{u}_1$ and $\mathbf{u}_2$

$$\mathbf{v} = t_1\mathbf{u}_1 + t_2\mathbf{u}_2$$

(See Figure 9.2).

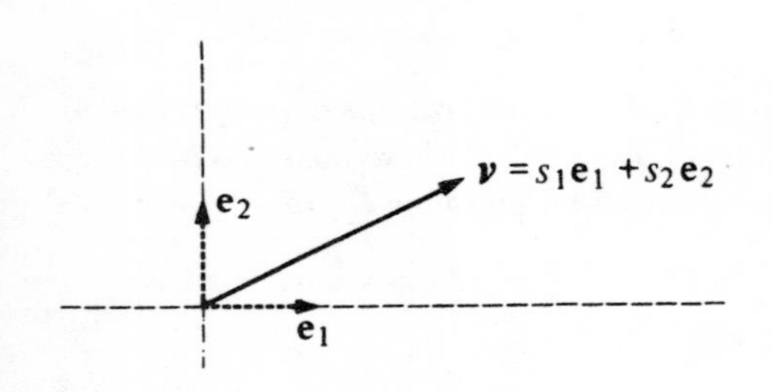

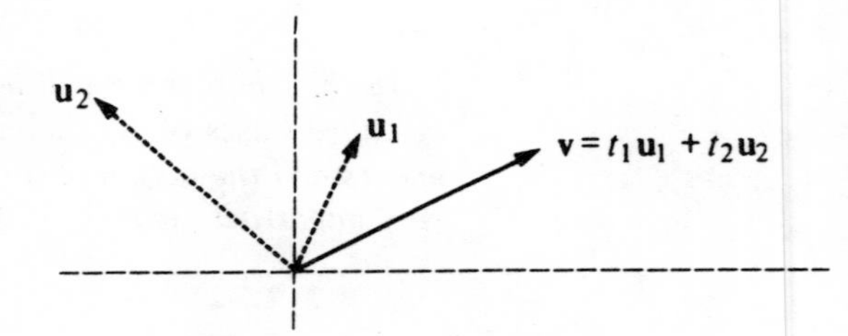

Figure 9.2

The problems we will discuss are:

1. If we know t_1 and t_2, how can we find s_1 and s_2?
2. If we know s_1 and s_2, how can we find t_1 and t_2?

The following is an example of a problem of the first type.

EXAMPLE 1 Let $\mathbf{v} = -\mathbf{u}_1 + 2\mathbf{u}_2$. We will see how to express $\mathbf{v}$ as a linear combination of $\mathbf{e}_1$ and $\mathbf{e}_2$.

Since $\quad \mathbf{u}_1 = 1\mathbf{e}_1 + 2\mathbf{e}_2 \quad$ and $\quad \mathbf{u}_2 = -3\mathbf{e}_1 + 3\mathbf{e}_2$

we can write
$$\begin{aligned}\mathbf{v} &= -\mathbf{u}_1 + 2\mathbf{u}_2 \\ &= -1(1\mathbf{e}_1 + 2\mathbf{e}_2) + 2(-3\mathbf{e}_1 + 3\mathbf{e}_2) \\ &= -7\mathbf{e}_1 + 4\mathbf{e}_2 \quad \blacksquare\end{aligned}$$

This is a simple and straightforward example, but more careful scrutiny of the ideas it contains will allow us to gain enough insight to solve problems of the second type, which are more difficult.

We begin by looking at the way we write vectors. When we write

$$\mathbf{v} = \begin{bmatrix} -7 \\ 4 \end{bmatrix}$$

what we really mean is that (in terms of the basis $\{\mathbf{e}_1, \mathbf{e}_2\}$)

$$\mathbf{v} = -7\mathbf{e}_1 + 4\mathbf{e}_2$$

As we saw in Example 1, the expression for $\mathbf{v}$ in terms of the basis $\mathbf{u}_1$ and $\mathbf{u}_2$ is

$$\mathbf{v} = -\mathbf{u}_1 + 2\mathbf{u}_2$$

We can write $\mathbf{v}$ in terms of its $\mathbf{u}_1\mathbf{u}_2$ coordinates as

$$\mathbf{v} = \begin{bmatrix} -1 \\ 2 \end{bmatrix}_{(\mathbf{u})}$$

EXAMPLE 2 We use the notation just introduced to express the vectors $\mathbf{u}_1$ and $\mathbf{u}_2$ in terms of their $\mathbf{u}_1\mathbf{u}_2$ coordinates as well as their $\mathbf{e}_1\mathbf{e}_2$ coordinates.

$$\mathbf{u}_1 = 1\mathbf{u}_1 + 0\mathbf{u}_2 = \begin{bmatrix} 1 \\ 0 \end{bmatrix}_{(\mathbf{u})} \quad \text{and} \quad \mathbf{u}_1 = 1\mathbf{e}_1 + 2\mathbf{e}_2 = \begin{bmatrix} 1 \\ 2 \end{bmatrix}$$

$$\mathbf{u}_2 = 0\mathbf{u}_1 + 1\mathbf{u}_2 = \begin{bmatrix} 0 \\ 1 \end{bmatrix}_{(\mathbf{u})} \quad \text{and} \quad \mathbf{u}_2 = -3\mathbf{e}_1 + 3\mathbf{e}_2 = \begin{bmatrix} -3 \\ 3 \end{bmatrix}$$

For an arbitrary vector $\quad \mathbf{w} = t_1\mathbf{u}_1 + t_2\mathbf{u}_2$

the coordinate vector for $\mathbf{w}$ in terms of the $\mathbf{u}_1\mathbf{u}_2$ coordinate system is

$$\mathbf{w} = \begin{bmatrix} t_1 \\ t_2 \end{bmatrix}_{(\mathbf{u})}$$

Since $\quad \mathbf{u}_1 = \begin{bmatrix} 1 \\ 2 \end{bmatrix} \quad \text{and} \quad \mathbf{u}_2 = \begin{bmatrix} -3 \\ 3 \end{bmatrix}$

the coordinate vector for $\mathbf{w}$ in terms of the $\mathbf{e}_1\mathbf{e}_2$ coordinate system is

$$t_1\begin{bmatrix} 1 \\ 2 \end{bmatrix} + t_2\begin{bmatrix} -3 \\ 3 \end{bmatrix} = \begin{bmatrix} 1t_1 - 3t_2 \\ 2t_1 + 3t_2 \end{bmatrix}$$

This can be written in matrix form as

$$\begin{bmatrix} 1 & -3 \\ 2 & 3 \end{bmatrix}\begin{bmatrix} t_1 \\ t_2 \end{bmatrix}$$

Therefore, we conclude that

$$\begin{bmatrix} \mathbf{e}_i \text{ coordinate} \\ \text{of } \mathbf{w} \end{bmatrix} = \begin{bmatrix} 1 & -3 \\ 2 & 3 \end{bmatrix} \begin{bmatrix} \mathbf{u}_i \text{ coordinate} \\ \text{of } \mathbf{w} \end{bmatrix}_{(\mathbf{u})}$$

We call the matrix
$$B = \begin{bmatrix} 1 & -3 \\ 2 & 3 \end{bmatrix}$$
the **basis change** matrix. ■

Using the basis change matrix, we do the problem of Example 1 again.

EXAMPLE 3 We express the vector $\mathbf{v} = -1\mathbf{u}_1 + 2\mathbf{u}_2$ in terms of the standard basis vectors $\mathbf{e}_1$ and $\mathbf{e}_2$.

$$\begin{bmatrix} \mathbf{e}_i \text{ coordinate} \\ \text{of } \mathbf{v} \end{bmatrix} = \begin{bmatrix} 1 & -3 \\ 2 & 3 \end{bmatrix} \begin{bmatrix} \mathbf{u}_i \text{ coordinate} \\ \text{of } \mathbf{v} \end{bmatrix}_{(\mathbf{u})}$$
$$= \begin{bmatrix} 1 & -3 \\ 2 & 3 \end{bmatrix} \begin{bmatrix} -1 \\ 2 \end{bmatrix}_{(\mathbf{u})} = \begin{bmatrix} -7 \\ 4 \end{bmatrix}$$

So
$$\mathbf{v} = -7\mathbf{e}_1 + 4\mathbf{e}_2$$
(which is the same answer we got before). This is shown graphically in Figure 9.3. ■

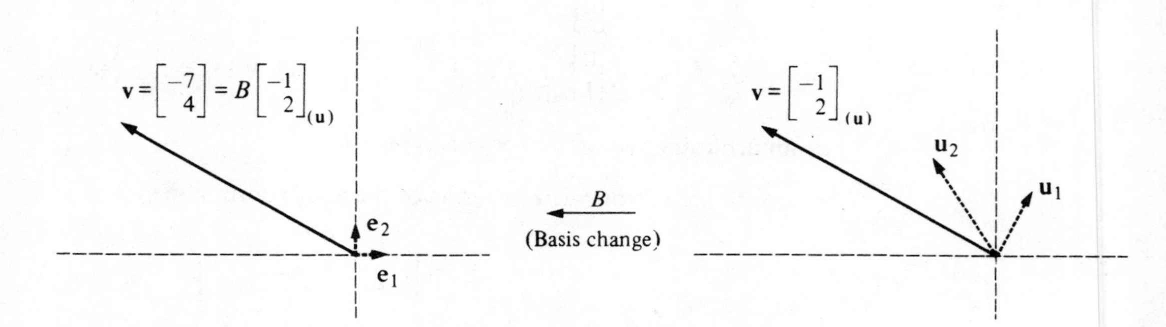

Figure 9.3

These ideas can be extended to higher dimensional spaces to give the following:

DEFINITION 9.1 Let V be a vector space with basis $\{\mathbf{u}_1, \mathbf{u}_2, \ldots, \mathbf{u}_n\}$. If $\mathbf{v} = t_1\mathbf{u}_1 + t_2\mathbf{u}_2 + \cdots + t_n\mathbf{u}_n$, then we say that $t_1, t_2, \ldots, t_n$ are the **coordinates of v relative to the basis** $\{\mathbf{u}_1, \mathbf{u}_2, \ldots, \mathbf{u}_n\}$. Moreover, we represent $\mathbf{v}$ by the coordinate vector

$$\mathbf{v} = \begin{bmatrix} t_1 \\ t_2 \\ \vdots \\ t_n \end{bmatrix}_{(\mathbf{u})}$$

(As before, coordinate vectors having no subscript will represent coordinate vectors relative to the standard basis $\{\mathbf{e}_1, \mathbf{e}_2, \ldots, \mathbf{e}_n\}$.)

THEOREM 9.1 Let $\{\mathbf{u}_1, \mathbf{u}_2, \ldots, \mathbf{u}_n\}$ be a basis of $\mathbb{R}^n$. Given any vector $\mathbf{v}$ of $\mathbb{R}^n$, it can be expressed as a linear combination of the basis vectors $\mathbf{u}_1, \mathbf{u}_2, \ldots, \mathbf{u}_n$ as

$$\mathbf{v} = t_1\mathbf{u}_1 + t_2\mathbf{u}_2 + \cdots + t_n\mathbf{u}_n$$

The vector $\mathbf{v}$ can also be expressed as a linear combination of the standard basis vectors $\mathbf{e}_1, \mathbf{e}_2, \ldots, \mathbf{e}_n$ as:

$$\mathbf{v} = s_1\mathbf{e}_1 + s_2\mathbf{e}_2 + \cdots + s_n\mathbf{e}_n$$

The relationship between the s_i and the t_i is

$$B\begin{bmatrix} t_1 \\ t_2 \\ \vdots \\ t_n \end{bmatrix}_{(\mathbf{u})} = \begin{bmatrix} s_1 \\ s_2 \\ \vdots \\ s_n \end{bmatrix}$$

where B, the basis change matrix, is the matrix that has as its ith column the coordinates of the vector of $\mathbf{u}_i$ relative to the basis $\mathbf{e}_1, \mathbf{e}_2, \ldots, \mathbf{e}_n$. That is, if

$$\mathbf{u}_i = b_{1i}\mathbf{e}_1 + b_{2i}\mathbf{e}_2 + \cdots + b_{ni}\mathbf{e}_n$$

then the ith column of B is

$$\begin{bmatrix} b_{1i} \\ b_{2i} \\ \vdots \\ b_{ni} \end{bmatrix}$$

EXAMPLE 4 Let

$$\mathbf{u}_1 = \begin{bmatrix} 1 \\ -1 \\ 3 \end{bmatrix} \qquad \mathbf{u}_2 = \begin{bmatrix} 1 \\ 2 \\ 1 \end{bmatrix} \qquad \mathbf{u}_3 = \begin{bmatrix} 1 \\ 1 \\ 2 \end{bmatrix}$$

Find the $\mathbf{e}_i$ coordinates of the vector

$$\mathbf{v} = 5\mathbf{u}_1 - 3\mathbf{u}_2 + 2\mathbf{u}_3$$

We form the matrix B

$$B = \begin{bmatrix} 1 & 1 & 1 \\ -1 & 2 & 1 \\ 3 & 1 & 2 \end{bmatrix}$$

Then we apply Theorem 9.1

$$B\begin{bmatrix} 5 \\ -3 \\ 2 \end{bmatrix}_{(\mathbf{u})} = \begin{bmatrix} 1 & 1 & 1 \\ -1 & 2 & 1 \\ 3 & 1 & 2 \end{bmatrix}\begin{bmatrix} 5 \\ -3 \\ 2 \end{bmatrix}_{(\mathbf{u})} = \begin{bmatrix} 4 \\ -9 \\ 16 \end{bmatrix}$$

So $$\mathbf{v} = 4\mathbf{e}_1 - 9\mathbf{e}_2 + 16\mathbf{e}_3 \qquad \blacksquare$$

Theorem 9.1 can be used to see how to find the $\mathbf{u}_i$ coordinates of a vector if you are given its $\mathbf{e}_i$ coordinates. Since

$$B\begin{bmatrix} \mathbf{u}_i \text{ coordinate} \\ \text{of} \\ \mathbf{v} \end{bmatrix}_{(\mathbf{u})} = \begin{bmatrix} \mathbf{e}_i \text{ coordinate} \\ \text{of} \\ \mathbf{v} \end{bmatrix}$$

multiplying both sides of this equation by B^{-1} on the left gives

$$\begin{bmatrix} \mathbf{u}_i \text{ coordinate} \\ \text{of} \\ \mathbf{v} \end{bmatrix}_{(\mathbf{u})} = B^{-1}\begin{bmatrix} \mathbf{e}_i \text{ coordinate} \\ \text{of} \\ \mathbf{v} \end{bmatrix}$$

So we have:

THEOREM 9.2

Using the same notation as in Theorem 9.1,

$$\begin{bmatrix} t_1 \\ t_2 \\ \vdots \\ t_n \end{bmatrix}_{(\mathbf{u})} = B^{-1}\begin{bmatrix} s_1 \\ s_2 \\ \vdots \\ s_n \end{bmatrix}$$

Note: The matrix B is invertible, see Exercise 3.

EXAMPLE 5 Let $V = \mathbb{R}^2$ and $\mathbf{u}_1 = \begin{bmatrix} 1 \\ 2 \end{bmatrix} \quad \mathbf{u}_2 = \begin{bmatrix} -3 \\ 3 \end{bmatrix}$

We find the $\mathbf{u}_1\mathbf{u}_2$ coordinates of the vector $\mathbf{v} = -7\mathbf{e}_1 + 4\mathbf{e}_2$.
From Theorem 9.2 we see that

$$\begin{bmatrix} \mathbf{u}_i \text{ coordinate} \\ \text{of } \mathbf{v} \end{bmatrix}_{(\mathbf{u})} = B^{-1}\begin{bmatrix} \mathbf{e}_i \text{ coordinate} \\ \text{of } \mathbf{v} \end{bmatrix} = B^{-1}\begin{bmatrix} -7 \\ 4 \end{bmatrix}$$

Since $B = \begin{bmatrix} 1 & -3 \\ 2 & 3 \end{bmatrix}$, using Theorem 8.7 we find that $B^{-1} = \begin{bmatrix} 1/3 & 1/3 \\ -2/9 & 1/9 \end{bmatrix}$

Thus $\begin{bmatrix} \mathbf{u}_i \text{ coordinate} \\ \text{of} \\ \mathbf{v} \end{bmatrix}_{(\mathbf{u})} = B^{-1}\begin{bmatrix} -7 \\ 4 \end{bmatrix} = \begin{bmatrix} 1/3 & 1/3 \\ -2/9 & 1/9 \end{bmatrix}\begin{bmatrix} -7 \\ 4 \end{bmatrix} = \begin{bmatrix} -1 \\ 2 \end{bmatrix}_{(\mathbf{u})}$

So

$$\mathbf{v} = -1\mathbf{u}_1 + 2\mathbf{u}_2 \quad \blacksquare$$

Exercise Set 9.1

1. Let $\mathbf{u}_1 = 2\mathbf{e}_1 + \mathbf{e}_2$ and $\mathbf{u}_2 = 5\mathbf{e}_1 + 2\mathbf{e}_2$. Find the $\mathbf{e}_1\mathbf{e}_2$ coordinates of each of the following vectors:

a. $\mathbf{v} = \mathbf{u}_2$ **b.** $\mathbf{v} = 3\mathbf{u}_1 + 7\mathbf{u}_2$

c. $\mathbf{v} = -4\mathbf{u}_1 + 11\mathbf{u}_2$

d. $\mathbf{v} = \begin{bmatrix} 3 \\ 7 \end{bmatrix}_{(\mathbf{u})}$ **e.** $\mathbf{v} = \begin{bmatrix} -9 \\ 11 \end{bmatrix}_{(\mathbf{u})}$

2. Let $\mathbf{u}_1\mathbf{u}_2$ be as in Exercise 1. Find the $\mathbf{u}_1\mathbf{u}_2$ coordinates of each of the following vectors.

a. $\mathbf{v} = \mathbf{e}_1$ **b.** $\mathbf{v} = \mathbf{e}_2$

c. $\mathbf{v} = 5\mathbf{e}_1 + 3\mathbf{e}_2$ **d.** $\mathbf{v} = -\mathbf{e}_1 + 7\mathbf{e}_2$

e. $\mathbf{v} = \begin{bmatrix} 3 \\ 7 \end{bmatrix}$ **f.** $\mathbf{v} = \begin{bmatrix} 2 \\ 7 \end{bmatrix}$

★ **3.** Suppose that the vectors $\mathbf{u}_1$ and $\mathbf{u}_2$ form a basis for $\mathbb{R}^2$ where

$\mathbf{u}_1 = b_{11}\mathbf{e}_1 + b_{21}\mathbf{e}_2$ and $\mathbf{u}_2 = b_{12}\mathbf{e}_1 + b_{22}\mathbf{e}_2$

Show that the matrix

$$B = \begin{bmatrix} b_{11} & b_{12} \\ b_{21} & b_{22} \end{bmatrix}$$

has an inverse. (*Hint*: Apply Theorem 7.5′.)

For Exercises 4–12 let

$$\mathbf{u}_1 = \begin{bmatrix} 1 \\ -1 \\ -1 \end{bmatrix} \quad \mathbf{u}_2 = \begin{bmatrix} 2 \\ 4 \\ 5 \end{bmatrix} \quad \mathbf{u}_3 = \begin{bmatrix} -2 \\ 2 \\ 3 \end{bmatrix}$$

Find the $\mathbf{e}_i$ coordinates of each of the vectors listed in Exercises 4–7.

4. $\mathbf{v} = 2\mathbf{u}_1 - 5\mathbf{u}_2 + 7\mathbf{u}_3$

5. $\mathbf{v} = \mathbf{u}_1 + \mathbf{u}_2 - 3\mathbf{u}_3$

6. $\mathbf{v} = \begin{bmatrix} 1 \\ 1 \\ -1 \end{bmatrix}_{(\mathbf{u})}$ **7.** $\mathbf{v} = \begin{bmatrix} 11 \\ -2 \\ 7 \end{bmatrix}_{(\mathbf{u})}$

Find the $\mathbf{u}_i$ coordinates of each of the vectors listed in Exercises 8–12.

8. $\mathbf{v} = \mathbf{e}_2$

9. $\mathbf{v} = 3\mathbf{e}_1 + 7\mathbf{e}_2 - 2\mathbf{e}_3$

10. $\mathbf{v} = -4\mathbf{e}_1 + 11\mathbf{e}_2 + \mathbf{e}_3$

11. $\mathbf{v} = \begin{bmatrix} -4 \\ 11 \\ 1 \end{bmatrix}$ **12.** $\mathbf{v} = \begin{bmatrix} -7 \\ 5 \\ 7 \end{bmatrix}$

9.2 Linear Transformations and Change of Basis

In Section 3.3 we saw that if $T: \mathbb{R}^n \to \mathbb{R}^n$ is a linear transformation, then T is completely determined if the vectors

$$T(\mathbf{e}_1), T(\mathbf{e}_2), \ldots, T(\mathbf{e}_n)$$

are known, where $\mathbf{e}_1, \mathbf{e}_2, \ldots, \mathbf{e}_n$ are the standard basis vectors.

Similarly, if $\{\mathbf{u}_1, \mathbf{u}_2, \ldots, \mathbf{u}_n\}$ is any basis for $\mathbb{R}^n$, using the fact that T is linear, we can easily see that T is completely determined if the vectors

$$T(\mathbf{u}_1), T(\mathbf{u}_2), \ldots, T(\mathbf{u}_n)$$

are known. The following example will illustrate this.

EXAMPLE 1 Let $T: \mathbb{R}^2 \to \mathbb{R}^2$ be reflection about the line $y = 2x$, and let

$$\mathbf{u}_1 = \begin{bmatrix} 1 \\ 2 \end{bmatrix} \quad \text{and} \quad \mathbf{u}_2 = \begin{bmatrix} -2 \\ 1 \end{bmatrix}$$

Then $\{\mathbf{u}_1, \mathbf{u}_2\}$ is a basis for $\mathbb{R}^2$.

Since $\mathbf{u}_1$ lies on the line $y = 2x$, $T(\mathbf{u}_1) = \mathbf{u}_1$, and since $\mathbf{u}_2$ is perpendicular to this line, $T(\mathbf{u}_2) = -\mathbf{u}_2$ (see Figure 9.4).

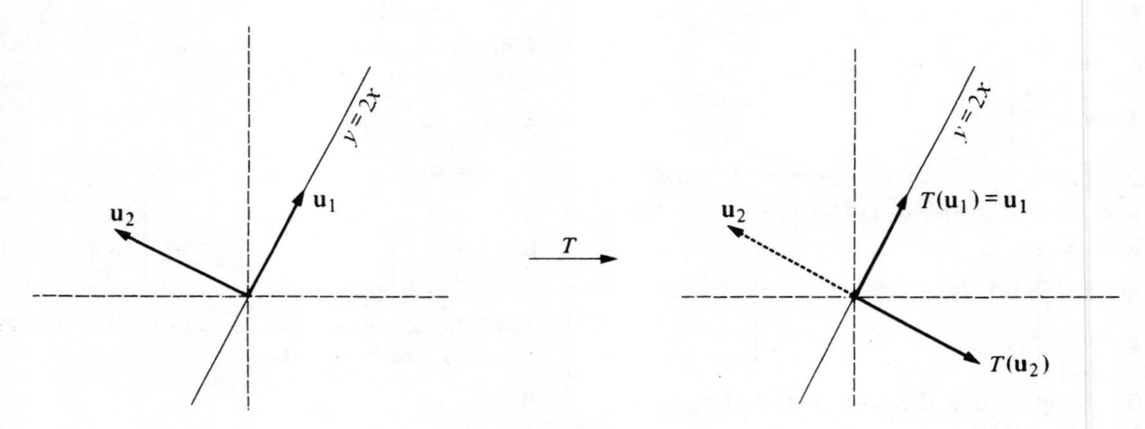

Figure 9.4

Because $\{\mathbf{u}_1, \mathbf{u}_2\}$ is a basis for $\mathbb{R}^2$, any vector $\mathbf{v}$ of $\mathbb{R}^2$ can be written

$$\mathbf{v} = a_1\mathbf{u}_1 + a_2\mathbf{u}_2$$

So

$$\begin{aligned} T(\mathbf{v}) &= T(a_1\mathbf{u}_1 + a_2\mathbf{u}_2) \\ &= a_1T(\mathbf{u}_1) + a_2T(\mathbf{u}_2) \qquad \text{since } T \text{ is linear} \\ &= a_1\mathbf{u}_1 + a_2(-\mathbf{u}_2) \end{aligned}$$

Using the coordinate vector notation introduced in Section 9.1, we can rewrite this as

$$T: \begin{bmatrix} a_1 \\ a_2 \end{bmatrix}_{(\mathbf{u})} \to \begin{bmatrix} a_1 \\ -a_2 \end{bmatrix}_{(\mathbf{u})} \quad \text{or} \quad T: \begin{bmatrix} a_1 \\ a_2 \end{bmatrix}_{(\mathbf{u})} \to a_1\begin{bmatrix} 1 \\ 0 \end{bmatrix}_{(\mathbf{u})} + a_2\begin{bmatrix} 0 \\ -1 \end{bmatrix}_{(\mathbf{u})} \quad \blacksquare$$

The ideas used in this example can be generalized as follows:

If $T: \mathbb{R}^n \to \mathbb{R}^n$ is a linear transformation and $\{\mathbf{u}_1, \mathbf{u}_2, \ldots, \mathbf{u}_n\}$ is a basis for $\mathbb{R}^n$, then any vector $\mathbf{v}$ of $\mathbb{R}^n$ can be written as

$$\mathbf{v} = a_1\mathbf{u}_1 + a_2\mathbf{u}_2 + \cdots + a_n\mathbf{u}_n$$

and, since T is a linear transformation,

$$T(\mathbf{v}) = T(a_1\mathbf{u}_1 + a_2\mathbf{u}_2 + \cdots + a_n\mathbf{u}_n) = a_1T(\mathbf{u}_1) + a_2T(\mathbf{u}_2) + \cdots + a_nT(\mathbf{u}_n)$$

Since the range of T is $\mathbb{R}^n$, we can express every vector in the image space of T as a linear combination of the basis vectors $\mathbf{u}_1, \mathbf{u}_2, \ldots, \mathbf{u}_n$. So, in particular, we can find scalars s_{ij} so that

$$\begin{aligned} T(\mathbf{u}_1) &= s_{11}\mathbf{u}_1 + s_{21}\mathbf{u}_2 + \cdots + s_{n1}\mathbf{u}_n \\ T(\mathbf{u}_2) &= s_{12}\mathbf{u}_1 + s_{22}\mathbf{u}_2 + \cdots + s_{n2}\mathbf{u}_n \\ &\vdots \\ T(\mathbf{u}_n) &= s_{1n}\mathbf{u}_1 + s_{2n}\mathbf{u}_2 + \cdots + s_{nn}\mathbf{u}_n \end{aligned}$$

We can express these relationships in terms of coordinate vectors relative to the $\mathbf{u}_i$ coordinate system (see Section 9.1) as

$$T(\mathbf{u}_1) = \begin{bmatrix} s_{11} \\ s_{21} \\ \vdots \\ s_{n1} \end{bmatrix}_{(\mathbf{u})}, \quad T(\mathbf{u}_2) = \begin{bmatrix} s_{12} \\ s_{22} \\ \vdots \\ s_{n2} \end{bmatrix}_{(\mathbf{u})}, \quad \ldots, \quad T(\mathbf{u}_n) = \begin{bmatrix} s_{1n} \\ s_{2n} \\ \vdots \\ s_{nn} \end{bmatrix}_{(\mathbf{u})}$$

Using this notation, the equation

$$T(a_1\mathbf{u}_1 + a_2\mathbf{u}_2 + \cdots + a_n\mathbf{u}_n) = a_1T(\mathbf{u}_1) + a_2T(\mathbf{u}_2) + \cdots + a_nT(\mathbf{u}_n)$$

can be rewritten as

$$T\begin{bmatrix} a_1 \\ a_2 \\ \vdots \\ a_n \end{bmatrix}_{(\mathbf{u})} = a_1\begin{bmatrix} s_{11} \\ s_{21} \\ \vdots \\ s_{n1} \end{bmatrix}_{(\mathbf{u})} + a_2\begin{bmatrix} s_{12} \\ s_{22} \\ \vdots \\ s_{n2} \end{bmatrix}_{(\mathbf{u})} + \cdots + a_n\begin{bmatrix} s_{1n} \\ s_{2n} \\ \vdots \\ s_{nn} \end{bmatrix}_{(\mathbf{u})}$$

This equation should look familiar. Except for the subscripts $(\mathbf{u})$, it is the same as the equation we use when finding the matrix that represents T (see Theorem 3.2 and Definition 3.8). Until now, when we have talked about a matrix that represents a linear transformation, we meant a matrix representing a linear transformation **relative to the standard basis**. Now we will define what we mean by a matrix relative to a different basis.

DEFINITION 9.2

Let $T: \mathbb{R}^n \to \mathbb{R}^n$ be a linear transformation and let $\{\mathbf{u}_1, \mathbf{u}_2, \ldots, \mathbf{u}_n\}$ be a basis for $\mathbb{R}^n$. Furthermore, let

$$T(\mathbf{u}_1) = s_{11}\mathbf{u}_1 + s_{21}\mathbf{u}_2 + \cdots + s_{n1}\mathbf{u}_n$$
$$T(\mathbf{u}_2) = s_{12}\mathbf{u}_1 + s_{22}\mathbf{u}_2 + \cdots + s_{n2}\mathbf{u}_n$$
$$\vdots \qquad \vdots \qquad \vdots \qquad \vdots$$
$$T(\mathbf{u}_n) = s_{1n}\mathbf{u}_1 + s_{2n}\mathbf{u}_2 + \cdots + s_{nn}\mathbf{u}_n$$

Then the **matrix of T relative to the basis** $\{\mathbf{u}_1, \mathbf{u}_2, \ldots, \mathbf{u}_n\}$ is

$$N = \begin{bmatrix} s_{11} & s_{12} & \cdots & s_{1n} \\ s_{21} & s_{22} & \cdots & s_{2n} \\ \vdots & \vdots & & \vdots \\ s_{n1} & s_{n2} & \cdots & s_{nn} \end{bmatrix}$$

(Sometimes we attach the subscript $(\mathbf{u})$ to the matrix to emphasize the fact that this is the matrix for T relative to the $\mathbf{u}_i$ coordinate system.)

The following theorem summarizes the above discussion.

THEOREM 9.3 Let $T: \mathbb{R}^n \to \mathbb{R}^n$ be the linear transformation represented by the matrix N relative to the basis $\{\mathbf{u}_1, \mathbf{u}_2, \ldots, \mathbf{u}_n\}$. If the $\mathbf{u}_i$ coordinate vector for a vector $\mathbf{v}$ of $\mathbb{R}^n$ is

$$\mathbf{v} = \begin{bmatrix} a_1 \\ a_2 \\ \vdots \\ a_n \end{bmatrix}_{(\mathbf{u})}$$

then the $\mathbf{u}_i$ coordinate vector for $T(\mathbf{v})$ is

$$T(\mathbf{v}) = N\begin{bmatrix} a_1 \\ a_2 \\ \vdots \\ a_n \end{bmatrix}_{(\mathbf{u})}$$

EXAMPLE 2 We find the matrix relative to the $\mathbf{u}_1\mathbf{u}_2$ coordinate system for the transformation T of Example 1. In Example 1 we saw that

$$T(\mathbf{u}_1) = 1\mathbf{u}_1 + 0\mathbf{u}_2$$
$$T(\mathbf{u}_2) = 0\mathbf{u}_1 - 1\mathbf{u}_2$$

So the matrix N that represents T relative to the $\mathbf{u}_1\mathbf{u}_2$ coordinate system is

$$N = \begin{bmatrix} 1 & 0 \\ 0 & -1 \end{bmatrix}$$

We can apply Theorem 9.3 to find $T(\mathbf{v})$ for the vector $\mathbf{v} = 3\mathbf{u}_1 - 4\mathbf{u}_2$.

$$T(\mathbf{v}) = N\begin{bmatrix} 3 \\ -4 \end{bmatrix}_{(\mathbf{u})} = \begin{bmatrix} 1 & 0 \\ 0 & -1 \end{bmatrix}\begin{bmatrix} 3 \\ -4 \end{bmatrix}_{(\mathbf{u})} = \begin{bmatrix} 3 \\ 4 \end{bmatrix}_{(\mathbf{u})} = 3\mathbf{u}_1 + 4\mathbf{u}_2$$

(*Note*: The matrix for T relative to the standard basis is

$$M = \begin{bmatrix} -(3/5) & (4/5) \\ (4/5) & (3/5) \end{bmatrix}$$

See Example 2 in Section 6.3.) ■

EXAMPLE 3 Let $T: \mathbb{R}^2 \to \mathbb{R}^2$ be counterclockwise rotation by 45°, and let

$$\mathbf{u}_1 = \begin{bmatrix} 1 \\ 0 \end{bmatrix} \quad \text{and} \quad \mathbf{u}_2 = \begin{bmatrix} 1/\sqrt{2} \\ 1/\sqrt{2} \end{bmatrix}$$

We will find the matrix for T relative to the basis $\{\mathbf{u}_1, \mathbf{u}_2\}$.

To find the matrix for T relative to $\mathbf{u}_1$ and $\mathbf{u}_2$, we must express each of the vectors $T(\mathbf{u}_1)$ and $T(\mathbf{u}_2)$ as linear combinations of $\mathbf{u}_1$ and $\mathbf{u}_2$ (see Figure 9.5).

It is easy to see that $\quad T(\mathbf{u}_1) = \mathbf{u}_2 = \begin{bmatrix} 0 \\ 1 \end{bmatrix}_{(\mathbf{u})}$

Finding $T(\mathbf{u}_2)$ is slightly more difficult. To obtain the $\mathbf{u}_1\mathbf{u}_2$ coordinates of $T(\mathbf{u}_2)$, we proceed as follows:

We know that $\quad T(\mathbf{u}_2) = 0\mathbf{e}_1 + \mathbf{e}_2 = \begin{bmatrix} 0 \\ 1 \end{bmatrix}$

We must solve the equation

$$T(\mathbf{u}_2) = s_{12}\mathbf{u}_1 + s_{22}\mathbf{u}_2$$

for s_{12} and s_{22}. We can write this equation in terms of $\mathbf{e}_1\mathbf{e}_2$ coordinate vectors as

$$\begin{bmatrix} 0 \\ 1 \end{bmatrix} = s_{12}\begin{bmatrix} 1 \\ 0 \end{bmatrix} + s_{22}\begin{bmatrix} 1/\sqrt{2} \\ 1/\sqrt{2} \end{bmatrix}$$

Figure 9.5

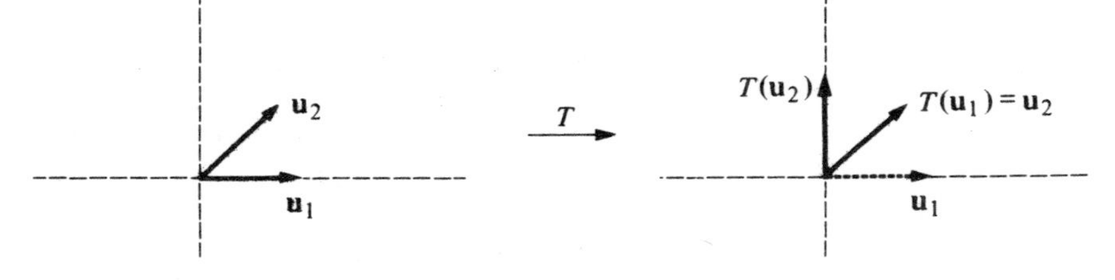

By looking at the coefficients in the second row, we see that $s_{22} = \sqrt{2}$, and from this it is easy to see that $s_{12} = -1$. So

$$\begin{bmatrix} 0 \\ 1 \end{bmatrix} = -1 \begin{bmatrix} 1 \\ 0 \end{bmatrix} + \sqrt{2} \begin{bmatrix} 1/\sqrt{2} \\ 1/\sqrt{2} \end{bmatrix}$$

which means that $T(\mathbf{u}_2) = -1\mathbf{u}_1 + \sqrt{2}\mathbf{u}_2 = \begin{bmatrix} -1 \\ \sqrt{2} \end{bmatrix}_{(\mathbf{u})}$

Thus the matrix for T relative to the $\mathbf{u}_i$ coordinate system is

$$N = \begin{bmatrix} 0 & -1 \\ 1 & \sqrt{2} \end{bmatrix} \quad \blacksquare$$

EXAMPLE 4 Using the same linear transformation T and basis $\{\mathbf{u}_1, \mathbf{u}_2\}$ as in Example 3, find

$$T(4\mathbf{u}_1 - 7\mathbf{u}_2)$$

Since the matrix for T relative to the basis $\{\mathbf{u}_1, \mathbf{u}_2\}$ is

$$N = \begin{bmatrix} 0 & -1 \\ 1 & \sqrt{2} \end{bmatrix}$$

we see that

$$T\begin{bmatrix} 4 \\ -7 \end{bmatrix}_{(\mathbf{u})} = N\begin{bmatrix} 4 \\ -7 \end{bmatrix}_{(\mathbf{u})} = \begin{bmatrix} 0 & -1 \\ 1 & \sqrt{2} \end{bmatrix}\begin{bmatrix} 4 \\ -7 \end{bmatrix}_{(\mathbf{u})} = \begin{bmatrix} 7 \\ 4 - 7\sqrt{2} \end{bmatrix}_{(\mathbf{u})}$$

So $$T(4\mathbf{u}_1 - 7\mathbf{u}_2) = 7\mathbf{u}_1 + (4 - 7\sqrt{2})\mathbf{u}_2 \quad \blacksquare$$

EXAMPLE 5 Let T be the linear transformation that, relative to the standard basis $\{\mathbf{e}_1, \mathbf{e}_2\}$, is represented by the matrix

$$M = \begin{bmatrix} -7 & -6 \\ 18 & 14 \end{bmatrix}$$

We wish to find the matrix N for T relative to the basis

$$\left\{\mathbf{u}_1 = \begin{bmatrix} -2 \\ 3 \end{bmatrix},\ \mathbf{u}_2 = \begin{bmatrix} 1 \\ -2 \end{bmatrix}\right\}$$

To do this, we must find numbers s_{ij} so that

$$T(\mathbf{u}_1) = s_{11}\mathbf{u}_1 + s_{21}\mathbf{u}_2$$

$$T(\mathbf{u}_2) = s_{12}\mathbf{u}_1 + s_{22}\mathbf{u}_2$$

Since M is the matrix for T relative to the $\mathbf{e}_1\mathbf{e}_2$ coordinate system, we can write these equations (relative to the basis $\{\mathbf{e}_1, \mathbf{e}_2\}$) as

$$\begin{bmatrix} -7 & -6 \\ 18 & 14 \end{bmatrix}\begin{bmatrix} -2 \\ 3 \end{bmatrix} = s_{11}\begin{bmatrix} -2 \\ 3 \end{bmatrix} + s_{21}\begin{bmatrix} 1 \\ -2 \end{bmatrix}$$

$$\begin{bmatrix} -7 & -6 \\ 18 & 14 \end{bmatrix}\begin{bmatrix} 1 \\ -2 \end{bmatrix} = s_{12}\begin{bmatrix} -2 \\ 3 \end{bmatrix} + s_{22}\begin{bmatrix} 1 \\ -2 \end{bmatrix}$$

Thus
$$\begin{bmatrix} -4 \\ 6 \end{bmatrix} = s_{11}\begin{bmatrix} -2 \\ 3 \end{bmatrix} + s_{21}\begin{bmatrix} 1 \\ -2 \end{bmatrix}$$

$$\begin{bmatrix} 5 \\ -10 \end{bmatrix} = s_{12}\begin{bmatrix} -2 \\ 3 \end{bmatrix} + s_{22}\begin{bmatrix} 1 \\ -2 \end{bmatrix}$$

It is easy to see that the solution of this system of equations is

$$s_{11} = 2 \qquad s_{21} = 0 \qquad \text{and} \qquad s_{12} = 0 \qquad s_{22} = 5$$

Therefore, the matrix for T relative to the basis $\{\mathbf{u}_1, \mathbf{u}_2\}$ is

$$N = \begin{bmatrix} 2 & 0 \\ 0 & 5 \end{bmatrix}$$

Thus, in terms of the basis $\{\mathbf{u}_1, \mathbf{u}_2\}$, the transformation T can be described very simply as

$$T(a_1\mathbf{u}_1 + a_2\mathbf{u}_2) = 2a_1\mathbf{u}_1 + 5a_2\mathbf{u}_2$$

Note: In the same way that we got a clearer picture of this linear transformation T by expressing it as a matrix relative to a basis different from the standard basis, we can also get more simple descriptions of other matrices by change of basis. Finding such bases is the subject of Section 9.4. ■

Exercise Set 9.2

1. Let $T: \mathbb{R}^2 \to \mathbb{R}^2$ be projection onto the line $y = 2x$ and

$$\mathbf{u}_1 = \begin{bmatrix} 1 \\ 2 \end{bmatrix} \quad \text{and} \quad \mathbf{u}_2 = \begin{bmatrix} -2 \\ 1 \end{bmatrix}$$

Find the matrix for T relative to the basis $\{\mathbf{u}_1, \mathbf{u}_2\}$. Is this matrix the same as the matrix for T relative to the $\mathbf{e}_1\mathbf{e}_2$ coordinate system?

2. Let $T: \mathbb{R}^2 \to \mathbb{R}^2$ be counterclockwise rotation of 90° and

$$\mathbf{u}_1 = \begin{bmatrix} 1 \\ 2 \end{bmatrix} \quad \text{and} \quad \mathbf{u}_2 = \begin{bmatrix} -2 \\ 1 \end{bmatrix}$$

Find the matrix that represents T relative to the $\{\mathbf{u}_1, \mathbf{u}_2\}$. Is this the same matrix for T as the matrix for T relative to the $\mathbf{e}_1\mathbf{e}_2$ coordinate system?

3. Let $\mathbf{u}_1$ and $\mathbf{u}_2$ be two linearly independent vectors of $\mathbb{R}^3$, and let $\mathbf{u}_3 = \mathbf{u}_1 \times \mathbf{u}_2$. If $T: \mathbb{R}^3 \to \mathbb{R}^3$ is the projection onto the plane spanned by $\mathbf{u}_1$ and $\mathbf{u}_2$, find the matrix for T relative to the basis $\{\mathbf{u}_1, \mathbf{u}_2, \mathbf{u}_3\}$.

4. Let $T: \mathbb{R}^2 \to \mathbb{R}^2$ be reflection about the line $y = 5x$, and

$$\mathbf{u}_1 = \begin{bmatrix} 1 \\ 5 \end{bmatrix} \quad \text{and} \quad \mathbf{u}_2 = \begin{bmatrix} -5 \\ 1 \end{bmatrix}$$

Find the matrix that represents T relative to the basis $\{\mathbf{u}_1, \mathbf{u}_2\}$.

5. Suppose that $\{\mathbf{u}_1, \mathbf{u}_2\}$ is a basis for $\mathbb{R}^2$, and $T: \mathbb{R}^2 \to \mathbb{R}^2$ is the linear transformation represented by the matrix M relative to the basis $\{\mathbf{u}_1, \mathbf{u}_2\}$. For each matrix M give a geometrical description of T.

a. $M = \begin{bmatrix} 0 & 1 \\ 1 & 0 \end{bmatrix}$ Assume that $\{\mathbf{u}_1, \mathbf{u}_2\}$ is an orthonormal basis.

b. $M = \begin{bmatrix} 0 & 0 \\ 0 & 1 \end{bmatrix}$ Assume that $\mathbf{u}_1$ is perpendicular to $\mathbf{u}_2$.

c. $M = \begin{bmatrix} 2 & 0 \\ 0 & 2 \end{bmatrix}$

Suppose $T: \mathbb{R}^2 \to \mathbb{R}^2$ is the linear transformation represented by the matrix M relative to the basis $\{\mathbf{e}_1, \mathbf{e}_2\}$. Find the matrix for T relative to the basis $\{\mathbf{u}_1, \mathbf{u}_2\}$ in each of the matrices M listed in Exercises 6–9.

6. $M = \begin{bmatrix} 4 & 18 \\ 3 & 1 \end{bmatrix}$

$\mathbf{u}_1 = \begin{bmatrix} 3 \\ 1 \end{bmatrix}$ and $\mathbf{u}_2 = \begin{bmatrix} -2 \\ 1 \end{bmatrix}$

7. $M = \begin{bmatrix} 7 & 3 \\ -9 & -5 \end{bmatrix}$

$\mathbf{u}_1 = \begin{bmatrix} 2 \\ -2 \end{bmatrix}$ and $\mathbf{u}_2 = \begin{bmatrix} 2 \\ -6 \end{bmatrix}$

8. $M = \begin{bmatrix} 1 & 0 \\ 0 & -2 \end{bmatrix}$

$\mathbf{u}_1 = \begin{bmatrix} 1 \\ -2 \end{bmatrix}$ and $\mathbf{u}_2 = \begin{bmatrix} 2 \\ 1 \end{bmatrix}$

9. $M = \begin{bmatrix} 1 & 2 \\ 1 & 1 \end{bmatrix}$

$\mathbf{u}_1 = \begin{bmatrix} 1 \\ -2 \end{bmatrix}$ and $\mathbf{u}_2 = \begin{bmatrix} 2 \\ 1 \end{bmatrix}$

9.3 Similar Matrices

In this section we continue our discussion of how to find matrices that represent linear transformations (from $\mathbb{R}^n$ to $\mathbb{R}^n$) relative to bases other than the standard basis. We begin by taking a more thorough look at the problem we solved in Example 5 in Section 9.2.

In that example we were given the matrix M, which represented the linear transformation T relative to the basis $\{\mathbf{e}_1, \mathbf{e}_2\}$, and we wanted to determine the matrix N that represented T relative to the basis

$$\left\{\mathbf{u}_1 = \begin{bmatrix} -2 \\ 3 \end{bmatrix}, \ \mathbf{u}_2 = \begin{bmatrix} 1 \\ -2 \end{bmatrix}\right\}$$

Figure 9.6 should make the situation clearer.

In the figure both rows of graphs show the action of T on an arbitrary vector $\mathbf{v}$. The vectors $\mathbf{v}$ and $T(\mathbf{v})$ in the top row are expressed relative to the $\mathbf{e}_i$ coordinate system and the linear transformation T is represented by the matrix M. The same vectors $\mathbf{v}$ and $T(\mathbf{v})$ are in the bottom row. Here, however, they are expressed relative to the $\mathbf{u}_i$ coordinate system, and the linear transformation T is represented by the matrix N.

The matrix $$B = \begin{bmatrix} -2 & 1 \\ 3 & -2 \end{bmatrix}$$

is the basis change matrix used to convert the $\mathbf{u}_i$ coordinates of a vector to the $\mathbf{e}_i$ coordinates of that vector.

We restate our problem: Given the matrix M that represents the linear transformation T relative to the basis $\{\mathbf{e}_1, \mathbf{e}_2\}$ and the relationship B

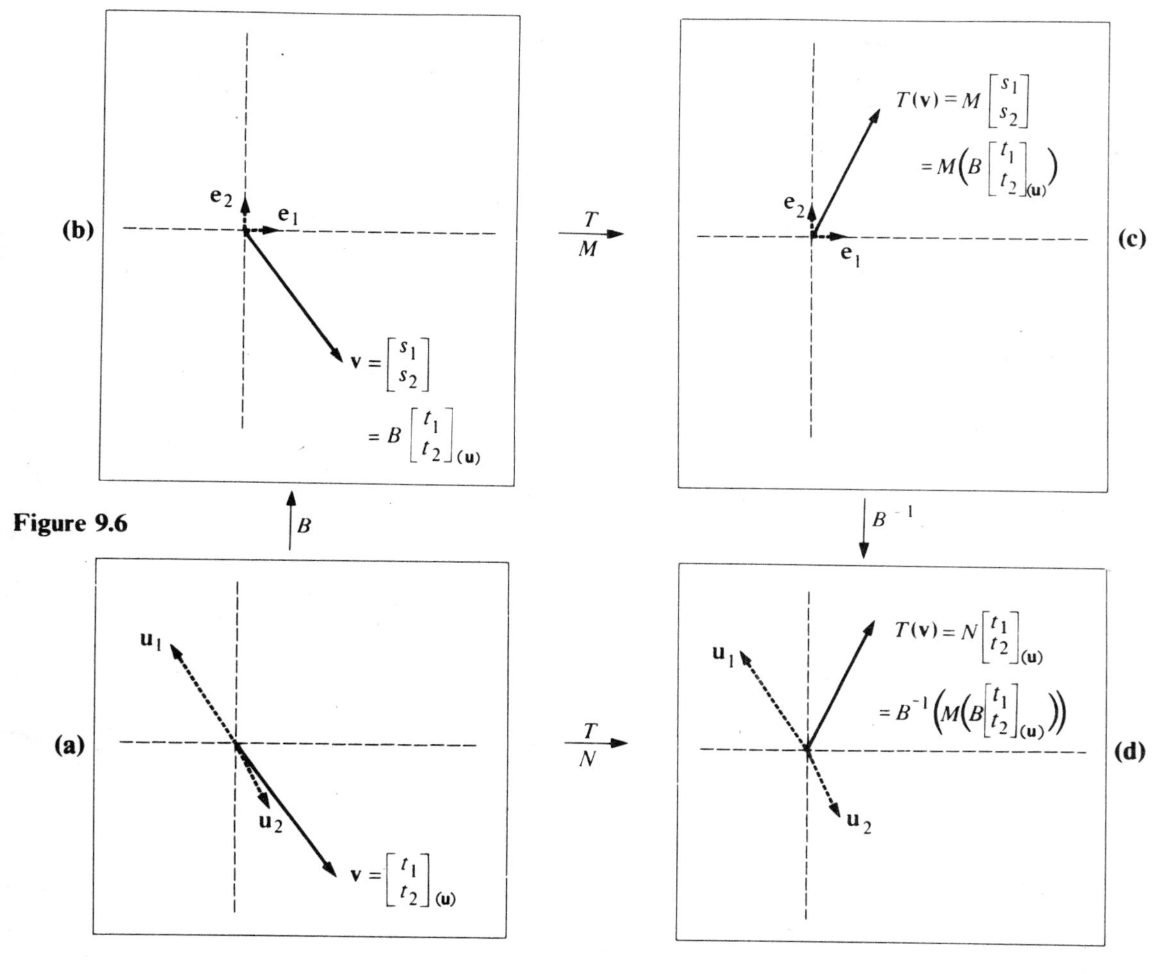

Figure 9.6

between the bases $\{\mathbf{u}_1, \mathbf{u}_2\}$ and $\{\mathbf{e}_1, \mathbf{e}_2\}$, find the matrix N that represents T relative to the basis $\{\mathbf{u}_1, \mathbf{u}_2\}$.

We use Figure 9.6. Let $\mathbf{v} = t_1\mathbf{u}_1 + t_2\mathbf{u}_2$ be an arbitrary vector in $\mathbb{R}^2$. We want to find the $\mathbf{u}_i$ coordinates of $T(\mathbf{v})$. The coordinate vector for $\mathbf{v}$ relative to the basis $\{\mathbf{u}_1, \mathbf{u}_2\}$ is

$$\mathbf{v} = \begin{bmatrix} t_1 \\ t_2 \end{bmatrix}_{(\mathbf{u})} \qquad \text{Figure 9.6a}$$

To find $T(\mathbf{v})$, we first convert $\mathbf{v}$ to its $\mathbf{e}_1\mathbf{e}_2$ coordinates. The coordinate vector for $\mathbf{v}$ relative to the basis $\{\mathbf{e}_1, \mathbf{e}_2\}$ is

$$\mathbf{v} = \begin{bmatrix} s_1 \\ s_2 \end{bmatrix} = B\begin{bmatrix} t_1 \\ t_2 \end{bmatrix}_{(\mathbf{u})} \qquad \text{Figure 9.6b}$$

We can now use the matrix M to find the coordinate vector for $T(\mathbf{v})$ relative to the basis $\{\mathbf{e}_1, \mathbf{e}_2\}$.

$$T(\mathbf{v}) = M\left(B\begin{bmatrix} t_1 \\ t_2 \end{bmatrix}_{(\mathbf{u})}\right) \qquad \text{Figure 9.6c}$$

We are trying to find the coordinate vector for $T(\mathbf{v})$ relative to the basis $\{\mathbf{u}_1, \mathbf{u}_2\}$. The matrix that converts the $\mathbf{e}_i$ coordinates of a vector to the $\mathbf{u}_i$ coordinates of that vector is B^{-1}. So the coordinate vector for $T(\mathbf{v})$ relative to the basis $\{\mathbf{u}_1, \mathbf{u}_2\}$ is

$$T(\mathbf{v}) = B^{-1}\left(M\left(B\begin{bmatrix} t_1 \\ t_2 \end{bmatrix}_{(\mathbf{u})}\right)\right) \qquad \text{Figure 9.6d}$$

Thus the matrix N that represents T relative to the basis $\{\mathbf{u}_1, \mathbf{u}_2\}$ is

$$N = B^{-1}(MB)$$

(Remember that this means that we first apply B, then M, and then B^{-1}.)

In our particular case, since

$$M = \begin{bmatrix} -7 & -6 \\ 18 & 14 \end{bmatrix}, \quad B = \begin{bmatrix} -2 & 1 \\ 3 & -2 \end{bmatrix} \quad \text{and} \quad B^{-1} = \begin{bmatrix} -2 & -1 \\ -3 & -2 \end{bmatrix}$$

we have

$$\begin{aligned} N &= B^{-1}(MB) \\ &= \begin{bmatrix} -2 & -1 \\ -3 & -2 \end{bmatrix}\left(\begin{bmatrix} -7 & -6 \\ 18 & 14 \end{bmatrix}\begin{bmatrix} -2 & 1 \\ 3 & -2 \end{bmatrix}\right) \\ &= \begin{bmatrix} -2 & -1 \\ -3 & -2 \end{bmatrix}\begin{bmatrix} -4 & 5 \\ 6 & -10 \end{bmatrix} = \begin{bmatrix} 2 & 0 \\ 0 & 5 \end{bmatrix} \end{aligned}$$

which is the same answer we got in Example 5 in Section 9.2. (*Note*: $B^{-1}(MB)$ is not equal to M.)

The same ideas that we just used to solve this problem work in general to give the following theorem.

THEOREM 9.4 Let $T: \mathbb{R}^n \to \mathbb{R}^n$ be a linear transformation, and M the matrix that represents T relative to the basis $\{\mathbf{e}_1, \mathbf{e}_2, \ldots, \mathbf{e}_n\}$. If $\{\mathbf{u}_1, \mathbf{u}_2, \ldots, \mathbf{u}_n\}$ is another basis for $\mathbb{R}^n$ and B is the basis change matrix (see Theorem 9.1) that converts the $\mathbf{u}_i$ coordinates of a vector to the $\mathbf{e}_i$ coordinates of that vector, then the matrix N that represents T relative to the basis $\{\mathbf{u}_1, \mathbf{u}_2, \ldots, \mathbf{u}_n\}$ is

$$N = B^{-1}(MB)$$

Note: Since matrix multiplication is associative, $B^{-1}(MB) = (B^{-1}M)B$. That is, to find N, you can either first find the product MB and then multiply this on the left by B^{-1}, or you can begin by finding the product $B^{-1}M$ and then multiply this on the right by B.

DEFINITION 9.3 Two matrices that represent the same linear transformation (relative to the same or different bases) are said to be **similar**. (See Theorem 9.4.)

As we have shown above

$$\begin{bmatrix} -7 & -6 \\ 18 & 14 \end{bmatrix} \quad \text{and} \quad \begin{bmatrix} 2 & 0 \\ 0 & 5 \end{bmatrix} \text{ are similar.}$$

EXAMPLE 1 Let the matrix M be the matrix that represents the linear transformation T relative to the basis $\{\mathbf{e}_1, \mathbf{e}_2, \mathbf{e}_3\}$

$$M = \begin{bmatrix} -1 & 3 & 2 \\ 2 & 1 & 2 \\ 1 & 0 & -1 \end{bmatrix}$$

We will find the matrix N that represents T relative to the basis

$$\left\{ \mathbf{u}_1 = \begin{bmatrix} 1 \\ -1 \\ 3 \end{bmatrix}, \quad \mathbf{u}_2 = \begin{bmatrix} 1 \\ 2 \\ 1 \end{bmatrix}, \quad \mathbf{u}_3 = \begin{bmatrix} 1 \\ 1 \\ 2 \end{bmatrix} \right\}$$

Figure 9.7 will help us when we apply Theorem 9.4. The basis change matrix is

$$B = \begin{bmatrix} 1 & 1 & 1 \\ -1 & 2 & 1 \\ 3 & 1 & 2 \end{bmatrix} \quad \text{and} \quad B^{-1} = \begin{bmatrix} 3 & -1 & -1 \\ 5 & -1 & -2 \\ -7 & 2 & 3 \end{bmatrix}$$

(See Example 6 in Section 7.3)

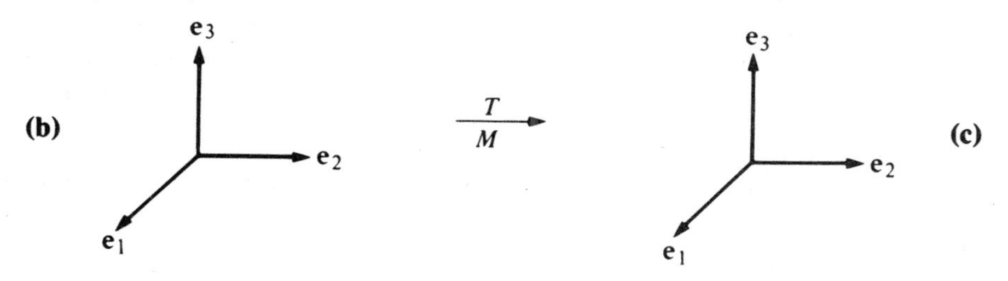

Figure 9.7

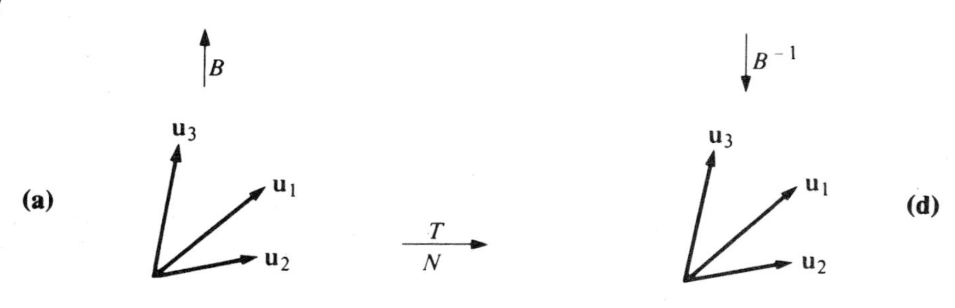

We compute N using Theorem 9.4.

$$N = B^{-1}(MB)$$

$$= \begin{bmatrix} 3 & -1 & -1 \\ 5 & -1 & -2 \\ -7 & 2 & 3 \end{bmatrix} \left(\begin{bmatrix} -1 & 3 & 2 \\ 2 & 1 & 2 \\ 1 & 0 & -1 \end{bmatrix} \begin{bmatrix} 1 & 1 & 1 \\ -1 & 2 & 1 \\ 3 & 1 & 2 \end{bmatrix} \right)$$

$$= \begin{bmatrix} 3 & -1 & -1 \\ 5 & -1 & -2 \\ -7 & 2 & 3 \end{bmatrix} \begin{bmatrix} 2 & 7 & 6 \\ 7 & 6 & 7 \\ -2 & 0 & -1 \end{bmatrix} = \begin{bmatrix} 1 & 15 & 12 \\ 7 & 29 & 25 \\ -6 & -37 & -31 \end{bmatrix} \quad ■$$

Exercise Set 9.3

Suppose $T: \mathbb{R}^n \to \mathbb{R}^n$ is the linear transformation represented by the matrix M relative to the $\mathbf{e}_i$ coordinate system. Use Theorem 9.4 to find the matrix for T relative to the $\mathbf{u}_i$ coordinate system for each matrix and basis given in Exercises 1–6.

1. $M = \begin{bmatrix} 4 & 18 \\ 3 & 1 \end{bmatrix}$

$\mathbf{u}_1 = \begin{bmatrix} 3 \\ 1 \end{bmatrix}$ and $\mathbf{u}_2 = \begin{bmatrix} -2 \\ 1 \end{bmatrix}$

2. $M = \begin{bmatrix} 7 & 3 \\ -9 & -5 \end{bmatrix}$

$\mathbf{u}_1 = \begin{bmatrix} 2 \\ -2 \end{bmatrix}$ and $\mathbf{u}_2 = \begin{bmatrix} 2 \\ -6 \end{bmatrix}$

3. $M = \begin{bmatrix} 1 & 0 \\ 0 & -2 \end{bmatrix}$

$\mathbf{u}_1 = \begin{bmatrix} 1 \\ -2 \end{bmatrix}$ and $\mathbf{u}_2 = \begin{bmatrix} 2 \\ 1 \end{bmatrix}$

4. $M = \begin{bmatrix} 1 & 2 \\ 1 & 1 \end{bmatrix}$

$\mathbf{u}_1 = \begin{bmatrix} 1 \\ -2 \end{bmatrix}$ and $\mathbf{u}_2 = \begin{bmatrix} 2 \\ 1 \end{bmatrix}$

5. $M = \begin{bmatrix} 15 & 35 & -5 \\ 8 & 6 & 8 \\ -1 & 13 & 19 \end{bmatrix}$

$\mathbf{u}_1 = \begin{bmatrix} 1 \\ 0 \\ -1 \end{bmatrix}, \quad \mathbf{u}_2 = \begin{bmatrix} 2 \\ 1 \\ 1 \end{bmatrix}, \quad \mathbf{u}_3 = \begin{bmatrix} 3 \\ -2 \\ 1 \end{bmatrix}$

6. $M = \begin{bmatrix} 1 & -1 & 4 \\ 3 & 2 & -1 \\ 2 & 2 & 0 \end{bmatrix}$

$\mathbf{u}_1 = \begin{bmatrix} 1 \\ 0 \\ -1 \end{bmatrix}, \quad \mathbf{u}_2 = \begin{bmatrix} 2 \\ 1 \\ 1 \end{bmatrix}, \quad \mathbf{u}_3 = \begin{bmatrix} 3 \\ -2 \\ 1 \end{bmatrix}$

7. Show that if M_1 and M_2 are similar matrices, then $\det(M_1) = \det(M_2)$.

8. Let $T: \mathbb{R}^n \to \mathbb{R}^n$ be a linear transformation and let N be the matrix that represents T relative to the basis $\{\mathbf{u}_1, \mathbf{u}_2, \ldots, \mathbf{u}_n\}$. Find an expression for the matrix M that represents T relative to the standard basis $\{\mathbf{e}_1, \mathbf{e}_2, \ldots, \mathbf{e}_n\}$ of $\mathbb{R}^n$. (Let the matrix B be the basis change matrix that changes the $\mathbf{u}_i$ coordinates of a vector to the $\mathbf{e}_i$ coordinates of that vector.)

9. Suppose that the $n \times n$ matrix M is similar to a matrix D of the form

$$D = \begin{bmatrix} d_1 & 0 & \cdots & 0 \\ 0 & d_2 & \cdots & 0 \\ \vdots & \vdots & & \vdots \\ 0 & \cdots & 0 & d_n \end{bmatrix}$$

Show that this means that there is a basis $\{\mathbf{u}_1, \mathbf{u}_2, \ldots, \mathbf{u}_n\}$ of $\mathbb{R}^n$ such that

$$M\mathbf{u}_i = \text{a scalar multiple of } \mathbf{u}_i$$

for $i = 1, 2, \ldots, n$.

10. Suppose that the matrices M and N are similar and that $N = B^{-1}MB$.

a. Show that $N^k = (B^{-1}MB)^k = B^{-1}M^kB$ and also that $M^k = BN^kB^{-1}$ for any integer k.

b. Let M, B, and $N = B^{-1}MB$ be as in Example 5 of Section 9.2.

$$M = \begin{bmatrix} -7 & -6 \\ 18 & 14 \end{bmatrix} \qquad B = \begin{bmatrix} -2 & 1 \\ 3 & -2 \end{bmatrix}$$

$$N = \begin{bmatrix} 2 & 0 \\ 0 & 5 \end{bmatrix}$$

Find M^2 by first computing N^2 and then using the fact (part **a**) that $M^2 = BN^2B^{-1}$. Compute M^2 directly to check your answer.

c. Let M, N, and B be as in part **b**. Find a matrix A so that $A^2 = M$. (*Hint:* Begin by finding a matrix R such that $R^2 = N$.)

11. Show that if M and N are similar matrices, then $\text{Rank}(M) = \text{Rank}(N)$

★**12.** **a.** If M and N are similar, explain how the vectors in the null space of M are related to the vectors of the null space of N.

b. Let

$$M = \begin{bmatrix} -10 & -30 \\ 4 & 12 \end{bmatrix} \qquad N = \begin{bmatrix} 2 & 0 \\ 0 & 0 \end{bmatrix}$$

Then M and N are similar under the basis change matrix

$$B = \begin{bmatrix} -5 & 3 \\ 2 & -1 \end{bmatrix}$$

Check the answer you obtained in part **a** by examining the relationship between the null spaces of the similar matrices M and N.

9.4 Characteristic (Eigen) Values and Vectors

As we saw in Section 9.3, a given linear transformation can be represented by a number of different matrices. Moreover, in many cases a change of basis allows us to give a very simple description of the action of a particular linear transformation. We saw this in the case of the linear transformation T discussed in Example 5 in Section 9.2. The matrix $M = \begin{bmatrix} -7 & -6 \\ 18 & 14 \end{bmatrix}$, which is the matrix for T relative to the $\mathbf{e}_i$ coordinate system, is quite complicated, whereas the description given by the matrix $N = \begin{bmatrix} 2 & 0 \\ 0 & 5 \end{bmatrix}$, which is the matrix for T relative to the $\mathbf{u}_i$ coordinate system, is very simple.

The question we discuss in this section is how to choose the proper basis so that a given linear transformation T has a simple description relative to this basis. To discuss this question we must first state precisely what we mean by simple description. By simple description we will mean that a basis $\{\mathbf{u}_1, \mathbf{u}_2, \ldots, \mathbf{u}_n\}$ can be chosen so that for each of the basis vectors $\mathbf{u}_i$

$$T(\mathbf{u}_i) = d_i \mathbf{u}_i \qquad \text{for some scalar } d_i$$

That is, $T(\mathbf{u}_i)$ lies on the same line as $\mathbf{u}_i$. In this case the matrix for T

relative to the basis $\{\mathbf{u}_1, \mathbf{u}_2, \ldots, \mathbf{u}_n\}$ would be the matrix

$$\begin{bmatrix} d_1 & 0 & 0 & \cdots & 0 & 0 \\ 0 & d_2 & 0 & \cdots & 0 & 0 \\ \vdots & \vdots & \vdots & & \vdots & \vdots \\ 0 & 0 & 0 & \cdots & 0 & d_n \end{bmatrix}$$

Such a matrix is called a **diagonal** matrix.

Vectors like these vectors $\mathbf{u}_i$ play an important role in the study of linear transformations and are given a special name.

DEFINITION 9.4 Let $T: \mathbb{R}^n \to \mathbb{R}^n$ be a linear transformation and $\mathbf{v} \neq \mathbf{0}$ a vector in $\mathbb{R}^n$ such that $T(\mathbf{v}) = d\mathbf{v}$ for some scalar d

Then $\mathbf{v}$ is called a **characteristic** (or **eigen**) **vector** for T and d is called the **characteristic** (or **eigen**) **value** associated with $\mathbf{v}$.

EXAMPLE 1 The vector $\mathbf{u} = \begin{bmatrix} 1 \\ -2 \end{bmatrix}$ is a characteristic vector with characteristic value 2

for the matrix
$$M = \begin{bmatrix} -7 & -6 \\ 18 & 14 \end{bmatrix}$$

since
$$M\mathbf{u} = \begin{bmatrix} -7 & -6 \\ 18 & 14 \end{bmatrix}\begin{bmatrix} 1 \\ -2 \end{bmatrix} = \begin{bmatrix} 5 \\ -10 \end{bmatrix} = 5\begin{bmatrix} 1 \\ -2 \end{bmatrix} = 5\mathbf{u}$$

See Example 5 in Section 9.2. ■

Now we can restate our question as: Given a linear transformation $T: \mathbb{R}^n \to \mathbb{R}^n$, can we find a basis $\{\mathbf{u}_1, \mathbf{u}_2, \ldots, \mathbf{u}_n\}$ for $\mathbb{R}^n$ so that each basis vector $\mathbf{u}_i$ is a characteristic vector for T?

This is a difficult question and we will not give a complete answer to it in this text. However, we will find large classes of linear transformations for which we can always find such a basis (Section 9.5). In this section we begin the discussion of this topic by learning how to find characteristic values and characteristic vectors. To do this we will make use of almost all of the ideas developed so far. The following discussion is a focal point of this text.

Let $T: \mathbb{R}^n \to \mathbb{R}^n$ be a linear transformation and let M be the matrix that represents T relative to the standard basis. If $\mathbf{v}$ is a characteristic vector for T with characteristic value d, then

$$T(\mathbf{v}) = d\mathbf{v} \qquad \text{or, in matrix form,} \qquad M(\mathbf{v}) = d\mathbf{v}$$

This equation can be written as

$$M(\mathbf{v}) - d\mathbf{v} = \mathbf{0}$$

which is the same as the equation

$$M(\mathbf{v}) - dI(\mathbf{v}) = \mathbf{0} \qquad \text{where } I \text{ is the } n \times n \text{ identity matrix}$$

This equation can be rewritten as

$$(M - dI)\mathbf{v} = \mathbf{0}$$

From this last equation we see that the vector $\mathbf{v}$ is a characteristic vector for T with characteristic value d if and only if $\mathbf{v}$ is in the null space of the matrix $M - dI$. Therefore, if we can find a scalar d so that $M - dI$ has a nontrivial (not $\langle \mathbf{0} \rangle$) null space, then any nonzero vector of the null space can be chosen as the characteristic vector $\mathbf{v}$. Thus, d is a characteristic value for T (or M) if and only if $M - dI$ is singular.

The determinant gives a very efficient way of determining if a matrix is singular (Theorem 8.11): The matrix $M - dI$ is singular if and only if $\det(M - dI) = 0$.

So we conclude that d is a characteristic value for T (or M) if and only if $\det(M - dI) = 0$; and if d is a characteristic value for T (or M), then any nonzero vector $\mathbf{v}$ of the null space of $M - dI$ is a characteristic vector for T (or M) with characteristic value d.

The following example illustrates how this can be used.

EXAMPLE 2 Find all characteristic vectors and characteristic values for the linear transformation T represented by the matrix M relative to the $\mathbf{e}_1\mathbf{e}_2$ basis where

$$M = \begin{bmatrix} -7 & -6 \\ 18 & 14 \end{bmatrix}$$

We begin by finding the characteristic values for M by solving the equation $\det(M - dI) = 0$ for the scalar d:

$$\begin{aligned} 0 &= \det(M - dI) \\ &= \left| \begin{bmatrix} -7 & -6 \\ 18 & 14 \end{bmatrix} - d \begin{bmatrix} 1 & 0 \\ 0 & 1 \end{bmatrix} \right| \\ &= \begin{vmatrix} -7 - d & -6 \\ 18 & 14 - d \end{vmatrix} \\ &= (-7 - d)(14 - d) - (-6)(18) \\ &= d^2 - 7d + 10 = (d - 2)(d - 5) \end{aligned}$$

So the characteristic values for M are

$$d = 2 \qquad \text{and} \qquad d = 5$$

We now find a characteristic vector associated with the characteristic value $d = 2$. To do this we must find the null space of $M - 2I$.

$$M - 2I = \begin{bmatrix} -7 & -6 \\ 18 & 14 \end{bmatrix} - 2\begin{bmatrix} 1 & 0 \\ 0 & 1 \end{bmatrix} = \begin{bmatrix} -9 & -6 \\ 18 & 12 \end{bmatrix}$$

We now reduce $M - 2I$ to find its null space

$$\begin{bmatrix} -9 & -6 \\ 18 & 12 \end{bmatrix} \xrightarrow{\text{reduces to}} \begin{bmatrix} -9 & -6 \\ 0 & 0 \end{bmatrix}$$

Thus the null space of $M - 2I$ is

$$\left\langle \begin{bmatrix} -2 \\ 3 \end{bmatrix} \right\rangle$$

and the vector $$\mathbf{u}_1 = \begin{bmatrix} -2 \\ 3 \end{bmatrix}$$

is a characteristic vector for T with characteristic value $d = 2$.

In the same way we can find a characteristic vector for T with characteristic value $d = 5$. We begin with the matrix $M - 5I$.

$$M - 5I = \begin{bmatrix} -7 & -6 \\ 18 & 14 \end{bmatrix} - 5\begin{bmatrix} 1 & 0 \\ 0 & 1 \end{bmatrix} = \begin{bmatrix} -12 & -6 \\ 18 & 9 \end{bmatrix}$$

We reduce the matrix $M - 5I$ to find its null space

$$\begin{bmatrix} -12 & -6 \\ 18 & 9 \end{bmatrix} \xrightarrow{\text{reduces to}} \begin{bmatrix} -12 & -6 \\ 0 & 0 \end{bmatrix}$$

So the null space of $M - 5I$ is $\left\langle \begin{bmatrix} 1 \\ -2 \end{bmatrix} \right\rangle$

and the vector $$\mathbf{u}_2 = \begin{bmatrix} 1 \\ -2 \end{bmatrix}$$

is a characteristic vector for T with characteristic value $d = 5$.

Since $\mathbf{u}_1$ and $\mathbf{u}_2$ are linearly independent vectors of $\mathbb{R}^2$, they form a basis for $\mathbb{R}^2$. The matrix for T relative to this basis is the diagonal matrix

$$D = \begin{bmatrix} 2 & 0 \\ 0 & 5 \end{bmatrix}$$

(*Note*: $D = B^{-1}MB$, where B is the basis change matrix

$$B = \begin{bmatrix} -2 & 1 \\ 3 & -2 \end{bmatrix}$$

Compare this with the results of Example 5 in Section 9.2.) ■

DEFINITION 9.5

Let $T: \mathbb{R}^n \to \mathbb{R}^n$ be a linear transformation and M be the matrix representing T relative to the standard basis. Then the equation

$$0 = \det(M - xI)$$

is called the **characteristic equation** of T (or M). (*Note:* $\det(M - xI)$ is a polynomial in x of degree n.)

The following theorem generalizes the results of Example 2.

THEOREM 9.5

Let $T: \mathbb{R}^n \to \mathbb{R}^n$ be a linear transformation and M the matrix representing T relative to the standard basis. Then

1. The scalar d is a characteristic value for T if and only if d is a solution of the characteristic equation $0 = \det(M - xI)$.
2. If d is a characteristic value for T, then $\mathbf{v}$ is a characteristic vector of T if and only if $\mathbf{v}$ is a nonzero vector of the null space of $M - dI$.

Note: Using Theorem 9.5, it is possible to find all characteristic vectors for a given linear transformation T. However, Theorem 9.5 does not say that it is always possible to choose a basis for $\mathbb{R}^n$ consisting entirely of characteristic vectors. In fact, there are many linear transformations for which this is not possible. The following is an example of such a linear transformation. Another example is given in Exercise 1.

EXAMPLE 3 Let T be the linear transformation represented by the matrix

$$M = \begin{bmatrix} 1 & 1 \\ 0 & 1 \end{bmatrix}$$

relative to the standard basis. We show that no pair of characteristic vectors for T is a basis of $\mathbb{R}^2$.

The characteristic equation for T is

$$0 = \det(M - xI) = \begin{vmatrix} 1 - x & 1 \\ 0 & 1 - x \end{vmatrix} = (1 - x)^2$$

Thus $d = 1$ is the only characteristic value for T.

Since
$$M - 1I = \begin{bmatrix} 0 & 1 \\ 0 & 0 \end{bmatrix}$$

the null space of $M - 1I$ is $\left\langle \begin{bmatrix} 1 \\ 0 \end{bmatrix} \right\rangle$

Therefore, all characteristic vectors of M lie along the x axis, and no pair of characteristic vectors spans $\mathbb{R}^2$. So no pair of characteristic vectors for T can be a basis for $\mathbb{R}^2$. ∎

Now we see how to apply Theorem 9.5 to linear transformations on $\mathbb{R}^3$.

EXAMPLE 4 Let $T: \mathbb{R}^3 \to \mathbb{R}^3$ be represented by the matrix

$$M = \begin{bmatrix} 8 & 9 & 9 \\ 3 & 2 & 3 \\ -9 & -9 & -10 \end{bmatrix}$$

relative to the standard basis. We find characteristic vectors $\mathbf{u}_1$, $\mathbf{u}_2$, and $\mathbf{u}_3$ so that $\{\mathbf{u}_1, \mathbf{u}_2, \mathbf{u}_3\}$ is a basis for $\mathbb{R}^3$.

We begin with the characteristic equation.

$$\begin{aligned} 0 &= \det(M - xI) \\ &= \begin{vmatrix} 8-x & 9 & 9 \\ 3 & 2-x & 3 \\ -9 & -9 & -10-x \end{vmatrix} \\ &= (8-x)[(2-x)(-10-x)+27] - 9[3(-10-x)+27] \\ &\quad + 9[-27 + 9(2-x)] \\ &= -x^3 + 0x^2 + 3x + 2 \\ &= -(x-2)(x+1)^2 \end{aligned}$$

So the characteristic values for M are

$$d = 2 \quad \text{and} \quad d = -1 \quad \text{(a double root)}$$

We find a characteristic vector for $d = 2$. Reducing $M - 2I$ gives

$$\begin{bmatrix} 6 & 9 & 9 \\ 3 & 0 & 3 \\ -9 & -9 & -12 \end{bmatrix} \xrightarrow{\text{reduces to}} \begin{bmatrix} 6 & 9 & 9 \\ 0 & 9 & 3 \\ 0 & 0 & 0 \end{bmatrix}$$

which has null space $\left\langle \begin{bmatrix} -3 \\ -1 \\ 3 \end{bmatrix} \right\rangle$

So a characteristic vector with characteristic value $d = 2$ is

$$\mathbf{u}_1 = \begin{bmatrix} -3 \\ -1 \\ 3 \end{bmatrix}$$

We now find characteristic vectors for $d = -1$. Reducing $M - (-1)I$ gives:

$$\begin{bmatrix} 9 & 9 & 9 \\ 3 & 3 & 3 \\ -9 & -9 & -9 \end{bmatrix} \xrightarrow{\text{reduces to}} \begin{bmatrix} 9 & 9 & 9 \\ 0 & 0 & 0 \\ 0 & 0 & 0 \end{bmatrix}$$

which has a two-dimensional null space consisting of all vectors of the form

$$\begin{bmatrix} -y - z \\ y \\ z \end{bmatrix}$$

A basis for the null space of $M - (-1)I$ is

$$\left\{ \begin{bmatrix} -1 \\ 1 \\ 0 \end{bmatrix}, \begin{bmatrix} -1 \\ 0 \\ 1 \end{bmatrix} \right\}$$

So we have two independent characteristic vectors with characteristic value -1

$$\mathbf{u}_2 = \begin{bmatrix} -1 \\ 1 \\ 0 \end{bmatrix} \quad \text{and} \quad \mathbf{u}_3 = \begin{bmatrix} -1 \\ 0 \\ 1 \end{bmatrix}$$

A simple calculation shows that the vectors $\mathbf{u}_1$, $\mathbf{u}_2$, and $\mathbf{u}_3$ are linearly independent, therefore they form a basis for $\mathbb{R}^3$. ■

EXAMPLE 5 Letting M be as in Example 4, we find a diagonal matrix D similar to M, and a matrix B such that

$$D = B^{-1}MB$$

From Example 4 we know that $\{\mathbf{u}_1, \mathbf{u}_2, \mathbf{u}_3\}$ is a basis for $\mathbb{R}^3$ and that

$$T(\mathbf{u}_1) = 2\mathbf{u}_1 \qquad T(\mathbf{u}_2) = -1\mathbf{u}_2 \qquad T(\mathbf{u}_3) = -1\mathbf{u}_3$$

So the matrix for T relative to the $\mathbf{u}_i$ coordinates is the diagonal matrix

$$D = \begin{bmatrix} 2 & 0 & 0 \\ 0 & -1 & 0 \\ 0 & 0 & -1 \end{bmatrix}$$

Since the matrix

$$B = \begin{bmatrix} -3 & -1 & -1 \\ -1 & 1 & 0 \\ 3 & 0 & 1 \end{bmatrix}$$

is the basis change matrix that converts the $\mathbf{u}_i$ coordinates of a vector to the $\mathbf{e}_i$ coordinates of that vector; using Theorem 9.5, we see that

$$D = B^{-1}MB \quad ■$$

Exercise Set 9.4

1. Show that the matrix $M = \begin{bmatrix} 0 & -1 \\ 1 & 0 \end{bmatrix}$ has no real characteristic values, and therefore is not similar to a diagonal matrix (with real coefficients). Compare your result with Example 5 of Section 3.2.

Suppose $T: \mathbb{R}^n \to \mathbb{R}^n$ is the linear transformation represented by the matrix M relative to the basis $\{\mathbf{e}_1, \mathbf{e}_2, \ldots, \mathbf{e}_n\}$.

For each of the matrices listed in Exercises 2–10, find a basis $\{\mathbf{u}_1, \mathbf{u}_2, \ldots, \mathbf{u}_n\}$ so that the matrix D that represents T relative to the $\mathbf{u}_i$ coordinates is diagonal. Also find D.

2. $M = \begin{bmatrix} 4 & 18 \\ 3 & 1 \end{bmatrix}$ **3.** $M = \begin{bmatrix} 7 & 3 \\ -9 & -5 \end{bmatrix}$

4. $M = \begin{bmatrix} -9 & 6 & 16 \\ -4 & 5 & 8 \\ -4 & 2 & 7 \end{bmatrix}$

5. $M = \begin{bmatrix} 1 & 5 & -4 \\ -2 & -10 & 8 \\ -2 & -7 & 5 \end{bmatrix}$

6. $M = \begin{bmatrix} 1 & 3 & 0 \\ 3 & 1 & 0 \\ 0 & 0 & -2 \end{bmatrix}$

7. $M = \begin{bmatrix} 4 & -3 & 0 \\ 4 & -1 & -2 \\ 1 & -3 & 3 \end{bmatrix}$

8. $M = \begin{bmatrix} 2 & 0 & 0 \\ 0 & 0 & 0 \\ 1 & 0 & 0 \end{bmatrix}$

9. $M = \begin{bmatrix} 1 & 0 & 0 & 0 \\ 0 & 1 & 5 & -5 \\ 0 & 0 & 2 & 0 \\ 0 & 0 & 0 & 3 \end{bmatrix}$

10. $M = \begin{bmatrix} 1 & 0 & -2 \\ 0 & 1 & -2 \\ 1 & 0 & 4 \end{bmatrix}$

11. Let $M = \begin{bmatrix} a_{11} & a_{12} & a_{13} \\ 0 & a_{22} & a_{23} \\ 0 & 0 & a_{33} \end{bmatrix}$

a. Show that the matrices $M - a_{11}I$, $M - a_{22}I$, and $M - a_{33}I$ are singular.

b. Find all characteristic values of M.

c. Find the characteristic equation for M.

d. Find a characteristic vector for M with characteristic value a_{11}.

12. Let M be a square matrix and $\mathbf{v}$ be a characteristic vector for M with characteristic value d. Suppose that P is an invertible matrix and $N = P^{-1}MP$.

a. Show that $\mathbf{w} = P^{-1}\mathbf{v}$ is a characteristic vector for N.

b. Find the characteristic value of N associated with $\mathbf{w}$.

13. Let M be a square matrix and

$$x^n + a_{n-1}x^{n-1} + \cdots + a_1x + a_0 = 0$$

be the characteristic equation for M. Show that $|a_0|$ is equal to $|\det(M)|$. (*Hint:* Let $x = 0$.)

14. a. Show that if d is a characteristic value for the matrix M, then d^2 is a characteristic value for M^2.

b. What is the characteristic vector of M^2 associated with d^2?

c. How can you use characteristic values of M to get some of the characteristic values of M^n? Do you get all the characteristic values of M^n this way?

d. How is the set of characteristic vectors for M related to the set of characteristic vectors for M^n?

e. Let $M = \begin{bmatrix} 0 & -1 & 0 \\ 1 & 0 & 0 \\ 0 & 0 & 2 \end{bmatrix}$

Find the characteristic values and characteristic

vectors of M as well as M^2. Do the answers you obtained agree with your answers to parts **c** and d?

15. The set C of all differentiable functions defined on the interval $[0, 1]$ is a vector space with the addition of two vectors (functions) being the usual addition of functions, and the multiplication of a vector (function) by a scalar being the multiplication of a function by a real number. The differential operator D defined by

$$D: C \to C$$
$$f \to f' \quad \text{where } f' \text{ is the derivative of } f$$

is a linear function.

a. Show that the function $f(x) = e^x$ is a characteristic vector for D. What is its characteristic value?

b. Let a be an arbitrary real number. Find a function f_a in C such that f_a is a characteristic vector for D with characteristic value a.

★**16.** Let M be a 2×2 matrix with two distinct real characteristic roots. Show that this implies that M is similar to a diagonal matrix. (*Note*: In Example 3 we showed that the condition that the characteristic roots be distinct is necessary.)

17. Suppose M and N are 2×2 matrices such that the characteristic roots of M are the same as the characteristic roots of N, and these roots are real numbers.

a. Show that M is similar to N if the roots are distinct. (*Hint:* Use the result of Exercise 16.)

b. Find an example in which M is not similar to N but M and N have the same characteristic roots.

★**18.** The property that if M and N are similar matrices, then $\det(M) = \det(N)$ is a very important property of the determinant. We will now discuss another important function that shares this property. The **trace** of a square matrix M, denoted $\mathrm{tr}(M)$, is defined to be the sum of the entries on the main diagonal of M. For example,

$$\mathrm{tr}\left(\begin{bmatrix} 3 & 0 & -1 \\ 2 & -5 & 7 \\ 1 & 1 & 2 \end{bmatrix}\right) = 3 - 5 + 2 = 0$$

a. Show that if A and B are 2×2 matrices, then $\mathrm{tr}(AB) = \mathrm{tr}(BA)$.

b. Show that if A and B are 3×3 matrices, then $\mathrm{tr}(AB) = \mathrm{tr}(BA)$.

c. Can you see how to show that if A and B are $n \times n$ matrices, then $\mathrm{tr}(AB) = \mathrm{tr}(BA)$?

d. Use part **c** to show that

$$\mathrm{tr}(P^{-1}MP) = \mathrm{tr}(M)$$

where M and P are $n \times n$ matrices and P^{-1} exists

This proves that similar matrices have the same trace.

e. Use part **d** to show that the following matrices are not similar

$$M = \begin{bmatrix} 1 & 3 \\ 2 & 2 \end{bmatrix} \qquad N = \begin{bmatrix} 2 & 2 \\ 1 & 3 \end{bmatrix}$$

9.5 Diagonalizable Matrices—The Spectral Theorem

As we saw in Section 9.4, the relationship between a linear transformation and the vector space on which it operates can be understood best by looking at the vector space relative to a basis consisting of characteristic vectors for the transformation. We also saw that choosing such a basis is not always possible (see Example 3 and Exercise 1 in Section 9.4). In this section we continue our investigation of the structure of a vector space relative to the action of a single linear transformation.

Let $T: \mathbb{R}^n \to \mathbb{R}^n$ be a linear transformation. We look at subspaces of $\mathbb{R}^n$ related to the characteristic values of T.

THEOREM 9.6 Let $T: \mathbb{R}^n \to \mathbb{R}^n$ be a linear transformation, d be a characteristic value of T, and S_d be the set containing the zero vector along with all characteristic vectors of T with characteristic value d, that is, $S_d = \{\mathbf{v} \mid T(\mathbf{v}) = d\mathbf{v}\}$. Then S_d is a subspace of $\mathbb{R}^n$.

To prove this theorem, we merely check the two conditions for a subset of a vector space to be a subspace (Theorem 3.1):

1. If $\mathbf{v}_1$ and $\mathbf{v}_2$ are in S_d, then

$$T(\mathbf{v}_1 + \mathbf{v}_2) = T(\mathbf{v}_1) + T(\mathbf{v}_2) = d\mathbf{v}_1 + d\mathbf{v}_2 = d(\mathbf{v}_1 + \mathbf{v}_2)$$

So $\mathbf{v}_1 + \mathbf{v}_2$ is a characteristic vector for T with characteristic value d. Hence $\mathbf{v}_1 + \mathbf{v}_2$ is in S_d.

2. If c is a scalar and $\mathbf{v}$ is in S_d, then

$$T(c\mathbf{v}) = cT(\mathbf{v}) = c(d\mathbf{v}) = d(c\mathbf{v})$$

Therefore, if $c \neq 0$, then $c\mathbf{v}$ is a characteristic vector for T with characteristic value d. If $c = 0$, then $c\mathbf{v} = \mathbf{0}$. Thus, $c\mathbf{v}$ is in S_d for any value of c. Hence S_d is a subspace of $\mathbb{R}^n$. (*Note:* One particular subspace S_d that we have studied extensively is S_0. This is the set of all vectors $\mathbf{v}$ for which $T\mathbf{v} = 0\mathbf{v} = \mathbf{0}$, that is, the null space of T.)

In studying the examples and exercises of the last section, you may have noticed that whenever you found two characteristic vectors with different characteristic values, they were linearly independent. This was not just coincidental, as the next theorem shows.

THEOREM 9.7 Let $T: \mathbb{R}^n \to \mathbb{R}^n$ be a linear transformation and $\mathbf{u}_1, \mathbf{u}_2, \ldots, \mathbf{u}_k$ be characteristic vectors for T with distinct characteristic values. Then the vectors $\mathbf{u}_1, \mathbf{u}_2, \ldots, \mathbf{u}_k$ are linearly independent.

Although the general proof of this theorem is not difficult, we will discuss the proof only for the cases when $k = 2$ and $k = 3$. From this discussion it should be clear why the theorem is true in general.

Suppose $\mathbf{u}_1$ and $\mathbf{u}_2$ are characteristic vectors for T with distinct characteristic values d_1 and d_2. We will show that if we assume that $\mathbf{u}_1$ and $\mathbf{u}_2$ are linearly dependent, this leads to a contradiction, and therefore

$\mathbf{u}_1$ and $\mathbf{u}_2$ must be linearly independent. If $\mathbf{u}_1$ and $\mathbf{u}_2$ were linearly dependent, $\mathbf{u}_1$ would be a scalar multiple of $\mathbf{u}_2$, that is, $\mathbf{u}_1 = c\mathbf{u}_2$ for some scalar c. Using this, we could evaluate $T(\mathbf{u}_1)$ in two different ways:

$$T(\mathbf{u}_1) = d_1\mathbf{u}_1$$

and $$T(\mathbf{u}_1) = T(c\mathbf{u}_2) = cT(\mathbf{u}_2) = c(d_2\mathbf{u}_2) = d_2(c\mathbf{u}_2) = d_2\mathbf{u}_1$$

From this we would obtain $\quad d_1\mathbf{u}_1 = d_2\mathbf{u}_1$

But since $\mathbf{u}_1 \neq \mathbf{0}$ and $d_1 \neq d_2$, this is impossible, and so $\mathbf{u}_1$ and $\mathbf{u}_2$ must have been linearly independent. Thus, we have shown that if two characteristic vectors have distinct characteristic values, they are linearly independent.

Now we look at the case for $k = 3$. Suppose that $\mathbf{u}_1, \mathbf{u}_2, \mathbf{u}_3$ are characteristic vectors for T with distinct characteristic values d_1, d_2, d_3. Again we use the method of contradiction to show that $\mathbf{u}_1, \mathbf{u}_2, \mathbf{u}_3$ are linearly independent. If $\mathbf{u}_1, \mathbf{u}_2, \mathbf{u}_3$ were linearly dependent, then we could find scalars s_1, s_2, s_3, not all zero, such that

$$\mathbf{0} = s_1\mathbf{u}_1 + s_2\mathbf{u}_2 + s_3\mathbf{u}_3$$

The scalar s_3 could not be zero, for otherwise we would have a nontrivial dependency relation between $\mathbf{u}_1$ and $\mathbf{u}_2$ which would contradict the result we have just proved for $k = 2$. So we can solve this equation for $\mathbf{u}_3$, and express $\mathbf{u}_3$ as a linear combination of the other vectors.

$$\mathbf{u}_3 = c_1\mathbf{u}_1 + c_2\mathbf{u}_2 \qquad \text{for scalars } c_1 \text{ and } c_2$$

Now we could evaluate $T(\mathbf{u}_3)$ in two different ways:

$$T(\mathbf{u}_3) = d_3\mathbf{u}_3 = d_3(c_1\mathbf{u}_1 + c_2\mathbf{u}_2) = d_3c_1\mathbf{u}_1 + d_3c_2\mathbf{u}_2$$

and $T(\mathbf{u}_3) = T(c_1\mathbf{u}_1 + c_2\mathbf{u}_2) = c_1T(\mathbf{u}_1) + c_2T(\mathbf{u}_2) = c_1d_1\mathbf{u}_1 + c_2d_2\mathbf{u}_2$

So we could conclude that

$$(d_3c_1)\mathbf{u}_1 + (d_3c_2)\mathbf{u}_2 = (c_1d_1)\mathbf{u}_1 + (c_2d_2)\mathbf{u}_2$$

Now $\mathbf{u}_1$ and $\mathbf{u}_2$ are two characteristic vectors with distinct characteristic values, and so, by what we have just proved, $\mathbf{u}_1$ and $\mathbf{u}_2$ are linearly independent. Therefore the coefficients of $\mathbf{u}_1$ (and of $\mathbf{u}_2$) on the left side of this equation must equal the coefficients of $\mathbf{u}_1$ (and of $\mathbf{u}_2$) on the right side of this equation. So

$$d_3c_1 = c_1d_1$$

But $c_1 \neq 0$ (otherwise $\mathbf{u}_3$ would be a scalar multiple of $\mathbf{u}_2$, which would contradict the fact that two characteristic vectors with distinct characteristic values are linearly independent). Therefore, we see that $d_3 = d_1$. Since we assumed that the characteristic values were distinct, this is a contradiction; therefore, the vectors $\mathbf{u}_1$, $\mathbf{u}_2$, $\mathbf{u}_3$ must have been linearly independent.

An important consequence of this theorem is Theorem 9.8.

THEOREM 9.8 If $T: \mathbb{R}^n \to \mathbb{R}^n$ is a linear transformation and the characteristic equation for T has n distinct roots $d_1, d_2, \ldots, d_n$, then there is a basis $\{\mathbf{u}_1, \mathbf{u}_2, \ldots, \mathbf{u}_n\}$ of $\mathbb{R}^n$ so that the matrix for T relative to this basis is the diagonal matrix

$$\begin{bmatrix} d_1 & 0 & 0 & \cdots & 0 & 0 \\ 0 & d_2 & 0 & \cdots & 0 & 0 \\ \vdots & \vdots & & \ddots & & \vdots \\ 0 & 0 & 0 & \cdots & 0 & d_n \end{bmatrix}$$

The proof of this theorem is quite easy. Using the technique discussed in Section 9.4, we can find characteristic vectors $\mathbf{u}_1, \mathbf{u}_2, \ldots, \mathbf{u}_n$ corresponding to the n distinct characteristic values $d_1, d_2, \ldots, d_n$. Using Theorem 9.7, we see that these vectors are linearly independent and are therefore a basis for $\mathbb{R}^n$.

EXAMPLE 1 Let M be a matrix with characteristic equation

$$0 = (x + 1)(x - 2)(x + 3)$$

From Theorem 9.8 we see that M is similar to the matrix

$$D = \begin{bmatrix} -1 & 0 & 0 \\ 0 & 2 & 0 \\ 0 & 0 & -3 \end{bmatrix}$$

(*Note:* We do not actually have to find the basis $\{\mathbf{u}_1, \mathbf{u}_2, \mathbf{u}_3\}$ to know that M is similar to D in this case.) ■

Now we discuss another class of matrices that are always similar to diagonal matrices.

DEFINITION 9.6 A matrix M with real coefficients such that $M = M^t$ (recall that M^t is the transpose of M) is called a **real symmetric matrix**.

Real symmetric matrices have three important properties. We state them in the following theorem.

THEOREM 9.9

Spectral Theorem

Let M be an $n \times n$ real symmetric matrix. Then

1. All characteristic roots of M are real numbers (there are no complex roots).

2. M is similar to a diagonal matrix—that is, there is a basis $\{\mathbf{u}_1, \mathbf{u}_2, \ldots, \mathbf{u}_n\}$ so that the matrix D representing the linear transformation induced by M relative to this basis is diagonal.

3. The basis $\{\mathbf{u}_1, \mathbf{u}_2, \ldots, \mathbf{u}_n\}$ can be chosen to be orthonormal.

This is a very powerful and important theorem, but its proof is somewhat beyond the scope of this book. We will indicate what is behind the proof of each part of this theorem and the significance of each part.

Part 1: This part says that just by looking at the form of the matrix, we can determine that it has no complex roots. As we saw in Exercise 1 in Section 9.4, if a matrix has complex roots, it cannot be diagonalized. The proof of this part of the Spectral Theorem involves the use of complex-valued vectors and matrices and the definition of the scalar product for vectors having complex entries.

Part 2: The proof of Part 2 is based on the fact that if $\mathbf{u}$ is a characteristic vector for the symmetric matrix M, then $\mathbf{u}^{\perp}$, the subset of all vectors perpendicular to $\mathbf{u}$, is a subspace of $\mathbb{R}^n$, and furthermore, that if $\mathbf{v}$ is a vector of $\mathbf{u}^{\perp}$, then $T(\mathbf{v})$ is also in $\mathbf{u}^{\perp}$. Since the dimension of $\mathbf{u}^{\perp}$ is $n - 1$, a reduction argument is used to complete the proof. (See Exercises 10 and 12.)

Part 3: The important thing about the fact that the new basis for $\mathbb{R}^n$ can be chosen to be orthonormal is that this change of basis preserves the dot product and, therefore, the size and shape of objects is not changed. The proof of this part depends primarily on the fact that if $\mathbf{u}_i$ and $\mathbf{u}_j$ are characteristic vectors for a real symmetric matrix with distinct characteristic values, then $\mathbf{u}_i$ and $\mathbf{u}_j$ are not only linearly independent, but they are orthogonal. This fact, coupled with the orthogonalization procedure discussed in Section 6.2, allows us to construct an orthonormal basis of characteristic vectors for M (see Exercise 11).

EXAMPLE 2 Let $T\colon \mathbb{R}^3 \to \mathbb{R}^3$ be the linear transformation represented by the matrix

$$M = \begin{bmatrix} 1 & 3 & 0 \\ 3 & 1 & 0 \\ 0 & 0 & 4 \end{bmatrix}$$

relative to the standard basis. Find an orthonormal basis $\{\mathbf{u}_1, \mathbf{u}_2, \mathbf{u}_3\}$ so that T is represented by a diagonal matrix.

Theorem 9.9 guarantees that it is possible to find such a basis. We begin, as in Section 9.4, by solving the characteristic equation:

$$\begin{aligned} 0 &= \det(M - xI) \\ &= \begin{vmatrix} 1-x & 3 & 0 \\ 3 & 1-x & 0 \\ 0 & 0 & 4-x \end{vmatrix} \\ &= (1-x)[(1-x)(4-x)] - 3[3(4-x)] \\ &= (x^2 - 2x - 8)(4-x) = -(x+2)(x-4)^2 \end{aligned}$$

So the characteristic values are

$$d = -2 \qquad \text{and} \qquad d = 4 \qquad \text{(a double root)}$$

To find a characteristic vector for $d = -2$, we reduce $M + 2I$.

$$M + 2I = \begin{bmatrix} 3 & 3 & 0 \\ 3 & 3 & 0 \\ 0 & 0 & 6 \end{bmatrix} \xrightarrow{\text{reduces to}} \begin{bmatrix} 3 & 3 & 0 \\ 0 & 0 & 1 \\ 0 & 0 & 0 \end{bmatrix}$$

So a characteristic vector for $d = -2$ is

$$\begin{bmatrix} 1 \\ -1 \\ 0 \end{bmatrix}$$

To find characteristic vectors for $d = 4$, we reduce the matrix $M - 4I$:

$$M - 4I = \begin{bmatrix} -3 & 3 & 0 \\ 3 & -3 & 0 \\ 0 & 0 & 0 \end{bmatrix} \xrightarrow{\text{reduces to}} \begin{bmatrix} -3 & 3 & 0 \\ 0 & 0 & 0 \\ 0 & 0 & 0 \end{bmatrix}$$

So the characteristic vectors with characteristic value 4 are all of the form

$$\begin{bmatrix} x \\ x \\ z \end{bmatrix}$$

We can easily find an orthogonal basis for this subspace of $\mathbb{R}^3$, namely

$$\begin{bmatrix} 1 \\ 1 \\ 0 \end{bmatrix} \quad \text{and} \quad \begin{bmatrix} 0 \\ 0 \\ 1 \end{bmatrix}$$

These two vectors are orthogonal to the vector

$$\begin{bmatrix} 1 \\ -1 \\ 0 \end{bmatrix}$$

So, by dividing each of these three characteristic vectors by its length, we obtain the orthonormal basis

$$\left\{ \mathbf{u}_1 = (1/\sqrt{2})\begin{bmatrix} 1 \\ -1 \\ 0 \end{bmatrix},\ \mathbf{u}_2 = (1/\sqrt{2})\begin{bmatrix} 1 \\ 1 \\ 0 \end{bmatrix},\ \mathbf{u}_3 = \begin{bmatrix} 0 \\ 0 \\ 1 \end{bmatrix} \right\}$$

The matrix for T relative to this basis is

$$D = \begin{bmatrix} -2 & 0 & 0 \\ 0 & 4 & 0 \\ 0 & 0 & 4 \end{bmatrix} \quad \blacksquare$$

Exercise Set 9.5

1. Let $T: \mathbb{R}^3 \to \mathbb{R}^3$ be reflection about the z axis.

a. Find S_1 and S_{-1} (see Theorem 9.6).

b. Show directly that if $\mathbf{v}$ is a nonzero vector of S_1 and $\mathbf{w}$ is a nonzero vector of S_{-1}, then $\mathbf{v}$ and $\mathbf{w}$ are linearly independent.

2. Determine which of the following statements are true and which are false. Justify your answers.

a. If M is a singular matrix, then 0 is a characteristic value for M.

b. Every matrix with characteristic equation

$$0 = (x - 1)(x + 1)(x - 5)$$

is similar to a diagonal matrix.

c. Every matrix with characteristic equation

$$0 = (x - 1)(x - 1)(x - 5)$$

is similar to a diagonal matrix.

d. There is a real symmetric matrix M that has characteristic equation

$$0 = x^2 + 1$$

e. There is a real symmetric matrix that has characteristic equation

$$0 = (x + 1)^2$$

★**3.** Suppose M is a 3×3 matrix with characteristic equation

$$0 = (x - 1)^2(x + 1)$$

Show that M is similar to a diagonal matrix if and only if the dimension of the subspace S_1 is 2.

For each of the matrices M listed in Exercises 4–9, find an orthogonal matrix B such that $D = B^{-1}MB$ is diagonal. Also find D.

4. $M = \begin{bmatrix} 1 & 2 \\ 2 & 1 \end{bmatrix}$ **5.** $M = \begin{bmatrix} 1 & 1 \\ 1 & 1 \end{bmatrix}$

6. $M = \begin{bmatrix} 1 & 1 & 1 \\ 1 & 1 & 1 \\ 1 & 1 & 1 \end{bmatrix}$

7. $M = \begin{bmatrix} -1 & -6 & 0 \\ -6 & 2 & -6 \\ 0 & -6 & 5 \end{bmatrix}$

8. $M = \begin{bmatrix} 2 & 2 & -2 \\ 2 & -1 & 4 \\ -2 & 4 & -1 \end{bmatrix}$

9. $M = \begin{bmatrix} 9 & 3 & -3 \\ 3 & 1 & -1 \\ -3 & -1 & 1 \end{bmatrix}$

★**10.** Suppose that M is an $n \times n$ symmetric matrix with characteristic vector $\mathbf{u}$. Let $\mathbf{v}$ be a vector perpendicular to $\mathbf{u}$. Show that $M(\mathbf{v})$ is also perpendicular to $\mathbf{u}$. (*Hint:* Use Theorem 8.1.)

★**11.** Let M be a real symmetric matrix and

$\mathbf{u}_1$, $\mathbf{u}_2$ be characteristic vectors with distinct characteristic values d_1 and d_2. Evaluate $M\mathbf{u}_1 \cdot \mathbf{u}_2$ two ways to obtain

$$M\mathbf{u}_1 \cdot \mathbf{u}_2 = d_1(\mathbf{u}_1 \cdot \mathbf{u}_2)$$

and

$$M\mathbf{u}_1 \cdot \mathbf{u}_2 = d_2(\mathbf{u}_1 \cdot \mathbf{u}_2)$$

(*Hint:* Use Theorem 8.1.) From this show that $\mathbf{u}_1$ is perpendicular to $\mathbf{u}_2$.

★**12.** (Proof of Spectral Theorem for 2×2 matrices.) Let

$$M = \begin{bmatrix} a & b \\ b & c \end{bmatrix}$$

be a real symmetric matrix.

a. Find the characteristic equation for M.

b. Using the quadratic formula, solve the characteristic equation for x, and show that both characteristic values of M are real numbers.

c. Show that if both characteristic values of M are equal, then M is a diagonal matrix.

d. If the characteristic values d_1 and d_2 of M are distinct, and if $\mathbf{u}_1$ is a characteristic vector with characteristic value d_1, then show that any nonzero vector $\mathbf{u}_2$ perpendicular to $\mathbf{u}_1$ is a characteristic vector with characteristic value d_2. (*Hint:* Use Exercises 10 and 11.)

e. Use the results you have proved in parts **a–d** to prove that there is an orthonormal basis $\{\mathbf{u}_1, \mathbf{u}_2\}$ of $\mathbb{R}^2$ so that the linear transformation represented by M has a diagonal matrix relative to the basis $\{\mathbf{u}_1, \mathbf{u}_2\}$.

Applications 9a Finding Trends in Statistical Data—Stochastic Matrices

In this section we see how linear algebra can be applied to predict trends in statistical data. We begin by considering a specific example. The data in this example are approximate, but are based on actual statistical information on automobile production and sales.

The problem we will discuss is this: Suppose that we start with a group of 10,000 owners of American-made cars (selected at random). We want to predict the number of people in this sample who will own Cadillacs after a given number of years, and to analyze trends in Cadillac ownership among these people.

Our initial data are these: About 4% of American cars on the road are Cadillacs; so we can assume that, of the 10,000 car owners in our sample, 400 (4% of 10,000) own Cadillacs and the remaining 9,600 own other makes of cars. Furthermore, the probability that a Cadillac owner will either keep his Cadillac or buy a new Cadillac is 0.78 (that is, 78% of the people who own Cadillacs will either keep their Cadillac or buy another Cadillac). The probability that a person who doesn't own a Cadillac will sell his car and buy a Cadillac is 0.01. Table 9.1 expresses these statistical data more concisely. The table expresses the various probabilities of buying or keeping a Cadillac or other type of car depending on whether the person owns a Cadillac or another type of car.

Using this table, we see that if we should start out with a sample of 100 Cadillac owners, then, after one year, (on the average) 78 of them will still own Cadillacs, and the remaining 22 will have bought other makes of cars. On the other hand, if we start with a sample of 100 owners of other

Table 9.1

Probability of Buying or Keeping	If Owns Cadillac	Other
Cadillac	0.78	0.01
Other	0.22	0.99

makes of cars, then after one year on the average 99 will still own other makes of cars and 1 will have bought a Cadillac. In general, if we begin with a sample of x Cadillac owners and y owners of other makes of cars, then after one year, on the average

$$0.78x + 0.01y \qquad \text{will own Cadillacs}$$

$$0.22x + 0.99y \qquad \text{will own other makes of cars}$$

We can use matrix notation to put this in a very concise form. Let

$$M = \begin{bmatrix} 0.78 & 0.01 \\ 0.22 & 0.99 \end{bmatrix}$$

Then
$$M\begin{bmatrix} x \\ y \end{bmatrix} = \begin{bmatrix} 0.78 & 0.01 \\ 0.22 & 0.99 \end{bmatrix}\begin{bmatrix} x \\ y \end{bmatrix} = \begin{bmatrix} 0.78x + 0.01y \\ 0.22x + 0.99y \end{bmatrix}$$

We interpret this as follows: In the vector $\begin{bmatrix} x \\ y \end{bmatrix}$, the first coordinate x tells us how many people in our sample own Cadillacs, and the second coordinate y tells us how many people in our sample own other makes of cars. The vector

$$M\begin{bmatrix} x \\ y \end{bmatrix} = \begin{bmatrix} 0.78x + 0.01y \\ 0.22x + 0.99y \end{bmatrix}$$

tells the number of Cadillac owners $(0.78x + 0.01y)$ and the number of owners of other makes of cars $(0.22x + 0.99y)$ after one year. The matrix M is called the **transition matrix**.

$$M: \begin{bmatrix} \text{Number who own Cadillacs} \\ \text{(now)} \\ \text{Number who own other makes} \\ \text{(now)} \end{bmatrix} \rightarrow \begin{bmatrix} \text{Number who own Cadillacs} \\ \text{(1 year later)} \\ \text{Number who own other makes} \\ \text{(1 year later)} \end{bmatrix}$$

Now we look at our particular sample of 10,000 people. We assumed that initially 400 of these people owned Cadillacs and 9,600 owned other makes of cars. We represent this initial data by the vector

$$\mathbf{v}_0 = \begin{bmatrix} 400 \\ 9{,}600 \end{bmatrix}$$

To find the number of Cadillac owners and non-Cadillac owners in our sample after one year, we multiply $\mathbf{v}_0$ by the transition matrix M, that is,

$$\mathbf{v}_1 = M\mathbf{v}_0 = \begin{bmatrix} 0.78 & 0.01 \\ 0.22 & 0.99 \end{bmatrix} \begin{bmatrix} 400 \\ 9{,}600 \end{bmatrix} = \begin{bmatrix} 408 \\ 9{,}592 \end{bmatrix}$$

To find out how many of our 10,000 car owners will own Cadillacs and how many will own other makes of cars after 2 years, we multiply $\mathbf{v}_1$ by the transition matrix M to obtain

$$\mathbf{v}_2 = M\mathbf{v}_1 = \begin{bmatrix} 0.78 & 0.01 \\ 0.22 & 0.99 \end{bmatrix} \begin{bmatrix} 408 \\ 9{,}592 \end{bmatrix} = \begin{bmatrix} 414.16 \\ 9{,}585.84 \end{bmatrix}$$

which rounds off to

$$\mathbf{v}_2 = M\mathbf{v}_1 = M^2\mathbf{v}_0 = \begin{bmatrix} 414 \\ 9{,}586 \end{bmatrix}$$

Continuing in this way we obtain Table 9.2.

Table 9.2

	$\mathbf{v}_0$	$M\mathbf{v}_0$	$M^2\mathbf{v}_0$	$M^3\mathbf{v}_0$	$M^4\mathbf{v}_0$	$M^5\mathbf{v}_0$
Cadillacs	400	408	414	419	423	426
Other Makes	9,600	9,592	9,586	9,581	9,577	9,574
	Initial	1 year	2 year	3 year	4 year	5 year

Now we get to the heart of the problem, can we predict any trends? On the one hand, we see that Cadillac ownership is increasing each year; but, on the other hand, we note that the increase in the number of Cadillac owners from year to year gets smaller and smaller (from 8 cars between the first two years to 3 cars between the fourth and fifth years). So, to find a trend (if there is one), we must see whether the number of Cadillacs will increase from year to year, whether it will stabilize at some point, or whether it will oscillate to such a degree that no trend can be predicted. In other words, we are asking if the sequence of vectors

$$\mathbf{v}_0, \mathbf{v}_1 = M\mathbf{v}_0, \mathbf{v}_2 = M^2\mathbf{v}_0, \ldots, \mathbf{v}_i = M^i\mathbf{v}_0, \ldots$$

converges or not.

We could continue computing more values of $M^i\mathbf{v}$. This is not only time consuming, but it does not tell us what the long-range behavior of this sequence will be. To get a better understanding of this situation, we will put this question in a more general framework.

The transition matrix M is an example of a special type of matrix called a stochastic matrix.

DEFINITION 9.7

An $n \times n$ matrix M is called a **stochastic matrix** if all of its entries m_{ij} satisfy the condition that $0 \leq m_{ij} \leq 1$ and the sum of the entries in each column of M is 1.

In general, transition matrices are stochastic matrices since the entries of a transition matrix are probabilities (and therefore are numbers between 0 and 1), and the probabilities in any particular column represent the probability of going from one state (the one corresponding to that column) to each of the possible states. Therefore, the sum of the entries in any column represents the probability of an event that is certain to occur, which is 1.

In our problem, we had a stochastic matrix M, and what we wanted to determine was whether the sequence of vectors

$$\mathbf{v}_0, M\mathbf{v}_0, M^2\mathbf{v}_0, \ldots, M^i\mathbf{v}_0, \ldots$$

converged or not for a particular vector $\mathbf{v}_0$. We also wanted to know to which vector it converged (if it did indeed converge).

We now analyze this problem for a general stochastic matrix

$$M = \begin{bmatrix} p & q \\ 1-p & 1-q \end{bmatrix} \quad \text{where } 0 \leq p \leq 1 \quad \text{and} \quad 0 \leq q \leq 1$$

We begin with the following simple result:

THEOREM 9.10

Let M be the 2×2 stochastic matrix

$$M = \begin{bmatrix} p & q \\ 1-p & 1-q \end{bmatrix} \quad \text{where } 0 \leq p \leq 1 \quad \text{and} \quad 0 \leq q \leq 1$$

Then the characteristic values of M are

$$1 \quad \text{and} \quad d = p - q$$

It is easy to see that this theorem is true since both of the matrices

$$M - 1I = \begin{bmatrix} p-1 & q \\ 1-p & -q \end{bmatrix}$$

and

$$M - dI = M - (p-q)I = \begin{bmatrix} q & q \\ 1-p & 1-p \end{bmatrix}$$

are singular (see Theorem 9.5). (If this proof doesn't appeal to you, solve the characteristic equation $0 = \det(M - xI)$ using the quadratic formula, and you will get the answer directly.)

If
$$M \neq \begin{bmatrix} 1 & 0 \\ 0 & 1 \end{bmatrix}$$

the characteristic values 1 and $d = p - q$ of M are distinct. So if $\mathbf{u}_1$ and $\mathbf{u}_2$ are characteristic vectors with characteristic values 1 and d, respectively, then (by Theorem 9.7) $\mathbf{u}_1$ and $\mathbf{u}_2$ are linearly independent. So $\{\mathbf{u}_1, \mathbf{u}_2\}$ is a basis for $\mathbb{R}^2$. Therefore, we can find scalars a_1 and a_2 so that

$$\mathbf{v}_0 = a_1\mathbf{u}_1 + a_2\mathbf{u}_2$$

Using the fact that $\mathbf{u}_1$ and $\mathbf{u}_2$ are characteristic vectors for M with characteristic values 1 and d, we have

$$M\mathbf{v}_0 = M(a_1\mathbf{u}_1 + a_2\mathbf{u}_2) = a_1M(\mathbf{u}_1) + a_2M(\mathbf{u}_2) = a_1\mathbf{u}_1 + d(a_2\mathbf{u}_2)$$

and, similarly
$$\begin{aligned} M^2\mathbf{v}_0 &= M(M\mathbf{v}_0) = M(a_1\mathbf{u}_1 + da_2\mathbf{u}_2) \\ &= a_1M\mathbf{u}_1 + da_2M\mathbf{u}_2 = a_1\mathbf{u}_1 + d^2(a_2\mathbf{u}_2) \end{aligned}$$

In general, we have

$$M^i\mathbf{v}_0 = a_1\mathbf{u}_1 + d^i(a_2\mathbf{u}_2) \qquad \text{where } d = p - q$$

See Figure 9.8 for a graph of these vectors.

Except in the cases in which $p = 0$ and $q = 1$, or $p = 1$ and $q = 0$, the absolute value of d is smaller than 1. So, since $|d| < 1$, d^i approaches zero as i tends to infinity. Therefore, as long as M is neither the matrix

$$\begin{bmatrix} 1 & 0 \\ 0 & 1 \end{bmatrix} \quad \text{nor} \quad \begin{bmatrix} 0 & 1 \\ 1 & 0 \end{bmatrix}$$

Figure 9.8

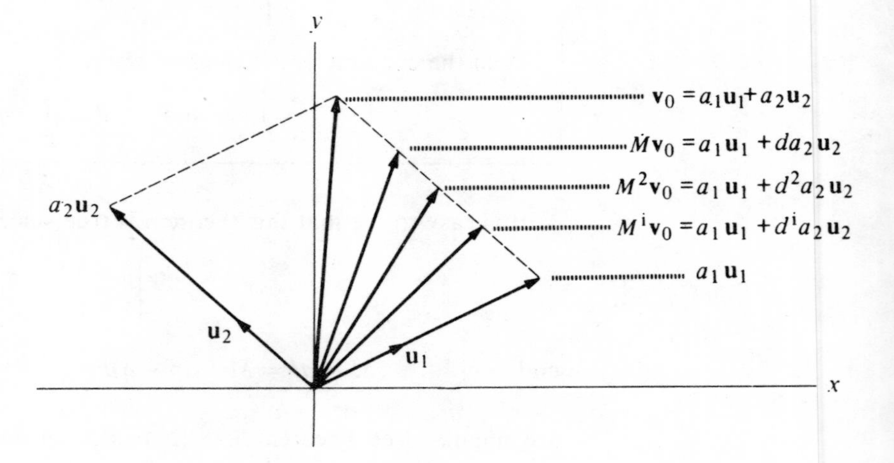

the sequence

$$\mathbf{v}_0, M\mathbf{v}_0, M^2\mathbf{v}_0, \ldots, M^i\mathbf{v}_0, \ldots$$

converges to the vector $a_1\mathbf{u}_1$ where $\mathbf{v}_0 = a_1\mathbf{u}_1 + a_2\mathbf{u}_2$ and $\mathbf{u}_1$ and $\mathbf{u}_2$ are characteristic vectors for M with characteristic values 1 and $p - q$.

There is a simpler way to find the vector $a_1\mathbf{u}_1$ than by expressing $\mathbf{v}_0$ as a linear combination of the vectors $\mathbf{u}_1$ and $\mathbf{u}_2$. It is based on the following simple observation. If $\mathbf{v} = \begin{bmatrix} x \\ y \end{bmatrix}$ is an arbitrary vector, then

$$M\begin{bmatrix} x \\ y \end{bmatrix} = \begin{bmatrix} p & q \\ 1-p & 1-q \end{bmatrix}\begin{bmatrix} x \\ y \end{bmatrix} = \begin{bmatrix} px + qy \\ (1-p)x + (1-q)y \end{bmatrix}$$

and the sum of the coefficients of $M\mathbf{v}$ is

$$(px + qy) + [(1-p)x + (1-q)y] = x + y$$

which is the same as the sum of the coefficients of the vector

$$\mathbf{v} = \begin{bmatrix} x \\ y \end{bmatrix}$$

So, for any vector $\mathbf{v}$, the sum of the coefficients of $\mathbf{v}$ is the same as the sum of the coefficients of $M\mathbf{v}$ (this does not mean that $\mathbf{v}$ and $M\mathbf{v}$ have the same length). Therefore, since the sum of the coefficients of any vector of the sequence

$$\mathbf{v}_0, M\mathbf{v}_0, M^2\mathbf{v}_0, \ldots, M^i\mathbf{v}_0, \ldots$$

is the same as the sum of the coefficients of $\mathbf{v}_0$, we see that the sum of the coefficients of $a_1\mathbf{u}_1$ is also the same as the sum of the coefficients of $\mathbf{v}_0$. Therefore, we have proved the following theorem:

THEOREM 9.11

Let M be a 2×2 stochastic matrix which is neither the identity matrix

$$I = \begin{bmatrix} 1 & 0 \\ 0 & 1 \end{bmatrix} \quad \text{nor the matrix} \quad \begin{bmatrix} 0 & 1 \\ 1 & 0 \end{bmatrix}$$

Then, for any vector $\mathbf{v}_0$ of $\mathbb{R}^2$, the sequence

$$\mathbf{v}_0, M\mathbf{v}_0, M^2\mathbf{v}_0, \ldots, M^i\mathbf{v}_0, \ldots$$

converges to a vector $\mathbf{v}_s$ (s stands for stability), where $\mathbf{v}_s$ is the characteristic vector for M with characteristic value 1 such that the sum of the coefficients of $\mathbf{v}_s$ is the same as the sum of the coefficients of $\mathbf{v}_0$.

We now apply these results to predict trends in Cadillac ownership. In this case, the initial vector $\mathbf{v}_0$ is the vector

$$\mathbf{v}_0 = \begin{bmatrix} 400 \\ 9{,}600 \end{bmatrix}$$

and the transition matrix M is the stochastic matrix

$$M = \begin{bmatrix} 0.78 & 0.01 \\ 0.22 & 0.99 \end{bmatrix}$$

From Theorem 9.11, we know that the sequence

$$\mathbf{v}_0, M\mathbf{v}_0, M^2\mathbf{v}_0, \ldots, M^i\mathbf{v}_0, \ldots$$

(which expresses the trends in Cadillac ownership among the people in our sample) converges to a vector $\mathbf{v}_s$ where $\mathbf{v}_s$ is the characteristic vector for M with characteristic value 1 such that the sum of the coefficients of $\mathbf{v}_s$ is equal to the sum of the coefficients of $\mathbf{v}_0$, which is $400 + 9{,}600 = 10{,}000$.

To find $\mathbf{v}_s$, we begin by finding the characteristic vectors for M with characteristic value 1. These are the nonzero vectors of the null space of

$$M - 1I = \begin{bmatrix} -0.22 & 0.01 \\ 0.22 & -0.01 \end{bmatrix}$$

(see Theorem 9.5). Thus, the characteristic vectors for M with characteristic value 1 are all nonzero vectors of the form

$$\begin{bmatrix} x \\ 22x \end{bmatrix}$$

Since the sum of the coefficients of $\mathbf{v}_s$ is 10,000, we see that x satisfies the equation

$$x + 22x = 10{,}000$$

Therefore
$$x = \frac{10{,}000}{23} = 434.78$$

and
$$22x = 9{,}565.22$$

We round this off to the nearest integer to obtain

$$\mathbf{v}_s = \begin{bmatrix} 435 \\ 9{,}565 \end{bmatrix}$$

Therefore, the trend in Cadillac ownership is that the number of Cadillacs owned by the people in our sample will increase until it reaches 435 at which point it will remain constant. (The Cadillac Corporation hopes their advertising can change this trend.)

We conclude this section by stating the generalization of Theorem 9.11 for $n \times n$ stochastic matrices.

THEOREM 9.12 Let M be a stochastic matrix such that for some power m of M, no entry in the matrix M^m is zero. Then, for any vector $\mathbf{v}$, the sequence

$$\mathbf{v}, M\mathbf{v}, M^2\mathbf{v}, \ldots, M^i\mathbf{v}, \ldots$$

converges to a vector $\mathbf{v}_s$, where $\mathbf{v}_s$ is the characteristic vector for M with characteristic value 1 such that the sum of the coefficients of $\mathbf{v}_s$ is equal to the sum of the coefficients of $\mathbf{v}$.

Exercises for Applications 9a

1. Let

$$M = \begin{bmatrix} 0.3 & 0.1 \\ 0.7 & 0.9 \end{bmatrix} \quad \text{and} \quad \mathbf{v}_0 = \begin{bmatrix} 2 \\ 5 \end{bmatrix}$$

a. Find the vectors $\mathbf{v}_0$, $M\mathbf{v}_0$, $M^2\mathbf{v}_0$, and $M^3\mathbf{v}_0$ and plot these vectors on the same pair of coordinate axes.

b. Find the vector $\mathbf{v}_s$ so that the sequence

$$\mathbf{v}_0, M\mathbf{v}_0, M^2\mathbf{v}_0, \ldots, M^i\mathbf{v}_0, \ldots$$

converges to $\mathbf{v}_s$.

2. Find characteristic vectors for the stochastic matrix

$$M = \begin{bmatrix} p & q \\ 1-p & 1-q \end{bmatrix}$$

3. Let M be an arbitrary 3×3 stochastic matrix. Show that 1 is a characteristic value for M.

4. Suppose two ponds are connected by a small stream as shown in Figure 9.9. In an average year about 15 percent of the fish who started the year in Pond A will have migrated to Pond B, and about 20 percent of the fish who started the year in Pond B will have migrated to Pond A. If the Fish and Wildlife Service plants 1,000 fish in Pond A, how many of these fish will be in each of the ponds (assuming that none die) after 1 year, after 2 years, after 3 years? What will be the stable population distribution of these fish?

Figure 9.9

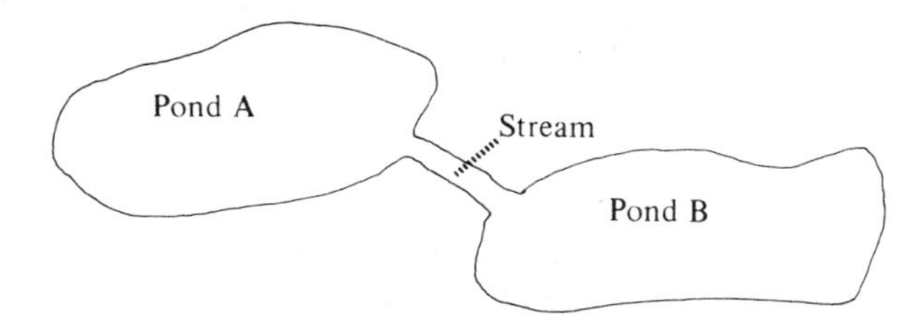

Applications 9b Graphing Equations of the Form $ax^2 + bxy + cy^2 = d$

In this section we consider a problem that you may have seen before. We will analyze equations of the form

$$ax^2 + bxy + cy^2 = d$$

We will see how the Spectral Theorem (9.9) can be used to graph the set of all points (x, y) that satisfy such an equation. If you have already studied such equations using rotation of axes, I think you will find the method described here far more efficient. In addition, this method, using the Spectral Theorem, can be easily generalized to equations in three unknowns of the form

$$a_1x^2 + a_2y^2 + a_3z^2 + b_1yz + b_2xz + b_3xy = d$$

We begin our study of the equation

$$ax^2 + bxy + cy^2 = d \qquad d \neq 0$$

In the special case where $b = 0$, this equation becomes

$$ax^2 + cy^2 = d$$

which can be written in a more familiar form as

$$\frac{x^2}{d/a} + \frac{y^2}{d/c} = 1 \qquad \text{when } a \neq 0, c \neq 0$$

The graph of an equation of this form is a conic section with its axes in the direction of the standard basis vectors $\mathbf{e}_1$ and $\mathbf{e}_2$ as is shown in Figure 9.10.

In what follows we will show that the graphs of equations of the form

$$ax^2 + bxy + cy^2 = d \qquad \text{where } b \neq 0$$

are also conic sections. However, their axes are not in the directions of the standard basis vectors $\mathbf{e}_1$ and $\mathbf{e}_2$. We will apply linear algebra to find suitable basis vectors $\mathbf{u}_1$ and $\mathbf{u}_2$ so that the axes of a conic section represented by this equation are in the direction of the vectors $\mathbf{u}_1$ and $\mathbf{u}_2$.

We introduce the symmetric matrix M

$$M = \begin{bmatrix} a & b/2 \\ b/2 & c \end{bmatrix}$$

The entries on the main diagonal of M are the coefficients of x^2 and y^2, and the off-diagonal entries are each half of the coefficient of xy.

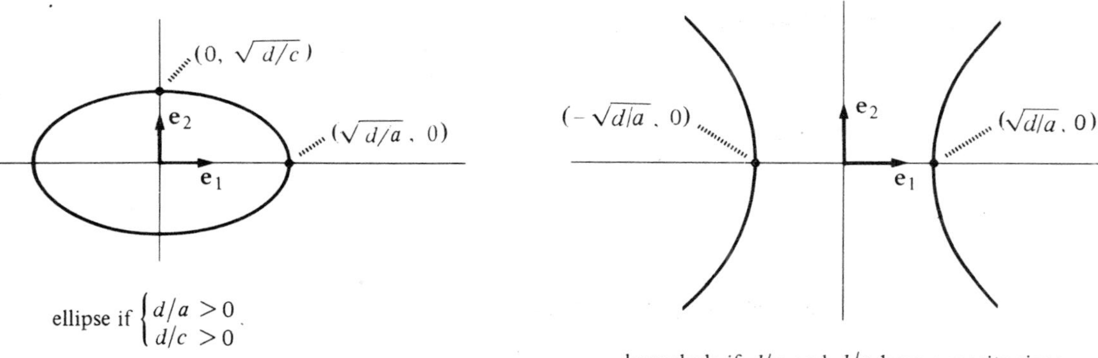

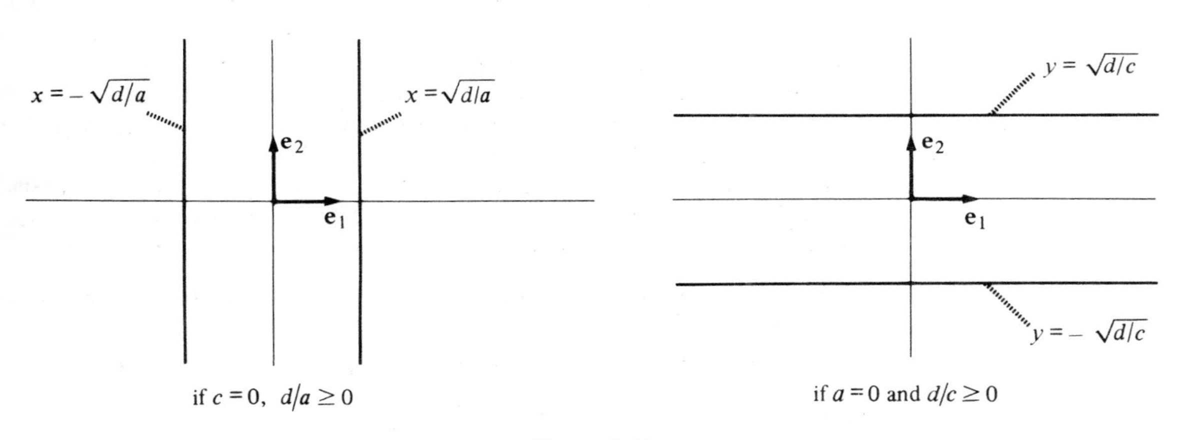

Figure 9.10

We note that

$$M\begin{bmatrix}x\\y\end{bmatrix} = \begin{bmatrix}a & b/2\\b/2 & c\end{bmatrix}\begin{bmatrix}x\\y\end{bmatrix} = \begin{bmatrix}ax + (b/2)y\\(b/2)x + cy\end{bmatrix}$$

So

$$\begin{aligned}\begin{bmatrix}x\\y\end{bmatrix}\cdot M\begin{bmatrix}x\\y\end{bmatrix} &= \begin{bmatrix}x\\y\end{bmatrix}\cdot\begin{bmatrix}ax + (b/2)y\\(b/2)x + cy\end{bmatrix}\\ &= ax^2 + (b/2)yx + (b/2)xy + cy^2\\ &= ax^2 + bxy + cy^2\end{aligned}$$

Thus the equation $\quad ax^2 + bxy + cy^2 = d$

can be written in matrix form as

$$\begin{bmatrix}x\\y\end{bmatrix}\cdot M\begin{bmatrix}x\\y\end{bmatrix} = d \qquad \text{where } M = \begin{bmatrix}a & b/2\\b/2 & c\end{bmatrix}$$

EXAMPLE 1 We express the equation

$$3x^2 + 2xy + 3y^2 = 8$$

in matrix form.

The matrix M is

$$M = \begin{bmatrix} 3 & 1 \\ 1 & 3 \end{bmatrix}$$

and the equation becomes

$$\begin{bmatrix} x \\ y \end{bmatrix} \cdot \begin{bmatrix} 3 & 1 \\ 1 & 3 \end{bmatrix} \begin{bmatrix} x \\ y \end{bmatrix} = 8 \quad \blacksquare$$

By looking at the matrix form of this equation, we can reinterpret what we mean by a solution of the equation. The numbers x and y that are solutions of the equation $ax^2 + bxy + cy^2 = d$ are the $\mathbf{e}_1\mathbf{e}_2$ coordinates of all vectors $\mathbf{v}$ such that $\mathbf{v} \cdot M\mathbf{v} = d$. Therefore, to solve this equation, we can first find the set of all vectors $\mathbf{v}$ for which $\mathbf{v} \cdot M\mathbf{v} = d$, and then find the $\mathbf{e}_1\mathbf{e}_2$ coordinates of these vectors.

The technique we will use to find these vectors $\mathbf{v}$ is change of basis. Let $\{\mathbf{u}_1, \mathbf{u}_2\}$ be a new basis for $\mathbb{R}^2$ and B be the basis change matrix that converts the $\mathbf{u}_i$ coordinates of a vector to the $\mathbf{e}_i$ coordinates of that vector. Then the relationship between the coordinate vector $\begin{bmatrix} x \\ y \end{bmatrix}$ that represents $\mathbf{v}$ relative to the basis $\{\mathbf{e}_1, \mathbf{e}_2\}$ and the vector $\begin{bmatrix} s \\ t \end{bmatrix}_{(\mathbf{u})}$, which represents the same vector $\mathbf{v}$ relative to the basis $\{\mathbf{u}_1, \mathbf{u}_2\}$ is

$$B\begin{bmatrix} s \\ t \end{bmatrix}_{(\mathbf{u})} = \begin{bmatrix} x \\ y \end{bmatrix} \qquad \text{See Theorem 9.1}$$

If the xy coordinates of $\mathbf{v}$ satisfy the equation

$$\begin{bmatrix} x \\ y \end{bmatrix} \cdot M \begin{bmatrix} x \\ y \end{bmatrix} = d \qquad \text{then, since} \qquad B\begin{bmatrix} s \\ t \end{bmatrix}_{(\mathbf{u})} = \begin{bmatrix} x \\ y \end{bmatrix}$$

the $\mathbf{u}_i$ coordinates of $\mathbf{v}$ satisfy the equation

$$B\begin{bmatrix} s \\ t \end{bmatrix}_{(\mathbf{u})} \cdot M\left(B\begin{bmatrix} s \\ t \end{bmatrix}_{(\mathbf{u})}\right) = d$$

Using Theorem 8.1 (to move B around the "dot"), this equation becomes

$$\begin{bmatrix} s \\ t \end{bmatrix}_{(\mathbf{u})} \cdot B^tMB\begin{bmatrix} s \\ t \end{bmatrix}_{(\mathbf{u})} = d$$

We summarize these results in the following theorem.

THEOREM 9.13 If the coordinate vector $\begin{bmatrix} x \\ y \end{bmatrix}$ that represents **v** relative to the $\mathbf{e}_i$ coordinates satisfies the equation

$$\begin{bmatrix} x \\ y \end{bmatrix} \cdot M \begin{bmatrix} x \\ y \end{bmatrix} = d$$

then the coordinate vector $\begin{bmatrix} s \\ t \end{bmatrix}_{(\mathbf{u})}$ that represents **v** relative to the $\mathbf{u}_i$ coordinates satisfies the equation

$$\begin{bmatrix} s \\ t \end{bmatrix}_{(\mathbf{u})} \cdot B^t M B \begin{bmatrix} s \\ t \end{bmatrix}_{(\mathbf{u})} = d$$

where B is the basis change matrix that converts the $\mathbf{u}_i$ coordinates of a vector to the $\mathbf{e}_i$ coordinates of that vector (see Theorem 9.1). The converse is also true; that is, if $\begin{bmatrix} s \\ t \end{bmatrix}_{(\mathbf{u})}$ is a solution of the second equation, then

$$B \begin{bmatrix} s \\ t \end{bmatrix}_{(\mathbf{u})} = \begin{bmatrix} x \\ y \end{bmatrix}$$

is a solution of the first equation.

The following example should help you understand Theorem 9.13 better.

EXAMPLE 2 Again we consider the equation

$$3x^2 + 2xy + 3y^2 = 8$$

As we saw in Example 1, the matrix form of this equation is

$$\begin{bmatrix} x \\ y \end{bmatrix} \cdot \begin{bmatrix} 3 & 1 \\ 1 & 3 \end{bmatrix} \begin{bmatrix} x \\ y \end{bmatrix} = 8 \tag{1}$$

If we choose the new basis

$$\left\{ \mathbf{u}_1 = \begin{bmatrix} 1/\sqrt{2} \\ 1/\sqrt{2} \end{bmatrix}, \quad \mathbf{u}_2 = \begin{bmatrix} -1/\sqrt{2} \\ 1/\sqrt{2} \end{bmatrix} \right\}$$

then the basis change matrix

$$B = \begin{bmatrix} 1/\sqrt{2} & -1/\sqrt{2} \\ 1/\sqrt{2} & 1/\sqrt{2} \end{bmatrix}$$

converts the $\mathbf{u}_i$ coordinates of a vector to the $\mathbf{e}_i$ coordinates of that vector.

Theorem 9.13 says that if and only if $\begin{bmatrix} s \\ t \end{bmatrix}_{(\mathbf{u})}$ is a solution of the equation

$$\text{(2)} \qquad \begin{bmatrix} s \\ t \end{bmatrix}_{(\mathbf{u})} \cdot B^t M B \begin{bmatrix} s \\ t \end{bmatrix}_{(\mathbf{u})} = 8$$

then $\begin{bmatrix} x \\ y \end{bmatrix} = B \begin{bmatrix} s \\ t \end{bmatrix}_{(\mathbf{u})}$ is a solution of Equation (1).

To explicitly write down equation (2) we will need to compute $B^t M B$.

$$B^t M B = \begin{bmatrix} 1/\sqrt{2} & 1/\sqrt{2} \\ -1/\sqrt{2} & 1/\sqrt{2} \end{bmatrix} \begin{bmatrix} 3 & 1 \\ 1 & 3 \end{bmatrix} \begin{bmatrix} 1/\sqrt{2} & -1/\sqrt{2} \\ 1/\sqrt{2}. & 1/\sqrt{2} \end{bmatrix} = \begin{bmatrix} 4 & 0 \\ 0 & 2 \end{bmatrix}$$

So equation (2) becomes

$$\begin{bmatrix} s \\ t \end{bmatrix}_{(\mathbf{u})} \cdot \begin{bmatrix} 4 & 0 \\ 0 & 2 \end{bmatrix} \begin{bmatrix} s \\ t \end{bmatrix}_{(\mathbf{u})} = 8$$

This equation can also be written as

$$4s^2 + 2t^2 = 8$$

Solutions to this last equation are quite easy to find; for example, $s = 1$ and $t = \sqrt{2}$ is a solution. Theorem 9.13 says that the vector

$$\begin{bmatrix} x \\ y \end{bmatrix} = B \begin{bmatrix} 1 \\ \sqrt{2} \end{bmatrix}_{(\mathbf{u})}$$

should be a solution to equation (1). We will verify that this is so.

$$B \begin{bmatrix} 1 \\ \sqrt{2} \end{bmatrix}_{(\mathbf{u})} = \begin{bmatrix} 1/\sqrt{2} & -1/\sqrt{2} \\ 1/\sqrt{2} & 1/\sqrt{2} \end{bmatrix} \begin{bmatrix} 1 \\ \sqrt{2} \end{bmatrix}_{(\mathbf{u})} = \begin{bmatrix} (1 - \sqrt{2})/\sqrt{2} \\ (1 + \sqrt{2})/\sqrt{2} \end{bmatrix}$$

Substituting the values

$$x = \frac{1 - \sqrt{2}}{\sqrt{2}} \quad \text{and} \quad y = \frac{1 + \sqrt{2}}{\sqrt{2}}$$

in our original equation gives

$$3x^2 + 2xy + 3y^2 = 3\left(\frac{1-\sqrt{2}}{\sqrt{2}}\right)^2 + 2\left(\frac{1-\sqrt{2}}{\sqrt{2}}\right)\left(\frac{1+\sqrt{2}}{\sqrt{2}}\right) + 3\left(\frac{1+\sqrt{2}}{\sqrt{2}}\right)^2 = 8$$

as we had expected.

We can do more than just find individual solutions of the equation

$$4s^2 + 2t^2 = 8$$

Since the basis change matrix B we chose is orthogonal, it preserves angles and lengths (see Section 8.1). Therefore, the curve consisting of all points whose $\mathbf{e}_1\mathbf{e}_2$ coordinates satisfy the equation

$$3x^2 + 2xy + 3y^2 = 8$$

has the same shape as the curve consisting of all points whose $\mathbf{u}_1\mathbf{u}_2$ coordinates satisfy the equation

$$4s^2 + 2t^2 = 8$$

The last equation represents an ellipse with major axis of length 2 in the $\mathbf{u}_2$ direction and minor axis of length $\sqrt{2}$ in the $\mathbf{u}_1$ direction. The curve satisfying this equation is graphed in Figure 9.11.

Figure 9.11

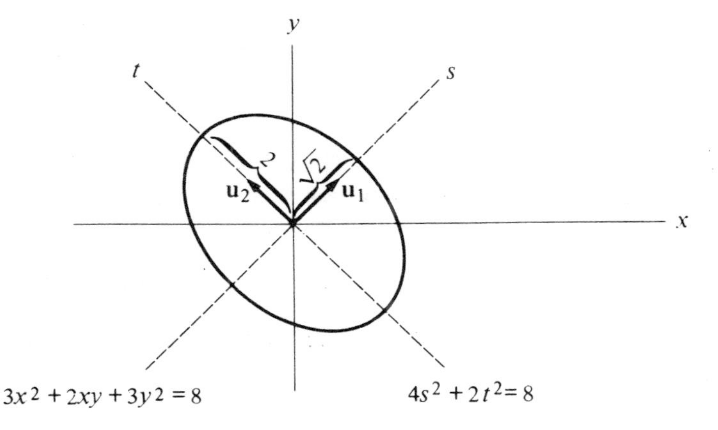

■

Now we consider the general problem of choosing a basis change matrix. From our previous discussion, we see that the basis change matrix B that we use to simplify equations of the form

$$ax^2 + bxy + cy^2 = 0$$

must have two properties:

1. B must be orthogonal.

2. B must be chosen so that $B^t M B$ is a diagonal matrix, where

$$M = \begin{bmatrix} a & b/2 \\ b/2 & c \end{bmatrix}$$

We note that if B is orthogonal, then $B^t = B^{-1}$ (Theorem 8.2). Furthermore since M is a symmetric matrix, we can apply the Spectral Theorem (Theorem 9.9) to guarantee the existence of such a matrix B. Therefore, we have proved the following theorem.

THEOREM 9.14 Given any equation of the form

(3) $$ax^2 + bxy + cy^2 = d$$

there is an orthonormal basis $\{\mathbf{u}_1, \mathbf{u}_2\}$ of $\mathbb{R}^2$ such that if the $\mathbf{e}_1\mathbf{e}_2$ coordinates of the vector $\mathbf{v}$ satisfy equation (3), then the $\mathbf{u}_1\mathbf{u}_2$ coordinates of $\mathbf{v}$ satisfy an equation of the form

(4) $$d_1 s^2 + d_2 t^2 = d$$

where d_1 and d_2 are the characteristic values of the matrix

$$M = \begin{bmatrix} a & b/2 \\ b/2 & c \end{bmatrix}$$

Combining Theorems 9.13 and 9.14, we obtain the following theorem, which summarizes this method for constructing the graph of an equation of the form $ax^2 + bxy + cy^2 = d$.

THEOREM 9.15 The curve traced out by the tips of all vectors of the form

$$\mathbf{v} = x\mathbf{e}_1 + y\mathbf{e}_2$$

where x and y satisfy the equation

$$ax^2 + bxy + cy^2 = d$$

is the same curve as the one traced out by the tips of all vectors of the form

$$\mathbf{v} = s\mathbf{u}_1 + t\mathbf{u}_2$$

where s and t satisfy the equation

$$d_1 s^2 + d_2 t^2 = d$$

The numbers d_1 and d_2 are the characteristic values of the matrix

$$M = \begin{bmatrix} a & b/2 \\ b/2 & c \end{bmatrix}$$

and the vectors $\mathbf{u}_1$ and $\mathbf{u}_2$ are unit characteristic vectors for M with characteristic values d_1 and d_2, respectively.

We apply this theorem in the following example.

EXAMPLE 3 We graph the equation $x^2 - 6xy + y^2 = 4$

We begin by forming the matrix M

$$M = \begin{bmatrix} 1 & -3 \\ -3 & 1 \end{bmatrix}$$

Using the method described in Section 9.4, we find that the characteristic equation for M is

$$0 = \det(M - xI) = (1 - x)^2 - 9 = (4 - x)(-2 - x)$$

So the characteristic values for M are 4 and -2.

Next we reduce the matrices $M - 4I$ and $M + 2I$ to find characteristic vectors for M. A simple calculation shows that

$$\mathbf{u}_1 = \begin{bmatrix} -1/\sqrt{2} \\ 1/\sqrt{2} \end{bmatrix} \quad \text{and} \quad \mathbf{u}_2 = \begin{bmatrix} 1/\sqrt{2} \\ 1/\sqrt{2} \end{bmatrix}$$

are unit characteristic vectors with characteristic values 4 and -2, respectively.

Therefore, using Theorem 9.15 we see that the curve we wish to graph is the curve traced out by the tips of all vectors of the form

$$\mathbf{v} = s\mathbf{u}_1 + t\mathbf{u}_2$$

where s and t satisfy the equation

$$4s^2 - 2t^2 = 4$$

We put the equation $4s^2 - 2t^2 = 4$ in standard form by dividing each side by 4 to obtain the equation

$$\frac{s^2}{1^2} - \frac{t^2}{(\sqrt{2})^2} = 1$$

It is easily seen that this equation represents a hyperbola. The graph is sketched in Figure 9.12.

Figure 9.12

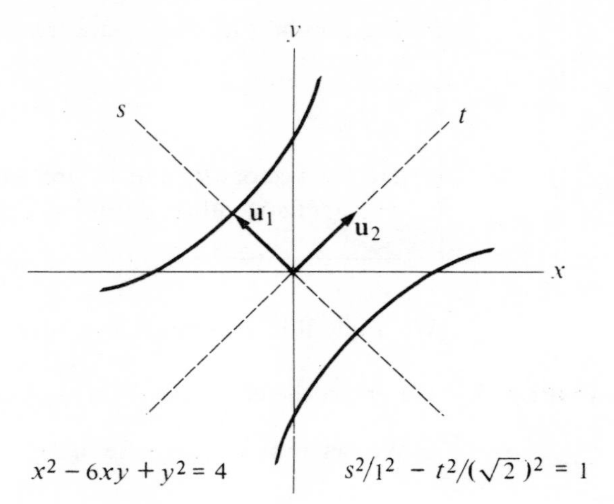

In some cases, when we study equations of the form

$$ax^2 + bxy + cy^2 = d$$

we are only interested in determining what type of conic section the equation represents and not actually in graphing it. We note that, after changing coordinates, the equation has the form

$$d_1 s^2 + d_2 t^2 = d \qquad \text{for convenience, assume } d > 0$$

If d_1 and d_2 are positive, this represents an ellipse, and if d_1 and d_2 have opposite signs, this represents a hyperbola. (If d_1 and d_2 are negative, no points satisfy this equation.) Another way of stating this is:

If $d_1 d_2 > 0$, the graph of the equation is an ellipse or the empty set.

If $d_1 d_2 < 0$, the graph of the equation is a hyperbola.

Now the product $d_1 d_2$ can be expressed as

$$d_1 d_2 = \det\left(\begin{bmatrix} d_1 & 0 \\ 0 & d_2 \end{bmatrix}\right)$$

and since

$$\begin{bmatrix} d_1 & 0 \\ 0 & d_2 \end{bmatrix} = B^t M B = B^{-1} M B \qquad \text{where } M = \begin{bmatrix} a & b/2 \\ b/2 & c \end{bmatrix}$$

we have
$$d_1 d_2 = \det(B^{-1} M B)$$

Since the determinant of the product of matrices is equal to the product of the determinants of the matrices (Theorem 8.9), we have

$$d_1 d_2 = \det(B^{-1}) \cdot \det(M) \cdot \det(B) = (\det(B))^{-1} \cdot \det(M) \cdot \det(B) = \det(M)$$

So we have proved the following theorem.

THEOREM 9.16

Consider the equation

$$ax^2 + bxy + cy^2 = d \qquad d > 0$$

Let

$$M = \begin{bmatrix} a & b/2 \\ b/2 & c \end{bmatrix}$$

Then

1. The graph of this equation represents an ellipse (or contains no points) if $\det(M) > 0$.

2. The graph of this equation represents a hyperbola if $\det(M) < 0$.

The equations in two unknowns discussed above represent only a basic introduction into the study of **quadratic forms**. A quadratic form in three variables is an expression of the form

$$a_{11}x_1^2 + a_{22}x_2^2 + a_{33}x_3^2 + b_{12}x_1x_2 + b_{13}x_1x_3 + b_{23}x_2x_3$$

This expression can be written in matrix form as

$$\begin{bmatrix} x_1 \\ x_2 \\ x_3 \end{bmatrix} \cdot \begin{bmatrix} a_{11} & (b_{12}/2) & (b_{13}/2) \\ (b_{12}/2) & a_{22} & (b_{23}/2) \\ (b_{13}/2) & (b_{23}/2) & a_{33} \end{bmatrix} \begin{bmatrix} x_1 \\ x_2 \\ x_3 \end{bmatrix}$$

Equations involving quadratic forms in three or more unknowns can be analyzed in the same way as equations in two unknowns. The details of this procedure are left as an exercise for the reader (see Exercise 3).

Other equations that can be analyzed using the method of Theorem 9.15 are equations of the form

$$ax^2 + bxy + cy^2 + dx + ey + f = 0$$

These equations represent conic sections that are not centered at the origin. To analyze such an equation, it is first necessary to find the center of the conic section it represents, then to rewrite the equation as an equation in which the coefficients of x and y are zero, and then to use the method of Theorem 9.15 to analyze this equation. This procedure is illustrated in the following example.

EXAMPLE 4 Graph the equation $3x^2 + 2xy + 3y^2 - 10x - 14y + 11 = 0$

We begin by changing coordinates. Let the $x'y'$ coordinate system be the one centered at the point $x = h$, $y = k$ and with the x' and y' axes parallel to the x and y axes, respectively (see Figure 9.13). If P is a point with xy coordinates (a, b), then its $x'y'$ coordinates are $(a - h, b - k)$. Suppose further that P is a point on the curve

$$3x^2 + 2xy + 3y^2 - 10x - 14y + 11 = 0$$

This means that $3a^2 + 2ab + 3b^2 - 10a - 14b + 11 = 0$

Now the $x'y'$ coordinates of P are $(a - h, b - k)$ and so the $x'y'$ coordinates of P satisfy the equation

$$3(x' + h)^2 + 2(x' + h)(y' + k) \\ + 3(y' + k)^2 - 10(x' + h) - 14(y' + k) + 11 = 0$$

which, after expansion and regrouping, becomes

$$3x'^2 + 2x'y' + 3y'^2 + (6h + 2k - 10)x' + (2h + 6k - 14)y' \\ + (3h^2 + 2hk + 3k^2 - 10h - 14k + 11) = 0$$

To reduce this equation to one that can be analyzed using Theorem 9.15, we must choose h and k so that the coefficients of x' and y' in the equation above are 0. Thus h and k must satisfy the equations

$$6h + 2k - 10 = 0 \quad \text{and} \quad 2h + 6k - 14 = 0$$

Therefore, $h = 1$ and $k = 2$; and this equation becomes

$$3x'^2 + 2x'y' + 3y'^2 - 8 = 0$$

Figure 9.13

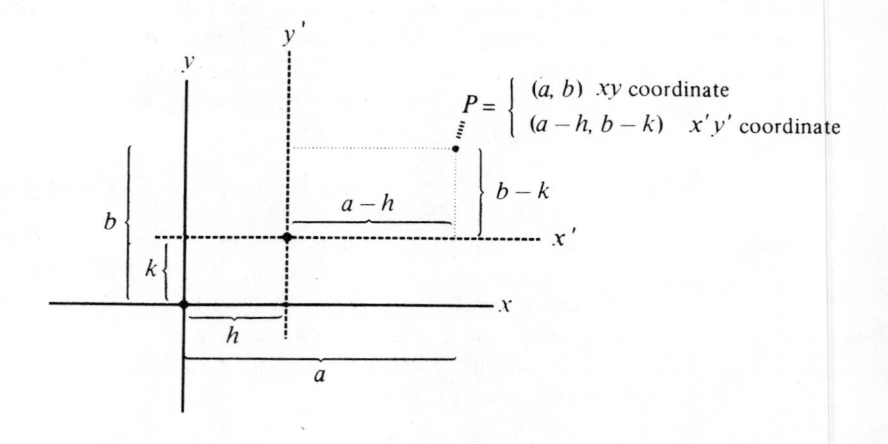

This is the same equation we analyzed in Example 2. So, using these results, we graph the equation in Figure 9.14.

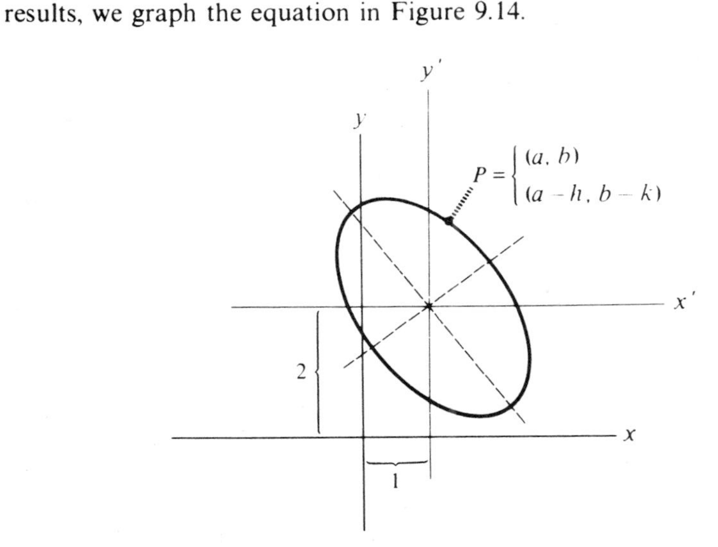

Figure 9.14 Graph of $3x^2 + 2xy + 3y^2 - 10x - 14y + 11 = 0$

■

Exercises for Applications 9b

1. Sketch the graph of the set of all points (x, y) that satisfy each of the following equations.

a. $2xy = 1$

b. $4x^2 + 2xy + 4y^2 = 15$

c. $x^2 + 6xy + y^2 = 8$

d. $-3x^2 + 4xy = 4$

e. $5x^2 - 4xy + 2y^2 = 24$

2. Sketch the graph of the set of all points (x, y) that satisfy each of the following equations.

a. $2xy - 4y + 6x - 13 = 0$

b. $x^2 + 6xy + y^2 - 4x + 4y - 12 = 0$

3. Sketch the graph of all points (x, y, z) that satisfy each of the following equations.

a. $3x^2 + 2xy + 3y^2 + z^2 = 8$

b. $-x^2 - 12xy + 2y^2 - 12yz + 5z^2 = 154$

Applications 9c Differential Equations

In this section we discuss several examples of ways in which linear algebra can be applied to finding solutions of differential equations. This section is not intended to be a survey of differential equations. The point of this section is to show you other examples of situations in which concepts of linear algebra play a critical role.

A differential equation is an equation that expresses a relationship between a function and some of its derivatives. The goal in studying differential equations is to find methods for determining which functions satisfy such an equation.

EXAMPLE 1 The equation

$$y'' + y' - 6y = 0$$

is a differential equation. The function

$$y = e^{2t}$$

is a solution of this equation. This follows since

$$y = e^{2t} \qquad y' = 2e^{2t} \qquad y'' = 4e^{2t}$$

Substituting these values in the original equation gives

$$y'' + y' - 6y = 4e^{2t} + 2e^{2t} - 6e^{2t} = 0 \quad \blacksquare$$

Although we found a solution of the differential equation in Example 1, we are not done because we have not determined if this is the only solution, and, if not, what the other solutions are. We will use linear algebra to study these questions.

We consider the general question of how to find all solutions of an equation of the form

$$ay'' + by' + cy = 0$$

To begin, we recall some ideas discussed in Section 3.2. In that section we gave conditions for a set of objects to be a vector space. It is easy to see that the set of all functions $f(t)$ that are twice differentiable (have first and second derivatives) is a vector space. The operations of addition and scalar multiplication are defined in the usual way; that is,

$$(f_1 + f_2)t = f_1(t) + f_2(t) \qquad \text{addition of vectors}$$

$$(cf)t = c(f(t)) \qquad \text{scalar multiplication}$$

Moreover, the transformation D such that

$$D: f \to f'$$

(where f' is the derivative of f) is a linear transformation since

$$D(f_1 + f_2) = (f_1 + f_2)' = f'_1 + f'_2 = D(f_1) + D(f_2)$$

and

$$D(cf) = (cf)' = cf' = cD(f)$$

In terms of this notation, we have

$$y' = Dy$$

$$y'' = D(y') = D(Dy) = D^2y$$

and the equation

$$ay'' + by' + cy = 0$$

can be written as

$$aD^2y + bDy + cy = 0$$

This can be expressed as

$$(aD^2 + bD + cI)y = 0$$

where I is the identity transformation. By looking at the differential equation from this point of view, we can restate our problem: We are looking for all functions (vectors) y such that the linear transformation

$$aD^2 + bD + cI$$

maps y to the zero function. In other words, we are looking for all functions (vectors) y that are in the null space of the transformation

$$aD^2 + bD + cI$$

Now we return to our specific example.

EXAMPLE 2 The differential equation $\quad y'' + y' - 6y = 0$

can be written as $\quad (D^2 + D - 6I)y = 0$

Now the expression $D^2 + D - 6I$ can be factored as

$$D^2 + D - 6I = (D + 3I)(D - 2I) = (D - 2I)(D + 3I)$$

Note: These factorizations work because D and I commute. Compare Exercise 9 in Section 7.1.

We can write our original equation as

$$(D + 3I)(D - 2I)y = 0 \qquad \text{or} \qquad (D - 2I)(D + 3I)y = 0$$

This means that any function y such that either

$$(D + 3I)y = 0 \qquad \text{or} \qquad (D - 2I)y = 0$$

is a solution of the original equation.

We consider the equation

$$0 = (D + 3I)y = y' + 3y$$

To solve this equation, we need a function $f(t)$ such that

$$f'(t) + 3f(t) = 0$$

that is, $\quad f'(t) = -3f(t)$

It is not difficult to see that all solutions to this equation are functions of the form

$$f(t) = c_1 e^{-3t} \qquad \text{where } c_1 \text{ is an arbitrary constant}$$

In the same way we can see that the solutions of the equation

$(D - 2I)y = 0$ are all of the form

$$g(t) = c_2 e^{2t} \qquad \text{where } c_2 \text{ is an arbitrary constant}$$

Since the null space of $\qquad D^2 + D - 6I$

is a vector space, all linear combinations

$$c_1 e^{-3t} + c_2 e^{2t}$$

of the functions $c_1 e^{-3t}$ and $c_2 e^{2t}$ are solutions of the original equation.

However, we still cannot be sure that we have found *all* solutions of our original equation. This is because equations involving linear transformations do not always behave as nicely as equations involving real numbers; in particular, if the product

$$(x - a)(x - b) = 0$$

then we know that one of the terms is zero. This is not the case with linear transformations. For example, if

$$T_1 \text{ is the linear transformation represented by } \begin{bmatrix} 0 & 1 \\ 0 & 0 \end{bmatrix}$$

$$T_2 \text{ is the linear transformation represented by } \begin{bmatrix} 0 & 3 \\ 0 & 0 \end{bmatrix}$$

and $\qquad \mathbf{v} = \begin{bmatrix} 0 \\ 1 \end{bmatrix}$

then $\quad T_1\mathbf{v} = \begin{bmatrix} 0 & 1 \\ 0 & 0 \end{bmatrix}\begin{bmatrix} 0 \\ 1 \end{bmatrix} = \begin{bmatrix} 1 \\ 0 \end{bmatrix} \quad$ and $\quad T_2\mathbf{v} = \begin{bmatrix} 0 & 3 \\ 0 & 0 \end{bmatrix}\begin{bmatrix} 0 \\ 1 \end{bmatrix} = \begin{bmatrix} 3 \\ 0 \end{bmatrix}$

Thus neither $T_1\mathbf{v}$ nor $T_2\mathbf{v}$ is $\mathbf{0}$. However,

$$T_1 T_2 \mathbf{v} = \begin{bmatrix} 0 & 1 \\ 0 & 0 \end{bmatrix}\begin{bmatrix} 0 & 3 \\ 0 & 0 \end{bmatrix}\begin{bmatrix} 0 \\ 1 \end{bmatrix} = \begin{bmatrix} 0 \\ 0 \end{bmatrix} = \mathbf{0}$$

Looking at our problem with this in mind, we see that there may be a function $y_0 = f_0(t)$ such that

$$(D + 3I)y_0 \neq 0 \qquad \text{and} \qquad (D - 2I)y_0 \neq 0$$

but $\qquad (D + 3I)(D - 2I)y_0 = 0$

This does not happen in this case as we will show in Example 3. Therefore, the set of all solutions of this equation is the set of all functions of the form

$$f(t) = c_1 e^{-3t} + c_2 e^{2t}$$

which can be written in vector notation as

$$\langle e^{-3t}, e^{2t} \rangle$$

Moreover, $\{e^{-3t}, e^{2t}\}$ is a basis for the null space since the vectors e^{-3t} and e^{2t} are linearly independent. This is easily seen since the only solution to the equation

$$c_1 e^{-3t} + c_2 e^{2t} = 0 \qquad \text{for all values of } t$$

is $c_1 = c_2 = 0$. ■

Now we state a general theorem about the relationship of the dimensions of the null spaces of two linear transformations and the nullity of their product. It is this theorem that we will use to show that the null space of $(D + 3I)(D - 2I)$ is $\langle e^{-3t}, e^{2t} \rangle$.

THEOREM 9.17

Let $T\colon V_1 \to V_2$ and $S\colon V_2 \to V_3$ be linear transformations, then

$$\text{nullity of } ST \leq \text{nullity of } S + \text{nullity of } T$$

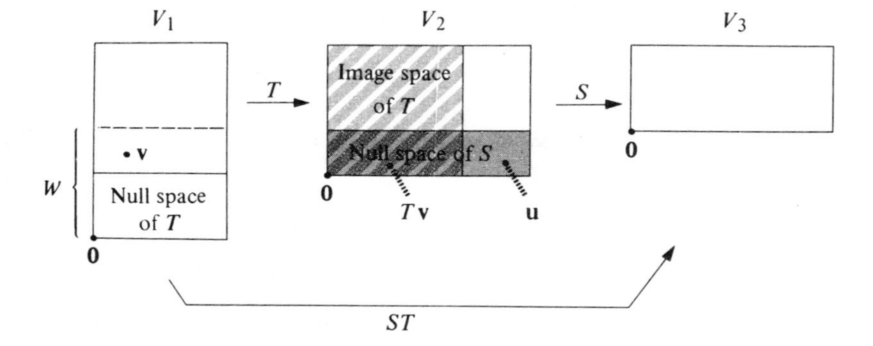

Figure 9.15

To see how to prove this consider Figure 9.15. Suppose **v** is a vector in the null space of ST. Then

$$0 = (ST)\mathbf{v} = S(T\mathbf{v})$$

which means that the vector $T\mathbf{v}$ is in the null space of S. Let W be the set of all vectors **v** of V_1 for which $T\mathbf{v}$ is in the null space of S—therefore W is the null space of ST. As Figure 9.16 illustrates, T maps W into the null space of S. (*Note:* There can be vectors **u** in the null space of S that are not the image of any vector of V_1 under T.) So, by Theorem 5.2,

$$\dim(W) = \text{nullity of } T + \dim(T(W))$$

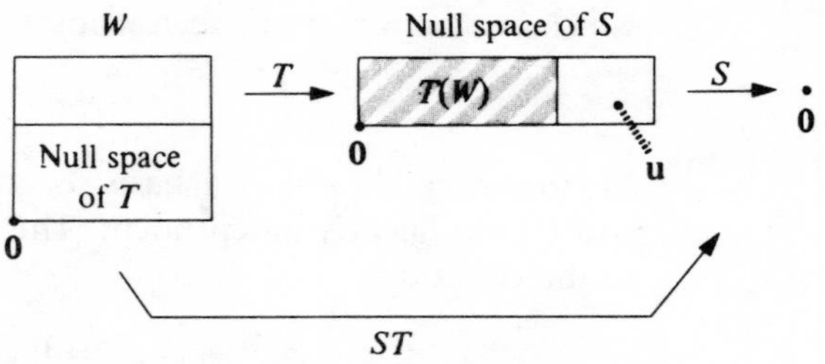

Figure 9.16

Since $T(W)$ is contained in the null space of S,

$$\dim(T(W)) \leq \dim(\text{null space of } S)$$

Therefore, we conclude that

$$\dim(W) \leq \text{nullity of } T + \text{nullity of } S$$

Since $W =$ (null space of ST,) this becomes

$$\text{Nullity of } ST \leq \text{nullity of } T + \text{nullity of } S$$

completing the proof.

EXAMPLE 3 We apply Theorem 9.17 to show that the null space of $D^2 + D - 6I$ is $\langle e^{-3t}, e^{2t} \rangle$.

Let $S = (D + 3I)$ and $T = (D - 2I)$

Then $$ST = D^2 + D - 6I$$

As we saw in Example 1, the null spaces of S and T are the 1-dimensional spaces

$$\text{Null space of } S = \langle e^{-3t} \rangle$$
$$\text{Null space of } T = \langle e^{2t} \rangle$$

So $$\text{Nullity of } S + \text{nullity of } T = 1 + 1 = 2$$

Hence, using Theorem 9.17, we have

$$\text{Nullity } ST \leq 2$$

Since $\langle e^{-3t}, e^{2t} \rangle$ is a two-dimensional subspace of the null space of ST,

$$\text{Nullity of } ST \geq \dim(\langle e^{-3t}, e^{2t} \rangle) = 2$$

Thus $$\text{Nullity of } ST = 2$$

and so $$\{e^{-3t}, e^{2t}\}$$

is a basis for the null space of ST. ■

In the general case of equations of the form

$$ay'' + by' + cy = 0$$

these ideas can be used to obtain solutions so long as the equation

$$ax^2 + bx + c = 0$$

has distinct real roots—that is, when $b^2 - 4ac > 0$. The cases of double roots and complex roots are handled somewhat differently and will not be discussed in this text.

Another type of differential equation problem that can be solved using linear algebra is the problem of solving systems of linear differential equations. For example, a solution of the system

$$y'_1 = a_{11}y_1 + a_{12}y_2$$
$$y'_2 = a_{21}y_1 + a_{22}y_2$$

would be a pair of functions $y_1 = f_1(t)$ and $y_2 = f_2(t)$ such that

$$f'_1(t) = a_{11}f_1(t) + a_{12}f_2(t)$$
$$f'_2(t) = a_{21}f_1(t) + a_{22}f_2(t)$$

for all values of t.

This equation can be written in matrix form as

$$\begin{bmatrix} y'_1 \\ y'_2 \end{bmatrix} = \begin{bmatrix} a_{11} & a_{12} \\ a_{21} & a_{22} \end{bmatrix} \begin{bmatrix} y_1 \\ y_2 \end{bmatrix}$$

or, more briefly, as

$$Y' = MY \qquad \text{where } Y = \begin{bmatrix} y_1 \\ y_2 \end{bmatrix} \qquad \text{and} \qquad M = \begin{bmatrix} a_{11} & a_{12} \\ a_{21} & a_{22} \end{bmatrix}$$

We now produce a solution for this equation. Unfortunately, the reasons why this is a solution won't become apparent until after you see the solution.

Let

$$\mathbf{v} = \begin{bmatrix} s_1 \\ s_2 \end{bmatrix}$$

be a characteristic vector for the matrix M with characteristic value c. Thus

$$M\mathbf{v} = \begin{bmatrix} a_{11} & a_{12} \\ a_{21} & a_{22} \end{bmatrix} \begin{bmatrix} s_1 \\ s_2 \end{bmatrix} = c \begin{bmatrix} s_1 \\ s_2 \end{bmatrix}$$

The solution to our original equation is

$$Y = \begin{bmatrix} y_1 \\ y_2 \end{bmatrix} = e^{ct} \begin{bmatrix} s_1 \\ s_2 \end{bmatrix} = \begin{bmatrix} s_1 e^{ct} \\ s_2 e^{ct} \end{bmatrix}$$

We see that Y is a solution to this equation since

$$Y' = \begin{bmatrix} (s_1 e^{ct})' \\ (s_2 e^{ct})' \end{bmatrix} = \begin{bmatrix} cs_1 e^{ct} \\ cs_2 e^{ct} \end{bmatrix} = c \begin{bmatrix} s_1 e^{ct} \\ s_2 e^{ct} \end{bmatrix} = cY$$

and

$$MY = M\left(e^{ct}\begin{bmatrix} s_1 \\ s_2 \end{bmatrix}\right) = (e^{ct})M\begin{bmatrix} s_1 \\ s_2 \end{bmatrix} = (e^{ct})\left(c\begin{bmatrix} s_1 \\ s_2 \end{bmatrix}\right) = c\left(e^{ct}\begin{bmatrix} s_1 \\ s_2 \end{bmatrix}\right) = cY$$

So
$$Y' = MY$$

as we claimed.

The following example should help you understand this better.

EXAMPLE 4 Solve the following system of linear differential equations:

$$\begin{aligned} y_1' &= -7y_1 - 6y_2 \\ y_2' &= 18y_1 + 14y_2 \end{aligned}$$

We rewrite the system as

$$Y' = \begin{bmatrix} -7 & -6 \\ 18 & 14 \end{bmatrix}\begin{bmatrix} y_1 \\ y_2 \end{bmatrix} = MY$$

As we saw in Example 2 of Section 9.4

$$\mathbf{u}_1 = \begin{bmatrix} -2 \\ 3 \end{bmatrix} \text{ is a characteristic vector for } M \text{ with value 2}$$

$$\mathbf{u}_2 = \begin{bmatrix} 1 \\ -2 \end{bmatrix} \text{ is a characteristic vector for } M \text{ with value 5}$$

Therefore two solutions of this equation are

$$Y = \begin{bmatrix} -2e^{2t} \\ 3e^{2t} \end{bmatrix} \quad \text{or} \quad \begin{cases} y_1 = -2e^{2t} \\ y_2 = 3e^{2t} \end{cases}$$

and

$$Y = \begin{bmatrix} 1e^{5t} \\ -2e^{5t} \end{bmatrix} \quad \text{or} \quad \begin{cases} y_1 = e^{5t} \\ y_2 = -2e^{5t} \end{cases}$$

We verify that the first solution satisfies the system. (Checking the second solution is left as an exercise for the interested reader.)

We have
$$y_1 = -2e^{2t}, \qquad y_1' = -4e^{2t}$$

and
$$y_2 = 3e^{2t}, \qquad y_2' = 6e^{2t}$$

Substituting these values into the left- and right-hand sides of the first equation gives

$$y_1' = -4e^{2t} \qquad \text{and} \qquad -7y_1 - 6y_2 = 14e^{2t} - 18e^{2t} = -4e^{2t}$$

So these values satisfy the first equation. Checking the second equation, we have

$$y_2' = 6e^{2t} \qquad \text{and} \qquad 18y_1 + 14y_2 = -36e^{2t} + 42e^{2t} = 6e^{2t}$$

So the second equation is also satisfied. Thus

$$y_1 = -2e^{2t} \qquad \text{and} \qquad y_2 = 3e^{2t}$$

is indeed a solution to the original system of linear differential equations. ■

There are many other applications of linear algebra to solving differential equations, but we will not discuss them in this text. It is hoped that this brief sampling will give you some insight into how the ideas of linear algebra can be used to understand differential equations.

Exercises for Applications 9c

1. Find all solutions of the following differential equations:

a. $y'' - 4y = 0$ **b.** $y'' - y' - 20y = 0$

c. $y'' + 2y' - 3y = 0$

2. Find a solution of each of the following systems of differential equations:

a. $y_1' = 7y_1 + 3y_2$
$y_2' = -9y_1 - 5y_2$

b. $y_1' = 4y_1 + 18y_2$
$y_2' = 3y_1 + 1y_2$

Review Exercises

1. Define the following terms:

a. Singular matrix **b.** Symmetric matrix

c. Similar matrices

d. The matrix of a linear transformation relative to the basis $\{\mathbf{u}_1, \mathbf{u}_2, \ldots, \mathbf{u}_n\}$

e. Characteristic value of a linear transformation

f. Characteristic vector of a linear transformation

g. Characteristic value of a matrix

h. Characteristic vector of a matrix

i. Characteristic equation of a matrix

2. For each of the linear transformations listed below find a diagonal matrix that represents that transformation. Also describe the basis associated with the diagonal matrix chosen.

a. $T: \mathbb{R}^2 \to \mathbb{R}^2$ is reflection about a line l containing the origin.

b. $T: \mathbb{R}^2 \to \mathbb{R}^2$ is projection onto a line l containing the origin.

c. $T: \mathbb{R}^2 \to \mathbb{R}^2$ is rotation by an angle of 180°.

d. $T: \mathbb{R}^3 \to \mathbb{R}^3$ is projection onto a plane containing the origin.

3. Let $T: \mathbb{R}^2 \to \mathbb{R}^2$ be rotation by an angle θ

where $0 < \theta < 180°$. Explain why there is no diagonal matrix that represents the linear transformation T.

4. Let T be a linear transformation and $\mathbf{v}_1, \mathbf{v}_2$ be characteristic vectors of T with distinct characteristic values. Show that $\mathbf{v}_1$ and $\mathbf{v}_2$ are linearly independent.

5. Find two matrices that are equivalent, but not similar.

6. Consider the basis $\{\mathbf{u}_1, \mathbf{u}_2\}$ of $\mathbb{R}^2$, where

$$\mathbf{u}_1 = 2\mathbf{e}_1 - 3\mathbf{e}_2 \quad \text{and} \quad \mathbf{u}_2 = -1\mathbf{e}_1 + 2\mathbf{e}_2$$

a. Find the $\mathbf{u}$ coordinates of the vector $\mathbf{v} = 5\mathbf{e}_1 - 7\mathbf{e}_2$.

b. Find the $\mathbf{e}$ coordinates of the vector $\mathbf{w} = 3\mathbf{u}_1 - \mathbf{u}_2$.

7. Consider the basis $\{\mathbf{u}_1, \mathbf{u}_2\}$ of $\mathbb{R}^2$, where

$$\mathbf{u}_1 = 2\mathbf{e}_1 - 3\mathbf{e}_2 \quad \text{and} \quad \mathbf{u}_2 = -\mathbf{e}_1 + 4\mathbf{e}_2$$

a. Find the $\mathbf{u}$ coordinates of the vector $\mathbf{v} = 6\mathbf{e}_1 + \mathbf{e}_2$.

b. Find the $\mathbf{e}$ coordinates of the vector $\mathbf{w} = -2\mathbf{u}_1 + 3\mathbf{u}_2$.

8. Let $T: \mathbb{R}^2 \to \mathbb{R}^2$ be projection onto the line $y = 5x$.

a. Find a basis for $\mathbb{R}^2$ so that the matrix for T is diagonal. Find this diagonal matrix.

b. What are the characteristic values for T?

9. Find a diagonal matrix similar to

$$\begin{bmatrix} 2 & 3 & 4 \\ 2 & 4 & 5 \\ 0 & 0 & 1 \end{bmatrix}$$

10. Find a basis for $\mathbb{R}^3$ under which the linear transformation represented by the matrix

$$\begin{bmatrix} 2 & -2 & 3 \\ 0 & 3 & -2 \\ 0 & -1 & 2 \end{bmatrix}$$

is diagonal. What is this diagonal matrix?

11. Suppose that M is a noninvertible 3×3 matrix. Complete the following statements:

a. $\text{Det}(M) =$ ______________.

b. $\text{Rank}(M)$ is/is not 3.

c. Nullity of M is/is not 0.

d. 0 is/is not a characteristic value for M.

12. For each of the following statements, fill in the blank with one of the words *every*, *some*, or *no*.

a. ____________ matrix with characteristic equation $x^2 + 1 = 0$ can be diagonalized.

b. ____________ matrix with characteristic equation $x^2 - 1 = 0$ can be diagonalized.

c. ____________ matrix with characteristic equation $(x - 1)^2 = 0$ can be diagonalized.

d. ____________ symmetric matrix with characteristic equation $(x - 1)^2 = 0$ can be diagonalized.

e. ____________ matrix with characteristic equation $x^2 + 1 = 0$ is symmetric.

13. Let $T: \mathbb{R}^2 \to \mathbb{R}^2$ be reflection about the line $y = x$.

a. Find a basis for the domain of T that is composed of characteristic vectors for T.

b. Find the matrix that represents T relative to the basis you found in part a.

c. Find the characteristic equation for T.

14. Answer the questions of Exercise 13 for the linear transformation: $T: \mathbb{R}^3 \to \mathbb{R}^3$ is reflection about the line containing the vector $\begin{bmatrix} 1 \\ 1 \\ 0 \end{bmatrix}$.

Answers and Solutions to Selected Exercises

Exercise Set 1.1, page 4

1. f is not a function since there is no element in the range of f corresponding to x if x is odd.

2. f is not a function since two distinct numbers $+\sqrt[4]{x}$ and $-\sqrt[4]{x}$ can correspond to the same element x.

3. f is not a function since 3^2 and 4^2 are not in the range of f.

4. f is a function. **5.** f is a function.

6. $f = g$. **7.** $f = g$.

8. **a.** f_2 **b.** f_2 **c.** f_1, f_2
d. f_3 **e.** f_2

9. **a.** f_1 **b.** f_1 **c.** f_1 **d.** f_1

10. $f = g$.

11. **a.** Function. **b.** Not function.
c. Not function (there are black cars).
d. Function.

Exercise Set 1.2, page 8

1. **a.** Not linear. **b.** Not linear
c. Linear **d.** Not linear

2. **a.** $f(x) = 5x$ **b.** $f(x) = 6x$
c. $f(x) = (5/2)x$
d. Every linear function from $\mathbb{R}$ to $\mathbb{R}$.
e. No linear function has this property.

4. **(1.)**
$$\begin{aligned} f(x_1 + x_2) &= a(x_1 + x_2) \\ &= ax_1 + ax_2 \\ &= f(x_1) + f(x_2). \end{aligned}$$
(2.) $f(cx) = a(cx) = c(ax) = cf(x)$.
Since f satisfies conditions 1 and 2 of Definition 1.3, f is linear.

5. **a.** **(1)**
$$\begin{aligned} (f+g)(x_1 + x_2) &= f(x_1 + x_2) + g(x_1 + x_2) \\ &= f(x_1) + f(x_2) + g(x_1) + g(x_2) \\ &\quad \text{since } f, g \text{ are linear} \\ &= (f(x_1) + g(x_1)) \\ &\quad + (f(x_2) + g(x_2)) \\ &= (f+g)(x_1) + (f+g)(x_2) \end{aligned}$$

(2)
$$\begin{aligned} (f+g)(cx) &= f(cx) + g(cx) \\ &= cf(x) + cg(x) \\ &\quad \text{since } f, g \text{ are linear} \\ &= c(f(x) + g(x)) \\ &= c(f+g)(x) \end{aligned}$$

So $(f+g)$ is linear.

b. **(1)**
$$\begin{aligned} fg(x_1 + x_2) &= f(g(x_1 + x_2)) \\ &= f(g(x_1) + g(x_2)) \\ &\quad \text{since } g \text{ is linear} \\ &= f(g(x_1)) + f(g(x_2)) \\ &\quad \text{since } f \text{ is linear} \\ &= fg(x_1) + fg(x_2) \end{aligned}$$

(2)
$$\begin{aligned} fg(cx) &= f(g(cx)) \\ &= f(cg(x)) && \text{since } g \text{ is linear} \\ &= cf(g(x)) && \text{since } f \text{ is linear} \\ &= cfg(x) \end{aligned}$$

Hence fg is linear.

6. $f(x) = x$, $g(x) = 2x$ works (as does almost any pair of linear functions).

7. $f(x) = xf(1) = x \cdot 0 = 0$

8. If $f = g$, then $f(b) = g(b)$ for every real number b. Conversely if $f(b) = g(b) = c$, then, since $b \neq 0$,

$$f(x) = \frac{c}{b}x \quad \text{and} \quad g(x) = \frac{c}{b}x \text{ (Theorem 1.1)}$$

Hence $f = g$.

9. **a.** From Theorem 1.1 we see that we can write $f(x) = cx$ and $g(x) = dx$ for some real numbers c, d

So we have

$$af(x) + bg(x) = acx + bdx = (ac + bd)x$$

is a linear function.

b. Use same ideas as in part a.

c. $h_1(x) = x^2 \qquad h_2(x) = x + 1$
$h_1(h_2(x)) = (x + 1)^2 = x^2 + 2x + 1$
$h_2(h_1(x)) = x^2 + 1$

Exercise Set 1.3, page 14

1.

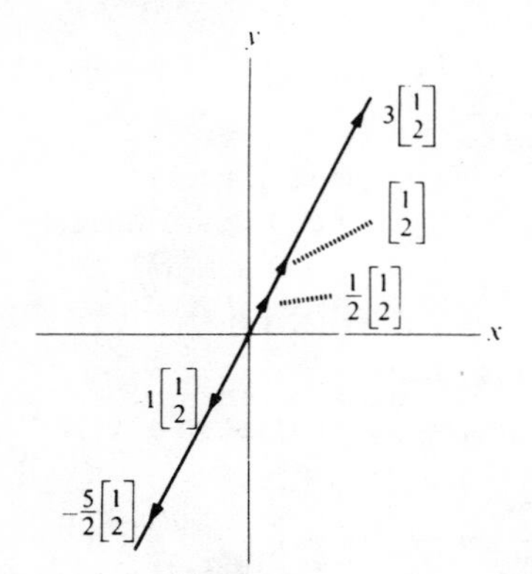

2.

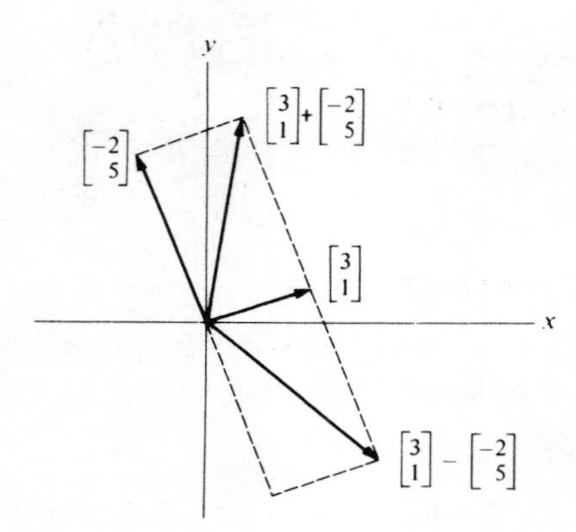

3.

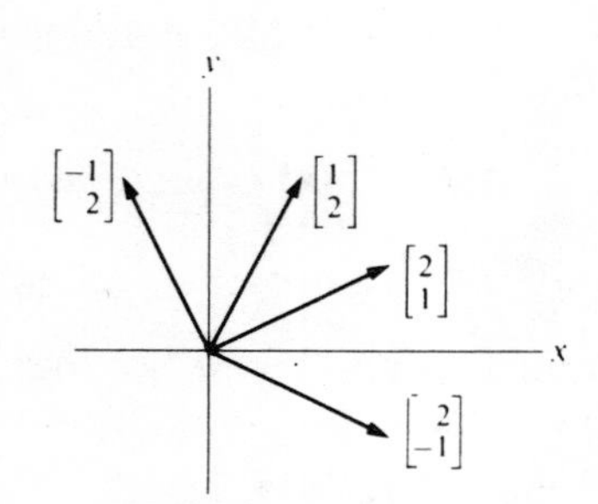

4. **a.** $\mathbf{x} = \begin{bmatrix} -3 \\ 2 \end{bmatrix}$. **b.** $c = -2$

c. $a = 1/7, b = -2/7$ **d.** $a = 3/7, b = 1/7$

5. **a.** (5, 4) **b.** (6, 3)

6. **a.** Linear **b.** Not linear

c. Linear **d.** Not linear **e.** Linear

7. **a.**

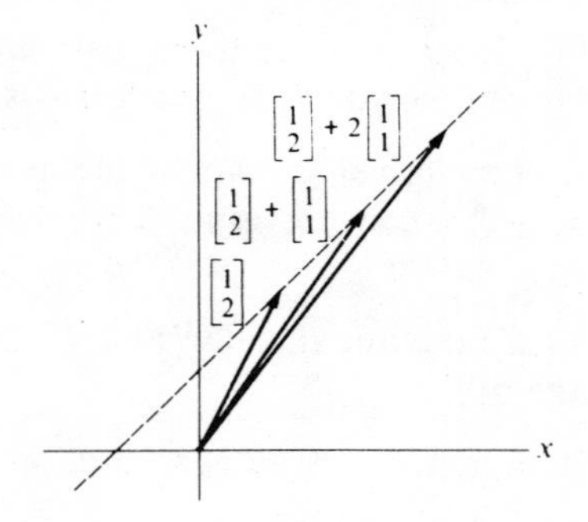

b.,c. The line parallel to the vector $\begin{bmatrix} 1 \\ 1 \end{bmatrix}$ that contains the point (1, 2).

8. The equation of the line is $y = 2x$.

10. $f\left(\begin{bmatrix} x \\ y \end{bmatrix}\right) = x$ and $g\left(\begin{bmatrix} x \\ y \end{bmatrix}\right) = 2x$.

Almost any pair of linear functions will work.

Exercise Set 1.4, page 19

1. $f\left(\begin{bmatrix} x \\ y \end{bmatrix}\right) = 5x - 2y$ **2.** $f\left(\begin{bmatrix} x \\ y \end{bmatrix}\right) = 17x$

3. $f\left(\begin{bmatrix} x \\ y \end{bmatrix}\right) = -13x + 3y$

4. $f\left(\begin{bmatrix} x \\ y \end{bmatrix}\right) = 12x - 3y$

5. $f\left(\begin{bmatrix} x \\ y \end{bmatrix}\right) = 5x - 2y$ **6.** $f\left(\begin{bmatrix} x \\ y \end{bmatrix}\right) = 7x - 3y$

7. $f\left(\begin{bmatrix} x \\ y \end{bmatrix}\right) = 3x, \qquad g\left(\begin{bmatrix} x \\ y \end{bmatrix}\right) = 2y$

8. **a.** $\begin{bmatrix} 1 \\ 0 \end{bmatrix} = 2/5\begin{bmatrix} 1 \\ 2 \end{bmatrix} - 1/5\begin{bmatrix} -3 \\ 4 \end{bmatrix}$

and $\begin{bmatrix} 0 \\ 1 \end{bmatrix} = 3/10\begin{bmatrix} 1 \\ 2 \end{bmatrix} + 1/10\begin{bmatrix} -3 \\ 4 \end{bmatrix}$

b. $\begin{bmatrix}1\\0\end{bmatrix} = 1/3\begin{bmatrix}5\\-3\end{bmatrix} - 1/3\begin{bmatrix}2\\-3\end{bmatrix}$

and $\begin{bmatrix}0\\1\end{bmatrix} = 2/9\begin{bmatrix}5\\-3\end{bmatrix} - 5/9\begin{bmatrix}2\\-3\end{bmatrix}$

c. $\begin{bmatrix}1\\0\end{bmatrix} = -1/24\begin{bmatrix}1\\5\end{bmatrix} + 5/24\begin{bmatrix}5\\1\end{bmatrix}$

and $\begin{bmatrix}0\\1\end{bmatrix} = 5/24\begin{bmatrix}1\\5\end{bmatrix} - 1/24\begin{bmatrix}5\\1\end{bmatrix}$

9. and 10. All linear combinations of $\begin{bmatrix}1\\2\end{bmatrix}$ and $\begin{bmatrix}-3\\-6\end{bmatrix}$ lie on the line $y = 2x$. Since $\begin{bmatrix}1\\0\end{bmatrix}$ does not lie along this line, it is not a linear combination of these two vectors.

Exercise Set 1.5, page 23

1. $\begin{bmatrix}1\\0\end{bmatrix} = \mathbf{w}_1$ and $\begin{bmatrix}0\\1\end{bmatrix} = 2\mathbf{w}_1 - \mathbf{w}_2$

2. $\mathbf{w}_1, \mathbf{w}_2$ is not a basis.

3. $\begin{bmatrix}1\\0\end{bmatrix} = \frac{1}{5}\mathbf{w}_1 + \frac{2}{5}\mathbf{w}_2$ and $\begin{bmatrix}0\\1\end{bmatrix} = -\frac{1}{3}\mathbf{w}_2$

4. $\begin{bmatrix}1\\0\end{bmatrix} = -\frac{3}{2}\mathbf{w}_1 - 2\mathbf{w}_2$ and $\begin{bmatrix}0\\1\end{bmatrix} = \frac{1}{2}\mathbf{w}_1 + \mathbf{w}_2$

5. Not a basis.

6. (1) $f\left(\begin{bmatrix}x\\y\end{bmatrix}\right) = 3x + 7y$

(2) No such linear function.

(3) $f\left(\begin{bmatrix}x\\y\end{bmatrix}\right) = \frac{1}{5}x + \frac{1}{3}y$

(5) $f\left(\begin{bmatrix}x\\y\end{bmatrix}\right) = ax + (-\frac{1}{3} - \frac{4}{3}a)y.$

7. a. $\mathbf{v}_1 = \begin{bmatrix}1\\-2\end{bmatrix}$ and $\mathbf{v}_2 = \begin{bmatrix}1\\-1\end{bmatrix}$

c. $\begin{bmatrix}1\\1\end{bmatrix} = -2\begin{bmatrix}1\\-2\end{bmatrix} + 3\begin{bmatrix}1\\-1\end{bmatrix}$

d.
$$\begin{aligned} f(\mathbf{w}) &= f(-2\mathbf{v}_1 + 3\mathbf{v}_2) \\ &= -2f(\mathbf{v}_1) + 3f(\mathbf{v}_2) \\ &= -2(0) + 3(1) = 3 \end{aligned}$$

$$f\left(\begin{bmatrix}1\\1\end{bmatrix}\right) = 2(1) + 1(1) = 3.$$

8. a. Using Theorem 1.2, we write $f\left(\begin{bmatrix}x\\y\end{bmatrix}\right) = ax + by$, where a and b are not both 0. We choose $\mathbf{v}_0 = \begin{bmatrix}-b\\a\end{bmatrix}$.

b. $f(c\mathbf{v}_0) = cf(\mathbf{v}_0) = c(\mathbf{0}) = 0.$

c. If $\mathbf{w}$ is not parallel to $\mathbf{v}_0$, then the vectors $\mathbf{v}_0$ and $\mathbf{w}$ form a basis of $\mathbb{R}^2$. So every vector $\mathbf{v}$ of $\mathbb{R}^2$ can be written as a linear combination of $\mathbf{v}_0$ and $\mathbf{w}$, that is,

$$\mathbf{v} = x\mathbf{v}_0 + y\mathbf{w}$$

and since f is linear

$$\begin{aligned} f(\mathbf{v}) &= f(x\mathbf{v}_0 + y\mathbf{w}) \\ &= xf(\mathbf{v}_0) + yf(\mathbf{w}) \\ &= x(0) + y(0) = 0 \end{aligned}$$

for every vector $\mathbf{v}$ of $\mathbb{R}^2$.

9. a. $N =$ the set of all vectors of the form $\begin{bmatrix}2t\\3t\end{bmatrix}$ for any real number t.

b. The line $y = (3/2)x$

d. The line $y = (3/2)x - 1/2.$

e. The line $y = (3/2)x - 1.$

Exercises for Applications 1, page 31

1. a.
$$\begin{aligned} f_i\left(\begin{bmatrix}a_1\\b_1\end{bmatrix} + \begin{bmatrix}a_2\\b_2\end{bmatrix}\right) &= f_i\left(\begin{bmatrix}a_1 + a_2\\b_1 + b_2\end{bmatrix}\right) \\ &= \begin{bmatrix}-(b_1 + b_2)\\(a_1 + a_2)\end{bmatrix} \\ &= \begin{bmatrix}-b_1\\a_1\end{bmatrix} + \begin{bmatrix}-b_2\\a_2\end{bmatrix} \\ &= f_i\left(\begin{bmatrix}a_1\\b_1\end{bmatrix}\right) + f_i\left(\begin{bmatrix}a_2\\b_2\end{bmatrix}\right) \end{aligned}$$

b. $f_i\left(c\begin{bmatrix}a\\b\end{bmatrix}\right) = f_i\left(\begin{bmatrix}ca\\cb\end{bmatrix}\right) = \begin{bmatrix}-cb\\ca\end{bmatrix}$

$= c\begin{bmatrix}-b\\a\end{bmatrix} = cf_i\left(\begin{bmatrix}a\\b\end{bmatrix}\right)$

2. Let $z = x_0 + y_0 i$. Then for any arbitrary complex number $a + bi$

$$z(a+bi) = (x_0 + y_0 i)(a + bi)$$
$$= (x_0 a - y_0 b) + (y_0 a + x_0 b)i$$

So f_z: $\mathbb{R}^2 \to \mathbb{R}^2$

$$\begin{bmatrix}a\\b\end{bmatrix} \to \begin{bmatrix}x_0 a - y_0 b\\y_0 a + x_0 b\end{bmatrix}$$

Using this definition of f_z, follow the same procedures used to solve Exercise 1.

3. $z = \cos(2\pi/5) + i\sin(2\pi/5) \doteq 0.309 + 0.951i$
All roots of $x^5 - 1$ are 1, z, z^2, z^3, z^4.

4. **a.** $\theta = \text{Arctan}(b/a)$

b. $z = a + bi$

$$= \sqrt{a^2+b^2}\left(\frac{a}{\sqrt{a^2+b^2}} + \frac{bi}{\sqrt{a^2+b^2}}\right)$$

c. $3 + 4i = 5(3/5 + (4/5)i)$
$2 + 2i = \sqrt{8}[1/\sqrt{2} + (1/\sqrt{2})i]$

5. f_z rotates by 90° counterclockwise and doubles length.
f_{z^2} rotates by 180° counterclockwise and quadruples length.

6. **a.** f_z is rotation by angle θ and stretch by a factor of r.
f_{z^n} is rotation by angle $n\theta$ and stretch by a factor of r^n.

b. $z_1 = 2$, $z_2 = 2[(-1/2) + (\sqrt{3}/2)i]$,
$z_3 = 2[(-1/2) - (\sqrt{3}/2)i]$

c. 64

7. The vector $\begin{bmatrix}a\\b\end{bmatrix}$ is the diagonal of the rectangle with sides $a\begin{bmatrix}1\\0\end{bmatrix}$ and $b\begin{bmatrix}0\\1\end{bmatrix}$. Since f is linear, $f\left(\begin{bmatrix}a\\b\end{bmatrix}\right) = af\left(\begin{bmatrix}1\\0\end{bmatrix}\right) + bf\left(\begin{bmatrix}0\\1\end{bmatrix}\right)$. This means that the image $f\left(\begin{bmatrix}a\\b\end{bmatrix}\right)$ of the diagonal of this rectangle is equal to the diagonal of the rectangle whose sides are $af\left(\begin{bmatrix}1\\0\end{bmatrix}\right)$ and $bf\left(\begin{bmatrix}0\\1\end{bmatrix}\right)$. Since the sides of this rectangle are each rotated by an angle θ, the diagonal is also rotated by the angle θ. That is, the mapping f is rotation by an angle θ.

8. **b.** $f\left(\begin{bmatrix}2\\1\end{bmatrix}\right) = \begin{bmatrix}-4\\-1\end{bmatrix} \neq \begin{bmatrix}-1\\2\end{bmatrix}$

c. f is not a linear function.

Exercise Set 2.1, page 38

2. **a.** $-12/13$, $18/13$ **b.** $5/11$, $23/11$

c. $-27/7$, $-18/7$

d. x is arbitrary and $y = 1/3x - 5/3$

e. No solution.

4. **a.** $k \neq -3$

b. $k = -3$, infinitely many solutions.

c. No value of k.

5. **a.** No value of m. **b.** $m = 3$

c. $m \neq 3$

Exercise Set 2.2, page 45

1. $5/13$, $-1/13$ **2.** $3/13$, $2/13$

3. $1/5$, 0, $-1/5$ **4.** 1, 1, 1

5. 1, -1, 3, -2

6. $x = (-5/11)z$, $y = (7/11)z$, $z = z$

7. $x = (-5/11)z + 16/11$
$y = (7/11)z + 4/11$
$z = z$

8. No solution. **9.** No solution.

10. $x = -1/5$, $y = 3/5$ **11.** No solution.

12. **a.** $c = 0$ **b.** $c \neq 0$ **c.** No

13. a. $x = \dfrac{ds - bt}{ad - bc} \quad y = \dfrac{at - cs}{ad - bc}$

b. All statements are true.

14. Each of the equations represents a plane through the origin. Hence these planes have at least one line in common. All the nonzero points on this line correspond to nontrivial solutions of this system of equations.

Exercises for Applications 2, page 62

1. a.

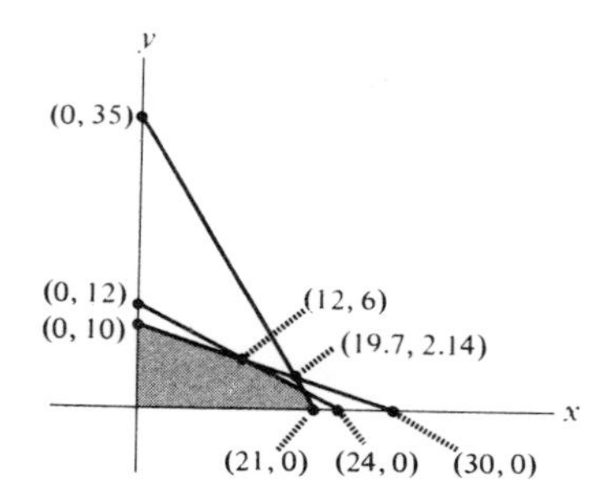

b.

Vertex	Value of f	
(0, 0)	0	
(0, 10)	270	
(12, 6)	294	Maximum
(19.7, 2.14)	274.48	
(21, 0)	231	

2. (1b)

Tableau 1

$$\left[\begin{array}{cccccc|c} 1 & \textcircled{3} & 1 & 0 & 0 & 0 & 30 \\ 1 & 2 & 0 & 1 & 0 & 0 & 24 \\ 5 & 3 & 0 & 0 & 1 & 0 & 105 \\ \hline -11 & -27 & 0 & 0 & 0 & 1 & 0 \end{array}\right]$$

Tableau 2

$$\left[\begin{array}{cccccc|c} 1/3 & 1 & 1/3 & 0 & 0 & 0 & 10 \\ \textcircled{1/3} & 0 & -2/3 & 1 & 0 & 0 & 4 \\ 4 & 0 & -1 & 0 & 1 & 0 & 75 \\ \hline -2 & 0 & 9 & 0 & 0 & 1 & 270 \end{array}\right]$$

Tableau 3

$$\left[\begin{array}{cccccc|c} 0 & 1 & 1 & -1 & 0 & 0 & 6 \\ 1 & 0 & -2 & 3 & 0 & 0 & 12 \\ 0 & 0 & -9 & -12 & 1 & 0 & 27 \\ \hline 0 & 0 & 5 & 6 & 0 & 1 & 294 \end{array}\right]$$

3. Constraint equations:

$$\begin{aligned} 6x + 2y + 3z &= \text{time in } D_1 \leq 150 \\ 4x + 3y + 3z &= \text{time in } D_2 \leq 200 \\ 2x + 3y + 0z &= \text{time in } D_3 \leq 30 \end{aligned}$$

where x, y, z are the number of units of P_1, P_2, and P_3 to be produced.

Objective function:

$$P = 100x + 110y + 120z$$

Tableau 1

$$\left[\begin{array}{ccccccc|c} x & y & z & u & v & w & P & \\ 6 & 2 & \textcircled{3} & 1 & 0 & 0 & 0 & 150 \\ 4 & 3 & 3 & 0 & 1 & 0 & 0 & 200 \\ 2 & 3 & 0 & 0 & 0 & 1 & 0 & 30 \\ \hline -100 & -110 & -120 & 0 & 0 & 0 & 1 & 0 \end{array}\right]$$

Tableau 2

$$\left[\begin{array}{ccccccc|c} 2 & 2/3 & 1 & 1/3 & 0 & 0 & 0 & 50 \\ -2 & 1 & 0 & -1 & 1 & 0 & 0 & 50 \\ 2 & \textcircled{3} & 0 & 0 & 0 & 1 & 0 & 30 \\ \hline 140 & -30 & 0 & 40 & 0 & 0 & 1 & 6000 \end{array}\right]$$

Tableau 3

$$\left[\begin{array}{ccccccc|c} 14/9 & 0 & 1 & 1/3 & 0 & -2/9 & 0 & 130/3 \\ -8/3 & 0 & 0 & -1 & 1 & -1/3 & 0 & 40 \\ 2/3 & 1 & 0 & 0 & 0 & 1/3 & 0 & 10 \\ \hline 160 & 0 & 0 & 40 & 0 & 10 & 1 & 6300 \end{array}\right]$$

Maximum profit occurs when

$$\begin{aligned} x &= \text{number of units of } P_1 = 0, \\ y &= \text{number of units of } P_2 = 10, \\ z &= \text{number of units of } P_3 = 130/3 \end{aligned}$$

Maximum profit = \$6,300.

Exercise Set 3.1, page 72

1. a.

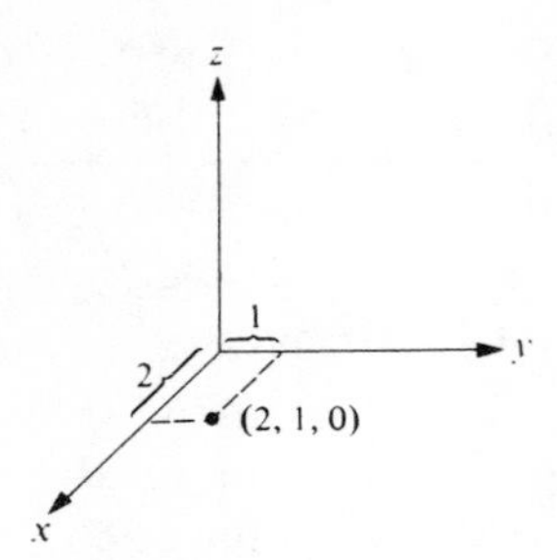

b.

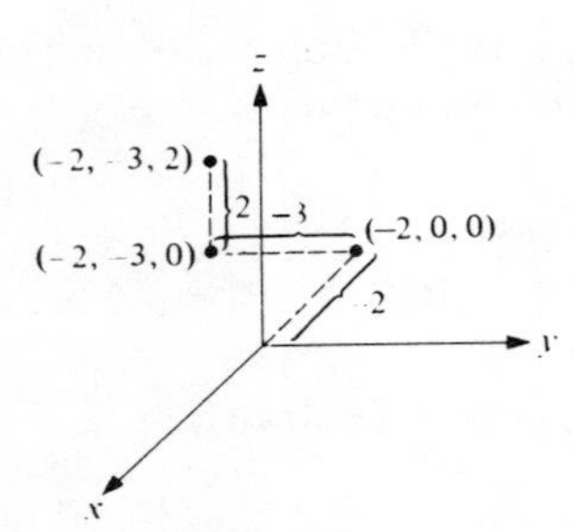

c.

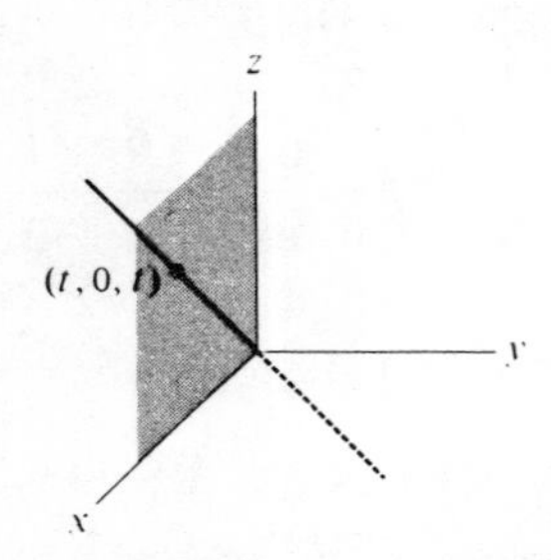

d. The plane parallel to the xy plane and 1 unit above the xy plane.

e. The plane perpendicular to the xy plane containing the line $y = 2x$, $z = 0$.

2. a. $\begin{bmatrix} -1 \\ -1 \\ 12 \end{bmatrix}$ **b.** $\begin{bmatrix} 8 \\ -7 \\ -11 \end{bmatrix}$ **c.** $\begin{bmatrix} -3 \\ 3 \\ 2 \end{bmatrix}$

d. $c_1 = 2$, $c_2 = -1$

3. a. $\begin{bmatrix} -1 \\ 6 \\ 9 \\ 0 \end{bmatrix}$ **b.** $\begin{bmatrix} 12 \\ 0 \\ 0 \\ -9 \end{bmatrix}$ **c.** $\begin{bmatrix} -3 \\ 2 \\ 3 \\ 2 \end{bmatrix}$

4. Linear. **5.** Linear. **6.** Linear.

7. Not linear. **8.** Not linear.

9. Linear **10.** Linear **11.** Linear.

12. a. The xy plane **b.** The xy plane

13. If T is linear, then

$$T\begin{bmatrix} x_1 \\ x_2 \end{bmatrix} = x_1 T\begin{bmatrix} 1 \\ 0 \end{bmatrix} + x_2 T\begin{bmatrix} 0 \\ 1 \end{bmatrix}$$

If $T\begin{bmatrix} 1 \\ 0 \end{bmatrix} = \begin{bmatrix} a_{11} \\ a_{21} \end{bmatrix}$ and $T\begin{bmatrix} 0 \\ 1 \end{bmatrix} = \begin{bmatrix} a_{12} \\ a_{22} \end{bmatrix}$ then the first equation becomes

$$T\begin{bmatrix} x_1 \\ x_2 \end{bmatrix} = x_1 \begin{bmatrix} a_{11} \\ a_{21} \end{bmatrix} + x_2 \begin{bmatrix} a_{12} \\ a_{22} \end{bmatrix} = \begin{bmatrix} x_1 a_{11} + x_2 a_{12} \\ x_1 a_{21} + x_2 a_{22} \end{bmatrix}$$

To show that every function of this form is linear, simply verify that functions of this form satisfy conditions 1 and 2 of Definition 3.4.

Exercise Set 3.2, page 79

3. See Exercise 1 in Applications 1

4. a. Subspace **b.** Not subspace

c. Not subspace **d.** Subspace

e. Not subspace

5. (1) $I(p_1(x) + p_2(x))$

$$= \int_0^1 (p_1(x) + p_2(x))\, dx$$
$$= \int_0^1 p_1(x)\, dx + \int_0^1 p_2(x)\, dx$$
$$= I(p_1(x)) + I(p_2(x))$$

(2) $I(cp(x)) = \int_0^1 cp(x)\, dx$

$$= c\int_0^1 p(x)\, dx$$
$$= cI(p(x))$$

6. (1) Let $p(x)$, $q(x)$ be polynomials of degree at most 3. Then the degree of $p(x) + q(x)$ is also at most 3. So $p(x) + q(x)$ is in S.

(2) Since $cp(x)$ is a polynomial of the same degree as $p(x)$, if $p(x)$ is in S, so is $cp(x)$.

7. a. (1) $X(p_1(x) + p_2(x)) = x(p_1(x) + p_2(x))$
$= xp_1(x) + xp_2(x)$
$= Xp_1(x) + Xp_2(x)$

(2) $X(cp(x)) = xcp(x) = cxp(x) = cXp(x)$.

8. a. If $p(x)$ is in N, then $p(0) = 0$, that is, the constant term of $p(x)$ is 0. Since the sum of two polynomials with constant term zero has constant term zero and the product of a real number and a polynomial with constant term zero also has constant term zero, N is a subspace of $\mathbb{R}_3[x]$.

b. This set is not a subspace. If $p(x)$ is in this set, then $p(0) = 1$. Since $2p(0) = 2$, this means that $2p(x)$ is not in this set, which implies that this set is not a subspace.

10. U_1 = all vectors of the form $\begin{bmatrix} x \\ 0 \end{bmatrix}$

U_2 = all vectors of the form $\begin{bmatrix} 0 \\ y \end{bmatrix}$

Then $\begin{bmatrix} 1 \\ 0 \end{bmatrix}$ and $\begin{bmatrix} 0 \\ 1 \end{bmatrix}$ are in $U_1 \cup U_2$, but $\begin{bmatrix} 1 \\ 0 \end{bmatrix} + \begin{bmatrix} 0 \\ 1 \end{bmatrix} = \begin{bmatrix} 1 \\ 1 \end{bmatrix}$ is not in $U_1 \cup U_2$.

11. a. Vector space

b. Not a vector space

12. a. Subspace **b.** Not subspace

c. Subspace

Exercise Set 3.3, page 91

1. a. $\begin{bmatrix} 0 & 1 \\ 1 & 0 \end{bmatrix}$ **b.** $\begin{bmatrix} -1 & 0 \\ 0 & 1 \end{bmatrix}$

c. $\begin{bmatrix} 1/\sqrt{2} & -1/\sqrt{2} \\ 1/\sqrt{2} & 1/\sqrt{2} \end{bmatrix}$ **d.** $\begin{bmatrix} 2 & 0 \\ 0 & 2 \end{bmatrix}$

e. $\begin{bmatrix} 1 & 1 & -1 \\ 2 & 0 & 0 \\ 1 & 0 & -4 \end{bmatrix}$ **f.** $\begin{bmatrix} 1 & 1 & -2 & 3 \end{bmatrix}$

g. $\begin{bmatrix} 2 & -1 \\ 0 & 0 \\ -1 & -1 \\ 0 & 1 \end{bmatrix}$

2. a. $\begin{bmatrix} 16 \\ 4 \end{bmatrix}$ **b.** $\begin{bmatrix} 13 \\ -5 \end{bmatrix}$ **c.** $\begin{bmatrix} 20 \\ 11 \\ 10 \end{bmatrix}$

d. $\begin{bmatrix} 11 \\ -5 \\ -8 \\ 6 \end{bmatrix}$ **e.** $\begin{bmatrix} 46 \\ -20 \\ -16 \\ 65 \end{bmatrix}$ **f.** $\begin{bmatrix} 15 \\ 24 \\ 27 \\ 19 \\ 8 \end{bmatrix}$

g. $[-15]$ **h.** $\begin{bmatrix} -8 \\ -8 \\ 32 \\ 12 \\ -12 \end{bmatrix}$

3. a. $\begin{bmatrix} 1 \\ 0 \end{bmatrix} + \left(\text{any multiple of } \begin{bmatrix} 3 \\ 1 \end{bmatrix}\right)$

b. $\begin{bmatrix} 0 \\ 1 \end{bmatrix} + \left(\text{any multiple of } \begin{bmatrix} 3 \\ 1 \end{bmatrix}\right)$

c. $\begin{bmatrix} -2 \\ 0 \end{bmatrix} + \left(\text{any multiple of } \begin{bmatrix} 3 \\ 1 \end{bmatrix}\right)$

d. Any multiple of $\begin{bmatrix} 3 \\ 1 \end{bmatrix}$ **e.** No solution.

4. Maps all vectors in $\mathbb{R}^2$ onto vectors on the line $y = -4x$. For any line in $\mathbb{R}^2$ that is parallel to the vector $\begin{bmatrix} 3 \\ 1 \end{bmatrix}$ all the vectors ending on that line are mapped to the same point.

5. a. Stretch each vector by a factor of 3.

b. Reflection about the origin.

c. Projection onto the y axis.

d. Rotation of 90° counterclockwise.

e. Rotation of 90° clockwise.

f. Rotation by 45° counterclockwise and stretch by a factor of 2.

6. a. $\begin{bmatrix} 0 & 0 & 0 \\ 0 & 0 & 0 \\ 0 & 0 & 0 \end{bmatrix}$ is called the zero matrix.

b. $\begin{bmatrix} 1 & 0 & 0 \\ 0 & 1 & 0 \\ 0 & 0 & 1 \end{bmatrix}$ is called the identity matrix.

7. c. $D: \begin{bmatrix} c_0 \\ c_1 \\ c_2 \\ c_3 \end{bmatrix} \to \begin{bmatrix} c_1 \\ 2c_2 \\ 3c_3 \\ 0 \end{bmatrix}$

d. $\tilde{D} = \begin{bmatrix} 0 & 1 & 0 & 0 \\ 0 & 0 & 2 & 0 \\ 0 & 0 & 0 & 3 \\ 0 & 0 & 0 & 0 \end{bmatrix}$

8. a. $[1 \quad 0]$ **b.** $[0 \quad 2 \quad 0]$
c. $[2 \quad 1 \quad 0]$

9. b. The matrix that represents T_j is the jth row of M.

11. a. The matrix for counterclockwise rotation by 90° is $\begin{bmatrix} 0 & -1 \\ 1 & 0 \end{bmatrix}$.
The matrix for reflection about the line $y = x$ is $\begin{bmatrix} 0 & 1 \\ 1 & 0 \end{bmatrix}$.
The net effect of rotation followed by reflection is

$$\begin{bmatrix} 0 & 1 \\ 1 & 0 \end{bmatrix}\left(\begin{bmatrix} 0 & -1 \\ 1 & 0 \end{bmatrix}\begin{bmatrix} x \\ y \end{bmatrix}\right) = \begin{bmatrix} 0 & 1 \\ 1 & 0 \end{bmatrix}\begin{bmatrix} -y \\ x \end{bmatrix} = \begin{bmatrix} x \\ -y \end{bmatrix}.$$

Therefore, the net effect of these two linear transformations is

$$\begin{bmatrix} x \\ y \end{bmatrix} \to \begin{bmatrix} x \\ -y \end{bmatrix}$$

which is reflection about the x axis.

Exercises for Applications 3, page 104

1. a. 21 **b.** 3 **c.** 5, 18
d. 21 **e.** 22 **f.** 7

2. a. 19 14 1 4 14 17 13 14 19 19 14 1 4
b. 24 19 6 9 19 22 18 19 24 24 19 6 9
c. 3 20 7 2 20 15 13 20 3 3 20 7 2
d. 16 21 24 9 11 22 18 15 15 5 9 16 1 24

4. LOVE IS GRAND

5. LOVE IS GRAND

6. LOVE IS GRAND

7. a. 7, 22, 5, 0.

b. $M_1^{-1} = \begin{bmatrix} 1 & 7 \\ 0 & 15 \end{bmatrix}$, M_2^{-1} does not exist,
$M_3^{-1} = \begin{bmatrix} 23 & 10 \\ 15 & 21 \end{bmatrix}$, M_4^{-1} does not exist.

c. $M_2 \begin{bmatrix} 0 \\ 0 \end{bmatrix} = M_2 \begin{bmatrix} 13 \\ 0 \end{bmatrix} = \begin{bmatrix} 0 \\ 0 \end{bmatrix}$

$M_4 \begin{bmatrix} 0 \\ 0 \end{bmatrix} = M_4 \begin{bmatrix} 21 \\ 3 \end{bmatrix} = \begin{bmatrix} 0 \\ 0 \end{bmatrix}$

8. YOU GET A GOLD STAR

9. $M^{-1} = (ad - bc)^{-1}\begin{bmatrix} d & -b \\ -c & a \end{bmatrix}$

Exercise Set 4.1, page 115

1. b. The horizontal line $y = 2$

2. a. $y = 3x + 2$ **b.** $y = -3x + 7$
c. $y = (4/3)x + 1/3$

3. a. $\mathbf{v}(t) = t\begin{bmatrix} 2 \\ -3 \end{bmatrix} + \begin{bmatrix} 1 \\ 1 \end{bmatrix}$

b. $\mathbf{v}(t) = t\begin{bmatrix} 3 \\ -5 \end{bmatrix} + \begin{bmatrix} 1 \\ 3 \end{bmatrix}$

c. $\mathbf{v}(t) = t\begin{bmatrix} 1 \\ 1 \\ 1 \end{bmatrix} + \begin{bmatrix} 1 \\ -1 \\ 1 \end{bmatrix}$

d. $\mathbf{v}(t) = t\begin{bmatrix} 3 \\ 6 \\ -2 \end{bmatrix} + \begin{bmatrix} 1 \\ 3 \\ 2 \end{bmatrix}$

4. a. $\mathbf{v}(t_1, t_2) = \begin{bmatrix} 0 \\ 2 \\ 3 \end{bmatrix} + t_1\begin{bmatrix} 1 \\ -1 \\ 2 \end{bmatrix} + t_2\begin{bmatrix} 2 \\ 3 \\ 4 \end{bmatrix}$

b. $\mathbf{v}(t_1, t_2) = t_1\begin{bmatrix} 1 \\ 3 \\ 2 \end{bmatrix} + t_2\begin{bmatrix} 2 \\ 2 \\ 4 \end{bmatrix}$

c. $\mathbf{v}(t_1, t_2) = \begin{bmatrix} 1 \\ -2 \\ 3 \end{bmatrix} + t_1 \begin{bmatrix} -1 \\ -7 \\ -4 \end{bmatrix} + t_2 \begin{bmatrix} -1 \\ -4 \\ 0 \end{bmatrix}$

5. a. Same line b. Same line
c. Distinct parallel lines
d. Distinct nonparallel planes

8. They have one point in common:

$$(1/14)\begin{bmatrix} 1 \\ -1 \\ 2 \end{bmatrix} + \begin{bmatrix} 2 \\ 0 \\ 3 \end{bmatrix}$$

10. $\mathbf{v}(t_1, t_2) = t_1 \begin{bmatrix} -b/a \\ 1 \\ 0 \end{bmatrix} + t_2 \begin{bmatrix} -c/a \\ 0 \\ 1 \end{bmatrix}$

11. $2x + 3y - z = 0$

Exercise Set 4.2, page 123

1. a. $W = \left\langle \begin{bmatrix} 3 \\ 1 \\ 5 \end{bmatrix}, \begin{bmatrix} 1 \\ -1 \\ 3 \end{bmatrix} \right\rangle = \left\langle \begin{bmatrix} 3 \\ 1 \\ 5 \end{bmatrix}, \begin{bmatrix} 1 \\ 1 \\ 1 \end{bmatrix} \right\rangle$

b. $W = \left\langle \begin{bmatrix} 2 \\ 0 \\ 4 \end{bmatrix}, \begin{bmatrix} 1 \\ 1 \\ 1 \end{bmatrix}, \begin{bmatrix} 3 \\ 1 \\ 5 \end{bmatrix} \right\rangle$

c. One vector can span only a line. W is a plane.

2. a. No b. Yes
c. Yes (see Section 1.5) d. Yes
e. No f. Yes

3.

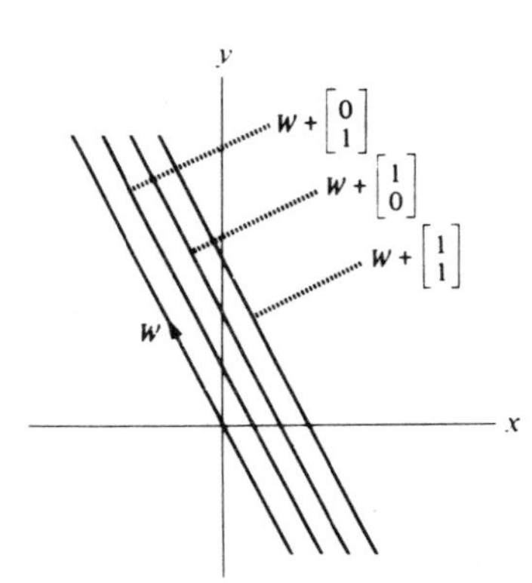

4. $W = \begin{bmatrix} 3 \\ -2 \\ 0 \end{bmatrix} + W = xy$ plane

$\begin{bmatrix} 0 \\ 0 \\ 1 \end{bmatrix} + W =$ the horizontal plane one unit above the xy plane

$\begin{bmatrix} 0 \\ 0 \\ 2 \end{bmatrix} + W = \begin{bmatrix} 3 \\ 5 \\ 2 \end{bmatrix} + W =$ the horizontal plane two units above the xy plane

5. a, c are in the coset. b, d are not.

6. a. $\mathbf{v}_1 = \begin{bmatrix} 1 \\ -2 \end{bmatrix}, \mathbf{v}_2 = \begin{bmatrix} 2 \\ -1 \end{bmatrix}$

b. $\mathbf{v} = \begin{bmatrix} 1 \\ -2 \end{bmatrix}, c = 2$ c. No

9. Let $p(x) = p_2(x) - p_1(x)$. Then $0 = p(1) = p(2) = p(3) = p(4)$, and $p(x)$ is a polynomial of degree at most 3. Since every nonzero polynomial of degree 3 has at most 3 roots, and nonzero polynomials of smaller degree have fewer roots, the fact that $p(x)$ has 4 roots implies that $p(x) = 0$. Hence $p_1(x) = p_2(x)$.

10. a. $L_i(j) = \begin{cases} 0 & \text{if } i \neq j \\ 1 & \text{if } i = j \end{cases}$

Exercise Set 4.3, page 134

1. Any two of the vectors listed is a basis.

2. The three vectors listed are a basis.

3. Dependency equation: $2\mathbf{w}_1 - \mathbf{w}_2 + \mathbf{w}_3 + 0\mathbf{w}_4 = \mathbf{0}$. To form a basis, delete one of the first three vectors.

4. Dependency equation: $-\mathbf{w}_1 - \mathbf{w}_2 + \mathbf{w}_3 + 0\mathbf{w}_4 = \mathbf{0}$. To form a basis delete one of the first three vectors.

5. Basis: $\{\mathbf{w}_1, \mathbf{w}_2, \mathbf{w}_4, \mathbf{w}_5\}$.

6. a. Plane b. $\mathbb{R}^3$
c. Line d. Plane

7. a. $\left\{\begin{bmatrix}1\\-1\\2\end{bmatrix}\right\}$ b. $\left\{\begin{bmatrix}2\\0\\3\end{bmatrix}, \begin{bmatrix}-1\\0\\3\end{bmatrix}\right\}$

c. $\left\{\begin{bmatrix}1\\0\\0\end{bmatrix}, \begin{bmatrix}0\\1\\0\end{bmatrix}, \begin{bmatrix}0\\0\\1\end{bmatrix}\right\}$

8. If $\mathbf{v}$ and $\mathbf{w}$ lie on the same line through the origin, then $\mathbf{v} = c\mathbf{w}$ for some scalar c. So the equation $1\mathbf{v} - c\mathbf{w} = \mathbf{0}$ is a nontrivial dependency relation between $\mathbf{v}$ and $\mathbf{w}$.

9. Since $\mathbf{w}_1, \mathbf{w}_2, \mathbf{w}_3$ are linearly independent, the two vectors $\mathbf{w}_1, \mathbf{w}_2$ are linearly independent. Therefore, $\langle \mathbf{w}_1, \mathbf{w}_2 \rangle$ is a plane. If $\langle \mathbf{w}_1, \mathbf{w}_2, \mathbf{w}_3 \rangle$ were a plane, then it would be the plane $\langle \mathbf{w}_1, \mathbf{w}_2 \rangle$, that is, $\mathbf{w}_3$ would lie in the plane spanned by $\mathbf{w}_1$ and $\mathbf{w}_2$. Therefore,

$$\mathbf{w}_3 = c_1\mathbf{w}_1 + c_2\mathbf{w}_2$$

or

$$\mathbf{0} = c_1\mathbf{w}_1 + c_2\mathbf{w}_2 - 1\mathbf{w}_3$$

giving a nontrivial dependency equation that would contradict the assumption that $\{\mathbf{w}_1, \mathbf{w}_2, \mathbf{w}_3\}$ is a linearly independent set.

10. Since $\begin{bmatrix}1\\3\end{bmatrix}$ and $\begin{bmatrix}2\\-4\end{bmatrix}$ are not parallel, they are linearly independent. Hence, they span a plane. Since these two vectors lie in $\mathbb{R}^2$, the plane they span lies in $\mathbb{R}^2$. The only plane contained in $\mathbb{R}^2$ is $\mathbb{R}^2$ itself. Therefore,

$$\left\langle \begin{bmatrix}1\\3\end{bmatrix}, \begin{bmatrix}2\\-4\end{bmatrix} \right\rangle = \mathbb{R}^2 = \left\langle \begin{bmatrix}1\\0\end{bmatrix}, \begin{bmatrix}0\\1\end{bmatrix} \right\rangle$$

11. a. To see if $f(x) = x^2$ and $g(x) = 1/(x-2)$ are linearly independent on $[0, 1]$, we must see if we can find a nontrivial solution to the equation

$$c_1 f(x) + c_2 g(x) = 0 \qquad \text{for all } x \text{ in } [0, 1]$$

or

$$c_1 x^2 + c_2(1/(x-2)) = 0 \qquad \text{for all } x \text{ in } [0, 1]$$

If this equation is to hold for all values of x between 0 and 1, it must hold for $x = 0$. Substituting $x = 0$ in the equation above gives

$$c_1(0) + c_2(1/(0-2)) = 0 \qquad \text{or} \qquad c_2 = 0$$

and substituting $x = 1$ in the equation gives

$$c_1(1) + c_2(1/(1-2)) = 0 \qquad \text{or} \qquad c_1 - c_2 = 0$$

From these two equations, we see that $0 = c_1 = c_2$. Therefore, these functions are linearly independent.

b. $c_1 \sin(0) + c_2 \cos(0) = 0$ implies that $c_2 = 0$

$c_1 \sin(\pi/2) + c_2 \cos(\pi/2) = 0$ implies that $c_1 = 0$

c. For $x = \pi/2$

$c_1(1) + c_2 \cos(2(\pi/2)) + c_3 \cos^2(\pi/2) = 0$ implies that $c_2 = c_1$

For $x = \pi/4$

$c_1(1) + c_2 \cos(2(\pi/4)) + c_3 \cos^2(\pi/4) = 0$ implies that $c_3 = -2c_1$

From these two equations we obtain

$$c_1(1) + c_1 \cos(2x) - 2c_1 \cos^2(x) = 0$$

Rearranging terms and letting $c_1 = 1$ gives the relation

$$\cos(2x) = 2\cos^2(x) - 1$$

12. Exercise 10 in Section 4.2 says that the Lagrange interpolation polynomials span $\mathbb{R}_3[x]$. We show that L_1, L_2, L_3, L_4 are linearly independent: Suppose

$$c_1 L_1(x) + c_2 L_2(x) + c_3 L_3(x) + c_4 L_4(x) = 0$$

for all x

Letting $x = 1$, this becomes $c_1(1) + c_2(0) + c_3(0) + c_4(0) = 0$ or $c_1 = 0$. Similarly, by letting $x = 2$, 3, and 4, we see that $0 = c_1 = c_2 = c_3 = c_4$. Thus, the Lagrange interpolation polynomials are linearly independent and form a basis for $\mathbb{R}_3[x]$.

Exercise Set 4.4, page 139

1. a. Basis b. Not basis
c. Not basis d. Not basis
e. Not basis

2. a. Since the set $\left\{\begin{bmatrix}1\\0\end{bmatrix}, \begin{bmatrix}0\\1\end{bmatrix}\right\}$ is a basis for $\mathbb{R}^2$ and contains two vectors, every subset of $\mathbb{R}^2$ containing more than two vectors is a set of linearly dependent vectors (Lemma 4.1). Hence

$\{\mathbf{v}_1, \mathbf{v}_2, \mathbf{v}\}$ is a linearly dependent set. In other words, there are scalars c_1, c_2, c which are not all 0 such that

$$\mathbf{0} = c_1\mathbf{v}_1 + c_2\mathbf{v}_2 + c\mathbf{v}$$

Since $\mathbf{v}_1$ and $\mathbf{v}_2$ are linearly independent, $c \neq 0$. So we can solve for $\mathbf{v}$ to obtain

$$\mathbf{v} = (-c_1/c)\mathbf{v}_1 + (-c_2/c)\mathbf{v}_2$$

In other words $\mathbf{v}$ is a linear combination of $\mathbf{v}_1$ and $\mathbf{v}_2$. Since $\mathbf{v}$ was arbitrary, every vector of $\mathbb{R}^2$ is a linear combination of $\mathbf{v}_1$ and $\mathbf{v}_2$. In other words, $\mathbf{v}_1$ and $\mathbf{v}_2$ span $\mathbb{R}^2$. Since $\mathbf{v}_1$ and $\mathbf{v}_2$ were assumed to be linearly independent, this means that $\{\mathbf{v}_1, \mathbf{v}_2\}$ is a basis for $\mathbb{R}^2$.

3. Suppose $\{\mathbf{v}_1, \mathbf{v}_2, \mathbf{v}_3\}$ spans $\mathbb{R}^3$ but is linearly dependent. Then one of the vectors $\mathbf{v}_1, \mathbf{v}_2$, or $\mathbf{v}_3$ is redundant and may be removed. This would mean that $\mathbb{R}^3$ could be spanned by two vectors, which means that $\mathbb{R}^3$ has a basis containing only two vectors. Since

$$\left\{\begin{bmatrix}1\\0\\0\end{bmatrix}, \begin{bmatrix}0\\1\\0\end{bmatrix}, \begin{bmatrix}0\\0\\1\end{bmatrix}\right\}$$

is also a basis for $\mathbb{R}^3$, this contradicts Theorem 4.7 which says that all bases of a given vector have the same number of elements. Therefore $\mathbf{v}_1$, $\mathbf{v}_2$, and $\mathbf{v}_3$ must have been linearly independent.

4. Types of subspaces of $\mathbb{R}^4$:

0-dimensional—$\{\mathbf{0}\}$
1-dimensional—lines containing the origin
2-dimensional—planes containing the origin
3-dimensional—copies of $\mathbb{R}^3$ containing the origin
4-dimensional—$\mathbb{R}^4$

5. a. $\mathbf{v}_0 = \begin{bmatrix}1\\0\\0\end{bmatrix}$, $\mathbf{v}_1 = \begin{bmatrix}1\\0\\-1\end{bmatrix}$,

$\mathbf{v}_2 = \begin{bmatrix}2\\-3\\-2\end{bmatrix}$, $\mathbf{v}_3 = \begin{bmatrix}-2\\-9\\-4\end{bmatrix}$

b. Yes **c.** $-6\mathbf{v}_0 - 2\mathbf{v}_1 + 3\mathbf{v}_2 = \mathbf{v}_3$

6. No. If the columns of the matrix representing T are linearly dependent then the vectors

$$T\begin{bmatrix}1\\0\\0\end{bmatrix}, \quad T\begin{bmatrix}0\\1\\0\end{bmatrix} \quad \text{and} \quad T\begin{bmatrix}0\\0\\1\end{bmatrix}$$

are linearly dependent.

7. To show that $\{\mathbf{v}_1, \mathbf{v}_2, \mathbf{v}_3\}$ is a basis for $\mathbb{R}^3$, it suffices to show that these three vectors are linearly independent. Suppose that

$$\mathbf{0} = c_1\mathbf{v}_1 + c_2\mathbf{v}_2 + c_3\mathbf{v}_3$$

Then

$$\begin{aligned}\mathbf{0} &= T(\mathbf{0})\\ &= T(c_1\mathbf{v}_1 + c_2\mathbf{v}_2 + c_3\mathbf{v}_3)\\ &= c_1T(\mathbf{v}_1) + c_2T(\mathbf{v}_2) + c_3T(\mathbf{v}_3)\end{aligned}$$

Since $T(\mathbf{v}_1)$, $T(\mathbf{v}_2)$, $T(\mathbf{v}_3)$ are linearly independent, the equation above means that $c_1 = c_2 = c_3 = 0$ which means that $\mathbf{v}_1$, $\mathbf{v}_2$, and $\mathbf{v}_3$ are linearly independent.

8. Let $W_i = \langle\mathbf{u}_i, \mathbf{v}_i, \mathbf{w}_i\rangle$ for $i = 1, 2$. Then the vectors $\mathbf{u}_1, \mathbf{v}_1, \mathbf{w}_1, \mathbf{u}_2, \mathbf{v}_2, \mathbf{w}_2$ are linearly dependent since they are six vectors in a five-dimensional space. Hence there are scalars c_i, d_i, and e_i (not all 0) so that

$$\mathbf{0} = c_1\mathbf{u}_1 + d_1\mathbf{v}_1 + e_1\mathbf{w}_1 + c_2\mathbf{u}_2 + d_2\mathbf{v}_2 + e_2\mathbf{w}_2$$

Then the vector

$$\begin{aligned}\mathbf{v} &= c_1\mathbf{u}_1 + d_1\mathbf{v}_1 + e_1\mathbf{w}_1\\ &= -c_2\mathbf{u}_2 - d_2\mathbf{v}_2 - e_2\mathbf{w}_2\end{aligned}$$

is in both W_1 and W_2.

9. Let $\mathbf{w}$ be a nonzero vector in both W_1 and W_2. We can find vectors $\mathbf{v}_1$ and $\mathbf{v}_2$ so that $W_1 = \langle\mathbf{w}, \mathbf{v}_1\rangle$ and $W_2 = \langle\mathbf{w}, \mathbf{v}_2\rangle$. Therefore, $W = \langle\mathbf{w}, \mathbf{v}_1, \mathbf{v}_2\rangle$. Using the fact that $W_1 \neq W_2$, it is not hard to show that these three vectors form a basis for W.

11. a. Basis **b.** Basis **c.** Not basis
d. Not basis **e.** Basis

Exercise Set 5.1, page 151

1. a. $\begin{bmatrix}2 & -1\\3 & 2\end{bmatrix}\begin{bmatrix}x\\y\end{bmatrix} = \begin{bmatrix}-1\\9\end{bmatrix}$

b. Null space $= \langle\mathbf{0}\rangle$, Image space $= \mathbb{R}^2$.

c. Solution $x = 1$, $y = 3$.

2. a. $\begin{bmatrix} 3 & 2 \\ -6 & -4 \end{bmatrix}\begin{bmatrix} x \\ y \end{bmatrix} = \begin{bmatrix} -4 \\ 8 \end{bmatrix}$

b. Null space $= \left\langle \begin{bmatrix} -2 \\ 3 \end{bmatrix} \right\rangle$,

Image space $= \left\langle \begin{bmatrix} 3 \\ -6 \end{bmatrix} \right\rangle$

c. Solution vectors: $x\begin{bmatrix} -2 \\ 3 \end{bmatrix} + \begin{bmatrix} 0 \\ -2 \end{bmatrix}$

d.

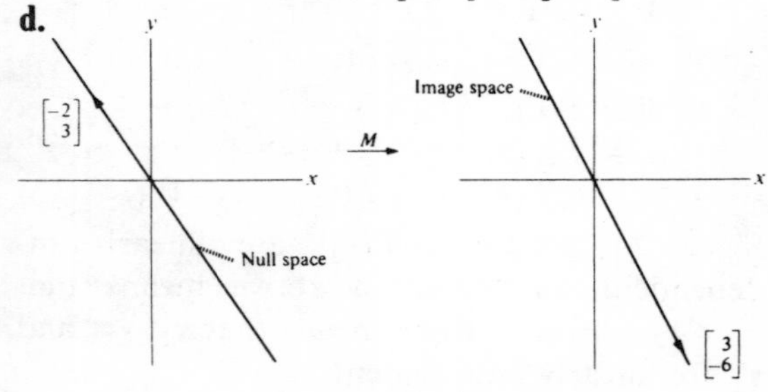

3. a. $\begin{bmatrix} 1 & 2 & 3 \\ -1 & 3 & 2 \end{bmatrix}\begin{bmatrix} x \\ y \\ z \end{bmatrix} = \begin{bmatrix} 11 \\ 14 \end{bmatrix}$

b. Null space $= \left\langle \begin{bmatrix} -1 \\ -1 \\ 1 \end{bmatrix} \right\rangle$,

Image space $= \mathbb{R}^2$

c. Solution vectors: $z\begin{bmatrix} -1 \\ -1 \\ 1 \end{bmatrix} + \begin{bmatrix} 1 \\ 5 \\ 0 \end{bmatrix}$

d.

4. a. $\begin{bmatrix} 1 & 1 \\ 1 & -1 \\ -1 & 2 \end{bmatrix}\begin{bmatrix} x \\ y \end{bmatrix} = \begin{bmatrix} 3 \\ 1 \\ 1 \end{bmatrix}$

b. Null space $\langle \mathbf{0} \rangle$

Image space $= \left\langle \begin{bmatrix} 1 \\ 1 \\ -1 \end{bmatrix}, \begin{bmatrix} 1 \\ -1 \\ 2 \end{bmatrix} \right\rangle$

c. No solutions.

d.

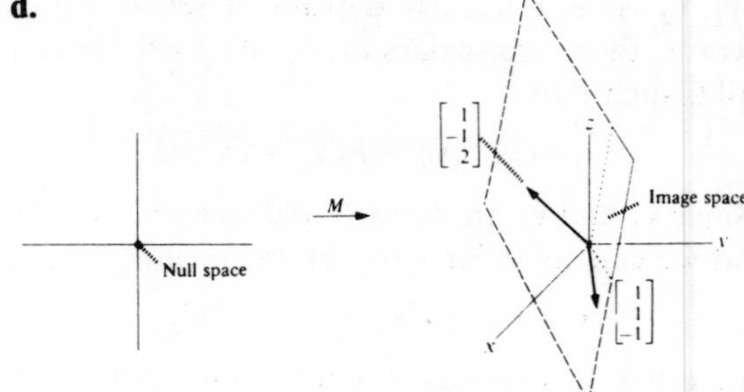

5. Null space $= \left\langle \begin{bmatrix} 1 \\ 1 \end{bmatrix} \right\rangle$, Image space $= \left\langle \begin{bmatrix} 1 \\ 2 \end{bmatrix} \right\rangle$

6. Null space $= \langle \mathbf{0} \rangle$, Image space $= \mathbb{R}^2$

7. Null space $= \langle \mathbf{0} \rangle$, Image space $= \mathbb{R}^3$

8. Null space: $\left\langle \begin{bmatrix} 1 \\ 1 \\ 1 \\ 0 \end{bmatrix}, \begin{bmatrix} -6 \\ -11 \\ 0 \\ 7 \end{bmatrix} \right\rangle$,

Image spaces: $\left\langle \begin{bmatrix} 1 \\ -3 \\ -7 \end{bmatrix}, \begin{bmatrix} 2 \\ 1 \\ 0 \end{bmatrix} \right\rangle$

9. Null space $= \langle \mathbf{0} \rangle$,

image space $= \left\langle \begin{bmatrix} 1 \\ -2 \\ 3 \\ 4 \end{bmatrix}, \begin{bmatrix} 2 \\ 3 \\ 1 \\ -2 \end{bmatrix} \right\rangle$

10 (5) $\mathbf{w}_1 =$ any vector that is not a scalar multiple of $\begin{bmatrix} 1 \\ 2 \end{bmatrix}$. $\mathbf{w}_2 = c\begin{bmatrix} 1 \\ 2 \end{bmatrix}$ for any scalar c.

(6) There is no $\mathbf{w}_1$ since the image space of M is $\mathbb{R}^2$.
There is no $\mathbf{w}_2$ since the null space of M is $\langle \mathbf{0} \rangle$.

(7) There is no $\mathbf{w}_1$ since the image space of M is $\mathbb{R}^3$.
There is no $\mathbf{w}_2$ since the null space of M is $\langle \mathbf{0} \rangle$.

(8) $\begin{bmatrix} 0 \\ 0 \\ 7 \end{bmatrix} = \mathbf{w}_1$ (Any vector not in the image space will do.)

$\begin{bmatrix} 2 \\ 1 \\ 0 \end{bmatrix} = \mathbf{w}_2$ (Any vector in the image space will do.)

(9) $\mathbf{w}_1 =$ any vector of $\mathbb{R}^4$ that is not a linear combination of the vectors

$$\left\langle \begin{bmatrix} 1 \\ -2 \\ 3 \\ 4 \end{bmatrix}, \begin{bmatrix} 2 \\ 3 \\ 1 \\ -2 \end{bmatrix} \right\rangle, \text{ for example } \mathbf{w}_1 = \begin{bmatrix} 1 \\ 0 \\ 0 \\ 0 \end{bmatrix}.$$

There is no vector $\mathbf{w}_2$ since the null space of M is $\langle \mathbf{0} \rangle$.

Exercise Set 5.2, page 157

1. Image space $= \mathbb{R}^2$, Null space $= \langle \mathbf{0} \rangle$

2. Image space $= \left\langle \begin{bmatrix} 1 \\ -3 \end{bmatrix} \right\rangle$,

Null space $= \left\langle \begin{bmatrix} 3 \\ 1 \end{bmatrix} \right\rangle$

3. Image space $= \mathbb{R}^3$, Null space $= \langle \mathbf{0} \rangle$

4. Image space $= \left\langle \begin{bmatrix} -1 \\ 0 \\ 0 \end{bmatrix} \right\rangle$,

Null space $= \left\langle \begin{bmatrix} 0 \\ 1 \\ 0 \end{bmatrix}, \begin{bmatrix} 0 \\ 0 \\ 1 \end{bmatrix} \right\rangle$

5. Image space $= \mathbb{R}^3$, Null space $= \langle \mathbf{0} \rangle$

8. By row reducing M we find that the nullity of M is 2. Therefore, rank of $M = 4 - 2 = 2$. To find a basis for the image space of M, we need only find two linearly independent columns of M. For example, a basis for the image space of M is

$$\left\langle \begin{bmatrix} 1 \\ 3 \\ 1 \end{bmatrix}, \begin{bmatrix} -1 \\ 2 \\ 4 \end{bmatrix} \right\rangle$$

9. Null space of D is the set of all polynomials of the form $p(x) = c$ for some constant c.

10. b. All polynomials with constant term 0.

Exercise Set 5.3, page 164

1. $\begin{bmatrix} 1 & 1 \\ 0 & 0 \end{bmatrix}$ rank $= 1$, nullity $= 1$

2. $\begin{bmatrix} 1 & 0 \\ 0 & 1 \end{bmatrix}$ rank $= 2$, nullity $= 0$

3. $\begin{bmatrix} 1 & 0 & -2 \\ 0 & 1 & 1 \\ 0 & 0 & 0 \end{bmatrix}$ rank $= 2$, nullity $= 1$

4. $\begin{bmatrix} 1 & 0 & 0 \\ 0 & 1 & 0 \\ 0 & 0 & 1 \end{bmatrix}$ rank $= 3$, nullity $= 0$

5. $\begin{bmatrix} 1 & 0 & 0 & -1 & 0 \\ 0 & 1 & 0 & 5/2 & 1/2 \\ 0 & 0 & 1 & 13/4 & 3/4 \end{bmatrix}$ rank $= 3$, nullity $= 2$

6. $\begin{bmatrix} 0 & 1 \\ 0 & 0 \end{bmatrix}$ rank $= 1$, nullity $= 1$

7. $\begin{bmatrix} 1 & 0 \\ 0 & 1 \\ 0 & 0 \end{bmatrix}$ rank $= 2$, nullity $= 0$

8. $\begin{bmatrix} 1 & -3/2 & 5/2 \\ 0 & 0 & 0 \\ 0 & 0 & 0 \end{bmatrix}$ rank $= 1$, nullity $= 2$

9. Completely reduced; rank $= 2$, nullity $= 3$

10. $\begin{bmatrix} 1 & 0 & 3 & -1 \\ 0 & 1 & 2 & 0 \\ 0 & 0 & 0 & 0 \end{bmatrix}$ rank $= 2$, nullity $= 2$

11. Completely reduced; rank $= 2$, nullity $= 0$

12. a. False

$$\left.\begin{aligned} x + y + z &= 1 \\ 2x + 2y + 2z &= 0 \end{aligned}\right\} \text{has no solution}$$

b. True

c. False. The image space of $\begin{bmatrix} 1 & 1 \\ 1 & 1 \end{bmatrix}$ is $\left\langle \begin{bmatrix} 1 \\ 1 \end{bmatrix} \right\rangle$ and the image space of the equivalent completely reduced matrix $\begin{bmatrix} 1 & 1 \\ 0 & 0 \end{bmatrix}$ is $\left\langle \begin{bmatrix} 1 \\ 0 \end{bmatrix} \right\rangle$.

13. a. Since every vector in the image space of P has nonzero entries only in the first k positions, the dimension of the image space of P is at most k. Moreover, the space

$$H = \langle \mathbf{h}_1, \mathbf{h}_2, \ldots, \mathbf{h}_k \rangle$$

is a subspace of the image space of P.
We will show that $\{\mathbf{h}_1, \mathbf{h}_2, \ldots, \mathbf{h}_k\}$ is linearly independent. This will show that H has dimension k. From this we obtain

$$k = \dim(H) \leq \dim(\text{image space of } P) \leq k$$

Therefore, the rank of P is k.

To show that the vectors $\mathbf{h}_1, \mathbf{h}_2, \ldots, \mathbf{h}_k$ are linearly independent we must show that the only solution to the equation $c_1\mathbf{h}_1 + \cdots + c_k\mathbf{h}_k = 0$ is $c_i = 0$ for all i. We examine the matrix whose columns are the vectors h_i for $i = 1, \ldots, k$.

$$\begin{bmatrix} \mathbf{h}_1 & \mathbf{h}_2 & \cdots & \mathbf{h}_k \end{bmatrix} = \begin{bmatrix} a_{11} & a_{12} & a_{13} & \cdots & a_{1k} \\ 0 & a_{22} & a_{23} & \cdots & a_{2k} \\ \vdots & \vdots & \vdots & & \vdots \\ 0 & 0 & 0 & \cdots & a_{kk} \end{bmatrix}$$

From this we see that the solution of the dependency equation for the $\mathbf{h}_i$'s is

$$a_{kk}c_k = 0. \text{ So } c_k = 0.$$

$$a_{(k-1)(k-1)}c_{k-1} + a_{(k-1)(k)}c_k = 0. \text{ So } c_{k-1} = 0$$

etc.

Hence, all the c_i's are zero, and therefore, the $\mathbf{h}_i$'s are linearly independent.

14. a. $p_i = m_i$ and $p_j = am_i + bm_j$
$m_i = p_i$ and $m_j = (-a/b)p_i + (1/b)p_j$

c. By part b we see that the rows of P span the same vector space as is spanned by the rows of M. Since P is completely reduced, its nonzero rows are linearly independent. Hence, the rows of P form a basis for this vector space.

d. Row rank of M = Dim(vector space spanned by rows of M)
= Number of nonzero rows of P (by part c)
= Rank of P (by Theorem 5.5)

Exercise Set 5.4, page 171

1. $\left\{\begin{bmatrix}1\\1\end{bmatrix}, \begin{bmatrix}1\\0\end{bmatrix}\right\}$ **2.** $\left\{\begin{bmatrix}1\\0\end{bmatrix}, \begin{bmatrix}0\\1\end{bmatrix}\right\}$

3. $\left\{\begin{bmatrix}1\\0\\0\end{bmatrix}, \begin{bmatrix}0\\1\\0\end{bmatrix}, \begin{bmatrix}0\\0\\1\end{bmatrix}\right\}$

4. $\left\{\begin{bmatrix}1\\1\\1\\0\end{bmatrix}, \begin{bmatrix}-6\\-11\\0\\7\end{bmatrix}, \begin{bmatrix}1\\0\\0\\0\end{bmatrix}, \begin{bmatrix}0\\1\\0\\0\end{bmatrix}\right\}$

5. $\left\{\begin{bmatrix}1\\0\end{bmatrix}, \begin{bmatrix}0\\1\end{bmatrix}\right\}$

6. Preimage basis: $\left\{\begin{bmatrix}-5\\4\\1\\0\end{bmatrix}, \begin{bmatrix}4\\-3\\0\\1\end{bmatrix}, \begin{bmatrix}1\\0\\0\\0\end{bmatrix}, \begin{bmatrix}0\\1\\0\\0\end{bmatrix}\right\}$

$$M\mathbf{v} = \begin{bmatrix}-1\\3\\2\end{bmatrix} = 7\begin{bmatrix}1\\1\\2\end{bmatrix} - 4\begin{bmatrix}2\\1\\3\end{bmatrix}$$

Therefore

$$\mathbf{v}_0 = 7\begin{bmatrix}1\\0\\0\\0\end{bmatrix} - 4\begin{bmatrix}0\\1\\0\\0\end{bmatrix}$$

Moreover,

$$\mathbf{v} - \mathbf{v}_0 = \begin{bmatrix}-6\\5\\2\\1\end{bmatrix} = 2\begin{bmatrix}-5\\4\\1\\0\end{bmatrix} + 1\begin{bmatrix}4\\-3\\0\\1\end{bmatrix}$$

Hence

$$\mathbf{v} = 7\begin{bmatrix}1\\0\\0\\0\end{bmatrix} - 4\begin{bmatrix}0\\1\\0\\0\end{bmatrix} + 2\begin{bmatrix}-5\\4\\1\\0\end{bmatrix} + 1\begin{bmatrix}4\\-3\\0\\1\end{bmatrix}$$

Exercises for Applications 5, page 183

1. a. 1 1 0 1 0 1 0

b. 1 0 1 1 1 0 0 0 1 0 1 1 0 1

2. a. 1 0 0 1 **b.** 1 1 1 0 **c.** 0 0 0 1

3. Message transmitted: 0 1 1 0 1 1 0
Message received: 0 1 0 0 1 0 0

Syndrome vector: $\begin{bmatrix}1\\1\\1\end{bmatrix}$

"Corrected" message: 1 1 0 0

4. 0 0 1 1 0 0 1 1 0 0 1 0 1 0 0

5. 0 1 0 0 1 0 0 0 1 0 0

6. Column with entries

0 0 0 0 0 0 0 0 0 1 0 1 0 1 0

7. If the symbol 1 was in the message to be sent, and it was encoded as 1 1 but was received as 0 1, there would be no way of determining whether the vector transmitted was 1 1 or 0 0.

8. $\mathbb{F}^{26} \xrightarrow{M} \mathbb{F}^{31} \xrightarrow{H} \mathbb{F}^{5}$

Exercise Set 6.1, page 193

1. a. $\sqrt{2}$ b. $\sqrt{14/2}$ c. $\sqrt{7}$

2. a. $6/\sqrt{85}$ b. $-1/\sqrt{2}$
c. $-7/\sqrt{319}$

3. a. $\pm\begin{bmatrix}1/3\\-2/3\\2/3\end{bmatrix}$ b. $\pm\begin{bmatrix}5/3\\-10/3\\10/3\end{bmatrix}$

c. $\begin{bmatrix}2\\1\\0\end{bmatrix}$ and $\begin{bmatrix}-2\\0\\1\end{bmatrix}$

5. a. $\begin{bmatrix}0\\-1\end{bmatrix}$ b. $\begin{bmatrix}2/5\\1/5\end{bmatrix}$ c. $\begin{bmatrix}-3/5\\0\\6/5\end{bmatrix}$

6. See Section 8.2.

7. a. The formula for the roots of $p(x)$ is

$$x = \frac{-(2b) \pm \sqrt{(2b)^2 - 4ac}}{2a}$$

Since $p(x) \geq 0$ for all x, the graph of $p(x)$ is never below the x axis. This means that either $p(x)$ has no real roots or that it has a double root. Therefore, the quantity $(2b)^2 - 4ac \leq 0$, that is $b^2 - ac \leq 0$.

b. Use the properties of the dot product.

c. Since the quadratic equation you found in part b is always positive, the coefficients satisfy the inequality given in part a. In this case $a = \mathbf{u} \cdot \mathbf{u}$, $b = \mathbf{u} \cdot \mathbf{v}$, $c = \mathbf{v} \cdot \mathbf{v}$. Hence

$$0 \geq b^2 - ac = (\mathbf{u} \cdot \mathbf{v})^2 - (\mathbf{u} \cdot \mathbf{u})(\mathbf{v} \cdot \mathbf{v}).$$

So we conclude that

$$-1 \leq \frac{\mathbf{u} \cdot \mathbf{v}}{|\mathbf{u}|\,|\mathbf{v}|} \leq 1$$

8. Suppose $\mathbf{0} = c_1\mathbf{u}_1 + c_2\mathbf{u}_2 + c_3\mathbf{u}_3$. Then taking the dot product of both sides of this equation with $\mathbf{u}_1$ gives

$$\mathbf{0} \cdot \mathbf{u}_1 = c_1\mathbf{u}_1 \cdot \mathbf{u}_1 + c_2\mathbf{u}_2 \cdot \mathbf{u}_2 + c_3\mathbf{u}_3 \cdot \mathbf{u}_3$$

or $0 = c_1|\mathbf{u}_1|^2$. Since $\mathbf{u}_1 \neq \mathbf{0}$, $c_1 = 0$.

In the same way, show that $0 = c_2 = c_3$ showing that $\{\mathbf{u}_1, \mathbf{u}_2, \mathbf{u}_3\}$ is a linearly independent set, and therefore a basis for $\mathbb{R}^3$.

9. If $\mathbf{v}_1$, $\mathbf{v}_2$ are in S, then, since

$$(\mathbf{v}_1 + \mathbf{v}_2) \cdot \mathbf{w} = \mathbf{v}_1 \cdot \mathbf{w} + \mathbf{v}_2 \cdot \mathbf{w} = 0 + 0 = 0$$

for every $\mathbf{w}$ in W, $(\mathbf{v}_1 + \mathbf{v}_2)$ is in P.

If $\mathbf{v}$ is in S and c is a scalar, then

$(c\mathbf{v}) \cdot \mathbf{w} = c(\mathbf{v} \cdot \mathbf{w}) = c(0) = 0$ for every $\mathbf{w}$ in W.

Hence $c\mathbf{v}$ is in S. Therefore S is a subspace. (*Note:* $\mathbf{0}$ is in S means that S is not the empty set.)

10. The quantity $(\mathbf{v} \cdot \mathbf{w})$ is a scalar.

Exercise Set 6.2, page 201

1. Not in W. 2. Not in W.

3. In W. 4. In W.

5. $\left\{(1/\sqrt{6})\begin{bmatrix}1\\-1\\2\end{bmatrix},\ (1/\sqrt{210})\begin{bmatrix}5\\13\\4\end{bmatrix}\right\}$

6. $\left\{(1/\sqrt{5})\begin{bmatrix}1\\0\\2\end{bmatrix},\ (1/\sqrt{205})\begin{bmatrix}12\\-5\\-6\end{bmatrix}\right\}$

7. $\left\{(1/\sqrt{7})\begin{bmatrix}1\\1\\-1\\2\end{bmatrix}, (1/\sqrt{266})\begin{bmatrix}12\\5\\9\\-4\end{bmatrix}, (1/\sqrt{322})\begin{bmatrix}10\\-13\\-7\\-2\end{bmatrix}\right\}$

8. a. The plane containing the origin that is perpendicular to **w**.

b. The line containing the origin that is perpendicular to $\langle \mathbf{w}_1, \mathbf{w}_2\rangle$.

d. Let $\{\mathbf{u}_1, \mathbf{u}_2, \ldots, \mathbf{u}_k\}$ be an orthonormal basis for W. For any vector **v** of V let

$$\mathbf{w} = (\mathbf{v}\cdot\mathbf{u}_1)\mathbf{u}_1 + (\mathbf{v}\cdot\mathbf{u}_2)\mathbf{u}_2 + \cdots + (\mathbf{v}\cdot\mathbf{u}_k)\mathbf{u}_k$$

and let $\mathbf{p} = \mathbf{v} - \mathbf{w}$

Then **w** is in W and $\mathbf{v} = \mathbf{w} + \mathbf{p}$. Moreover, since

$$\mathbf{p}\cdot\mathbf{u}_i = (\mathbf{v}-\mathbf{w})\cdot u_i = \mathbf{v}\cdot\mathbf{u}_i - (\mathbf{v}\cdot\mathbf{u}_i)(\mathbf{u}_i\cdot\mathbf{u}_i) = 0$$

the vector **p** is in $W^\perp$.

e. Since every vector in W is perpendicular to every vector in $W^\perp$, and **d** is in both W and $W^\perp$, we see that **d** is perpendicular to itself. In other words $\mathbf{d}\cdot\mathbf{d} = 0$. Therefore, $\mathbf{d} = \mathbf{0}$.

f. If $\mathbf{w}_1 + \mathbf{p}_1 = \mathbf{w}_2 + \mathbf{p}_2$, then $\mathbf{w}_1 - \mathbf{w}_2 = \mathbf{p}_2 - \mathbf{p}_1$. Since W is a subspace, $\mathbf{w}_1 - \mathbf{w}_2$ is in W. Since $W^\perp$ is a subspace, $\mathbf{p}_2 - \mathbf{p}_1$ is in $W^\perp$. However, since $\mathbf{w}_1 - \mathbf{w}_2 = \mathbf{p}_2 - \mathbf{p}_1$, the vector $\mathbf{w}_1 - \mathbf{w}_2$ is in both W and $W^\perp$. Therefore, $\mathbf{w}_1 - \mathbf{w}_2 = \mathbf{0}$. So $\mathbf{w}_1 = \mathbf{w}_2$ and $\mathbf{p}_1 = \mathbf{p}_2$.

10. These ideas are discussed in detail in Section 8.1.

Exercise Set 6.3, page 207

1. Use the fact that

$$T\begin{bmatrix}x\\y\end{bmatrix} = \left(\frac{ax+by}{a^2+b^2}\right)\begin{bmatrix}a\\b\end{bmatrix} = \begin{bmatrix}c_{11}x + c_{12}y\\c_{21}x + c_{22}y\end{bmatrix}$$

where $c_{11} = a^2/(a^2+b^2)$,
$c_{12} = c_{21} = ab/(a^2+b^2)$,
$c_{22} = b^2/(a^2+b^2)$.

2. a. $\begin{bmatrix}1/2 & -1/2\\-1/2 & 1/2\end{bmatrix}$ b. $\begin{bmatrix}0 & -1\\-1 & 0\end{bmatrix}$

c. $\begin{bmatrix}1/10 & 3/10\\3/10 & 9/10\end{bmatrix}$ d. $\begin{bmatrix}-4/5 & 3/5\\3/5 & 4/5\end{bmatrix}$

e. $\begin{bmatrix}1/11 & -1/11 & 3/11\\-1/11 & 1/11 & -3/11\\3/11 & -3/11 & 9/11\end{bmatrix}$

f. $\begin{bmatrix}-9/11 & -2/11 & 6/11\\-2/11 & -9/11 & -6/11\\6/11 & -6/11 & 7/11\end{bmatrix}$

4. a. $\sqrt{28/15}$ b. $\cos\theta = -3/\sqrt{14}$

5. a. $\begin{bmatrix}0\\0\\0\end{bmatrix}$ b. $\begin{bmatrix}0\\0\\1\end{bmatrix}$ c. $\begin{bmatrix}1\\0\\0\end{bmatrix}$

d. $\begin{bmatrix}-1\\0\\0\end{bmatrix}$ e. $\begin{bmatrix}10\\-7\\-4\end{bmatrix}$ f. $\begin{bmatrix}10\\11\\-4\end{bmatrix}$

6. $\mathbf{v}\times\mathbf{w} = -\mathbf{w}\times\mathbf{v}$

7. $(\mathbf{e}_1\times\mathbf{e}_2)\times\mathbf{e}_2 = \mathbf{e}_3\times\mathbf{e}_2 = -\mathbf{e}_1$
and $\mathbf{e}_1\times(\mathbf{e}_2\times\mathbf{e}_2) = \mathbf{e}_1\times\mathbf{0} = \mathbf{0}$

8. Rank = 1, nullity = $n - 1$

9. Rank = n, nullity = 0

10. Let $\mathbf{w} = \begin{bmatrix}x\\y\end{bmatrix}$. Then

$$\mathbf{u}_1 = \frac{1}{|\mathbf{w}|}\mathbf{w} = \frac{1}{\sqrt{x^2+y^2}}\begin{bmatrix}x\\y\end{bmatrix}$$

and $$\mathbf{u}_2 = \frac{1}{\sqrt{x^2+y^2}}\begin{bmatrix}-y\\x\end{bmatrix}$$

11. The vector $\mathbf{v}\times\mathbf{w}$ is perpendicular to the plane spanned by **v** and **w**. Moreover, the vector $\mathbf{u}\times(\mathbf{v}\times\mathbf{w})$ is perpendicular to $(\mathbf{v}\times\mathbf{w})$. Thus, $\mathbf{u}\times(\mathbf{v}\times\mathbf{w})$ lies in the plane perpendicular to $\mathbf{v}\times\mathbf{w}$, which is the plane spanned by **v** and **w**.

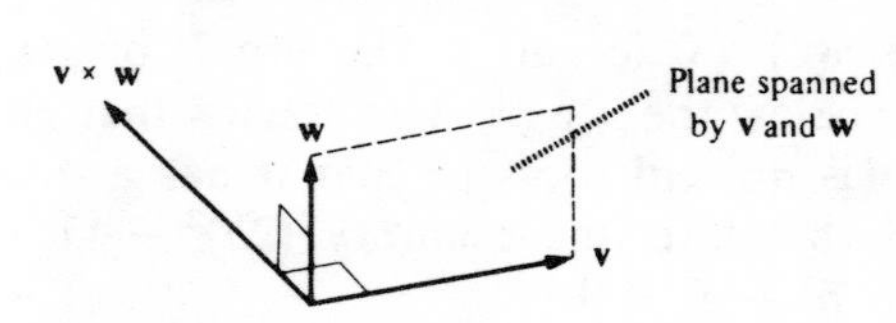

Exercises for Applications 6, page 216

1. $y = -2.02x + 1.01$

2. $y = 1.99x + 3.07$ **3.** $y = 0.51x + 2.84$

Exercise Set 7.1, page 225

1. $\begin{bmatrix} 0 & 0 \\ 2 & 0 \end{bmatrix}$ **2.** $\begin{bmatrix} -1 & 0 \\ 0 & 1 \end{bmatrix}$ **3.** $\begin{bmatrix} 1 & 0 \\ 0 & -1 \end{bmatrix}$

4. $\begin{bmatrix} -3/10 & -1/10 \\ 9/10 & 3/10 \end{bmatrix}$

5, 6. $\begin{bmatrix} -7/10 & -9/10 \\ 1/10 & 7/10 \end{bmatrix}$

7. a. $T_0 \leftrightarrow \begin{bmatrix} 0 & 0 \\ 0 & 0 \end{bmatrix}$ **b.** $I \leftrightarrow \begin{bmatrix} 1 & 0 \\ 0 & 1 \end{bmatrix}$

8. Matrix for TT_1 and TT_2 is $\begin{bmatrix} 1/2 & 1/2 \\ 1/2 & 1/2 \end{bmatrix}$

9. $(T_1 + T_2)^2 = T_1^2 + T_1T_2 + T_2T_1 + T_2^2$. Since $T_1T_2 \neq T_2T_1$, the middle terms cannot be combined.

10. The domain of T_2 is $\mathbb{R}^2$ not $\mathbb{R}^3$.

12. It suffices to show that if $\mathbf{v}$ is in the null space of T_2, then $\mathbf{v}$ is in the null space of T_1T_2. Since

$$T_1T_2(\mathbf{v}) = T_1(T_2(\mathbf{v})) = T_1(\mathbf{0}) = \mathbf{0},$$

$\mathbf{v}$ is in the null space of T_1T_2.

13. $\mathbf{v}$ is in the null space of T_1T_2 if and only if $T_1T_2(\mathbf{v}) = \mathbf{0}$. This is true if and only if $T_1(T_2(\mathbf{v})) = \mathbf{0}$, which is equivalent to the statement that $T_2(\mathbf{v})$ is in the null space of T_1. Therefore,

$$\text{Null space of } T_1T_2 = \{\mathbf{v} \mid T_2(\mathbf{v}) \text{ is in the null space of } T_1\}$$

Exercise Set 7.2, page 230

1. $\begin{bmatrix} 5 & 4 & -1 \\ 2 & 1 & 1 \end{bmatrix}$ **2.** Not defined

3. $\begin{bmatrix} 12 & 2 & -2 \\ 6 & 1 & -6 \\ -2 & 0 & 6 \end{bmatrix}$ **4.** $\begin{bmatrix} 4 & -5 & 13 \\ 1 & 3 & 10 \\ 8 & -5 & 1 \\ 9 & -5 & 18 \end{bmatrix}$

5. $\begin{bmatrix} 34 & -1 \\ 21 & 6 \end{bmatrix}$ **6.** $\begin{bmatrix} 34 & -1 \\ 21 & 6 \end{bmatrix}$, Yes

7. $\begin{bmatrix} 3 & -1 \\ 1 & 13 \end{bmatrix}$ **8.** $\begin{bmatrix} 6 & 4 \\ 5 & 10 \end{bmatrix}$, No

9. $[31]$ **10.** Not defined

11. $[20 \quad 12 \quad -9 \quad -20]$

12., 13. See Exercise 14

16. Matrix for T_1: $\begin{bmatrix} 0 & -1 \\ 1 & 0 \end{bmatrix}$. Matrix for T_2: $\begin{bmatrix} 0 & 1 \\ 1 & 0 \end{bmatrix}$. Matrix for T_3: $\begin{bmatrix} 1/10 & -3/10 \\ -3/10 & 9/10 \end{bmatrix}$

20. Let $\mathbf{a} = \begin{bmatrix} a_1 \\ \vdots \\ a_n \end{bmatrix}$, $\mathbf{b} = \begin{bmatrix} b_1 \\ \vdots \\ b_n \end{bmatrix}$. Then

$$MN = \begin{bmatrix} \mathbf{a}\cdot\mathbf{a} & \mathbf{a}\cdot\mathbf{b} \\ \mathbf{b}\cdot\mathbf{a} & \mathbf{b}\cdot\mathbf{b} \end{bmatrix}.$$

(Note that the entries in MN are scalars). To find the null space of MN we reduce this matrix to obtain

$$MN \xrightarrow{\text{reduces to}} \begin{bmatrix} \mathbf{a}\cdot\mathbf{a} & \mathbf{a}\cdot\mathbf{b} \\ 0 & (\mathbf{a}\cdot\mathbf{b})^2 - (\mathbf{a}\cdot\mathbf{a})(\mathbf{b}\cdot\mathbf{b}) \end{bmatrix}$$

Now the null space of MN is $\{\mathbf{0}\}$ if and only if $(\mathbf{a}\cdot\mathbf{b})^2 - (\mathbf{a}\cdot\mathbf{a})(\mathbf{b}\cdot\mathbf{b}) = \mathbf{0}$. But

$$\begin{aligned}(\mathbf{a}\cdot\mathbf{b})^2 - (\mathbf{a}\cdot\mathbf{a})(\mathbf{b}\cdot\mathbf{b}) &= |\mathbf{a}|^2|\mathbf{b}|^2\cos^2\theta - |\mathbf{a}|^2|\mathbf{b}|^2 \\ &= |\mathbf{a}|^2|\mathbf{b}|^2(\cos^2\theta - 1) \\ &\neq 0 \text{ when } \theta \text{ is not an integral multiple of } \pi\end{aligned}$$

Therefore, the null space of MN is $\{\mathbf{0}\}$ when the vectors $\mathbf{a}$ and $\mathbf{b}$ are not parallel.

21. $M = \begin{bmatrix} -1/2 & -\sqrt{3}/2 \\ \sqrt{3}/2 & -1/2 \end{bmatrix}$

$c = -1/2 + (\sqrt{3}/2)i$

Exercise Set 7.3, page 242

1. a. Inverse correspondence is clockwise rotation by 45°.

b. Inverse correspondence maps every vector $\begin{bmatrix} x_0 \\ 2x_0 \end{bmatrix}$ to all vectors of the form $\begin{bmatrix} x_0 \\ 2x_0 \end{bmatrix} + t\begin{bmatrix} 2 \\ -1 \end{bmatrix}$, where t is a scalar.

c. Inverse correspondence is reflection about the line $y = -x$.

d. Inverse correspondence is multiplication by $-1/2$.

2. $\begin{bmatrix} 3/4 & 1/2 \\ 1/4 & 1/2 \end{bmatrix}$ **3.** Not invertible

4. Not invertible **5.** Not invertible

6. $\begin{bmatrix} -1/12 & 1/6 & 1/4 \\ -13/12 & 1/6 & 1/4 \\ 5/12 & 1/6 & -1/4 \end{bmatrix}$

7. $\begin{bmatrix} 3 & -3 & -1 & 7 \\ 7 & -6 & -3 & 14 \\ 5 & -4 & -2 & 9 \\ -5 & 5 & 2 & -11 \end{bmatrix}$

8. Matrix for T_1^{-1}: $\begin{bmatrix} 0 & 1 \\ 1 & 0 \end{bmatrix}$

Matrix for T_2^{-1}: $\begin{bmatrix} 0 & 1 \\ -1 & 0 \end{bmatrix}$

Matrix for $(T_1T_2)^{-1}$: $\begin{bmatrix} 1 & 0 \\ 0 & -1 \end{bmatrix}$

Matrix for $T_1^{-1}T_2^{-1}$: $\begin{bmatrix} -1 & 0 \\ 0 & 1 \end{bmatrix}$

Matrix for $T_2^{-1}T_1^{-1}$: $\begin{bmatrix} 1 & 0 \\ 0 & -1 \end{bmatrix}$

9. $M_1M_2 = \begin{bmatrix} 4 & -8 \\ 3 & -5 \end{bmatrix}$, $M_1^{-1} = \begin{bmatrix} -1 & 2 \\ -1 & 1 \end{bmatrix}$,

$$M_2^{-1} = \begin{bmatrix} 3/4 & 1/2 \\ 1/4 & 1/2 \end{bmatrix},$$

$$(M_1M_2)^{-1} = M_2^{-1}M_1^{-1} = \begin{bmatrix} -5/4 & 2 \\ -3/4 & 1 \end{bmatrix},$$

$$M_1^{-1}M_2^{-1} = \begin{bmatrix} -1/4 & 1/2 \\ -1/2 & 0 \end{bmatrix}$$

10. $(M_1M_2M_3)^{-1} = M_3^{-1}M_2^{-1}M_1^{-1}$.

11. See Exercises 12 and 13 in Section 7.1 and Theorem 7.5.

12. D does not have an inverse. The inverse correspondence is

$$a_nx^n + a_{n-1}x^{n-1} + \cdots + a_1x + a_0 \xrightarrow{\text{corresponds to}} [(a_n/n+1)x^{n+1} + (a_{n-1}/n)x^n + \cdots + (a_1/2)x^2 + a_0x] + c$$

where c is any scalar. (The inverse correspondence maps a polynomial to its antiderivatives.)

13. c. $M^{-1} = (-1/10)(M - 7I)$

$$= (-1/10)\begin{bmatrix} -14 & -6 \\ 18 & 7 \end{bmatrix}$$

14. $M^{-1} = (-1/a_0)(M^{n-1} + a_{n-1}M^{n-2} + a_{n-2}M^{n-3} + \cdots + a_2M + a_1)$

15. We can assume that M is a square matrix (Theorem 7.4). We let M be an $n \times n$ matrix.

a. Since M has a column of zeros, its image space is spanned by $n - 1$ vectors, which means that the dimension of the image space is smaller than n. Hence M is not invertible (Theorem 7.5′).

b. If the ith row of M consists of only zeros, then the vector $\mathbf{e}_i$ that has a 1 in the ith position and zeros elsewhere is a nonzero vector in the null space of M. Hence, since the nullity of M is not 0, M is not invertible (Theorem 7.5′).

17. a. The inverse operations are:

i. Multiplication of a row by $1/c$ (multiplying by c and then by $1/c$ leaves the row unchanged.)

ii. The operation of interchanging rows of a matrix is its own inverse (doing it twice gives the original matrix).

iii. The inverse of the row operation

$$\begin{bmatrix} \text{row } i \\ \text{row } j \end{bmatrix} \begin{matrix} (1) \\ (1) \end{matrix} \longrightarrow \begin{bmatrix} \text{row } i \\ (\text{row } i) + (\text{row } j) \end{bmatrix}$$

is the operation

$$\begin{bmatrix} \text{row } i \\ \text{row } j \end{bmatrix} \quad \begin{matrix} (-1) \\ (1) \end{matrix} \longrightarrow \begin{bmatrix} \text{row } i \\ -(\text{row } i) + (\text{row } j) \end{bmatrix}$$

(that is, the matrix obtained by applying the first operation and then the second operation is the original matrix).

Exercises for Applications 7, page 250

1. Touch lights 1, 3, 5, 6, 7, 9.

2. a.

•	
•	•

b. $C = \begin{bmatrix} 1 & 1 & 0 & 0 \\ 0 & 1 & 1 & 1 \\ 1 & 0 & 0 & 1 \\ 0 & 0 & 1 & 1 \end{bmatrix}$

c. $C^{-1} = \begin{bmatrix} 1 & 1 & 0 & 1 \\ 0 & 1 & 0 & 1 \\ 1 & 1 & 1 & 0 \\ 1 & 1 & 1 & 1 \end{bmatrix}$

d. Touch lights 2 and 3.

3. a. $C = \begin{bmatrix} 1 & 1 & 0 & 0 \\ 0 & 1 & 1 & 0 \\ 1 & 0 & 0 & 1 \\ 0 & 0 & 1 & 1 \end{bmatrix}$

b. Since $\mathbf{c}_1 + \mathbf{c}_2 + \mathbf{c}_3 + \mathbf{c}_4 = \mathbf{0}$, the vector

$$\begin{bmatrix} 1 \\ 1 \\ 1 \\ 1 \end{bmatrix}$$

is in the null space of C. Hence C is not invertible by Theorem 7.5.

c. To get from the first pattern to the second we would need to solve the equation

$$C\mathbf{x} = \begin{bmatrix} 1 \\ 0 \\ 0 \\ 0 \end{bmatrix}$$

Since the sum of the entries in each column of C is 0 modulo 2, the sum of the entries in every vector of the image space of C is also 0 modulo 2. The sum of the entries in the vector

$$\begin{bmatrix} 1 \\ 0 \\ 0 \\ 0 \end{bmatrix}$$

is 1. So this vector is not in the image space of C, and the equation above cannot be solved.

d. Either by touching lights 1 and 2, or by touching lights 3 and 4.

e. Null space $= \left\langle \begin{bmatrix} 1 \\ 1 \\ 1 \\ 1 \end{bmatrix} \right\rangle$,

Image space $= \left\langle \begin{bmatrix} 1 \\ 0 \\ 1 \\ 0 \end{bmatrix}, \begin{bmatrix} 1 \\ 1 \\ 0 \\ 0 \end{bmatrix}, \begin{bmatrix} 0 \\ 1 \\ 0 \\ 1 \end{bmatrix} \right\rangle$

Exercise Set 8.1, page 259

1. a. Orthogonal **b.** Orthogonal
c. Not orthogonal **d.** Not orthogonal
e. Orthogonal

2. a. Orthogonal **b.** Orthogonal
c. Not orthogonal **d.** Not orthogonal

5.

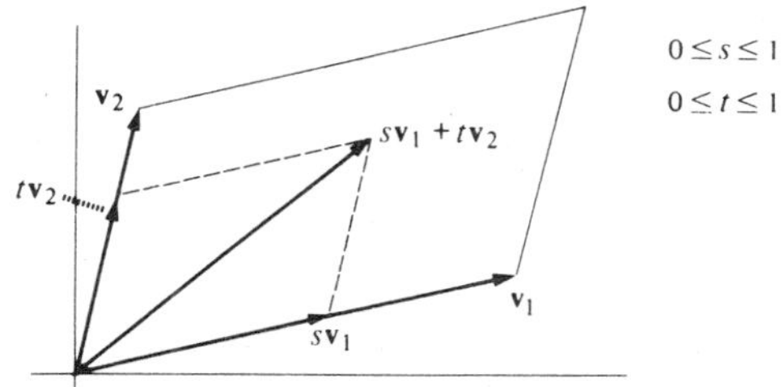

6. a. $\mathbf{e}_i \cdot M\mathbf{e}_j = \mathbf{e}_i \cdot (j\text{th column of } M) = i\text{th}$ entry in jth column of M

b. ij entry of $M = \mathbf{e}_i \cdot M\mathbf{e}_j = \mathbf{e}_i \cdot N\mathbf{e}_j = ij$ entry of N

7. $MN\mathbf{v}_1 \cdot \mathbf{v}_2 = M(N\mathbf{v}_1) \cdot \mathbf{v}_2 = N\mathbf{v}_1 \cdot M^t\mathbf{v}_2$
$= \mathbf{v}_1 \cdot N^tM^t\mathbf{v}_2$

On the other hand

$$MN\mathbf{v}_1 \cdot \mathbf{v}_2 = (MN)\mathbf{v}_1 \cdot \mathbf{v}_2 = \mathbf{v}_1 \cdot (MN)^t\mathbf{v}_2$$

Therefore, using the result of Exercise 6, we have $(MN)^t = N^tM^t$

8. **b.** Identity linear transformation

c. The line perpendicular to **v**

11. **a.** Symmetric **b.** Neither

c. Antisymmetric **d.** Symmetric

e. Antisymmetric

Exercise Set 8.2, page 269

1. **a.** 18 **b.** 37 **c.** -27 **d.** 1

2. 37

3. **a.** $(1/18)\begin{bmatrix} 6 & -3 \\ 4 & 1 \end{bmatrix}$ **b.** $(1/37)\begin{bmatrix} 7 & 1 \\ -2 & 5 \end{bmatrix}$

c. $(-1/27)\begin{bmatrix} 7 & -13 \\ -1 & -2 \end{bmatrix}$ **d.** $\begin{bmatrix} 3/5 & 4/5 \\ -4/5 & 3/5 \end{bmatrix}$

4. From Theorem 8.5 we have: (M_1M_2) is invertible if and only if $\det(M_1M_2) \neq 0$. Since $\det(M_1M_2) = \det(M_1)\det(M_2)$, we see that $\det(M_1M_2) \neq 0$ implies that $\det(M_1) \neq 0$ and $\det(M_2) \neq 0$, which means that M_1 and M_2 are invertible if and only if M_1M_2 is invertible.

5. $\det(M_1) = 5$, $\det(M_2) = -7$, $\det(M_1M_2) = -35$

6. **a.**

$$\begin{bmatrix} 1 & 0 \\ a & b \end{bmatrix}\begin{bmatrix} m_{11} & m_{12} \\ m_{21} & m_{22} \end{bmatrix} = \begin{bmatrix} m_{11} & m_{12} \\ am_{11} + bm_{21} & am_{12} + bm_{22} \end{bmatrix}$$

c. $\det(P) = \det(BM) = \det(B)\det(M)$

$$= \left(\det\begin{bmatrix} 1 & 0 \\ a & b \end{bmatrix}\right)(\det(M)) = b\det(M)$$

7. Interchanging rows in a 2×2 matrix can be accomplished by multiplication on the left by the matrix

$$C = \begin{bmatrix} 0 & 1 \\ 1 & 0 \end{bmatrix}$$

Since $\det(C) = -1$, $\det(CM) = -1\det(M)$

8. $\det(M) = \det\begin{bmatrix} a & b \\ c & d \end{bmatrix} = ad - bc = ad - cb$

$$= \det\begin{bmatrix} a & c \\ b & d \end{bmatrix} = \det(M^t)$$

9. **b.** $M\mathbf{v}(t) = \begin{bmatrix} 2\cos t \\ \sin t \end{bmatrix}$. The relationship between the coordinates of this vector is:

$$1 = \left(\frac{2\cos t}{2}\right)^2 + (\sin t)^2$$

$$= [\tfrac{1}{2}(x \text{ coordinate})]^2 + (y \text{ coordinate})^2$$

or

$$\frac{x^2}{2^2} + y^2 = 1$$

c. The area of the circle $x^2 + y^2 = 1$ is π. By Theorem 8.4, the area of the image of the circle under the matrix M is $\det(M)$ times as large as the area of the circle. Since $M = \begin{bmatrix} 2 & 0 \\ 0 & 1 \end{bmatrix}$, $\det(M) = 2$. Therefore the area of the ellipse is 2π.

Exercise Set 8.3, page 277

1. **a.** 0 **b.** -137 **c.** -127

d. -22 **e.** 72 **f.** 147

g. $(2)(-1)(-3) = 6$

h. $(3)(-4)(2) = -24$ **i.** -6 **j.** 6

2. **a.** Not invertible

b. $(-1/137)\begin{bmatrix} 9 & -25 & 1 \\ -38 & -1 & 11 \\ -7 & -11 & -16 \end{bmatrix}$

c. $(-1/127)\begin{bmatrix} 15 & 1 & -26 \\ 29 & -15 & 9 \\ -23 & -10 & 6 \end{bmatrix}$

d. $(-1/22)\begin{bmatrix} 22 & -14 & 24 \\ 11 & -9 & 17 \\ -11 & 3 & -13 \end{bmatrix}$

g. $\begin{bmatrix} 1/2 & 3/2 & 31/6 \\ 0 & -1 & -4 \\ 0 & 0 & -1/3 \end{bmatrix}$

3. f. $\begin{bmatrix} 2 & -1 & 3 & 2 \\ -1 & 4 & 0 & 5 \\ 2 & 3 & -2 & 7 \\ -1 & -1 & 4 & -5 \end{bmatrix}$ (1) (−1) (1) ② ① ②

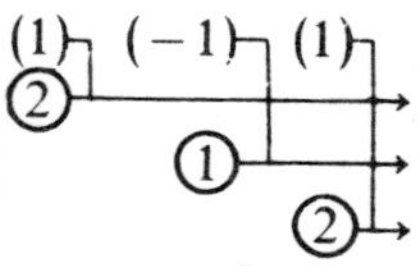

$$\begin{bmatrix} 2 & -1 & 3 & 2 \\ 0 & 7 & 3 & 12 \\ 0 & 4 & -5 & 5 \\ 0 & -3 & 11 & -8 \end{bmatrix} = P$$

$(2)(1)(2)\det(M) = \det(P)$

$$= 2\begin{vmatrix} 7 & 3 & 12 \\ 4 & -5 & 5 \\ -3 & 11 & -8 \end{vmatrix} = (2)(294)$$

So $\det(M) = \dfrac{(2)(294)}{4} = 147$

4. a. $d_1 d_2$ $\quad d_1 d_2 d_3$ $\quad d_1 d_2 d_3 d_4$

5. a. If rows r_i and r_j of M are the same, then the row operation

$$M = \begin{vmatrix} \vdots \\ r_i \\ \vdots \\ r_j \\ \vdots \end{vmatrix} \begin{matrix} -1 \\ ① \end{matrix} \rightarrow \begin{vmatrix} & r_i & \\ 0 & \cdots & 0 \end{vmatrix} = P$$

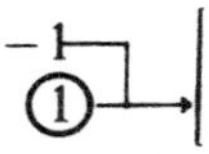

yields a singular matrix P. So $\det(M) = (1)\det(P) = 0$.

b. If two columns of M are the same, then the image space of M has dimension at most $n - 1$. By Theorem 8.11 we see that $\det(M) = 0$.

6. b. Let $B = \begin{bmatrix} 0 & 0 & 1 \\ 1 & 0 & 0 \\ 0 & 1 & 0 \end{bmatrix}$. Then $\det(B) = 1$ and $M^* = BM$. Therefore,

$$\det(M^*) = \det(BM) = \det(B)\det(M) = 1 \cdot \det(M).$$

8. b. $C = \begin{bmatrix} b & 0 & a \\ 0 & 1 & 0 \\ 0 & 0 & 1 \end{bmatrix}$

9. b. It suffices to show that if f_1, f_2, f_3 are linearly dependent on (a, b), then $\det(W(x)) = 0$ for each x in (a, b).

If f_1, f_2, f_3 are linearly dependent on (a, b), then there are scalars c_1, c_2, c_3, not all zero, such that

$$c_1 f_1(x) + c_2 f_2(x) + c_3 f_3(x) = 0 \quad \text{for every } x \text{ in } (a, b).$$

By part a, this implies that

$$c_1 f'_1(x) + c_2 f'_2(x) + c_3 f'_3(x) = 0 \quad \text{for every } x \text{ in } (a, b)$$

and

$$c_1 f''_1(x) + c_2 f''_2(x) + c_3 f''_3(x) = 0 \quad \text{for every } x \text{ in } (a, b)$$

(for the same values of c_1, c_2, c_3). These three equations are equivalent to the matrix equation

$$W(x)\mathbf{c} = \begin{bmatrix} f_1(x) & f_2(x) & f_3(x) \\ f'_1(x) & f'_2(x) & f'_3(x) \\ f''_1(x) & f''_2(x) & f''_3(x) \end{bmatrix} \begin{bmatrix} c_1 \\ c_2 \\ c_3 \end{bmatrix} = \begin{bmatrix} 0 \\ 0 \\ 0 \end{bmatrix}$$

for all x in (a, b)

In other words, if f_1, f_2, f_3 are linearly dependent, then the nonzero vector $\mathbf{c}$ is in the null space of $W(x)$ for all x in (a, b). This means that for each x in (a, b), the matrix $W(x)$ is singular and therefore $\det(W(x)) = 0$ for every x in (a, b).

10. a. Use Theorem 8.12.

b. $M = \begin{bmatrix} 1 & 0 \\ 0 & 2 \end{bmatrix}$ is one of many examples.

Exercise Set 8.4, page 288

1. a. 3 ① −2 ①

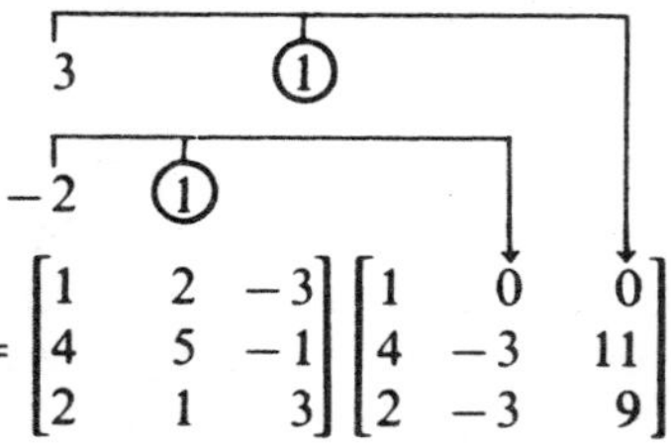

$$M = \begin{bmatrix} 1 & 2 & -3 \\ 4 & 5 & -1 \\ 2 & 1 & 3 \end{bmatrix} \begin{bmatrix} 1 & 0 & 0 \\ 4 & -3 & 11 \\ 2 & -3 & 9 \end{bmatrix} = P$$

$$\det(M) = 1 \cdot \det(P) = 1\begin{bmatrix} -3 & 11 \\ -3 & 9 \end{bmatrix} = 6$$

b.

$-3 \quad (2)$

$5 \quad (2)$

$$M = \begin{bmatrix} 2 & -5 & 3 \\ 1 & -1 & 1 \\ 4 & 5 & 2 \end{bmatrix} \begin{bmatrix} 2 & 0 & 0 \\ 1 & 3 & -1 \\ 4 & 30 & -8 \end{bmatrix} = P$$

$(2)(2)\det(M) = \det(P) = 12$. So $\det(M) = 3$

2. a. The volume of the parallelepiped with edges $\mathbf{v}_1, \mathbf{v}_2, \mathbf{v}_3$ is given by the formula

Volume = (area of base)(height)

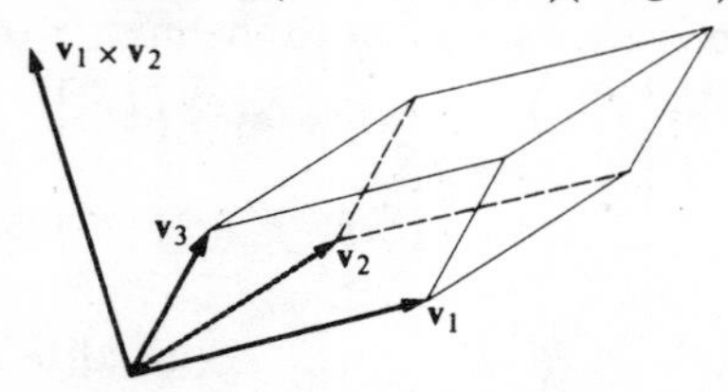

Area of base $= |\mathbf{v}_1||\mathbf{v}_2||\sin\theta| = |\mathbf{v}_1 \times \mathbf{v}_2|$.

Height = length of projection of $\mathbf{v}_3$ in direction perpendicular to base

= length of projection of $\mathbf{v}_3$ in direction of $\mathbf{v}_1 \times \mathbf{v}_2$

$$= \left| \frac{\mathbf{v}_3 \cdot (\mathbf{v}_1 \times \mathbf{v}_2)}{(\mathbf{v}_1 \times \mathbf{v}_2)\cdot(\mathbf{v}_1 \times \mathbf{v}_2)} |\mathbf{v}_1 \times \mathbf{v}_2| \right|$$

$$= \frac{|\mathbf{v}_3 \cdot (\mathbf{v}_1 \times \mathbf{v}_2)|}{|\mathbf{v}_1 \times \mathbf{v}_2|}$$

Therefore, the volume of this parallelepiped is

Volume = (area of base)(height)

$= |\mathbf{v}_3 \cdot (\mathbf{v}_1 \times \mathbf{v}_2)|$

Exercises for Applications 8, page 300

1. a. The region $\mathscr{R}$ is bounded by the lines

$$x + y = 0 \qquad x + y = 1$$
$$x - y = 0 \qquad x - y = 1$$

Each of these lines is the image, under C, of some curve in the uv plane. To find these curves we substitute $(u - v)/2$ for x and $(u + v)/2$ for y. The boundaries of the region $\mathscr{S}$ are

$$\frac{u-v}{2} + \frac{u+v}{2} = 0 \qquad \frac{u-v}{2} + \frac{u+v}{2} = 1$$

$$\frac{u-v}{2} - \frac{u+v}{2} = 0 \qquad \frac{u-v}{2} - \frac{u+v}{2} = 1$$

which simplify to

$$u = 0 \qquad u = 1 \qquad v = 0 \qquad v = -1$$

The region $\mathscr{S}$ is the region shown below.

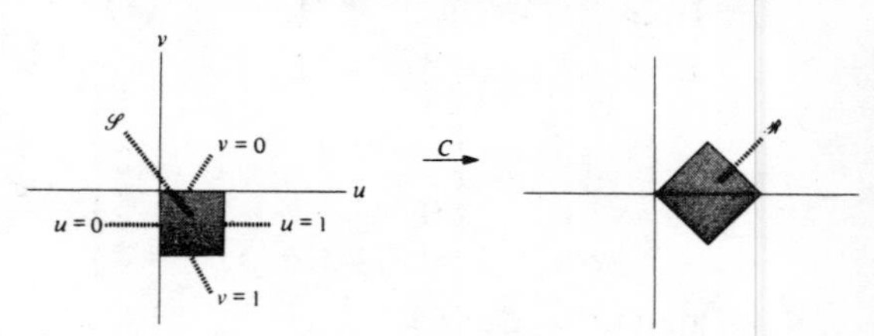

b. $M = \begin{bmatrix} 1/2 & -1/2 \\ 1/2 & 1/2 \end{bmatrix}$, $\quad \det(M) = 1/2$

c. $A = \iint_{\mathscr{R}} 1\, dx\, dy = \iint_{\mathscr{S}} |\det(M)|\, du\, dv$

$$= \int_{-1}^{0} \int_{0}^{1} (1/2)\, du\, dv = 1/2$$

d. Area of $\mathscr{R} = \begin{vmatrix} 1/2 & -1/2 \\ 1/2 & 1/2 \end{vmatrix} = 1/2$ (see Section 8.2).

2. To use the Jacobian we substitute

$$x + y = (u - v)/2 + (u + v)/2 = u,$$
$$x - y = (u - v)/2 - (u + v)/2 = -v$$

So $\iint_{\mathscr{R}} (x + y)^3 (x - y)^2\, dy\, dx$

$$= \iint_{\mathscr{S}} u^3 (-v)^2 |\det(M)|\, du\, dv$$

$$= \int_{-1}^{0} \int_{0}^{1} u^3 v^2 (1/2)\, du\, dv$$

$$= 1/24$$

Exercise Set 9.1, page 309

1. a. $5\mathbf{e}_1 + 2\mathbf{e}_2$ **b.** $41\mathbf{e}_1 + 17\mathbf{e}_2$
c. $47\mathbf{e}_1 + 18\mathbf{e}_2$ **d.** $41\mathbf{e}_1 + 17\mathbf{e}_2$
e. $37\mathbf{e}_1 + 13\mathbf{e}_2$

2. a. $-2\mathbf{u}_1 + \mathbf{u}_2$ **b.** $5\mathbf{u}_1 - 2\mathbf{u}_2$
c. $5\mathbf{u}_1 - 1\mathbf{u}_2$ **d.** $37\mathbf{u}_1 - 15\mathbf{u}_2$

e. $29\mathbf{u}_1 - 11\mathbf{u}_2$ **f.** $31\mathbf{u}_1 - 12\mathbf{u}_2$

3. The basis change matrix B is a square matrix with linearly independent columns. Therefore, the nullity of B is zero and so B is invertible (Theorem 7.5′).

4. $-22\mathbf{e}_1 - 8\mathbf{e}_2 - 6\mathbf{e}_3$ **5.** $9\mathbf{e}_1 - 3\mathbf{e}_2 - 5\mathbf{e}_3$

6. $5\mathbf{e}_1 + 1\mathbf{e}_2 + 1\mathbf{e}_3$ **7.** $-7\mathbf{e}_1 - 5\mathbf{e}_2 + 0\mathbf{e}_3$

For Exercises 8–12

$$B^{-1} = (1/6)\begin{bmatrix} 2 & -16 & 12 \\ 1 & 1 & 0 \\ -1 & -7 & 6 \end{bmatrix}$$

8. $-(8/3)\mathbf{u}_1 + (1/6)\mathbf{u}_2 - (7/6)\mathbf{u}_3$

9. $(-65/3)\mathbf{u}_1 + (5/3)\mathbf{u}_2 - (32/3)\mathbf{u}_3$

10. $(-86/3)\mathbf{u}_1 + (7/6)\mathbf{u}_2 - (67/6)\mathbf{u}_3$

11. $(-86/3)\mathbf{u}_1 + (7/6)\mathbf{u}_2 - (67/6)\mathbf{u}_3$

12. $(-5/3)\mathbf{u}_1 - (1/3)\mathbf{u}_2 + (7/3)\mathbf{u}_3$

Exercise Set 9.2, page 315

1. Matrix for T relative to $\mathbf{u}$ coordinates $\begin{bmatrix} 1 & 0 \\ 0 & 0 \end{bmatrix}$. No.

2. Matrix for T relative to $\mathbf{u}$ coordinates $\begin{bmatrix} 0 & -1 \\ 1 & 0 \end{bmatrix}$. Yes.

3. $\begin{bmatrix} 1 & 0 & 0 \\ 0 & 1 & 0 \\ 0 & 0 & 0 \end{bmatrix}$ **4.** $\begin{bmatrix} 1 & 0 \\ 0 & -1 \end{bmatrix}$

5. a. Reflection about the line containing $\mathbf{u}_1 + \mathbf{u}_2$.

b. Projection in the direction of $\mathbf{u}_2$.

c. Stretch each vector by a factor of 2.

6. $\begin{bmatrix} 10 & 0 \\ 0 & -5 \end{bmatrix}_{(\mathbf{u})}$ **7.** $\begin{bmatrix} 4 & 0 \\ 0 & -2 \end{bmatrix}_{(\mathbf{u})}$

8. $\begin{bmatrix} -7/5 & 6/5 \\ 6/5 & 2/5 \end{bmatrix}_{(\mathbf{u})}$ **9.** $\begin{bmatrix} -1/5 & -2/5 \\ -7/5 & 11/5 \end{bmatrix}_{(\mathbf{u})}$

Exercise Set 9.3, page 320

1.–4. See Exercises 6–9 in Section 9.2.

For Exercises 5 and 6

$$B = \begin{bmatrix} 1 & 2 & 3 \\ 0 & 1 & -2 \\ -1 & 1 & 1 \end{bmatrix}$$

$$B^{-1} = \begin{bmatrix} 0.3 & 0.1 & -0.7 \\ 0.2 & 0.4 & 0.2 \\ 0.1 & -0.3 & 0.1 \end{bmatrix}$$

5. $N = \begin{bmatrix} 20 & 0 & 0 \\ 0 & 30 & 0 \\ 0 & 0 & -10 \end{bmatrix}$

6. $N = \begin{bmatrix} -1.9 & -2 & 1.7 \\ 1.4 & 5 & 3.8 \\ -1.3 & -1 & -0.1 \end{bmatrix}$

8. $M = BNB^{-1}$ **9.** $M\mathbf{u}_i = d_i\mathbf{u}_i$

10. c. Let $R = \begin{bmatrix} \sqrt{2} & 0 \\ 0 & \sqrt{5} \end{bmatrix}$. Then $R^2 = N$. Therefore

$$M = BNB^{-1} = B(R^2)B^{-1} = BR(B^{-1}B)RB^{-1} = (BRB^{-1})^2$$

Therefore, we choose $A = BRB^{-1}$ or

$$A = \begin{bmatrix} (4\sqrt{2} - 3\sqrt{5}) & (2\sqrt{2} - 2\sqrt{5}) \\ (-6\sqrt{2} + 6\sqrt{5}) & (-3\sqrt{2} + 4\sqrt{5}) \end{bmatrix}.$$

11. Since M and N are similar, they represent the same linear transformation T. Therefore

$$\text{Rank}(M) = \text{dimension of the image space of } T = \text{rank}(N)$$

12. a. The vectors in the null space of $N = B^{-1}MB$ are all obtained by applying the matrix B^{-1} to all the vectors in the null space of M.

b. Null space of $M = \left\langle \begin{bmatrix} 3 \\ -1 \end{bmatrix} \right\rangle$; null space of $N = \left\langle \begin{bmatrix} 0 \\ 1 \end{bmatrix} \right\rangle$ and

$$B^{-1}\begin{bmatrix} 3 \\ -1 \end{bmatrix} = \begin{bmatrix} 1 & 3 \\ 2 & 5 \end{bmatrix}\begin{bmatrix} 3 \\ -1 \end{bmatrix} = \begin{bmatrix} 0 \\ 1 \end{bmatrix}$$

Exercise Set 9.4, page 328

1. The characteristic equation for M is $x^2 + 1 = 0$, which has no real roots.

2. $\mathbf{u}_1 = \begin{bmatrix} 3 \\ 1 \end{bmatrix}$, $\mathbf{u}_2 = \begin{bmatrix} -2 \\ 1 \end{bmatrix}$, $D = \begin{bmatrix} 10 & 0 \\ 0 & -5 \end{bmatrix}$

3. $\mathbf{u}_1 = \begin{bmatrix} 1 \\ -1 \end{bmatrix}$, $\mathbf{u}_2 = \begin{bmatrix} 1 \\ -3 \end{bmatrix}$, $D = \begin{bmatrix} 4 & 0 \\ 0 & -2 \end{bmatrix}$

4. $\mathbf{u}_1 = \begin{bmatrix} 1 \\ -1 \\ 1 \end{bmatrix}$, $\mathbf{u}_2 = \begin{bmatrix} 2 \\ 0 \\ 1 \end{bmatrix}$, $\mathbf{u}_3 = \begin{bmatrix} 1 \\ 2 \\ 0 \end{bmatrix}$,

$D = \begin{bmatrix} 1 & 0 & 0 \\ 0 & -1 & 0 \\ 0 & 0 & 3 \end{bmatrix}$

5. $\mathbf{u}_1 = \begin{bmatrix} -1 \\ 1 \\ 1 \end{bmatrix}$, $\mathbf{u}_2 = \begin{bmatrix} -1 \\ 2 \\ 2 \end{bmatrix}$, $\mathbf{u}_3 = \begin{bmatrix} -2 \\ 4 \\ 3 \end{bmatrix}$,

$D = \begin{bmatrix} 0 & 0 & 0 \\ 0 & -1 & 0 \\ 0 & 0 & -3 \end{bmatrix}$

6. $\mathbf{u}_1 = \begin{bmatrix} 1 \\ 1 \\ 0 \end{bmatrix}$, $\mathbf{u}_2 = \begin{bmatrix} 1 \\ -1 \\ 0 \end{bmatrix}$, $\mathbf{u}_3 = \begin{bmatrix} 0 \\ 0 \\ 1 \end{bmatrix}$,

$D = \begin{bmatrix} 4 & 0 & 0 \\ 0 & -2 & 0 \\ 0 & 0 & -2 \end{bmatrix}$

7. $\mathbf{u}_1 = \begin{bmatrix} 1 \\ 1 \\ 1 \end{bmatrix}$, $\mathbf{u}_2 = \begin{bmatrix} 3 \\ 2 \\ 3 \end{bmatrix}$, $\mathbf{u}_3 = \begin{bmatrix} 3 \\ 1 \\ 4 \end{bmatrix}$,

$D = \begin{bmatrix} 1 & 0 & 0 \\ 0 & 2 & 0 \\ 0 & 0 & 3 \end{bmatrix}$

8. $\mathbf{u}_1 = \begin{bmatrix} 0 \\ 1 \\ 0 \end{bmatrix}$, $\mathbf{u}_2 = \begin{bmatrix} 0 \\ 0 \\ 1 \end{bmatrix}$, $\mathbf{u}_3 = \begin{bmatrix} 2 \\ 0 \\ 1 \end{bmatrix}$,

$D = \begin{bmatrix} 0 & 0 & 0 \\ 0 & 0 & 0 \\ 0 & 0 & 2 \end{bmatrix}$

9. $\mathbf{u}_1 = \begin{bmatrix} 1 \\ 0 \\ 0 \\ 0 \end{bmatrix}$, $\mathbf{u}_2 = \begin{bmatrix} 0 \\ 1 \\ 0 \\ 0 \end{bmatrix}$, $\mathbf{u}_3 = \begin{bmatrix} 0 \\ 5 \\ 1 \\ 0 \end{bmatrix}$,

$\mathbf{u}_4 = \begin{bmatrix} 0 \\ 5 \\ 0 \\ -2 \end{bmatrix}$, $D = \begin{bmatrix} 1 & 0 & 0 & 0 \\ 0 & 1 & 0 & 0 \\ 0 & 0 & 2 & 0 \\ 0 & 0 & 0 & 3 \end{bmatrix}$

10. $\mathbf{u}_1 = \begin{bmatrix} 0 \\ 1 \\ 0 \end{bmatrix}$, $\mathbf{u}_2 = \begin{bmatrix} -2 \\ -2 \\ 1 \end{bmatrix}$, $\mathbf{u}_3 = \begin{bmatrix} -1 \\ -1 \\ 1 \end{bmatrix}$,

$D = \begin{bmatrix} 1 & 0 & 0 \\ 0 & 2 & 0 \\ 0 & 0 & 3 \end{bmatrix}$

11. **a.** The matrix $M - a_{ii} I$ can be reduced to a matrix that contains at least one row of zeros. Hence, $M - a_{ii} I$ is singular (the vector $\mathbf{e}_i$ is in the null space of $M - a_{ii} I$).

b. Characteristic values: a_{11}, a_{22}, a_{33}.

c. Characteristic equation:

$$(c - a_{11})(x - a_{22})(x - a_{33}) = 0.$$

d. Characteristic vector $\begin{bmatrix} 1 \\ 0 \\ 0 \end{bmatrix}$ has characteristic value a_{11}.

12. **a.**
$$\begin{aligned} N(P^{-1}\mathbf{v}) &= (P^{-1}MP)(P^{-1}\mathbf{v}) \\ &= P^{-1}(M\mathbf{v}) = P^{-1}(d\mathbf{v}) \\ &= d(P^{-1}\mathbf{v}). \end{aligned}$$

b. Characteristic value for $\mathbf{w}$ is d.

13.
$$\det(M - xI) = x^n + a_{n-1}x^{n-1} + \cdots + a_1 x + a_0$$

Letting $x = 0$ this equation becomes

$$\det(M) = a_0$$

14. **a.**
$$\begin{aligned} M^2\mathbf{v} &= M(M\mathbf{v}) = M(d\mathbf{v}) = d(M\mathbf{v}) \\ &= d(d\mathbf{v}) = d^2\mathbf{v}. \end{aligned}$$

b. The characteristic vector for M^2 is $\mathbf{v}$.

c. If d is a characteristic value for M, then d^n is a characteristic vector for M^n. This does not always give all characteristic values for M^n.

d. The set of characteristic vectors of M is a subset of the set of characteristic vectors of M^n.

e. Set of characteristic values of $M = \{2\}$.

Characteristic vector $= \begin{bmatrix} 0 \\ 0 \\ 1 \end{bmatrix}$. Set of characteristic values of $M^2 = \{-1, \ -1, \ 2^2 = 4\}$.

Characteristic vectors $= \begin{bmatrix} 1 \\ 0 \\ 0 \end{bmatrix}, \begin{bmatrix} 0 \\ 1 \\ 0 \end{bmatrix}, \begin{bmatrix} 0 \\ 0 \\ 1 \end{bmatrix}$.

15. a. Characteristic value = 1

b. $f_a(x) = e^{ax}$

16. See Theorem 9.7.

17. b. $M = \begin{bmatrix} 1 & 0 \\ 0 & 1 \end{bmatrix}$ is not similar to $N = \begin{bmatrix} 1 & 1 \\ 0 & 1 \end{bmatrix}$, but these two matrices have the same set of characteristic values, $\{1, 1\}$.

18. c. Let

$$A = \begin{bmatrix} a_{11} & a_{12} & \cdots & a_{1n} \\ \vdots & \vdots & & \vdots \\ a_{i1} & a_{i2} & \cdots & a_{in} \\ \vdots & \vdots & & \vdots \\ a_{n1} & a_{n2} & \cdots & a_{nn} \end{bmatrix}$$

$$B = \begin{bmatrix} b_{11} & \cdots & b_{1i} & \cdots & b_{1n} \\ \vdots & & b_{2i} & & \vdots \\ & & \vdots & & \\ b_{n1} & \cdots & b_{ni} & \cdots & b_{nn} \end{bmatrix}$$

The entry in the ith row and the ith column of the matrix AB is $(a_{i1}b_{1i} + a_{i2}b_{2i} + \cdots + a_{in}b_{ni})$. The trace of AB is the sum of all such expressions for $i = 1, 2, \ldots, n$, that is,

$\text{tr}(AB) =$ sum of all products of the form $a_{ij}b_{ji}$, for $i, j = 1, \ldots, n$.

Similarly, we see that

$\text{tr}(BA) =$ sum of all products of the form $b_{st}a_{ts}$ for $s, t = 1, \ldots, n$.

Thus, each term in the expression for the trace of AB is a term in the expression for the trace of BA, and conversely. Hence, $\text{tr}(AB) = \text{tr}(BA)$.

Exercise Set 9.5, page 335

1. a. $S_1 = \left\langle \begin{bmatrix} 0 \\ 0 \\ 1 \end{bmatrix} \right\rangle$,

$S_{-1} = \left\langle \begin{bmatrix} 1 \\ 0 \\ 0 \end{bmatrix}, \begin{bmatrix} 0 \\ 1 \\ 0 \end{bmatrix} \right\rangle$

2. a. True. If M is singular, then the null space of M is not $\{\mathbf{0}\}$. Let $\mathbf{v} \neq \mathbf{0}$ be in the null space of M. Since $M\mathbf{v} = 0\mathbf{v}$, $\mathbf{v}$ is a characteristic vector of M with characteristic value 0.

b. True. Theorem 9.8.

c. False. $M = \begin{bmatrix} 5 & 0 & 0 \\ 0 & 1 & 1 \\ 0 & 0 & 1 \end{bmatrix}$ is not similar to a diagonal matrix.

d. False. The characteristic roots of a real symmetric matrix must be real numbers. The roots of $x^2 + 1$ are the complex numbers $\pm i$.

e. True. $M = \begin{bmatrix} -1 & 0 \\ 0 & -1 \end{bmatrix}$

3. Assume $\dim(S_1) = 2$. We must show that there exist three linearly independent characteristic vectors for M. For each characteristic value d (in this case $d = 1$ or -1) of M, any nonzero vector of the null space of $M - dI$ is a characteristic vector for M. This means that for each value of d, we can find at least one characteristic vector. In particular, we can find characteristic vectors $\mathbf{v}_1$ and $\mathbf{v}_{-1}$ with characteristic values $+1$ and -1, respectively. The fact that $\dim(S_1) = 2$ says that we can find a second characteristic vector $\mathbf{w}_1$ with characteristic value 1 such that $\mathbf{v}_1$ and $\mathbf{w}_1$ are linearly independent. We show that $\{\mathbf{v}_1, \mathbf{w}_1, \mathbf{v}_{-1}\}$ is linearly independent. If

$$\mathbf{0} = c_1\mathbf{v}_1 + c_2\mathbf{w}_1 + c_3\mathbf{v}_{-1}$$

then applying M to this equation yields

$$\mathbf{0} = c_1\mathbf{v}_1 + c_2\mathbf{w}_1 - c_3\mathbf{v}_{-1}$$

From these two equations, we see that $c_3 = 0$. So

$$\mathbf{0} = c_1\mathbf{v}_1 + c_2\mathbf{w}_1$$

But since $\mathbf{v}_1$ and $\mathbf{w}_1$ are linearly independent, this means that $c_1 = c_2 = 0$. So $\{\mathbf{v}_1, \mathbf{w}_1, \mathbf{v}_{-1}\}$ is linearly independent.

Conversely, if M is similar to a diagonal matrix, then the entries on the diagonal of this matrix are the characteristic values of M, namely 1, 1, -1, and the basis associated with this diagonal matrix consists of three linearly independent characteristic vectors with characteristic values 1, 1, and -1. Hence, S_1 is spanned by the two characteristic vectors with characteristic value 1 and has dimension 2.

4. $B = \begin{bmatrix} 1/\sqrt{2} & 1/\sqrt{2} \\ -1/\sqrt{2} & 1/\sqrt{2} \end{bmatrix}, \quad D = \begin{bmatrix} -1 & 0 \\ 0 & 3 \end{bmatrix}$

5. $B = \begin{bmatrix} -1/\sqrt{2} & 1/\sqrt{2} \\ 1/\sqrt{2} & 1/\sqrt{2} \end{bmatrix}, \quad D = \begin{bmatrix} 0 & 0 \\ 0 & 2 \end{bmatrix}$

6. $B = \begin{bmatrix} -1/\sqrt{2} & 0 & 1/\sqrt{3} \\ 0 & -1/\sqrt{2} & 1/\sqrt{3} \\ 1/\sqrt{2} & 1/\sqrt{2} & 1/\sqrt{3} \end{bmatrix},$

$D = \begin{bmatrix} 0 & 0 & 0 \\ 0 & 0 & 0 \\ 0 & 0 & 3 \end{bmatrix}$

7. $B = \begin{bmatrix} -2/3 & 2/3 & 1/3 \\ 1/3 & 2/3 & -2/3 \\ 2/3 & 1/3 & 2/3 \end{bmatrix},$

$D = \begin{bmatrix} 2 & 0 & 0 \\ 0 & -7 & 0 \\ 0 & 0 & 11 \end{bmatrix}$

8. $B = \begin{bmatrix} 0 & 4/3\sqrt{2} & 1/3 \\ 1/\sqrt{2} & 1/3\sqrt{2} & -2/3 \\ 1/\sqrt{2} & -1/3\sqrt{2} & 2/3 \end{bmatrix},$

$D = \begin{bmatrix} 3 & 0 & 0 \\ 0 & 3 & 0 \\ 0 & 0 & -6 \end{bmatrix}$

9. $B = \begin{bmatrix} 0 & 2/\sqrt{22} & -3/\sqrt{11} \\ 1/\sqrt{2} & -3/\sqrt{22} & -1/\sqrt{11} \\ 1/\sqrt{2} & 3/\sqrt{22} & 1/\sqrt{11} \end{bmatrix},$

$D = \begin{bmatrix} 0 & 0 & 0 \\ 0 & 0 & 0 \\ 0 & 0 & 9 \end{bmatrix}$

10. $M\mathbf{v} \cdot \mathbf{u} = \mathbf{v} \cdot M^t\mathbf{u}$ by Theorem 8.1
$= \mathbf{v} \cdot M\mathbf{u}$ since M is symmetric
$= \mathbf{v} \cdot c\mathbf{u}$ where c is the characteristic value associated with $\mathbf{u}$
$= c(\mathbf{v} \cdot \mathbf{u}) = 0$

11. $M\mathbf{u}_1 \cdot \mathbf{u}_2 = (d_1\mathbf{u}_1) \cdot \mathbf{u}_2 = d_1(\mathbf{u}_1 \cdot \mathbf{u}_2)$
$M\mathbf{u}_1 \cdot \mathbf{u}_2 = \mathbf{u}_1 \cdot M^t\mathbf{u}_2 = \mathbf{u}_1 \cdot M\mathbf{u}_2$
$= \mathbf{u}_1 \cdot (d_2\mathbf{u}_2) = d_2(\mathbf{u}_1 \cdot \mathbf{u}_2).$

Since $d_1 \neq d_2$ and $d_1(\mathbf{u}_1 \cdot \mathbf{u}_2) = d_2(\mathbf{u}_1 \cdot \mathbf{u}_2)$, we see that $\mathbf{u}_1 \cdot \mathbf{u}_2 = 0$.

12. **a.** Characteristic equation:

$$x^2 - (a + c)x + (ac - b^2) = 0$$

b. Characteristic roots:

$$x = \frac{(a+c) \pm \sqrt{(a+c)^2 - 4(ac - b^2)}}{2}$$

$$= \frac{(a+c) \pm \sqrt{(a-c)^2 + 4b^2}}{2}$$

Since the quantity inside the radical is the sum of two squares, it is always nonnegative. Therefore, the characteristic roots of M are both real.

c. If both characteristic roots are equal, then

$$0 = (a - c)^2 + 4b^2$$

The sum of two squares is equal to zero only if each of the squares is itself zero. Hence

$(a - c)^2 = 0$ or $a = c$

and $4b^2 = 0$ or $b = 0.$

So $M = \begin{bmatrix} a & 0 \\ 0 & a \end{bmatrix}$

Exercises for Applications 9a, page 343

1. **a.** $\mathbf{v}_0 = \begin{bmatrix} 2 \\ 5 \end{bmatrix}, \quad \mathbf{v}_1 = \begin{bmatrix} 1.1 \\ 5.9 \end{bmatrix},$

$\mathbf{v}_2 = \begin{bmatrix} 0.92 \\ 6.08 \end{bmatrix}, \quad \mathbf{v}_3 = \begin{bmatrix} 0.884 \\ 6.116 \end{bmatrix}$

b. $\mathbf{v}_s = \begin{bmatrix} 7/8 \\ 49/8 \end{bmatrix} = \begin{bmatrix} 0.875 \\ 6.125 \end{bmatrix}$

2. $\mathbf{v}_1 = \begin{bmatrix} q \\ 1-p \end{bmatrix}, \quad \mathbf{v}_{p-q} = \begin{bmatrix} -1 \\ 1 \end{bmatrix}$

3. $M =$

$$\begin{bmatrix} p_1 & p_2 & p_3 \\ q_1 & q_2 & q_3 \\ 1-p_1-q_1 & 1-p_2-q_2 & 1-p_3-q_3 \end{bmatrix}$$

$M - 1I =$

$$\begin{bmatrix} p_1 - 1 & p_2 & p_3 \\ q_1 & q_2 - 1 & q_3 \\ 1-p_1-q_1 & 1-p_2-q_2 & -p_3-q_3 \end{bmatrix}$$

Since the sum of the elements of each column of $M = 1I$ is zero, it is easy to see that M is singular. Hence, 1 is a characteristic value for M.

4. Transition matrix: $\begin{bmatrix} 0.85 & 0.20 \\ 0.15 & 0.80 \end{bmatrix}$

No. Years	Pond A	Pond B
0	1,000	0
1	850	150
2	753	247
3	689	311
4	648	352
5	621	379

Stable population: 571 in Pond A, 429 in Pond B.

Exercises for Applications 9b, page 355

1. a. $B = \begin{bmatrix} 1/\sqrt{2} & -1/\sqrt{2} \\ 1/\sqrt{2} & 1/\sqrt{2} \end{bmatrix}, \quad s^2 - t^2 = 1$

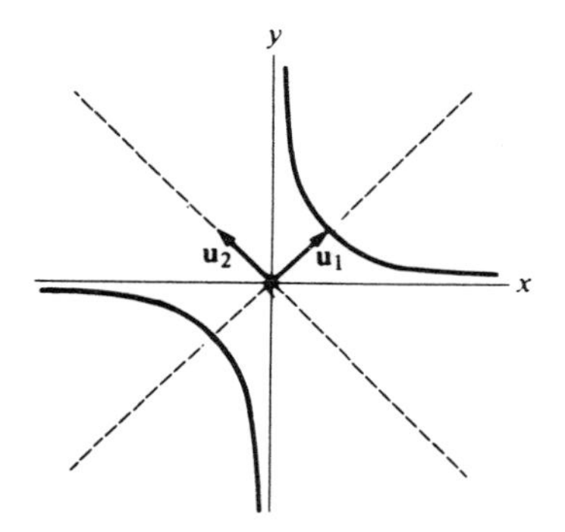

b. $B = \begin{bmatrix} 1/\sqrt{2} & 1/\sqrt{2} \\ -1/\sqrt{2} & 1/\sqrt{2} \end{bmatrix}$

$s^2/(\sqrt{5})^2 + t^2/(\sqrt{3})^2 = 1$

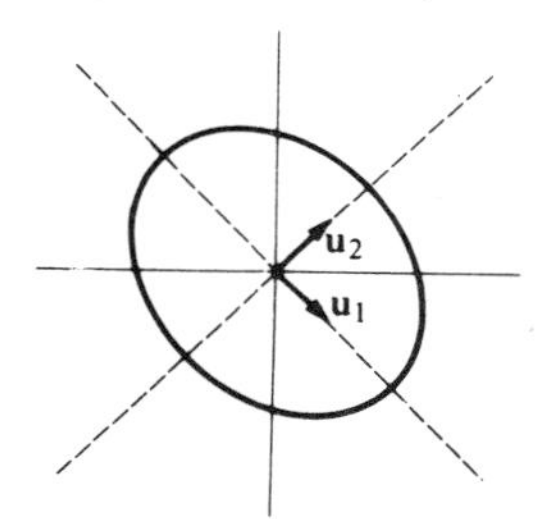

c. $B = \begin{bmatrix} -1/\sqrt{2} & 1/\sqrt{2} \\ 1/\sqrt{2} & 1/\sqrt{2} \end{bmatrix}, \quad \dfrac{-s^2}{2^2} + \dfrac{t^2}{(\sqrt{2})^2} = 1$

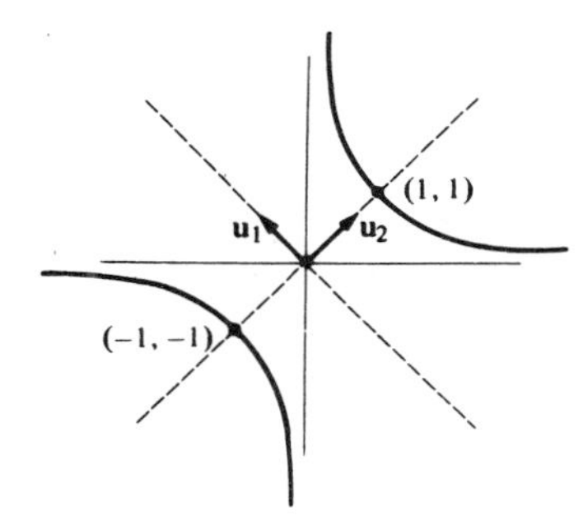

d. $B = \begin{bmatrix} -2/\sqrt{5} & 1/\sqrt{5} \\ 1/\sqrt{5} & 2/\sqrt{5} \end{bmatrix}$

$-s^2/1^2 + t^2/2^2 = 1$

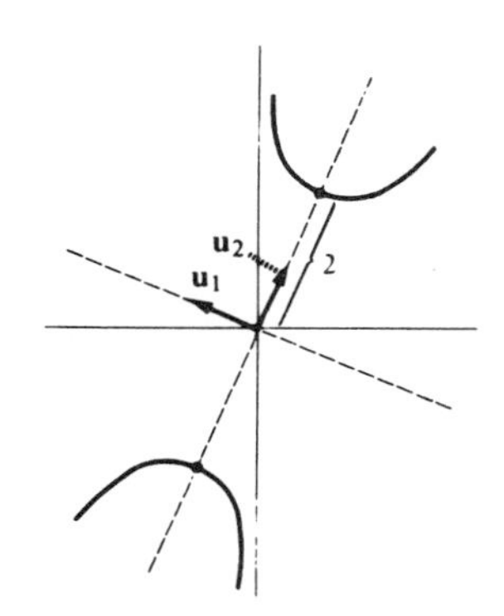

e. $B = \begin{bmatrix} -2/\sqrt{5} & 1/\sqrt{5} \\ 1/\sqrt{5} & 2/\sqrt{5} \end{bmatrix}$

$s^2/2^2 + t^2/(2\sqrt{6})^2 = 1$

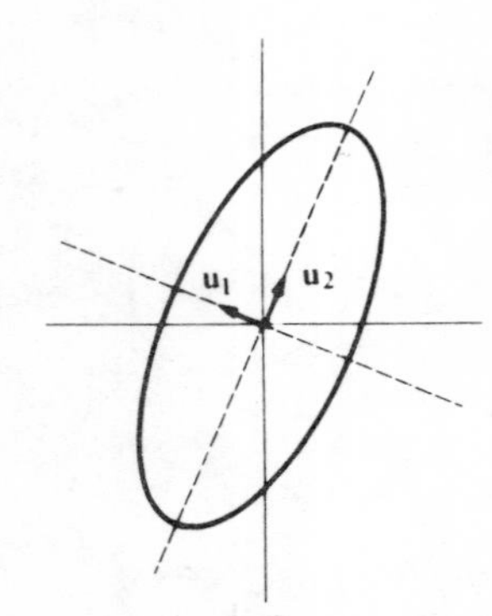

2. a. $x = x' + 2,\ 2x'y' = 1,\ y = y' - 3$

Graph same as graph in 1a., but centered at $(2, -3)$.

b. $x = x' - 1,\ y = y' + 1,$

$x'^2 + y'^2 + 6x'y' = 8$

Graph same as graph in 1c., centered at $(-1, 1)$.

3. a. $B = \begin{bmatrix} -1/\sqrt{2} & 1/\sqrt{2} & 0 \\ 1/\sqrt{2} & 1/\sqrt{2} & 0 \\ 0 & 0 & 1 \end{bmatrix}$

$2s^2 + 4t^2 + 1u^2 = 8$

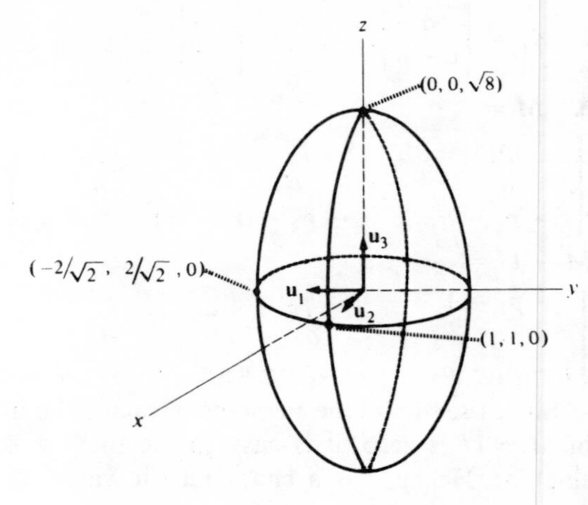

b. $B = \begin{bmatrix} -2/3 & 2/3 & 1/3 \\ 1/3 & 2/3 & -2/3 \\ 2/3 & 1/3 & 2/3 \end{bmatrix},$

$2s^2 - 7t^2 + 11u^2 = 154$

Exercises for Applications 9c, page 363

1. a. $c_1e^{2t} + c_2e^{-2t}$ **b.** $c_1e^{5t} + c_2e^{-4t}$

c. $c_1e^{t} + c_2e^{-3t}$

2. a. $y_1 = c_1e^{4t} + c_2e^{-2t}$

$y_2 = -c_1e^{4t} - 3c_2e^{-2t}$

See Exercise 3 in Section 9.4

b. $y_1 = 3c_1e^{10t} - 2c_2e^{-5t}$

$y_2 = c_1e^{10t} + c_2e^{-5t}$

See Exercise 2 in Section 9.4.

INDEX